Physics for Technology

Textbook Series in Physical Sciences

Lou Chosen, Executive Editor, Taylor & Francis

Understanding Nanomaterials
Malkiat S. Johal

Concise Optics: Concepts, Examples, and Problems
Ajawad I. Haija, M. Z. Numan, W. Larry Freeman

A Mathematica Primer for Physicists
Jim Napolitano

Understanding Nanomaterials, Second Edition
Malkiat S. Johal, Lewis E. Johnson

Physics for Technology, Second Edition
Daniel H. Nichols

For more information about this series, please visit: www.crcpress.com/Textbook-Series-in-Physical-Sciences/book-series/TPHYSCI

Physics for Technology

With Applications in Industrial Control Electronics

Second Edition

Daniel H. Nichols

CRC Press is an imprint of the
Taylor & Francis Group, an **informa** business

CRC Press
Taylor & Francis Group
6000 Broken Sound Parkway NW, Suite 300
Boca Raton, FL 33487-2742

First issued in paperback 2020

ISBN-13: 978-0-8153-8292-8 (hbk)
ISBN-13: 978-0-367-78059-3 (pbk)

Library of Congress Cataloging-in-Publication Data

Names: Nichols, Daniel H., author.
Title: Physics for technology : with applications in industrial control electronics / Daniel H. Nichols.
Description: Second edition. | Boca Raton : CRC Press, Taylor & Francis Group, 2018. | Series: Textbook series in physical sciences | Includes index.
Identifiers: LCCN 2018027638 (print) | LCCN 2018051265 (ebook) | ISBN 9781351207270 (eBook General) | ISBN 9781351207263 (eBook Adobe Reader) | ISBN 9781351207256 (eBook ePub) | ISBN 9781351207249 (eBook Mobipocket) | ISBN 9780815382928 (hardback : alk. paper)
Subjects: LCSH: Physics. | Physics—Textbooks. | Industrial electronics. | Electronic control.
Classification: LCC QC23.2 (ebook) | LCC QC23.2 .N53 2018 (print) | DDC 530.02/462—dc23
LC record available at https://lccn.loc.gov/2018027638

Visit the Taylor & Francis Web site at
http://www.taylorandfrancis.com

and the CRC Press Web site at
http://www.crcpress.com

To

Kathy and Henry

Contents

Preface

This book is intended as an introductory physics text for students majoring in technology. A background in algebra is assumed. It was written in an effort to try to make clear the relevance of physics for a career in technology. To achieve this, lengthy derivations and theoretical discussions on physics have been replaced with practical applications of physics in industry. Unlike most texts in physics where a general concept is explained leading to specific examples, the opposite approach has been taken here. Whenever possible, a discussion starts off with something the student is familiar with, for example, a speedometer, thermometer, etc., and then generalizes it to the physics behind these devices and concepts they address. This approach makes clear the relevance and motivation to study physics. Laboratory methods and data analysis are included setting this book apart from other introductory texts. This book is intended for a one-semester course in physics using a judicious selection of chapters.

WHAT'S NEW IN THIS EDITION?

This second edition of Physics for Technology was written in an attempt to bridge the gap in advances in technology since the first edition. This was achieved by incorporating sensors into the general physics discussions along with labs using data acquisition through the use of sensors interfaced to microcontrollers, along with smartphone physics labs. This book is effectively a text and lab book in one. Specifically, the new additions to the first edition are:

- Updating each chapter with current technology.
- Four new chapters on electric force, electricity, data acquisition with microcontrollers, and smartphone labs.
- Over 300 additional end of the chapter problems added.

D.H. Nichols

Acknowledgements

I would like to thank my wife Kathy and son Henry for their patience and support during the writing of this book. I would like to also acknowledge phyphox.org for their excellent smartphone physics app, and adafruit.com and sparkfun.com for the use of figures from their website.

Author

Dan Nichols has been teaching Physics and Electronics for more than 20 years. Dan received a bachelors from Indiana University and a PhD from Temple University, both in Physics. When he is not teaching at the University of Alaska Anchorage Matanuska-Susitna College, he is in his lab building robots, rockets, weather monitoring equipment, hydroponics systems, and assorted IOT projects. If you can't find Dan in either the classroom or in his lab, you will find him hiking with his beautiful wife Kathy and his wonderful son Henry in the Alaskan wilderness.

1

Units and Measurements

Units are based on standards wherein we define measurable quantities, such as length, mass, time, and so on. The units vary depending on the size of things we are measuring. For example, we would not measure the distance across the United States in inches; that is much too small a unit. Instead we would use a unit such as miles or kilometers. It is very handy to be able to convert back and forth between like units, such as miles, inches, and kilometers. This chapter will present some of the various ways of measuring, recording, and converting between units (Figure 1.1).

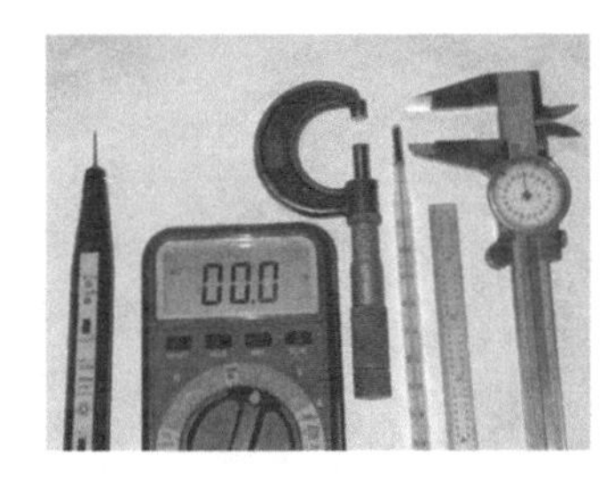

FIGURE 1.1 Measuring instruments.

1.1 SCIENTIFIC NOTATION

Many measurements are written in *scientific notation*, which is a method of writing big and small numbers in a compact way. In this notation, all numbers are written in a power of ten.

$$10 = 1.0 \times 10^{1}$$

$$100 = 1.0 \times 10^{2}$$

$$1{,}000 = 1.0 \times 10^{3}$$

The number above the ten is called an *exponent*, which tells how many zeros come after the one.

For numbers such as 1/10 or 1/100, we use a minus sign in the exponent.

$$\frac{1}{10} = 1.0 \times 10^{-1}$$

$$\frac{1}{100} = 1.0 \times 10^{-2}$$

$$\frac{1}{1{,}000} = 1.0 \times 10^{-3}$$

The number in the exponent now tells how many zeros follow the one at the bottom of the fraction.

Numbers such as 638,000 or 0.00012 can be written in scientific notation.

$$638{,}000 = 6.38 \times 10^{5} \leftarrow \text{Decimal moved to the left five places}$$

$$0.00012 = 1.2 \times 10^{-4} \leftarrow \text{Decimal moved to the right four places}$$

Place the decimal after the first nonzero digit on the left and count the number of places the decimal was moved over. Use that number in the exponent. If the decimal is moved to the left, the exponent is positive. If the decimal is moved to the right, the exponent is negative.

1.2 UNITS OF LENGTH

In the English system, length units are miles, yards, feet, and inches. Another system is the SI system. SI stands for the Système International (International System, in English). The metric system is part of the SI system. The basic unit of length in SI is meter. All units of length are based on it.

In the early part of the twentieth century, the standard of measuring length was defined using a metal bar that was kept in a vault in France. To calibrate a meter stick, one would have to travel to France and compare the two. This was very inconvenient! It therefore became necessary to devise a generic method of measuring length so that the whole world could have access to it. To alleviate this problem, the meter is defined as the distance that light travels in a vacuum during the time interval of 1/299,792,458 s. This may seem like a strange way to define a meter, but it eliminates the need to have a single object stored in a vault somewhere, which acts as the definition of what "1 meter long" actually is (Figure 1.2).

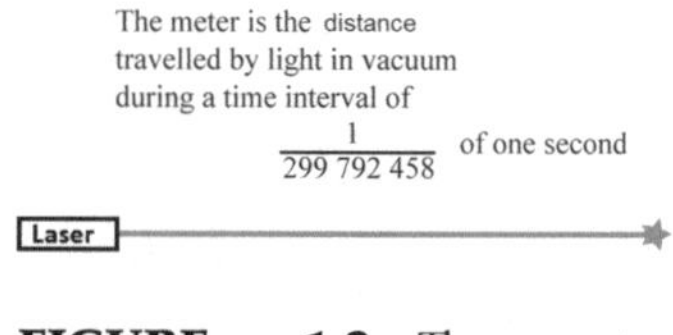

FIGURE 1.2 The meter defined.

In the metric system, length units are kilometers, meters, centimeters, and millimeters. The metric system is much easier to use than the English system of units because everything is grouped into powers of ten (Figure 1.3).

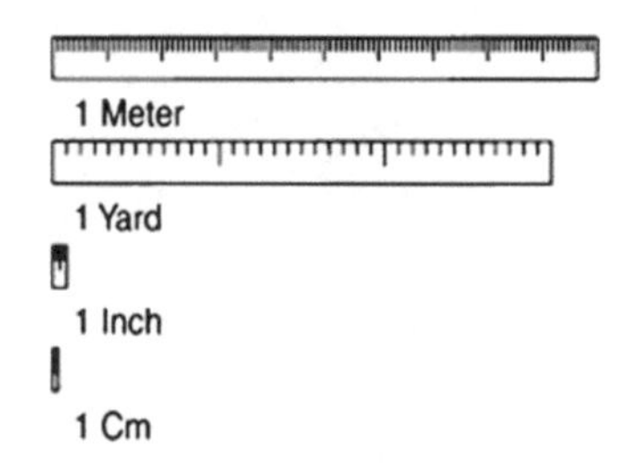

FIGURE 1.3 Comparison of various length units.

- The MKS system is a part of the SI system. All units are recorded and calculated in meters, kilograms, and seconds (MKS).
- In the English system, calculations are done using pounds, slugs, feet, and seconds. This system is sometimes called the FPS system, or the foot-pound-second system.

In the metric system, all units are grouped into powers of ten using prefixes (Table 1.1).

TABLE 1.1
Table of Metric Prefixes

1 giga = 1 G = 1×10^{9}	
1 mega = 1 M = 1×10^{6}	
1 kilo = 1 k = 1×10^{3}	
Unity = 1 = 1×10^{0}	
1 milli = 1 m = 1×10^{-3}	
1 micro = 1 μ = 1×10^{-6}	
1 nano = 1 n = 1×10^{-9}	
1 pico = 1 p = 1×10^{-12}	
1 km = 1,000 m	1 km is approximately 0.6 miles.
1 m = 100 cm	1 cm is approximately 0.4 in
1 mm = $\frac{1}{1,000}$ m	1 mm is approximately the thickness of a dime.
1 μm = $\frac{1}{1,000,000}$ m	1 μm is approximately $\frac{1}{100}$ the width of a human hair.

1.2.1 Mechanical Measuring Instruments

The choice of the measuring instrument depends on the accuracy needed. A standard machinist's rule can measure accurately to the smallest division on the scale, typically 0.01 in. or 0.1 mm (Figure 1.4).

For better accuracy, a Vernier caliper can measure both inside and outside diameters accurately to 0.001 in. (Figure 1.5).

A high-precision micrometer can measure accurately to 0.0001 in. or 0.002 mm (Figure 1.6).

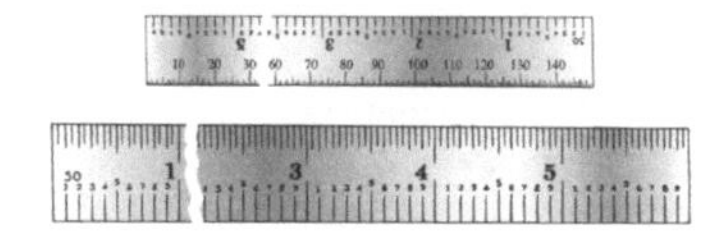

FIGURE 1.4 A machinist's scale is accurate to 0.01 in.

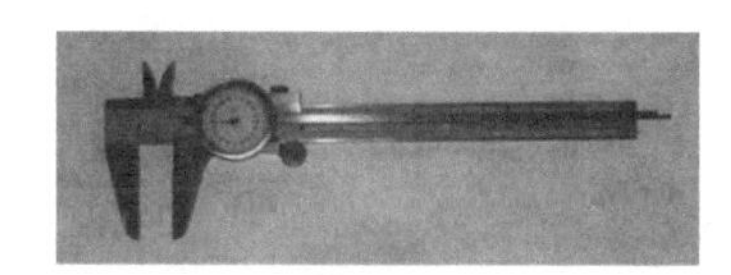

FIGURE 1.5 A vernier caliper measures accurately to 0.001 in.

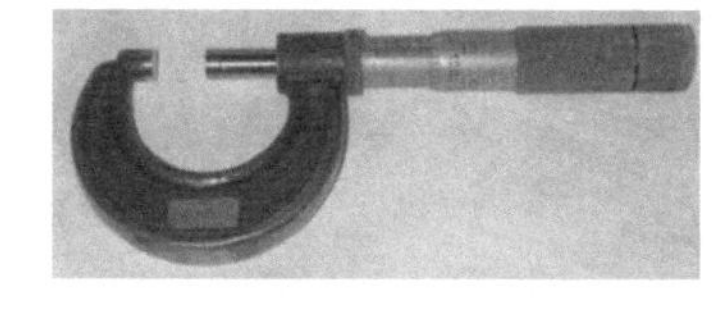

FIGURE 1.6 A high-precision micrometer measures accurately to 0.0001 in.

1.2.2 Recording and Calculating with Measurements

When we record a measurement, the number we write down should convey how accurately the measurement was made. For example, 2.1 in. and 2.10 in. mean the same thing mathematically but not in terms of measurements. The number 2.10 in. means the number was measured to one-hundredth of an inch. The number 2.1 in. means the number was measured to one-tenth of an inch.

Suppose a measurement is made with a voltmeter that displays the measurement as 2.1 V. If the real value of the measurement is 2.15 V or greater, the meter would read 2.2 V. If the value is less than 2.05 V, the meter would read 2.0 V. Therefore, 2.1 V means the measurement was made with an uncertainty of ±0.05 V, and 2.10 V means the measurement was made with an uncertainty of ±0.005 V.

The uncertainty is $\pm\frac{1}{2}$ the smallest scale division.

1.2.3 Significant Digits

A *significant digit* is a digit in a number that has been measured. When calculating with numbers, we must take into account the accuracy of the numbers in the calculation. The following rules have been adopted to convey how accurately something has been measured:

Number	Significant Digits
2.1 in.	2
2.10	3
0.01 in.	1
350	2
2.01×10^4	3
$\bar{3}00$	2
$40\bar{0}$	3

1. All nonzero numbers are significant.
2. Zeros are significant if
   a. They have a bar placed over them.
   b. They lie between significant digits.
   c. They follow the decimal and a significant digit.

1.2.4 Rounding Off

When rounding off a number, if the first digit dropped is less than five, round down; if it is five or greater, round up.

Example 1.1

Round off 254.735 to the nearest 100, 10, 1, 0.1, and 0.01.

254.735 = 300	Nearest 100
254.735 = 250	Nearest 10
254.735 = 255	Nearest 1
254.735 = 254.7	Nearest 0.1
254.735 = 254.74	Nearest 0.01

1.2.5 Addition/Subtraction

When quantities are added or subtracted, first round the numbers to the correct number of significant digits, then round the result to the decimal place of the number with the fewest decimal places in the addition or subtraction.

$$2\,\text{in.} + 1.4\,\text{in.} = 3\,\text{in.}$$

$$2.0\,\text{in.} + 1.4\,\text{in.} = 3.4\,\text{in.}$$

$$2\,\text{in.} - 1.4\,\text{in.} = 1\,\text{in.}$$

1.2.6 Multiplication/Division

When quantities are multiplied or divided, round the result so that it does not have more significant digits than the number with the least significant digits in the multiplication or division.

$$(28.0\,\text{ft})(6\,\text{ft}) = 200\,\text{ft}^2$$

$$(28.0\,\text{ft})(6.0\,\text{ft}) = 170\,\text{ft}^2$$

$$\frac{28.0\,\text{ft}}{6\,\text{ft}} = 5$$

$$\frac{28.0\,\text{ft}}{6.0\,\text{ft}} = 4.7$$

An exception to this is the case when conversions are used. For example, 1 ft = 12 in. is exact and should be treated as exact, that is, having an infinite number of significant digits.

$$0.2563\,\text{ft}^2\left(\frac{12\,\text{in.}}{1\,\text{ft}}\right)^2$$

$$= 0.2563\,\cancel{\text{ft}^2}\left(\frac{144\,\text{in.}^2}{1\,\cancel{\text{ft}^2}}\right)$$

$$= 36.91\,\text{in.}^2$$

1.2.7 Calibration

Calibration means to compare a measurement device against a known standard. Calibrating a measurement device means to adjust its output so that it displays correctly what it is measuring. A correctly calibrated instrument is very important, especially in the medical field, where an incorrectly calibrated instrument could cost someone's life.

1.2.8 Unit Conversions

To convert between the two systems, use the following conversions:

1 in. = 2.54 cm
1 m = 100 cm
1 cm = 10 mm
1 mile = 5,280 ft
1 yd = 3 ft
1 ft = 12 in.

Suppose you are given a measurement in feet and want to know how many inches that equals. Use the bracket method.

1.2.9 The Bracket Method

The bracket method is a way to convert between any type of like units. Convert 4.20 ft into in.

To use the bracket method, apply the following steps:

Step 1 Write down the conversion.

$$1\,\text{ft} = 12\,\text{in.}$$

Step 2 Write down what is to be converted with a bracket next to it.

$$4.20\,\text{ft}\left(\frac{\text{unit to convert to}}{\text{original unit}}\right)$$

Step 3 Fill in the bracket with the conversion, placing the unit that is on the outside of the bracket at the bottom of the bracket and the unit it is equal to at the top part of the bracket.

$$4.20\,\cancel{\text{ft}}\left(\frac{12\,\text{in.}}{1\,\cancel{\text{ft}}}\right) = 50.4\,\text{in.}$$

Example 1.2

Convert 30.0 in. into ft.

$$1\text{ft} = 12\text{in.}$$

$$30.0\,\cancel{\text{in.}}\left(\frac{1\text{ft}}{12\,\cancel{\text{in.}}}\right) = 2.50\,\text{ft}$$

Notice that the unit on the outside of the bracket is in inches; therefore, the unit in the bottom of the bracket must also be in inches in order for inches to cancel out, leaving feet. It is not necessary to try to remember what goes at the top or bottom; just remember to place the unit that is on the outside of the bracket at the bottom of the bracket. Or if the unit on the outside is at the bottom, place the same unit inside the bracket at the top.

Example 1.3

Convert 25 cm into m.

$$1\,\text{m} = 100\,\text{cm}$$

$$25\,\cancel{\text{cm}}\left(\frac{1\text{m}}{100\,\cancel{\text{cm}}}\right) = 0.25\text{m}$$

Example 1.4

Convert 25 mph into miles/min.

$$1\,\text{h} = 60\,\text{min}$$

$$25\frac{\text{miles}}{\cancel{\text{h}}}\left(\frac{1\,\cancel{\text{h}}}{60\,\text{min}}\right) = 0.42\frac{\text{miles}}{\text{min}}$$

The bracket method can be used when there is no direct conversion from one unit to another. To perform such a conversion, use the bracket method twice.

1.2.10 Multiple Conversions

Multiple conversions are used when no direct conversion exists. For example, convert 32.2 yd into in. We know that:

$$1\,\text{yd} = 3\,\text{ft}$$

$$1\,\text{ft} = 12\,\text{in.}$$

We do not have a direct conversion from yards to inches. We can, however, convert yards to feet, then feet into inches.

$$32.2\,\cancel{\text{yd}}\left(\frac{3\,\cancel{\text{ft}}}{1\,\cancel{\text{yd}}}\right)\left(\frac{12\,\text{in.}}{1\,\cancel{\text{ft}}}\right) = 1{,}160\,\text{in.}$$

In the first bracket, yards is canceled out, leaving feet. Therefore, the unit feet is on the outside of the second bracket, and we should put feet at the bottom of the second bracket.

Example 1.5

Convert 2.00 m into in. using the following conversions:

$$1\,\text{m} = 100\,\text{cm}$$

$$1\,\text{in.} = 2.54\,\text{cm}$$

We are not given a direct conversion from meters to inches, so we will have to perform a double conversion.

$$2.00\,\cancel{\text{m}}\left(\frac{100\,\cancel{\text{cm}}}{1\,\cancel{\text{m}}}\right)\left(\frac{1\,\text{in.}}{2.54\,\cancel{\text{cm}}}\right) = 78.7\,\text{in.}$$

1.2.11 Measuring Length Electronically

There are many ways to measure length or distance. Your accuracy requirements determine which type of instrument to use.

Electronic Micrometer: An electronic micrometer is accurate to 0.0001 in. (Figures 1.7 and 1.8).

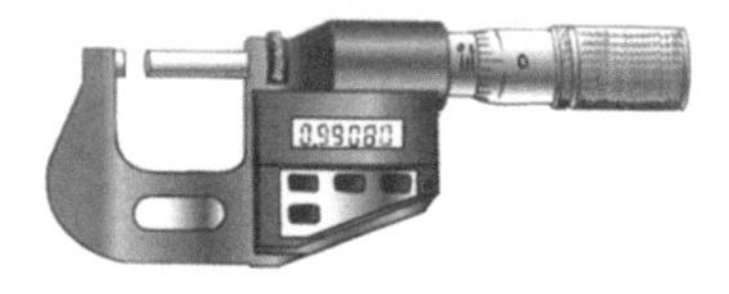

FIGURE 1.7 An electronic micrometer.

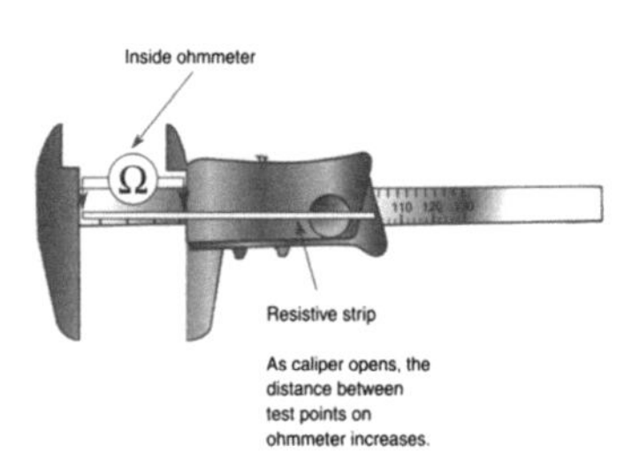

FIGURE 1.8 Electronic micrometers and calipers contain resistive strips. As the barrel or caliper is moved back, the distance between measuring points on the resistive strip increases, and therefore the resistance measured will increase. This resistance is converted into a distance and displayed.

Ultrasonic Tape Measure: An ultrasonic tape measure, which is accurate to a few tenths of an inch over a distance of tens of feet, can be bought in any hardware store (Figures 1.9 and 1.10).

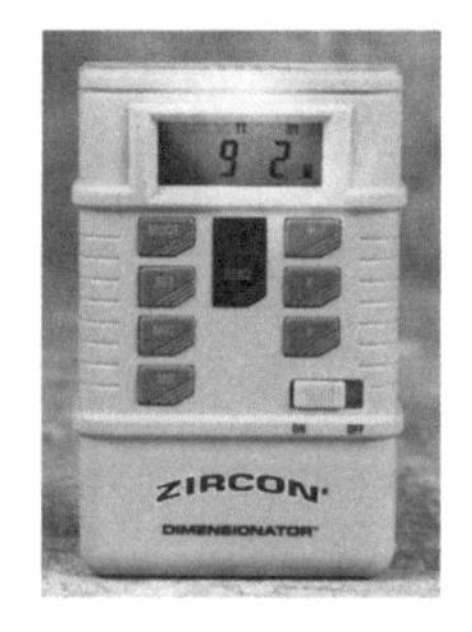

FIGURE 1.9 Ultrasonic tape measure (courtesy of Zircon).

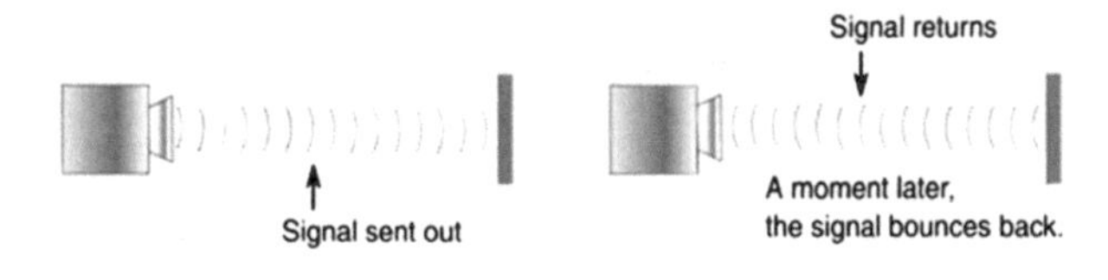

FIGURE 1.10 The ultrasonic tape measure works by sending a sound wave and timing how long it takes to return after bouncing off an object whose distance is being measured. The longer it takes, the farther away the object.

Laser Ranging works by sending a light beam and timing how long it takes to return after bouncing off the object being measured. The longer it takes, the farther away the object. (Figures 1.11–1.13).

Capacitive Sensors: As the spacing between plates on a capacitor decreases, the capacitance increases. This change in capacitance can be used to determine the relative change in plate separation. By anchoring one

FIGURE 1.11 A commercial laser ranger (courtesy of Zircon).

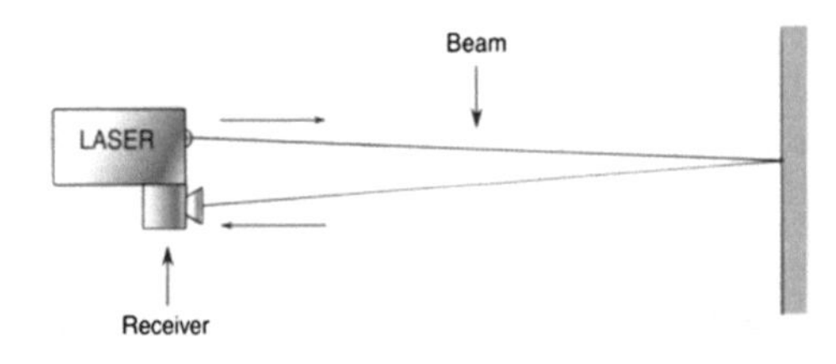

FIGURE 1.12 Laser ranging works by sending a light beam and timing how long it takes to return after bouncing off the object being measured. The longer it takes, the farther away the object. Background light does not interfere with the measurement because the detector is sensitive only to the frequency of the laser.

FIGURE 1.13 Adafruit's VL53L0X is a *Time of Flight* distance sensor, 30–1,000 mm, courtesy of adafruit.com.

of the plates to a fixed object and the other to an object in motion, the object in motion can be monitored by measuring the capacitance (Figure 1.14).

A capacitor can also measure displacement using a change in the dielectric's position between the two plates. As the dielectric is inserted into the capacitor, its capacitance will rise. This method is sometimes used to measure the height of a fluid in a tube (Figure 1.15).

Smartphones have a grid of fine wires under the glass. This grid is used to determine where your finger is on the screen by measuring the change in the capacitance at each point in the grid that your finger touches. There are plenty of apps that use this capability of the phone to measure small lengths (Figures 1.16 and 1.17).

For large distances, smartphones use the Global Positioning System, GPS. GPS uses a system of satellites in space to triangulate your position on the earth. Locations on the ground can be recorded and used to determine distance (Figure 1.18).

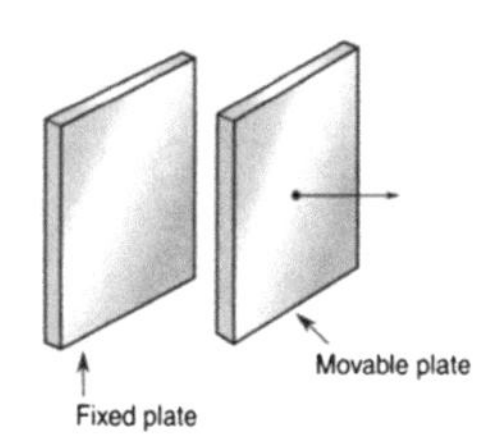

FIGURE 1.14 A variable capacitor is used to measure displacement.

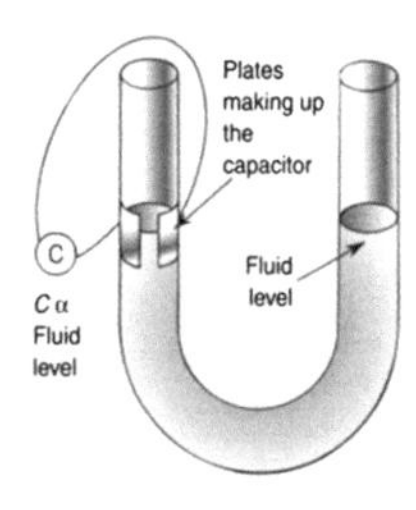

FIGURE 1.15 A capacitor is used to measure the level of a fluid.

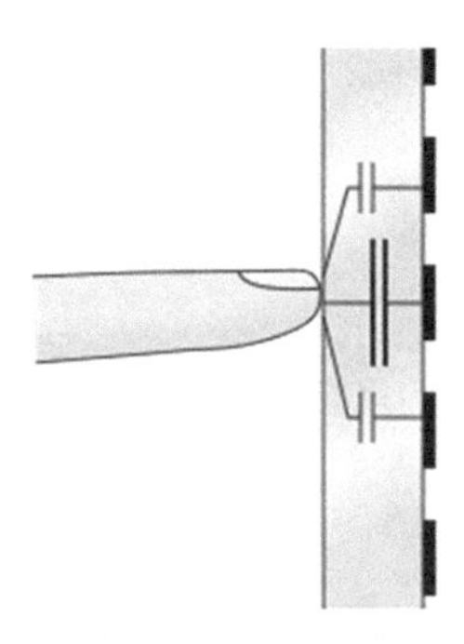

FIGURE 1.16 Smartphones use a grid of fine wires under the glass to determine where your finger is on the screen by measuring the change in the capacitance at each point in the grid that your finger touches.

FIGURE 1.17 A smartphone app to measure length.

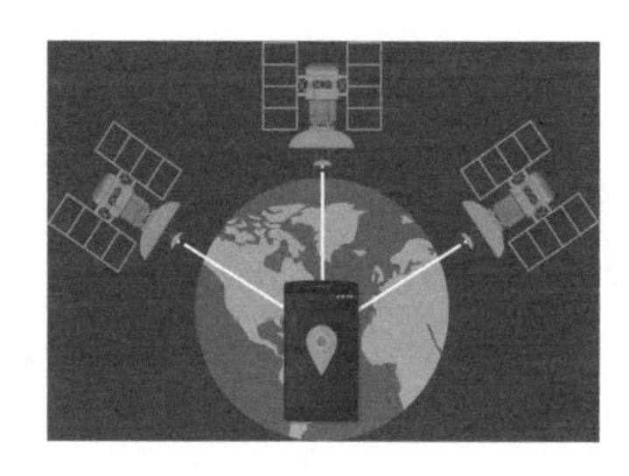

FIGURE 1.18 GPS uses a system of satellites.

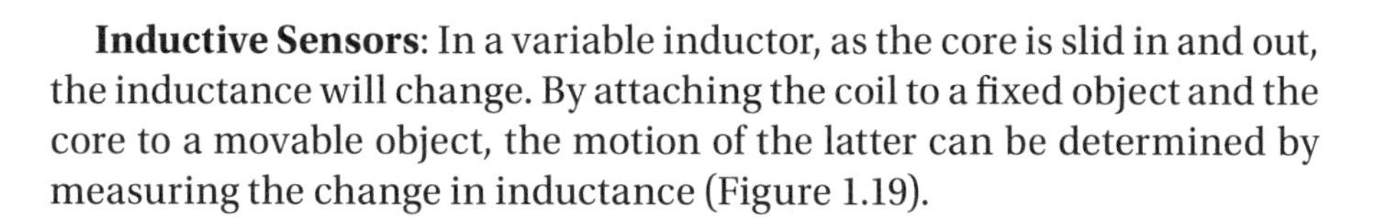

Inductive Sensors: In a variable inductor, as the core is slid in and out, the inductance will change. By attaching the coil to a fixed object and the core to a movable object, the motion of the latter can be determined by measuring the change in inductance (Figure 1.19).

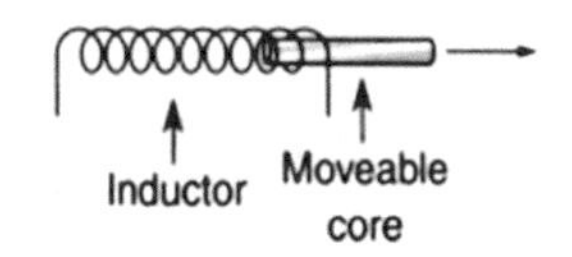

FIGURE 1.19 A variable inductor is used to measure length. As the core is moved out, the inductance will decrease.

1.3 AREA

If you have ever looked at heat sinks mounted on integrated circuit (IC) chips, the fins are designed to increase the surface area of the chip. A large surface area will diffuse heat faster than a small one.

Area has units of $(\text{length})^2$ (spoken as "square meters" or "square feet", etc.) (Figure 1.20).

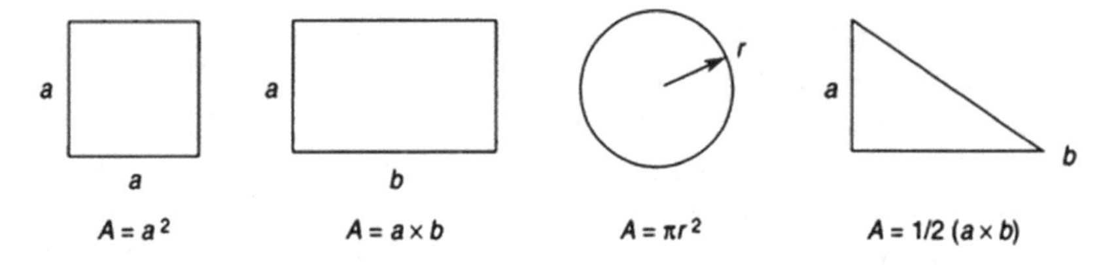

FIGURE 1.20 Areas of various shapes.

$$1\,\text{m}^2 = 10{,}000\,\text{cm}^2$$

$$1\,\text{ft}^2 = 144\,\text{in.}^2$$

$$1\,\text{cm}^2 = 0.155\,\text{in.}^2$$

$$1\,\text{ft}^2 = 929\,\text{cm}^2$$

Example 1.6

What is the area of a square with sides that are of 2.0 in. in length?

$$A = a^2$$

$$A = (2.0\,\text{in.})^2 = 4.0\,\text{in.}^2$$

Example 1.7

What is the area of a rectangle with a height of 2.5 in. and a width of 3.0 in.?

$$A = a \times b$$

$$A = (2.5\,\text{in.})(3.0\,\text{in.})$$

$$A = 7.5\,\text{in.}^2$$

Example 1.8

What is the area of a circle with a radius of 3.00 cm?

$$A = \pi r^2$$

$$A = \pi (3.00\,\text{cm})^2$$

$$A = 28.3\,\text{cm}^2$$

Example 1.9

What is the area of a right triangle with a height of 11 cm and a base of 12 cm?

$$A = \frac{1}{2}ab$$
$$A = \frac{1}{2}(11\text{cm})(12\text{cm})$$
$$A = 66\text{cm}^2$$

1.3.1 Surface Area of Three-Dimensional Objects

The surface area of a three-dimensional shape is the sum of the areas of each of its sides (Figure 1.21).

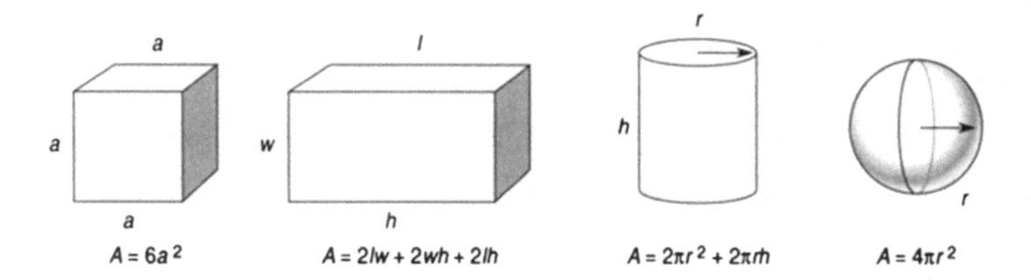

FIGURE 1.21 Surface areas of various shapes.

Example 1.10

What is the surface area of a cube with sides that are 2.0 cm long?

$$A = 6a^2$$
$$A = 6(2.0\text{cm})^2$$
$$A = 24\text{cm}^2$$

Example 1.11

What is the surface area of a rectangular prism with a length of 2.5 in., a width of 3.5 in., and a height of 5.1 in.?

$$A = 2lw + 2wh + 2lh$$
$$A = 2(2.5\text{in.})(3.5\text{in.}) + 2(3.5\text{in.})(5.1\text{in.}) + 2(2.5\text{in.})(5.1\text{in.})$$
$$A = 79\text{in.}^2$$

Example 1.12

What is the surface area of a sphere with a radius of 3.00 cm?

$$A = 4\pi r^2$$
$$A = 4\pi(3.00\text{cm})^2$$
$$A = 113\text{cm}^2$$

Example 1.13

What is the surface area of a Pringles® potato chip can with a radius of 3.00 cm and a height of 9.00 cm, including the ends?

$$A = 2\pi r^2 + 2\pi rh$$

$$A = 2\pi(3.00\,\text{cm})^2 + 2\pi(3.00\,\text{cm})(9.00\,\text{cm})$$

$$A = 226\,\text{cm}^2$$

1.3.2 The Bracket Method Revisited

When converting units that involve powers, the bracket must be raised to the same power as the unit being converted. In case of area conversions, the bracket must be squared.

Remember, when raising a bracket to a power, everything in the bracket is raised to that power.

Example 1.14

Convert 3 m^2 to cm^2.

$$3\,\text{m}^2\left(\frac{100\,\text{cm}}{1\text{m}}\right)^2$$

$$= 3\,\cancel{\text{m}^2}\,\frac{100^2\,\text{cm}^2}{\cancel{\text{m}^2}}$$

$$= 30{,}000\,\text{cm}^2$$

Example 1.15

Convert 12.3 cm^2 to m^2.

$$12.3\,\text{cm}^2\left(\frac{1\text{m}}{100\,\text{cm}}\right)^2$$

$$= 12.3\,\cancel{\text{cm}^2}\,\frac{1\text{m}^2}{100^2\,\cancel{\text{cm}^2}}$$

$$= 1.23\times10^{-3}\text{m}^2$$

Example 1.16

Convert 13.2 ft^2 to in^2.

$$13.2\,\text{ft}^2\left(\frac{12\text{in.}}{1\text{ft}}\right)^2$$

$$= 13.2\,\cancel{\text{ft}^2}\,\frac{12^2\,\text{in.}^2}{\cancel{\text{ft}^2}}$$

$$= 1.9\bar{0}0\,\text{in.}^2$$

1.4 VOLUME CALCULATIONS

A volume calculation involves a length × length × length, or (length)^3, such as the volume of a box where volume is determined by width times length times height (Figures 1.22 and 1.23).

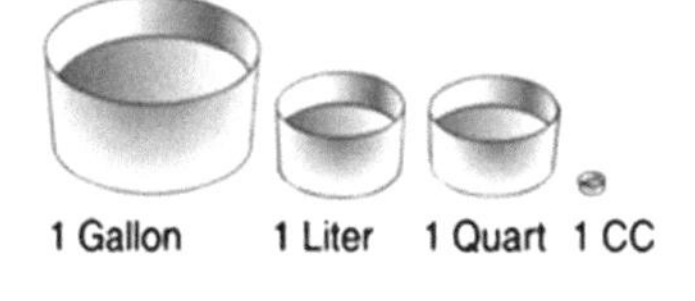

FIGURE 1.22 A comparison of various units.

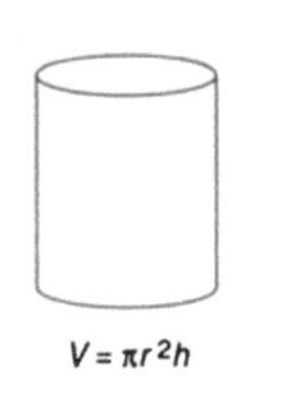
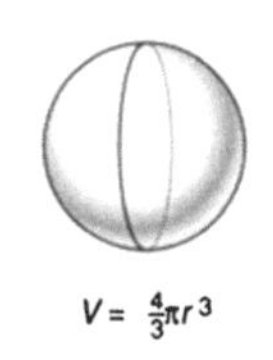

FIGURE 1.23 Volumes of various shapes.

1.4.1 Volume Units

SI
1 liter (L) = 1,000 cm^3
1 m^3 = 10^3 L
1 cm^3 = 1,000 mm^3

English
1 gal = 4 qt
1 gal = 8 pints
1 gal = 128 fl. oz.
1 ft^3 = 7.48 gal

English to SI
1 ft^3 = 28.3 L
ft^3 = 1 m^3
1 gal = 3.785 L

Example 1.17

Find the radius of a baseball with a volume of 33.5 in.^{3}

$$V = \frac{4}{3}\pi \cdot r^3$$

$$r^3 = \left(\frac{3 \cdot V}{4 \cdot \pi}\right)$$

$$r = \left(\frac{3 \cdot V}{4 \cdot \pi}\right)^{1/3}$$

$$r = \left(\frac{3 \cdot 33.5\text{in.}^3}{4 \cdot \pi}\right)^{1/3}$$

$$r = 2.00\,\text{in.}$$

1.4.2 Volume Unit Conversions—The Bracket Method

When converting units that involve powers, the bracket must be raised to the same power as the unit being converted. In case of volume conversions, the bracket must be cubed.

Remember, when raising a bracket to a power, everything in the bracket is raised to that power.

Example 1.18

Convert 25 in.^{3} into ft^3..

$$1\text{ft} = 12\text{in.}$$

$$25\text{in.}^3 \left(\frac{1\text{ft}}{12\text{in.}}\right)^3$$

$$= 25\,\cancel{\text{in}^3}\,\frac{\text{ft}^3}{12^3\,\cancel{\text{in}^3}}$$

$$= 0.014\,\text{ft}^3$$

Example 1.19

Convert 1.2 m^3 into cm^3.

$$1\text{m} = 100\,\text{cm}$$

$$1.2\text{m}^3 \left(\frac{100\,\text{cm}}{1\text{m}}\right)^3$$

$$= 1.2\,\cancel{\text{m}^3}\,\frac{100^3\,\text{cm}^3}{\cancel{\text{m}^3}}$$

$$= 1.2 \times 10^6\,\text{cm}^3$$

Example 1.20

Convert 2.5 in.^{3} into ft^3.

$$1\text{ft} = 12\text{in.}$$

$$2.50\text{in.}^3 \left(\frac{1\text{ft}}{12\text{in.}}\right)^3$$

$$= 2.50\,\cancel{\text{in.}^3}\,\frac{\text{ft}^3}{12^3\,\cancel{\text{in.}^3}}$$

$$= 1.4 \times 10^{-3}\text{ft}^3$$

1.4.3 Measuring Volume

One way to measure the volume of an object is to submerge it in water and measure how much water is displaced (Figure 1.24).

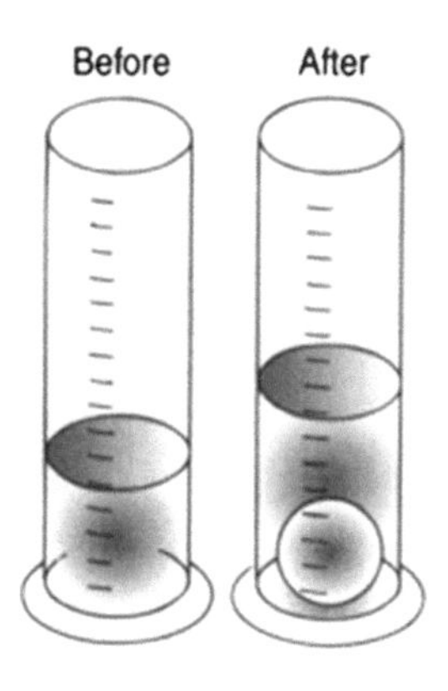

FIGURE 1.24 To measure the volume of an object, place the object in a full container of water and measure the amount of water displaced. The volume of the object equals the volume of water displaced.

1.5 MASS AND WEIGHT

Mass and weight are sometimes referred to as the same thing. This is not quite correct, however, as will be explained in more detail in a later chapter. For now, we will use simple definitions of mass and weight.

Mass: The *mass* of an object is the measure of how much force is required to change the motion of the object. Its basic unit of measurement is the gram. A small paper clip has the mass of about 1 g. The kilogram is currently defined in terms of the mass of a platinum-iridium cylinder stored in a vault in France. Imagine having to travel to France to determine whether your scale is correct or not! This definition of the kilogram that is going to be replaced with a method based on fundamental constants of nature, like the speed of light or the charge on an electron. The definition of the kilogram will be based on a constant called Planck's constant, a value from an area of physics called quantum mechanics that describes the physics of atoms, see www.nist.gov for more details.

Weight: The *weight* of an object refers to how heavy it is or how strongly it is attracted to the earth because of gravity.

Density: *Density* is a measure of how much mass or weight there is in a given volume. For example, a little bit of lead weighs a lot. It has a high density.

$$\rho = \frac{m}{V}$$

$$\text{Density}\left(\frac{\text{g}}{\text{cm}^3}\right) = \frac{\text{mass}(\text{g})}{\text{volume}\left(\text{cm}^3\right)}$$

$$\rho = \frac{w}{V}$$

$$\text{Density}\left(\frac{\text{lb}}{\text{ft}^3}\right) = \frac{\text{weight}(\text{lb})}{\text{volume}\left(\text{ft}^3\right)}$$

These equations show that a lot of mass or weight contained in a small volume has a high density (see Table 1.2).

TABLE 1.2
Table of Densities[a]

Solid	g/cm^3	lb$_m$/ft^3
Aluminum	2.699	168.5
Aluminum, A1-Clad 17 ST alloy	2.96	185
Brass, yellow, cast	8.44	527
Brick (average)	1.9	120
Bronze, gun metal	8.78	548
Cement, dry	1.5	94
Cement, set	2.7–3.0	170–190
Concrete (average)	2.4	150
Copper, cast	8.30–8.95	518–559
Copper, hard-drawn	8.89	555
Douglas fir, dry	0.446	27.8
Glass, common	2.4–2.8	150–175
Gold, cast	19.3	1,200
Ice (0°C)	0.917	57.2
Lead	11.3	705
Oak, white, dry	0.710	44.3
Paper	0.70–1.15	44–72
Pine, white, dry	0.373	23.3
Platinum	21.37	1,334
Sand	1.5	94
Silver, cast	10.5	655
Solder, plumbing	9.4	590
Steel, 1% carbon	7.83	489
Steel, stainless	7.75	484
Tungsten	18.6–19.1	1,160–1,190
Tungsten carbide	14.0	874
Uranium	18.7	1,170

[a] The temperature is 20°C [68°F] unless stated otherwise.
Note: $1\ \text{g/cm}^3 = 1\ \text{kg/L} = 1{,}000\ \text{kg/m}^3$

Example 1.21

The density of water is 62.4 lb/ft^3. A 55 gal drum contains approximately 7.33 ft^3 of volume (Figure 1.25). How much does a 55 gal drum of water weigh?

$$\rho = \frac{w}{V}$$

$$w = \rho \cdot V$$

$$w = 62.4\,\text{lb}/\cancel{\text{ft}^3} \cdot 7.33\ \cancel{\text{ft}^3}$$

$$w = 457.4\,\text{lb}$$

FIGURE 1.25 A 55 gal drum.

Example 1.22

Suppose you buy a mobile home, a double-wide trailer measuring 80.0 ft by 9.0 ft by 20.0 ft, and have it trucked to your lot (Figure 1.26). If the density of air at the time of the move is 0.0810 lb/ft^3, what is the weight of the air being transported on the truck?

$$V = l \cdot w \cdot h$$
$$V = 80.0\,\text{ft} \cdot 9.0\,\text{ft} \cdot 20.0\,\text{ft}$$
$$V = 14,400\,\text{ft}^3$$
$$\rho = \frac{w}{V}$$
$$w = \rho \cdot V$$
$$w = 0.0810\frac{\text{lb}}{\text{ft}^3} \cdot 14,400\,\text{ft}^3$$
$$w = 1166.4\,\text{lb}$$

Who would ever think air could weigh so much!

20.0 ft
9.00 ft
80.0 ft

FIGURE 1.26 A double-wide trailer.

1.6 TIME

The standard measurement of time is the second. Time is measured by a clock. A *clock* is anything that repeats itself at a constant rate.

Atoms make for very accurate clocks. A cesium 133 atom, for example, emits microwave radiation at a very precise frequency when stimulated. A second is defined as the duration of 9,192,631,700 periods of oscillation of radiation coming from a cesium 133 atom. Listed below are a number of ways to measure time.

FIGURE 1.27 A pendulum clock.

1.6.1 Time Measurement by Mechanical Devices

Pendulum: A pendulum could be simply a weight on a string swinging back and forth, repeating its motion at a regular rate. A repetitive motion is known as an *oscillation*. One back-and-forth motion is one oscillation. To time something with a pendulum, simply count the number of complete back-and-forth motions or oscillations during any action being timed and multiply this number by the time for one oscillation of the pendulum (Figure 1.27).

Example 1.23

Make a pendulum from a weight tied onto a string 1 m long. If you pull the weight back a small amount, it will take 2 s to complete one oscillation for a pendulum of this length. Time your heart rate by counting the number of beats for 30 oscillations; this will give you your heart rate in beats/min (Figure 1.28).

FIGURE 1.28 Timing your heart rate.

Weight on a Spring: A weight on a spring, bouncing up and down, completes an oscillation at a regular rate. To time something, simply count

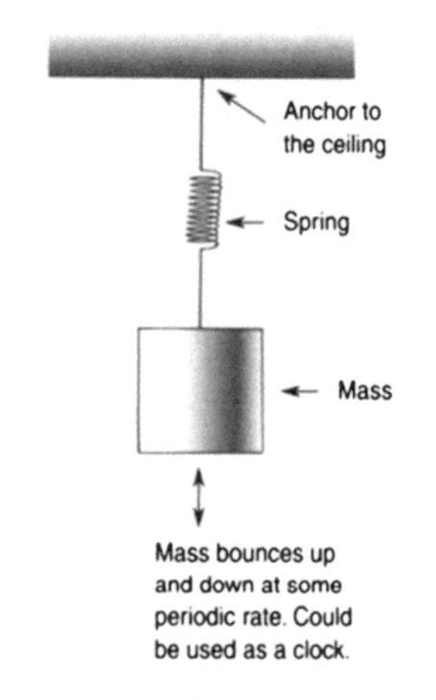

FIGURE 1.29 A spring clock.

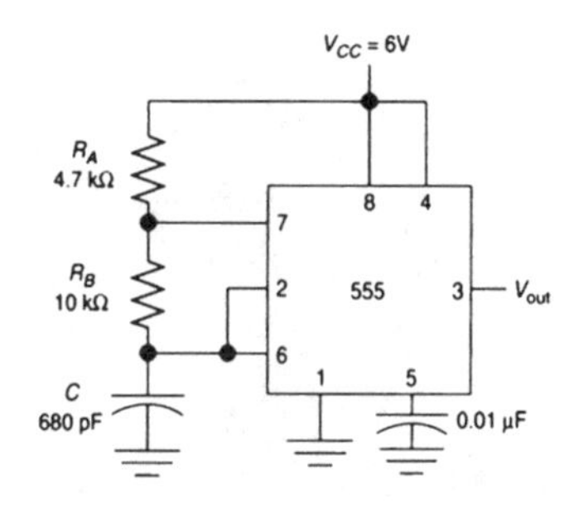

FIGURE 1.30 A 555-timer chip.

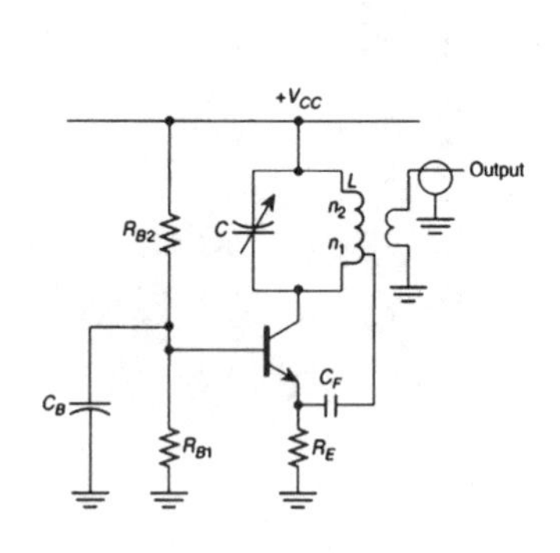

FIGURE 1.31 An LC oscillator.

the number of oscillations it takes for some action to be completed and multiply this by the time for one oscillation (Figure 1.29).

1.6.2 Electronic Measurement Devices

RC Oscillator: The resistive capacitor oscillator or RC oscillator works by alternating charging and discharging of a capacitor through a resistor happens at a determined rate, dependent on the time constant $\tau = R \times C$. This can be used to measure time. One of the electronic industry's standards that controls this charging and discharging is the 555-timer chip (Figure 1.30). It has a stability of 1%.

Example 1.24

An oscillator with the period of 1 ms may be used as a timer to cook a 1-min, soft-boiled egg. Simply wait for 60,000 pulses, and the egg is done.

LC Oscillator: The inductor capacitor or LC oscillator device has an accuracy 100 times better than the RC oscillator (Figure 1.31). It has a stability of 0.01%.

Crystal Oscillator: This device is many times more accurate than the LC oscillator (Figure 1.32). Over a normal temperature range, stabilities of a few parts per million are easily obtained.

Atomic Clock: Atomic clocks have the best accuracy of all the oscillators. Currently, the best atomic clock measures the movement of strontium atoms held in place by a laser. All atoms naturally vibrate at particular

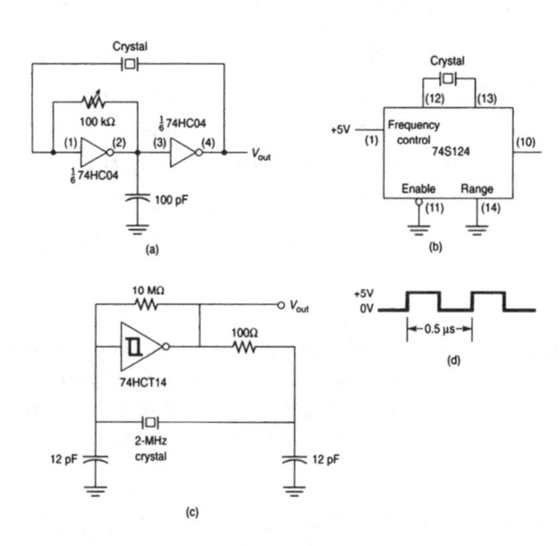

FIGURE 1.32 A crystal oscillator.

frequencies depending on the atom and its environment. Strontium, under the right conditions, has a consistent vibrational frequency of about 430 trillion times/s. Counting 430 trillion of these oscillations is, therefore, equivalent to the passage of one second of time. Currently, these clocks loose only one second every 15 billion years (Figure 1.33).

Smartphones rely on a connection with your carrier to ensure they are synchronized to the correct time. Without this connection, however, the time could be off by several seconds per day.

FIGURE 1.33 An atomic clock.

1.6.3 Calibrating Clocks

As stated above, calibrating a measurement device means to adjust its output against a known standard so that it accurately displays what it is measuring. In case of time, the best standard is an atomic clock. So, where do you get an atomic clock? The good news is that you don't need one. The National Institute of Standards and Technology (NIST) and radio station WWVB operate on short wave at 60 kHz and broadcast times and frequencies derived from their atomic clocks. This radio station can be used for calibration. For more information, look at the NIST web site at www.boulder.nist.gov/timefreq.

1.7 CHAPTER SUMMARY

Units used in this chapter:

Symbol	Unit
d—distance	m
r—radius	m
A—area	m^2
V—Volume	m^3
ρ—density	g/cm^3, kg/m^3, lb/ft^3

The bracket method is a systematic way to make unit conversions.

Step 1: Write down the conversions.

Step 2: Write down what is to be converted with a bracket next to it.

Step 3: Fill in the bracket with the conversion, placing the unit that is on the outside of the bracket at the bottom of the bracket and the unit it is equal to at the top part of the bracket.

When converting units that are squared or cubed, don't forget to raise the bracket to the same power.

1.7.1 Significant Digits

All nonzero numbers are significant. Zeros are significant if

a. They have a bar placed over them.
b. They lie between significant digits.
c. They follow the decimal and a significant digit.

1.7.2 Adding and Subtracting Significant Digits

When quantities are added or subtracted, round the result to the decimal place of the number with the fewest decimal places in the addition or subtraction.

1.7.3 Multiplying and Dividing Significant Digits

When quantities are multiplied or divided, round the result so that it does not have more significant digits than the number with the least significant digits in the multiplication or division.

1.7.4 Conversions

1.7.4.1 Length

$$1\,\text{in.} = 2.54\,\text{cm}$$

$$1\,\text{m} = 100\,\text{cm}$$

$$1\,\text{cm} = 10\,\text{mm}$$

$$1\,\text{mi} = 5{,}280\,\text{ft}$$

$$1\,\text{yd} = 3\,\text{ft}$$

$$1\,\text{ft} = 12\,\text{in.}$$

1.7.4.2 Area

$$1\,\text{m}^2 = 10{,}000\,\text{cm}^2$$

$$1\,\text{ft}^2 = 144\,\text{in.}^2$$

$$1\,\text{cm}^2 = 0.155\,\text{in.}^2$$

$$1\,\text{ft}^2 = 929\,\text{cm}^2$$

1.7.4.3 Volume

SI

$$1\,\text{liter}(\text{L}) = 1{,}000\,\text{cm}^3$$

$$1\,\text{m}^3 = 10^3\,\text{L}$$

$$1\,\text{cm}^3 = 1{,}000\,\text{mm}^3$$

English to SI

$$1\,\text{ft}^3 = 28.3\,\text{L}$$

$$25.3\,\text{ft}^3 = 1\,\text{m}^3$$

$$1\,\text{gal} = 3.785\,\text{L}$$

$$\rho = \frac{m}{V}$$

$$\text{Density} = \frac{\text{mass}(\text{g})}{\text{volume}(\text{cm}^3)}$$

$$\rho = \frac{w}{V}$$

$$\text{Density} = \frac{\text{weight}(\text{lb})}{\text{volume}(\text{ft}^3)}$$

1.7.4.4 Formulas for Area

Square: $A = a^2$

Rectangle: $A = a \times b$

Circle: $A = \pi r^2$

Right triangle: $A = 1/2(a \times b)$

1.7.4.5 Formulas for Volume

$$\text{Cube: } V = a^3$$

$$\text{Rectangular volume: } V = l \times w \times h$$

$$\text{Cylinder: } V = \pi r^2 h$$

$$\text{Sphere: } V = \frac{4}{3}\pi r^3$$

PROBLEM SOLVING TIPS

1. Remember to raise the bracket in a conversion to the same power that is being converted.
2. Don't forget to apply the rules of addition/subtraction and multiplication/division of significant digits when calculating.

PROBLEMS

1. Convert 1.2 miles into ft.
2. Calculate the distance to the moon in ft. The moon is 250,000 miles away.
3. Convert 25 miles into ft.
4. Convert 3.5 yd into ft.
5. Convert 10 ft into in.
6. Convert 12 in. into cm.
7. Convert 5.6 m into cm.
8. Convert 25 cm into mm.
9. Convert 350 μm into mm.
10. Convert 0.8 yd into in.
11. Convert 2.1 ft into cm.
12. Convert 25 $in.^2$ into cm^2.
13. Convert 59 $in.^3$ into m^3.
14. Convert 327 $in.^3$ into L.
15. Calculate the area of a rectangle with a length of 3.0 m and a width of 5.0 m.
16. Calculate the area of a circle with radius of 1.5 cm.
17. Calculate the area of a right triangle with a base of 2.1 in. and a height of 4.8 in.
18. Calculate the surface area of a cylinder with a radius of 23 cm and a height of 4.9 cm.

19. Calculate the volume of a sphere with a radius of 5.6 mm.
20. A car has a 5.6 L engine. How many cubic centimeters is this?
21. A pump fills a pool with 5,000 L of water. How many cubic inches is this?
22. Calculate how many seconds there are in 1 year.
23. A man lives to be 72 years old. How many minutes is this?
24. A clock pulse from an oscillator is on for 50 μs and off for 25 μs. If 100 pulses have elapsed, how long has the clock been running for?
25. A pendulum takes 1.2 s to complete one oscillation. If the pendulum completes 32 oscillations, how much time has elapsed?
26. A strain gauge is used to measure the weight of some object. The gauge emits a signal of 10 lb/mV. If the gauge is measuring 25 mV, how much does the object weigh?
27. An object has a density of 84.5 g/cm^3 and a mass of 13.2 g. What is its volume?
28. A radar gun measures the speed of a moving car by determining the difference in frequency of the outgoing and incoming waves it emits. The signal out of the gun is equal to $\frac{10\,\text{mph}}{\text{Hz}}$. Suppose the gun measures a frequency difference of 9 Hz. How fast is the car moving in mph?

Taking into account the rules for calculating with significant digits, calculate the following.

29. 3.21 in. + 3.1 in. + 4.675 in. = ?
30. 2.76×103 m − 2.4×102 m = ?
31. (6.67 in.) (2.1 in.) (3.32 in.) = ?
32. $\frac{0.012\,\text{m}}{0.00354\,\text{m}} = ?$
33. What is the density of a ball with a radius of 2 in and weighing 1.7 lb?
34. Using the rule of significant digits, how many significant digits does 2.05×100 have?
35. A heat sink consists of 4 fins measuring 1.0 in by 2.0 in. in area. What is the total surface area available for cooling? Remember each fin has two sides!
36. A box top measures 3.5 in. by 3.5 in. What is the area of this box top in cm^2?
37. If the period of a pendulum is 2 s, how many oscillations take place in 20 min?

38. Convert 1,800 revolutions/min into revolutions/s.
39. A battery lasts for 2 years. How many seconds is this?
40. A fish tank measures 18 in. wide × 12 in. tall × 10 in. long. If 1,800 in.^{3} of water is added, how deep is the water in the tank?
41. Air has a density of 0.0081 lb/ft^3. What does the air weigh in a room measuring 10 ft × 15 ft × 12 ft.
42. A clock is made from a 0 –5 V square wave oscillator with a period of 1 ms and a counter. If the clock is turned on and the counter registers a count of 5,000, how many seconds have elapsed?
43. An object of unknown volume is placed in a cylinder of water, raising the water level 3 in. The radius of the cylinder is 2 in. What is the volume of the object?
44. Using the rule of significant digits, how many significant digits does 2.8 × 203 have?
45. A heat sink consists of 8 fins and gives off 100 W of heat. How much power does each fin dissipate?
46. A rectangular computer measures 18 in. by 16 in. by 6.0 in. What is the volume of this computer in cm^3?
47. If the period of a mass oscillating on a spring is 1 s, how many oscillations take place in 2 min?
48. Convert 25 mph into miles/min.
49. A computer stays on for 3 years. How many seconds is this?
50. A rectangular box measures 12 cm × 14 cm × 20 cm. If cubes measuring 2 cm × 2 cm × 2 cm were placed in the box, how many would fit?
51. Water has a density of 62.4 lb/ft^3. What does a barrel of volume 5 ft^3 of water weigh?
52. A clock is made from a crystal oscillator with a period of 0.1 ms. If the clock is turned on and 200 pulses come out of it, how many seconds have elapsed?
53. A fish of unknown volume is placed in a fish tank of water measuring 12 in. long, 6 in. wide, and 8 in. tall and raises the water level 0.2 in. What is the volume of the fish?

2 Linear Motion

Linear motion means movement in a straight line. The equations that describe this motion are used in the programs that run elevators, computer numerical control (CNC) milling machines, military ballistics, and robotic arms that put together anything from automobiles to electronic circuit boards. This chapter will describe how to measure and calculate this motion.

2.1 RATES

A *rate* is a measure of how something changes over time. For example, the rate of water flowing out of a faucet is measured in gallons per minute (gal/min), and the rate of a machine gun firing is measured in rounds per second (rounds/s). Rates have units of some quantity over time (quantity/time). *Speed* is also a rate. Speed is a measure of how far something travels in a given amount of time. If a car travels 50 miles in 1 h, then the average speed of the car is 50 mph (Figure 2.1).

FIGURE 2.1 A moving car.

2.2 VECTOR

A quantity that has both magnitude and direction is called a *vector.* Traveling in a car at a certain speed in a particular direction is an example of a vector. The description contains both magnitude or size and direction.

2.3 SCALAR

A *scalar* is a quantity without a direction attached to it. Speed, mass, length, and time are scalar quantities.

2.4 COORDINATE SYSTEMS

A typical coordinate system defines up as positive and down as negative, and to the right as positive and to the left as negative (Figure 2.2).

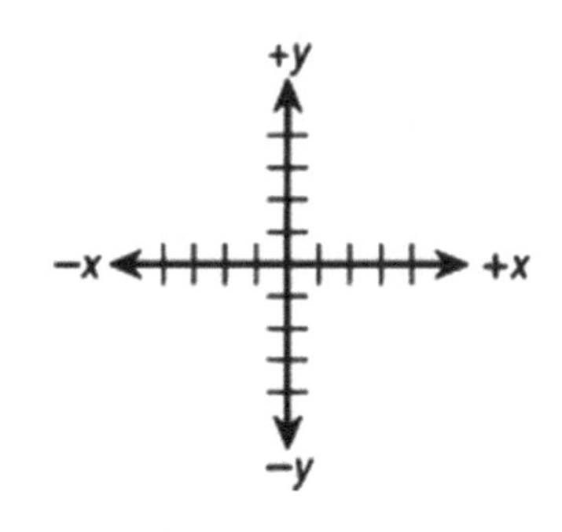

FIGURE 2.2 A typical coordinate system defines up as positive and down as negative, and to the right as positive and to the left as negative.

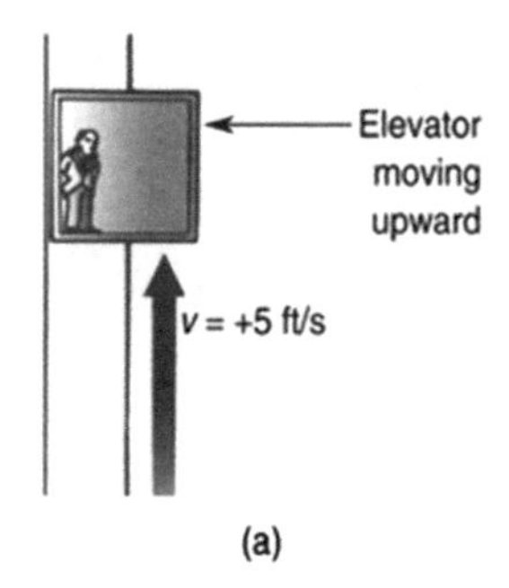

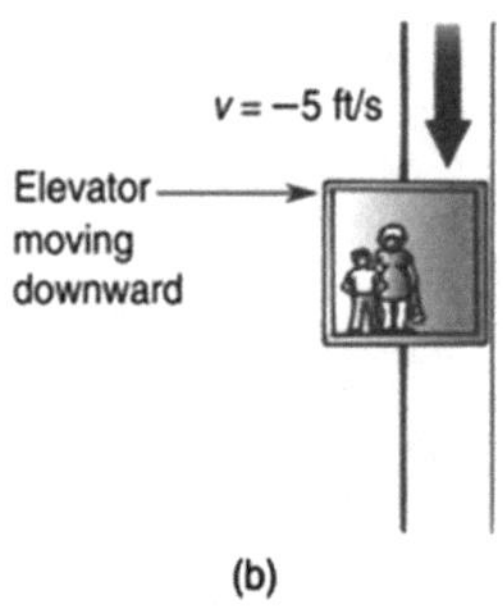

FIGURE 2.3 A elevator moving downward.

2.5 VELOCITY AND SPEED

Velocity is similar to speed, but it also includes a direction of travel.

Average velocity $\bar{v}$ can be calculated by:

$$\bar{v} = \frac{d}{t}$$

$$\text{Average velocity} = \frac{\text{distance}}{\text{time}}$$

In the metric system, velocity has units of m/s and in the English system, ft/s.

Example 2.1

A car moves at an average speed of 50 mph, traveling northwest. The average velocity of the car is 50 mph, northwest.

Example 2.2

An elevator moving upward at a speed of 5 ft/s has a velocity of +5 ft/s (Figure 2.3). If the same elevator is moving downward at a speed of 5 ft/s, its velocity is –5 ft/s. The sign indicates the direction. As you can see, velocity describes both speed and direction.

Example 2.3

A robot travels 20 ft to the right in 4 s (Figure 2.4). What is its average velocity?

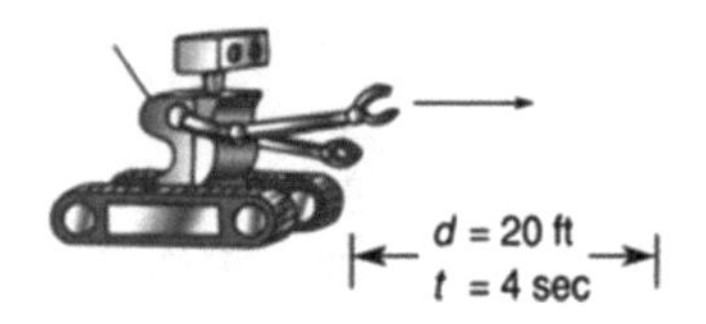

FIGURE 2.4 A robot moving.

$$d = 20\,\text{ft}$$

$$t = 4\,\text{s}$$

$$\bar{v} = ?$$

$$\bar{v} = \frac{d}{t}$$

$$\bar{v} = \left(\frac{20\,\text{ft}}{4\text{s}}\right)$$

$$\bar{v} = 5\frac{\text{ft}}{\text{s}}, \text{ to the right}$$

$$\text{or } \bar{v} = +5\frac{\text{ft}}{\text{s}}$$

Example 2.4

A robotic arm moving at an average speed of 35 ft/s travels 1.2 ft. How long did it take to travel this distance? (Figure 2.5)

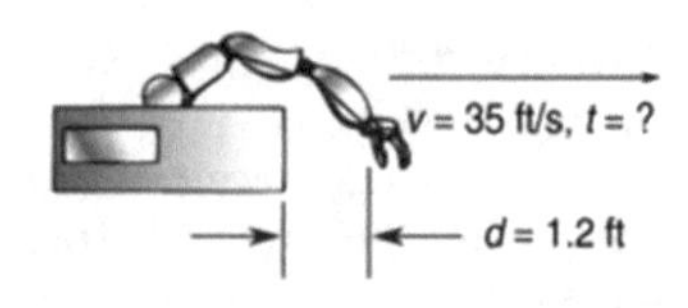

FIGURE 2.5 A robotic arm.

$$v = 35\frac{\text{ft}}{\text{s}}$$

$$d = 1.2 \text{ ft}$$

$$t = ?$$

$$\bar{v} = \frac{d}{t}$$

$$t = \frac{d}{\bar{v}}$$

$$t = \frac{1.2 \text{ ft}}{35 \text{ ft/s}}$$

$$t = 0.034 \text{ s}$$

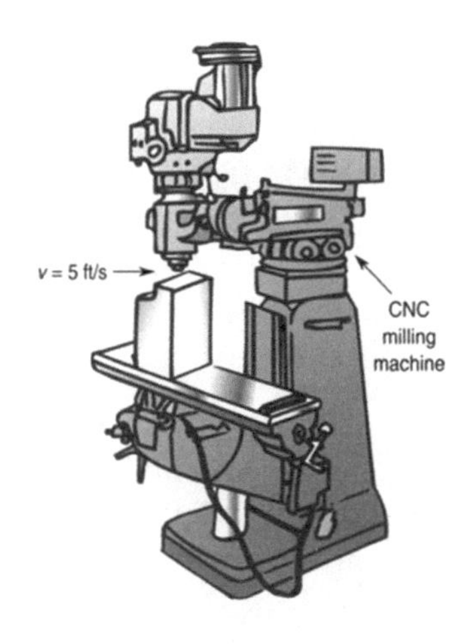

FIGURE 2.6 A mill.

Example 2.5

The head of a CNC milling machine moves across the table at an average speed of 5 ft/s. How far did the head travel in 0.2 s? (Figure 2.6)

$$\bar{v} = 5\frac{\text{ft}}{\text{s}}$$

$$t = 0.2 \text{s}$$

$$d = ?$$

$$\bar{v} = \frac{d}{t}$$

$$d = \bar{v}t$$

$$d = \left(5\frac{\text{ft}}{\cancel{\text{s}}}\right)0.2\cancel{\text{s}}$$

$$d = 1.0 \text{ft}$$

The NASA Parker Space Probe, scheduled to launch in 2018 will reach 430,000 mph as it dives into the sun. This will beat the last record held by NASA space probe Juno that was sling-shot toward Jupiter in 2013 (Figure 2.7).

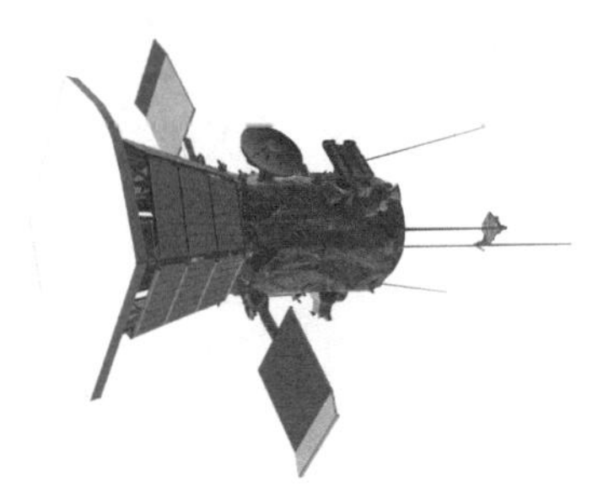

FIGURE 2.7 The NASA Parker Space Probe, scheduled to launch in 2018 will reach 430,000 mph as it dives into the sun.

2.6 INSTANTANEOUS VELOCITY

Instantaneous velocity is the velocity of an object at one point in time. For instance, the velocity of a car 5 s after the traffic light turns green.

We shall use the symbols for initial and final velocities:

$$v_i = \text{velocity}_{\text{initial}}$$

$$v_f = \text{velocity}_{\text{final}}$$

FIGURE 2.8 A race car.

Example 2.6

A race car starts from rest at a stoplight; after passing the finish line, the velocity of the car is +440 ft/s. What are the initial and final velocities of the car? (Figure 2.8)

$$v_i = 0\frac{\text{ft}}{\text{s}}$$

$$v_f = +440\frac{\text{ft}}{\text{s}}$$

FIGURE 2.9 Instantaneous velocity.

Example 2.7

A ball is tossed upward at 12 ft/s and reaches some maximum height before changing its direction and falling back down. What is the initial velocity and the final velocity of the ball when it reaches maximum height? (Figure 2.9)

$$v_i = +12\frac{\text{ft}}{\text{s}}$$

$$v_f = 0\frac{\text{ft}}{\text{s}}$$

When the ball reached maximum height, it was not moving.

PROBLEM SOLVING TIP

A projectile traveling upward, only under the influence of gravity, reaches maximum height when it changes its direction of travel, and at this point its velocity is zero.

2.7 MEASURING SPEED, SPEEDOMETERS

A nonelectronic or mechanical car speedometer measures the speed of the car using a steel cable connected to the transmission (Figure 2.10). The cable spins at a rate proportional to the car's speed. The cable enters a speed cup, which consists of a rotating magnet inside an aluminum cup attached to a spring and a needle. As the magnet spins, it induces an electric current in the cup, which produces its own magnetic field. The two magnetic fields interact, producing a torque in the cup and causing it to rotate the needle (Figure 2.10).

An electronic speedometer shown in Figure 2.11, measures the rate of spin of a transmission gear to determine the speed of a car.

Radar Gun: A police radar gun consists of a microwave radio source emitting a beam of radio waves of a given frequency (Figure 2.12). The waves are

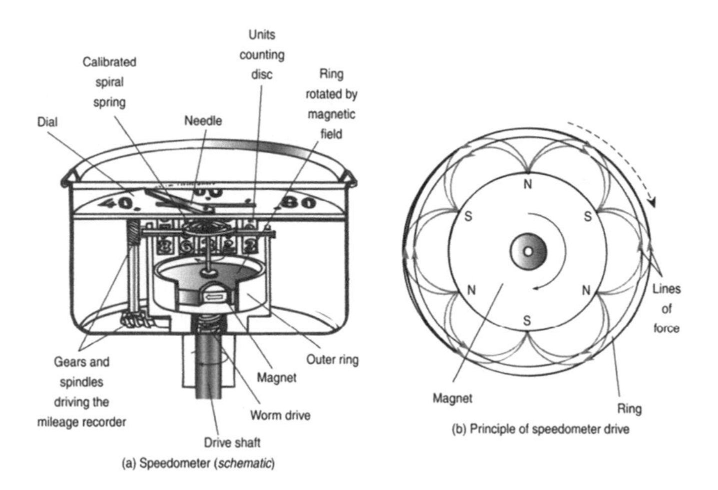

FIGURE 2.10 A mechanical car speedometer.

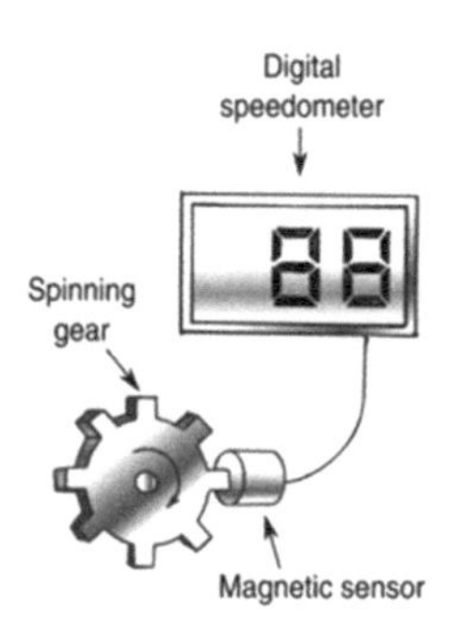

FIGURE 2.11 An electronic speedometer uses a magnetic sensor to determine the rate at which the transmission gears are spinning and converts this into a speed.

shot at moving vehicles. After reflecting off a car, the wave returns altered in frequency because of the car's motion. This is known as the *Doppler effect*, which will be studied in a later chapter. The reflected wave's frequency is compared with the frequency of the original wave sent out. The difference in the frequency of the two waves is a measure of how fast the car is moving.

GPS: Global Positioning System is a network of approximately 30 satellites orbiting the earth at an altitude 20,000 km above the earth. GPS works based on a method called triangulation to determine a location on earth. A GPS receiver on earth, such as your phone, picks up a signal from several of the satellites and measures the time travel of each of the signals. The radio signals from the satellites travel at the speed of light, $v = 3\times10^8\ \frac{\text{m}}{\text{s}}$, and using the equation $d = vt$, the distance to each satellite can be determined. Measuring your position at specific intervals of time and using $v = \frac{d}{t}$ will give you your speed (Figures 2.13 and 2.14).

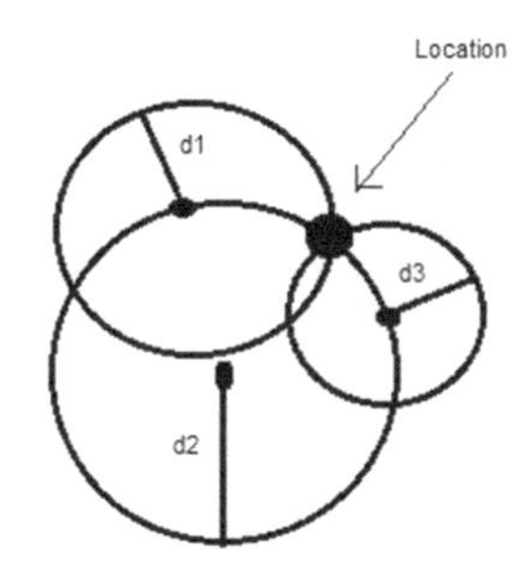

FIGURE 2.14 Each satellite gives a location with an uncertainty of radius d. Combining the possible locations from multiple satellite and finding the intersection give a more precise measurement.

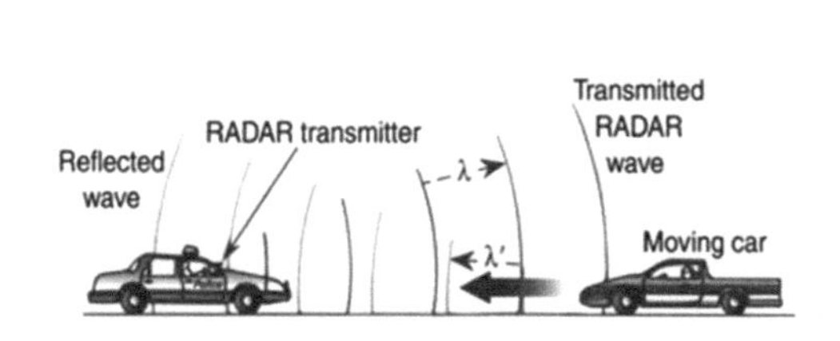

FIGURE 2.12 A police radar gun.

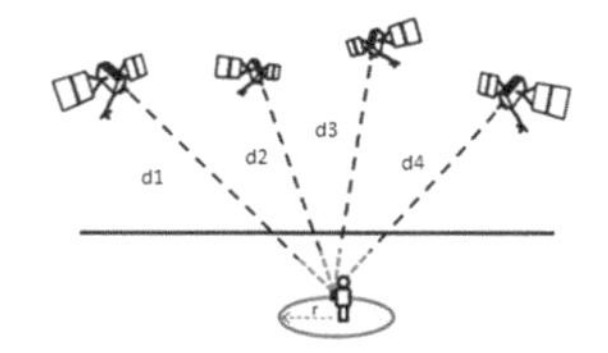

FIGURE 2.13 GPS uses a system of satellites in space to triangulate your position on the earth.

2.8 ACCELERATION

Acceleration is the rate at which velocity is changing. Normally, one thinks accelerating means speeding up and deaccelerating means slowing down. This is not always the case, for example, an object moving in a circle at a constant speed is accelerating because it is constantly changing direction. Acceleration has units of length/time². This makes sense because acceleration is the rate of a rate, so we end up with the unit length/time². The average acceleration is given by the formula:

$$\bar{a} = \frac{v_f - v_i}{t}$$

$$\text{Average acceleration} = \frac{\text{velocity final} - \text{velocity initial}}{\text{time}}$$

In the metric system, acceleration gas unit is m/s^2 and in the English system, ft/s^2.

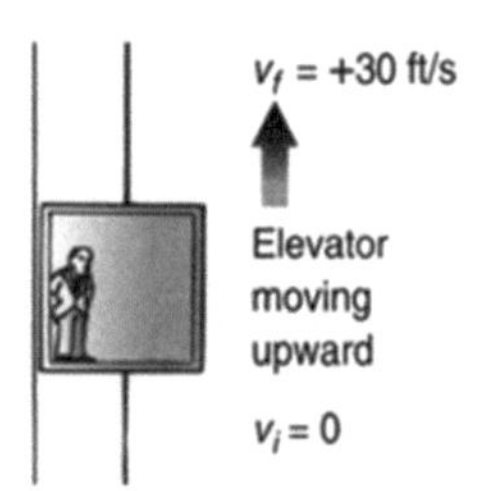

FIGURE 2.15 What is the average acceleration of the elevator?

Example 2.8

An object moving in a straight line, accelerating at +25 ft/s², means every second, the object increases its velocity +25 ft/s. If the object started from rest, after 1 s it would be moving at +25 ft/s; after 2 s, +50 ft/s; 3 s, +75 ft/s, and so on.

Example 2.9

An elevator moves from zero to +30 ft/s in 5 s. What is its average acceleration? (Figure 2.15)

$$\bar{a} = \frac{v_f - v_i}{t}$$

$$\bar{a} = \frac{30\frac{\text{ft}}{\text{s}} - 0\frac{\text{ft}}{\text{s}}}{5\text{s}}$$

$$\bar{a} = 6\ \frac{\text{ft}}{\text{s}^2}$$

As you can see, acceleration has units of distance/time².

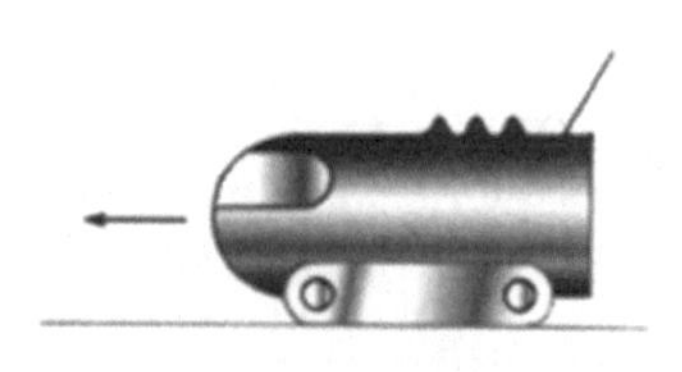

FIGURE 2.16 What is the acceleration of the machine?

Example 2.10

A machine moves across the floor, changing its velocity from –11 to –3 m/s, and it takes 4 s to do this. What is its acceleration? (Figure 2.16)

$$\bar{a} = \frac{v_f - v_i}{t}$$

$$\bar{a} = \frac{-3\frac{\text{m}}{\text{s}} - \left(-11\frac{\text{m}}{\text{s}}\right)}{4\text{s}}$$

$$\bar{a} = 2\ \frac{\text{m}}{\text{s}^2}$$

2.8.1 Acceleration Due to Gravity (Free-Fall Acceleration)

Drop an object. As the object falls, it picks up speed. This picking up of speed, or acceleration, is the acceleration due to gravity known as *g*.

$$g = -32.2\ \text{ft/s}^2$$

$$g = -9.8\ \text{m/s}^2 \text{ in the metric system}$$

Note: *g* is negative because it points downward.

In free-fall acceleration, *g* acts the same upon every object independent of their weights. For example, if a bowling ball and a marble were dropped at the same time, they both would hit the ground at the same time (Figure 2.17). This is because gravity acts the same upon every object regardless of its weight; this is neglecting air resistance, of course.*

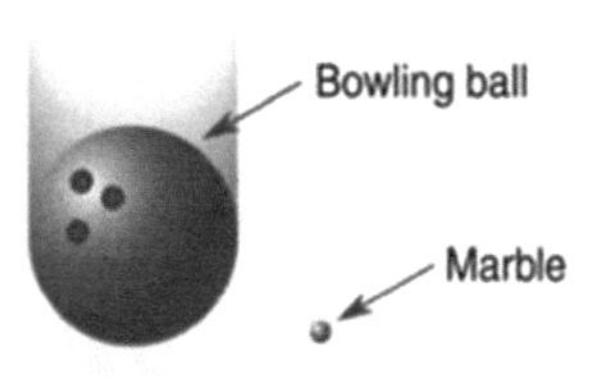

FIGURE 2.17 Both objects fall at the same rate and therefore hit the ground at the same time.

Example 2.11

Take a penny and a quarter, and drop them at the same time. Do they hit the ground at the same time? (Figure 2.18)

FIGURE 2.18 Which will hit the ground first, a quarter or a penny?

In all problems where objects are freely falling or being acted upon by gravity, acceleration $a = g$.

The free-fall acceleration can vary slightly over the earth's surface, depending on what is underneath the surface. For example, if there is a large ore deposit underneath the surface, the acceleration due to gravity will be slightly enhanced because ore is denser than ordinary soil. Mining companies use this property to determine where to dig for ore and other metals (Figure 2.19).

FIGURE 2.19 Determining ore deposits.

* The reason everything falls at the same rate is because inertial mass and gravitational mass are equivalent. These concepts will be discussed in later chapters.

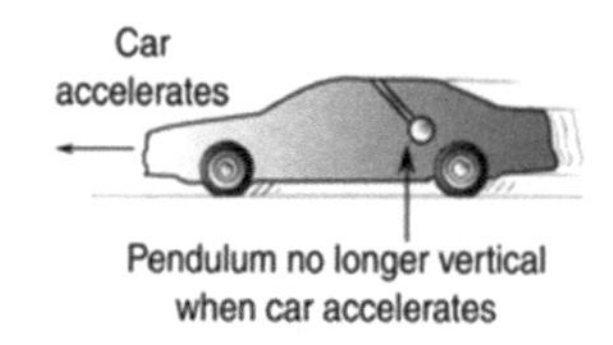

FIGURE 2.20 A pendulous accelerometer.

2.8.2 Accelerometers

A device that measures acceleration is called an *accelerometer.*

Pendulous Type: Pendulous-type accelerometers have suspended masses, with some type of transducer measuring the position of the mass (Figure 2.20). The mass will move only if accelerated, while the transducer will pick up this motion.

Inertia Switch: Many cars now have an *inertia switch,* which measures acceleration (Figure 2.21). If the acceleration is too large, such as in the event of a collision, the inertia switch shuts off the fuel pump and/or deploys the air bags.

Resistive Accelerometer: A *resistive accelerometer* is essentially a strain gauge with a mass mounted on it. During acceleration, the mass produces a stress on the resistive element. Stressing the resistor produces a change in its resistance (Figure 2.22).

MEMS Accelerometer: A *MEMS* (microelectromechanical systems) accelerometer is micro machined silicon wafer mass suspended by polysilicon springs. The wafer has a plate attached to it that acts as a plate of a capacitor, and the other plate is fixed. During an acceleration, the wafer is displaced and the capacitance between the plates changes. This change in capacitance produces a change in voltage that is proportional to the acceleration (Figures 2.23 and 2.24).

Instantaneous velocity is the velocity of an object at one point in time.

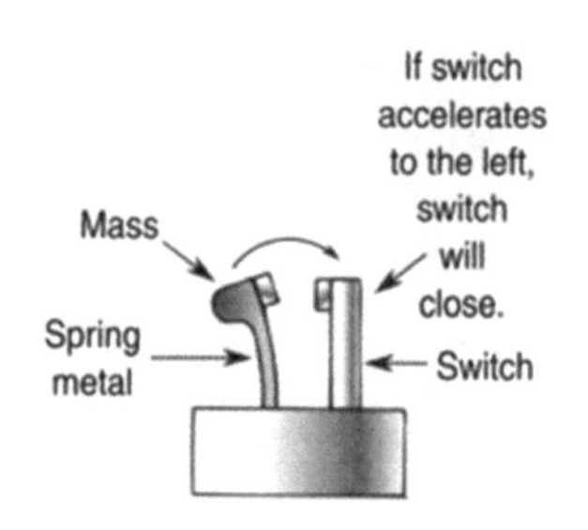

FIGURE 2.21 An inertia switch in an automobile shuts off the fuel supply in the event of a crash.

2.8.3 Accelerated Systems with a Constant Acceleration

For a system with a constant acceleration, as in the case of gravity, the average acceleration, $\bar{a}$, is the same at all times, and we can write $\bar{a}$ as a.

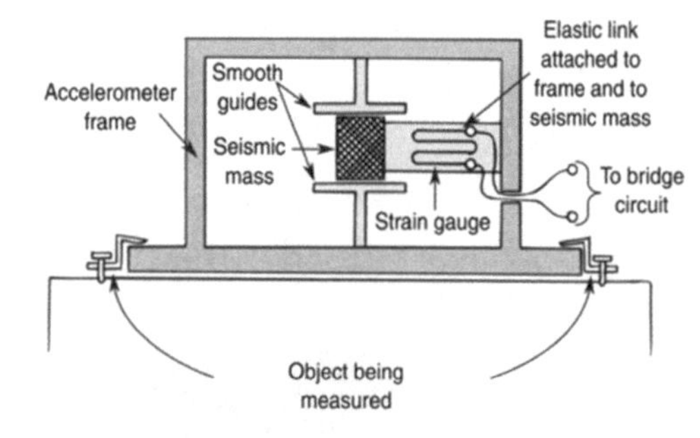

FIGURE 2.22 A resistive accelerometer. During acceleration, the mass produces a stress on the resistive element. Stressing the resistor produces a change in its resistance.

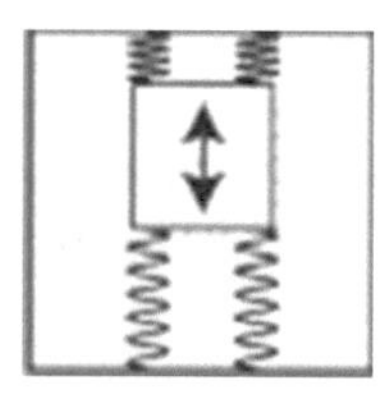

FIGURE 2.23 A *MEMS accelerometer* is micro machined silicon wafer mass suspended by polysilicon springs.

FIGURE 2.24 A commercial accelerometer capable of measuring up to 200 g. Courtesy of adafruit.com.

Putting together the equations for velocity and acceleration, along with some other useful relations, we obtain:

For any type of motion use:

$$\bar{v} = \frac{d}{t} \rightarrow \text{velocity}_{\text{average}} = \frac{\text{distance}}{\text{time}}$$

$$\bar{a} = \frac{v_f - v_i}{t} \rightarrow \text{acceleration}_{\text{average}} = \frac{V_{\text{final}} - V_{\text{initial}}}{t}$$

For motion with a constant acceleration use:

$$\bar{v} = \frac{v_f + v_i}{2} \rightarrow \text{velocity}_{\text{average}} = \frac{V_{\text{final}} + V_{\text{initial}}}{2}$$

$$d = \left(\frac{v_f + v_i}{2}\right) \cdot t \rightarrow \text{distance} = \left(\frac{V_{\text{final}} + V_{\text{initial}}}{2}\right) \cdot \text{time}$$

$$v_f = v_i + a \cdot t \rightarrow \text{velocity}_{\text{final}} = \text{velocity}_{\text{initial}} + \text{acceleration} \cdot \text{time}$$

$$y = \frac{1}{2} a \cdot t^2 + v_i \cdot t + y_i \rightarrow \text{position} = \frac{1}{2} \cdot \text{acceleration} \times \text{time}^2 + \text{velocity}_{\text{initial}} \cdot \text{time} + \text{position}_{\text{initial}}$$

$$2 \cdot a \cdot d = v_f^2 - v_i^2 \rightarrow 2 \cdot \text{acceleration} \cdot \text{distance} = \text{velocity}_{\text{final}}^2 - \text{velocity}_{\text{initial}}^2$$

Note: $y = \frac{1}{2} a \cdot t^2 + v_i \cdot t + y_i$ can be rewritten as $d = \frac{1}{2} a \cdot t^2 + v_i \cdot t$ where $d = y - y_i$

Example 2.12

A load is dropped from a crane at rest, and it falls for 2.0 s. How fast in ft/s is the load traveling after 2 s? Remember, if anything is falling, it is because gravity is pulling it, causing it to accelerate.

When a problem involves a constant acceleration, time, initial and final velocities the following equation may be used: $v_f = v_i + a\,t$ (Figure 2.25).

$$v_i = 0$$

$$t = 2.0\,\text{s}$$

$$a = -32.2 \frac{\text{ft}}{\text{s}^2}$$

$$v_f = ?$$

$$v_f = v_i + a \cdot t$$

$$v_f = 0 + \left(-32.2 \frac{\text{ft}}{\text{s}^2}\right) \cdot 2.0\,\text{s}$$

$$v_f = -64 \frac{\text{ft}}{\text{s}}$$

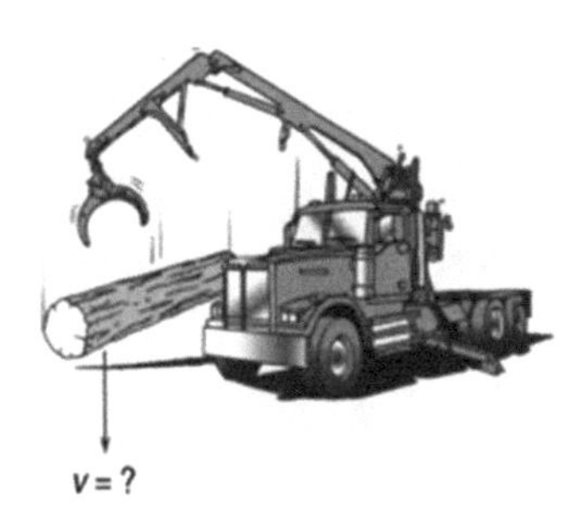

FIGURE 2.25 How fast will the log hit the ground?

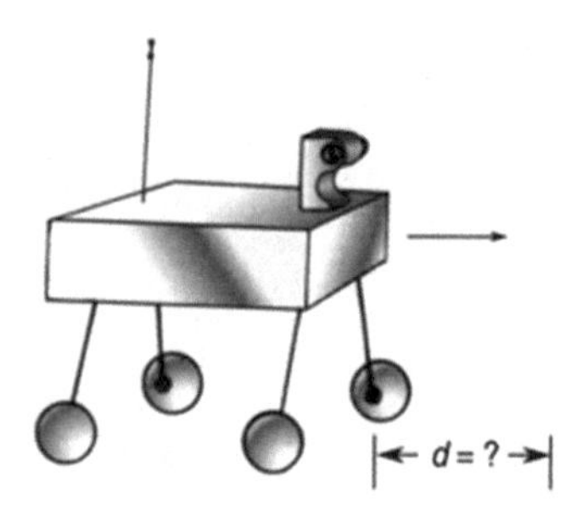

FIGURE 2.26 How far does the machine travel?

Example 2.13

A machine rolls across a platform, accelerating at a rate of $a = +3.00\,\text{ft/s}^2$ with an initial velocity of +6.00 ft/s, traveling for a time of 1.50 s. How far does the machine travel?

When a problem involves acceleration, distance, time, and velocity, the following equation may be used: $d = v \cdot t + \frac{1}{2} a \cdot t^2$, where $d = y - y_i$ (Figure 2.26).

$$v_i = +6.00 \text{ ft/s}$$

$$a = +3.00 \text{ ft/s}^2$$

$$t = 1.5 \text{ s}$$

$$d = ?$$

$$d = \frac{1}{2} a \cdot t^2 + v_i \cdot t$$

$$d = \frac{1}{2} \cdot \left(3.00 \frac{\text{ft}}{\text{s}^2}\right) \cdot (1.5 \text{ s})^2 + \left(6.00 \frac{\text{ft}}{\text{s}}\right) \cdot (1.5 \text{ s})$$

$$d = 12.4 \text{ ft}$$

FIGURE 2.27 How high is the bridge above the water?

Example 2.14

A man drops a rock off a bridge from rest, and it takes 4.0 s for the rock to hit the water below. In meters, how high is the bridge above the water? Remember, the rock is falling because of gravity ($g = -9.8\,\text{m/s}^2$) (Figure 2.27).

$$v_i = 0 \frac{\text{m}}{\text{s}} \text{ The rock strted from rest.}$$

$$a = -9.8 \text{ m/s}^2$$

$$t = 4.0 \text{ s}$$

$$d = ?$$

$$d = \frac{1}{2} a \cdot t^2 + v_i \cdot t$$

$$d = \frac{1}{2} \cdot \left(-9.8 \frac{\text{m}}{\text{s}^2}\right) \cdot (4.0 \text{ s})^2 + \left(0 \frac{\text{m}}{\text{s}}\right) \cdot (4.0 \text{s})$$

$$d = -78.4 \text{ m}$$

When problems involve motion with no reference to time, the following equations may be used:

$$v = \frac{v_i + v_f}{2}$$

$$\text{Average velocity} = \frac{\text{velocity}_{\text{initial}} + \text{velocity}_{\text{final}}}{2}$$

$$2 \cdot a \cdot d = v_f^2 - v_i^2$$

$$2 \cdot \text{acceleration} \cdot \text{distance} = (\text{velcocty}_{\text{final}})^2 - (\text{velocity}_{\text{initial}})^2$$

Example 2.15

What is the average velocity of an electron inside a CRT (cathode ray tube) shot out of an electron gun at 6.0×10^7 m/s and stopped by the screen? (Figure 2.28)

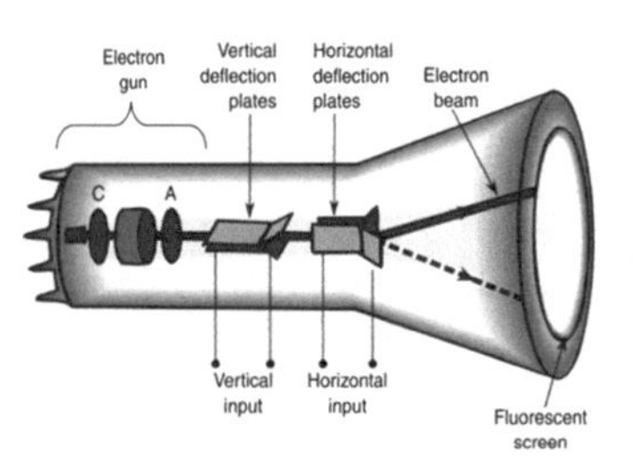

FIGURE 2.28 What is the average velocity of an electrons?

$$v_f = 0$$

$$v_i = +6.0 \times 10^7 \frac{\text{m}}{\text{s}}$$

$$\bar{v} = ?$$

$$\bar{v} = \frac{v_i + v_f}{2}$$

$$\bar{v} = \frac{6.0 \times 10^7 \frac{\text{m}}{\text{s}} + 0}{2}$$

$$\bar{v} = 3.0 \times 10^7 \frac{\text{m}}{\text{s}}$$

Example 2.16

A weight is dropped from a resting position from a pile driver, at a height of 25 ft, onto a pile below. How fast does the weight hit the pile? (Figure 2.29)

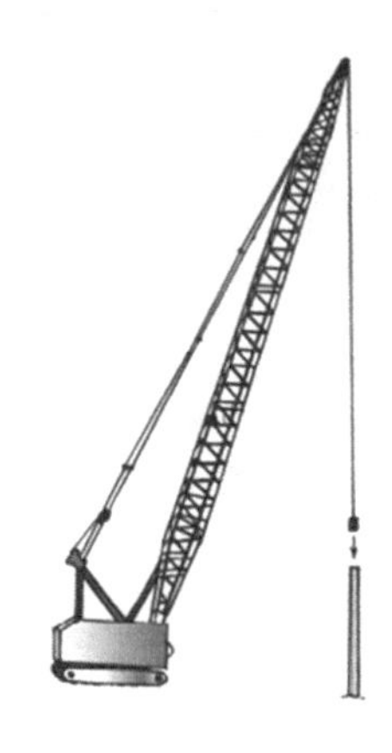

FIGURE 2.29 How fast does the weight hit the pile?

$$d = -25\,\text{ft}$$

$$v_i = 0$$

$$a = -32.2 \frac{\text{ft}}{\text{s}^2}$$

$$v_f = ?$$

$$2\bar{a}d = v_f^2 - v_i^2$$

$$2\bar{a}d + v_i^2 = v_f^2 - \cancel{v}_1^2 + \cancel{v}_i^2 \leftarrow \text{Solving for } v_f$$

$$v_f^2 = 2\bar{a}d + v_i^2$$

$$v_f = \sqrt{2\bar{a}d + v_i^2}$$

$$v_f = \sqrt{2\left(-32.2 \frac{\text{ft}}{\text{s}^2}\right)(-25\,\text{ft}) + (0)^2}$$

$$= 4\bar{0} \frac{\text{ft}}{\text{s}}$$

Example 2.17

A bullet is shot straight up into the air. The bullet reaches a maximum height of 3,000 ft. What was its initial velocity? (Figure 2.30)

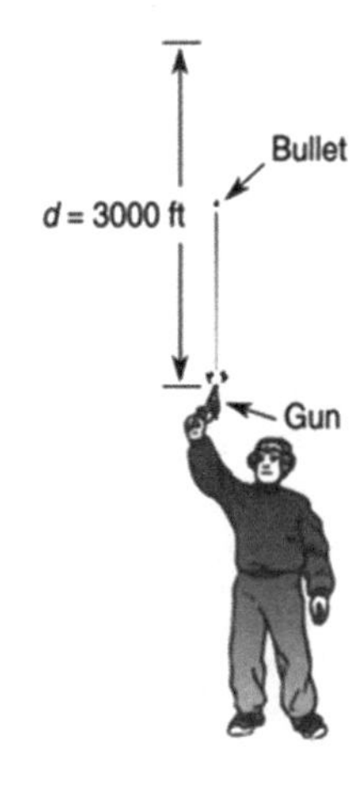

FIGURE 2.30 What was the initial velocity of the bullet?

$$d = 3{,}000 \text{ ft}$$
$$a = -32.2 \text{ ft/s}^2$$
$$v_f = 0 \leftarrow \text{At maximum height the bullet is still.}$$
$$v_i = ?$$
$$2 \cdot a \cdot d = v_f^2 - v_i^2$$
$$v_i^2 = v_f^2 - 2 \cdot a \cdot d$$
$$v_i = \sqrt{v_f^2 - 2 \cdot a \cdot d}$$
$$v_i = \sqrt{0 - 2 \cdot (-32.2 \text{ ft/s}^2) \cdot 3000 \text{ ft}}$$
$$v_i = 440 \text{ ft/s}$$

0.25m

FIGURE 2.31 How much time do you have to respond, after the trigger is pulled?

Example 2.18

You are held at gun point, but are not worried because you feel confident you will be able to jump out of the way when you see the trigger is pulled. The gun has a barrel of length 0.25 m and an exit velocity of 380 m/s. How long do you have to respond, after the trigger is pulled to get out of the way? Let's assume the bullet has a constant acceleration while in the barrel (Figure 2.31).

$$v_i = 0 \leftarrow \text{Bullet starts from rest.}$$
$$v_f = 380 \text{ m/s}$$
$$d = 0.25 \text{ m}$$
$$d = \left(\frac{v_f + v_i}{2}\right) \cdot t \leftarrow \text{For constant acceleration.}$$
$$t = d \cdot \left(\frac{2}{v_f + v_i}\right) \leftarrow \text{After solving for } t.$$
$$t = (0.25 \text{ m}) \cdot \left(\frac{2}{380 \text{ m/s} + 0}\right)$$
$$t = 1.32 \times 10^{-3} \text{ s}$$

Googling minimum human visual reaction time gives 0.25 s.
You better rethink your response to this situation!

2.8.4 Problem Solving Procedure

How do you determine which equation to use when solving a problem?

1. Make a formula sheet and have it in front of you.
2. As you read through the problem, write down the values of the quantities given.

3. Make a sketch of the problem, labeling everything to help you visualize what is being asked.
4. Work the problem in a vertical column, organizing the information and writing down everything that is relevant.
5. List all the relevant equations. If you have a problem figuring out which formulas apply, you can use the process of elimination to narrow down the possibilities. For example, if there is no mention of time in the problem, cross out all formulas dealing with time.
6. Work out any algebra first before plugging in numbers.
7. Make sure the quantities have the correct units when you are plugging them into formulas. If working in the MKS or FPS systems, stay in the systems when plugging in.
8. Check to see if the answer makes sense both numerically and unit-wise.

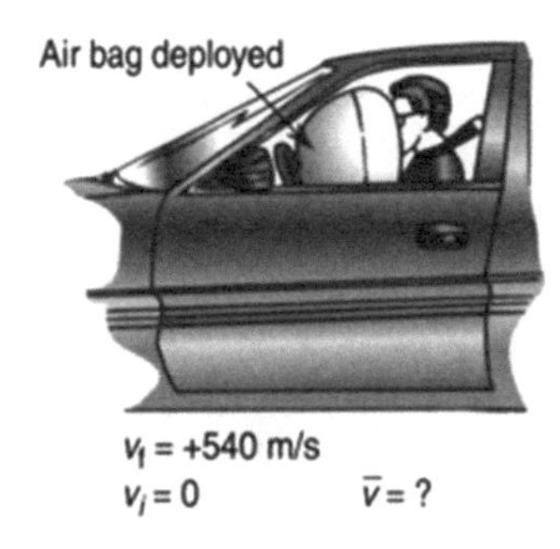

FIGURE 2.32 What is the average velocity of the air bag?

Example 2.19

An air bag is deployed from rest during an accident at +540 m/s. What is the average velocity of the air bag? (Figure 2.32)

Follow the steps above to solve the problem.

1. Have a formula sheet in front of you.

$$\text{For any type of motion}: \bar{v} = \frac{d}{t} \quad \bar{a} = \frac{v_f - v_i}{t}$$

For accelerated motion with a constant acceleration:

$$\bar{v} = \frac{v_f + v_i}{2}$$

$$d = \left(\frac{v_f + v_i}{2}\right) \cdot t$$

$$v_f = v_i + a \cdot t$$

$$d = \frac{1}{2} a \cdot t^2 + v_i \cdot t$$

$$2 \cdot a \cdot d = v_f^2 - v_i^2$$

2. As we read the problem, we first read "from rest," which means $v_i = 0$. Then we read "at +540 m/s," which means $v_f = +540$ m/s. Finally, we read "What is the average velocity?" which means $\bar{v} = ?$
3. Draw a sketch.
4. Work the problem in a vertical column, writing everything down.

$$v_f = +540 \text{ m/s}$$

$$v_i = 0 \text{ m/s}$$

$$\bar{v} = ?$$

5. List all the relevant equations. If you have a problem determining which formulas apply, you can use the process of elimination to narrow down the

possibilities. For example, in this problem, there is no mention of time, so cross out all formulas dealing with time.

$$\cancel{\bar{v} = \frac{d}{t}}$$

$$\cancel{\bar{a} = \frac{v_f - v_i}{t}}$$

$$\bar{v} = \frac{v_i + v_f}{2}$$

$$\cancel{d = \left(\frac{v_i + v_f}{2}\right) \cdot t}$$

$$\cancel{v_f = v_i + a \cdot t}$$

$$\cancel{d = v_i \cdot t + \frac{1}{2} a \cdot t^2}$$

$$2 \cdot a \cdot d = v_f^2 - v_i^2$$

From these remaining equations, only one fits our needs: $\bar{v} = \frac{v_i + v_f}{2}$

6. Work out the algebra before plugging in the numbers.

$$v_f = +540\text{m/s}$$
$$v_i = 0\text{m / s}$$
$$\bar{v} = ?$$
$$\bar{v} = \frac{v_i + v_j}{2}$$

Work out the algebra. No algebra needed here.

7. Plug in the numbers, checking first to see that all the units match. Everything is in MKS.

$$\bar{v} = \frac{0 + 540\text{m / s}}{2} = +270\text{m / s}$$

8. $\bar{v}$ should have units of m/s.

2.9 CHAPTER SUMMARY

Symbol	Unit
t—time	s
d—distance	m, ft
y—position	m
v—velocity	m/s
a—acceleration	m/s^2
g—acceleration due to gravity	–9.8 m/s^2 or –32.2 ft/s^2

Anything that moves in the industry—from robotic arms to elevators, conveyor belts, or air bags in a car —can be described by the equations in this chapter. The problem set at the end of the chapter contains relevant issues you may encounter when working in the industry.

For any type of motion use:

$$\bar{v} = \frac{d}{t} \rightarrow \text{velocity}_{\text{average}} = \frac{\text{distance}}{\text{time}}$$

$$\bar{a} = \frac{v_f - v_i}{t} \rightarrow \text{acceleration}_{\text{average}} = \frac{V_{\text{final}} - V_{\text{initial}}}{t}$$

For motion with a constant acceleration use:

$$\bar{v} = \frac{v_f + v_i}{2} \rightarrow \text{velocity}_{\text{average}} = \frac{V_{\text{final}} + V_{\text{initial}}}{2}$$

$$d = \left(\frac{v_f + v_i}{2}\right) \cdot t \rightarrow \text{distance} = \left(\frac{V_{\text{final}} + V_{\text{initial}}}{2}\right) \cdot \text{time}$$

$$v_f = v_i + a \cdot t \rightarrow \text{velocity}_{\text{final}} = \text{velocity}_{\text{initial}} + \text{acceleration} \cdot \text{time}$$

$$y = \frac{1}{2} a \cdot t^2 + v_i \cdot t + y_i \rightarrow \text{position} = \frac{1}{2} \cdot \text{acceleration} \cdot \text{time}^2 + \text{velocity}_{\text{initial}} \cdot \text{time} + y_{\text{initial}}$$

$$2 \cdot a \cdot d = v_f^2 - v_i^2 \rightarrow 2 \cdot \text{acceleration} \cdot \text{distance} = \text{velocity}_{\text{final}}^2 - \text{velocity}_{\text{initial}}^2$$

When problems involve motion with no reference to time, the following equations may be used:

$$2 \cdot a \cdot d = v_f^2 - v_i^2$$

$$\bar{v} = \frac{v_f + v_i}{2}$$

PROBLEM SOLVING TIPS

- As you read through the problem, write down the values of the quantities given.
- Have all your formulas in front of you and choose only the relevant ones. You can use the process of elimination to choose the formulas that are useful.

  For example, if a problem does not mention time, then time is not relevant to the solution; therefore, choose equations without t. From the remaining equations, determine the proper one from the context of the problem.
- If something is at rest, its velocity is zero.
- When a projectile reaches maximum height, it changes direction, and $v = 0$.
- Pay close attention to the units, remembering to stay in either the metric or the English system of units.

PROBLEMS

1. A robotic arm moves a distance of 3 m in 2 s. What is the average speed of the arm?
2. A radio transmission is broadcasting from an antenna. The radio wave travels for 3 s. How far has the wave traveled? Note: Radio waves travel at the speed of light, at a constant of 3×10^8 m/s.
3. An elevator travels upward three floors at an average speed of 0.5 floors/s. How long will it take to travel three floors?
4. An ultrasonic tape measure is used to measure the distance across a room. It takes 0.1 s for the sound wave to travel to the opposite wall and back. How far away is the wall? Note: The speed of sound at 72°F is 330 m/s.
5. A car starts from rest and accelerates for 10 s, reaching a speed of 100 ft/s. What is the acceleration of the car? How does this compare to the acceleration due to gravity?
6. An accelerometer is placed on a plane and emits a signal of 10 mV/(m/s^2). At takeoff, the accelerometer measures 25 mV. What is the acceleration of the plane? Hint: Let the units guide you. Add, subtract, multiply, or divide to end up with m/s^2 in the answer.
7. Laser ranging is a method used to determine the distance to an object. A laser pulse is sent out, and a clock records how long it takes the pulse to return after reflecting off of something. This

information, along with the speed of light, is used to determine the distance to the object. Suppose a pulse is sent out and takes 2 μs to return. Given that the speed of light is 3×10^8 m/s, how far away is the object?

8. An arrow is shot straight up into the air and reaches a height of 300 m. What was its initial velocity?
9. A conveyer belt transports parts on an assembly line. The conveyer belt starts from rest and accelerates at 1.2 ft/s^2 for a time of 4 s. How far did a part travel on the belt?
10. An electronic signal travels at 90% of the speed of light through a wire. If the signal travels for 1 ns, how far does it travel? Note that the speed of light is 3×10^8 m/s.
11. You are in a car traveling at 55 mph, and you step on the gas, accelerating at a rate of 600 mi/h^2. How fast are you traveling after 1 min?
12. Suppose the speedometer on your car broke and you want to determine the speed of your car on the highway. How would you use the mile markers on the roadside to determine your speed?
13. What is the difference between acceleration and velocity?
14. Which falls faster when dropped, a penny or a quarter? How fast do these objects pick up speed as they fall?
15. How might you build an accelerometer using chewing gum and some string?
16. A CD is ejected from a computer at 0.5 m/s and then comes to rest. What is its average velocity?
17. A radio signal travels at 3×10^8 m/s. If the signal travels to a satellite 30×10^6 m away, how long is the travel time?
18. A conveyer belt starts from rest, accelerates at 2 m/s^2 for 1.2 s. How far has a box on the conveyer belt been transported?
19. A computer sits on a table top 3 ft above the ground. By accident it falls off and hits the ground. What is the speed of impact?
20. A computer sits on the table top 3 ft above the ground. By accident it falls off and hits the ground. How long does it take before it hits the ground?
21. A robotic arm starts from rest and accelerates uniformly until it reaches a speed of 2.5 in./s at a distance 22 in. from where it started. How long does it take to travel the 22 in.?
22. In order to determine the height of a bridge, you drop a stone off the bridge and measure the time it takes to fall. If it takes 3 s to fall, how high is the bridge in meters?

23. Sometimes acceleration is measured in g, the acceleration due to gravity, 32.2 ft/s^2. Example, if an object accelerates at 64.4 ft/s^2 then its acceleration is 2g or two times gravity. Suppose a fighter jet changes speed from 600 to 700 mph in 1 s. How many g's did the pilot "pull"? Remember to convert mph to ft/s!
24. An arm on an IC (integrated circuit) chip insertion machine moves across a table at 12 cm/s. How long does it take to move 24 cm?
25. An elevator on the tenth floor malfunctions and begins to free fall downward at 32.2 ft/s^2 until it hits the ground floor. If each floor is 13 ft apart, how long do the riders in the elevator have to live?
26. A motor drives a robotic arm. The arm moves 10 cm for every second the motor is on. How long should the motor be turned on to move the arm 25 cm?
27. The cursor on a computer screen moves across the screen at 20 cm in 2 s. How long does it take the cursor to move 1 cm?
28. An electronic speedometer sends out a signal of 0.1 V/mph. How many volts does the speedometer put out when it is measuring 100 mph?
29. When disassembling a machine, a spring pops out of the machine rising 1 ft into the air. What was the initial velocity of the spring?
30. How long did it take the spring to rise 1 ft in the previous problem?
31. A copy machine scans a document. The scanner starts at rest and accelerates at 2 in./s^2 for 0.5 s. How fast is the scanning head moving at 0.5 s?
32. How far did the scanning head move in 0.5 s in the previous problem?
33. A screwdriver falls off a table and hits the ground 0.5 s later. How high is the table in feet?
34. How fast did the screwdriver hit the ground in the previous problem?
35. The cone on a speaker moves back and forth to produce sound. Sound travels at 345 m/s at room temperature. If you are standing 2 m away from the speaker, how long does it take the sound to reach you?

3

Force and Momentum

This chapter describes force and momentum and how they are related to each other. The concepts learned in this chapter, along with the equations of motion learned in the previous chapter, will provide a more complete understanding of the mechanics of motion.

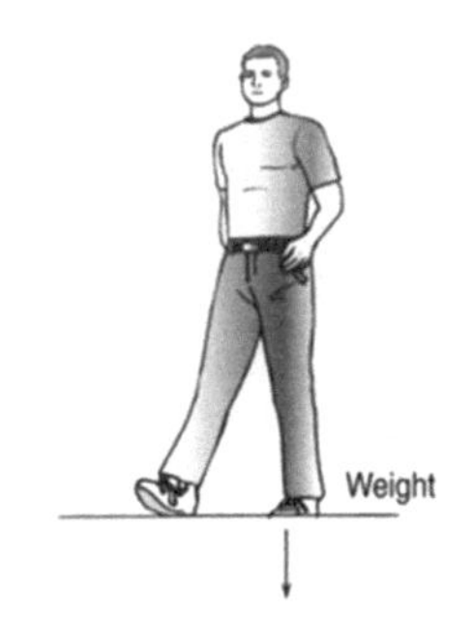

FIGURE 3.1 Your weight is an example of a force. Gravity is pulling you down toward the ground.

3.1 FORCE

A *force* is a push or pull in a certain direction (Figures 3.1 and 3.2).

Force is a vector because it has both magnitude and direction. For example, force = 180 lb downward or 10 lb to the right. Instead of saying up or down, or to the left or to the right, we can use a coordinate system to designate direction (Figure 3.3).

3.1.1 Measuring Force

A simple force gauge can be made from a spring (Figure 3.4). In order to stretch or compress a spring, a force is needed. The amount by which the spring changes length is determined by the force.

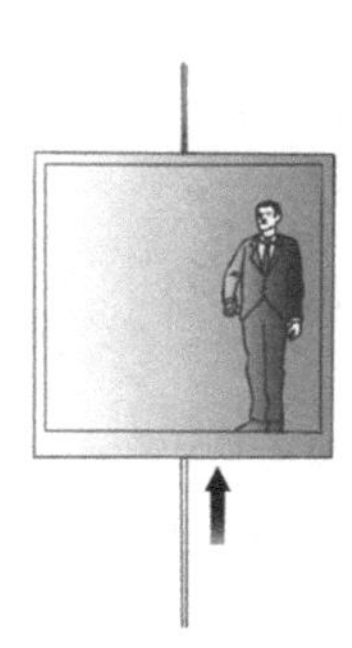

FIGURE 3.2 An elevator moves upward because a force below (a hydraulic cylinder) pushes it upward.

FIGURE 3.3 Instead of saying up or down, or to the left or to the right, we can use a coordinate system to designate direction.

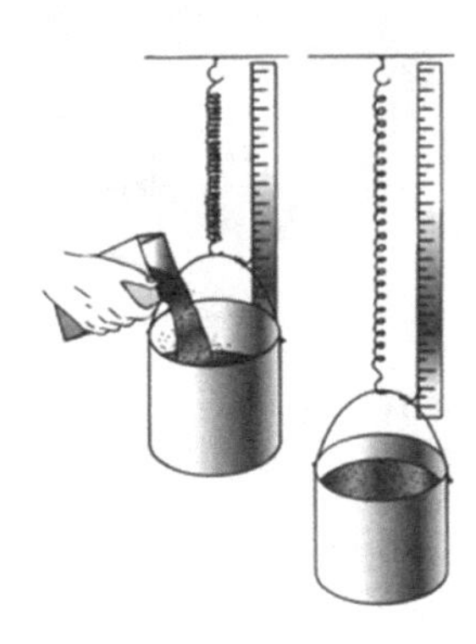

FIGURE 3.4 A spring force gauge.

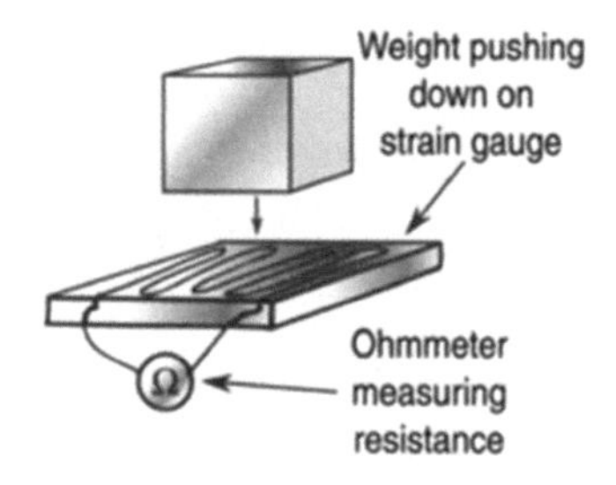

FIGURE 3.5 A resistive strain gauge converts an applied force, which deforms the gauge into a change in resistance. The resistance is then converted into the corresponding weight.

3.1.2 Electronic Scales

Electronic scales typically use a strain gauge to measure weight or force. Most strain gauges are resistive. A resistive strain gauge consists of a very thin resistor mounted on a material that will flex as force is applied. A force applied to the resistor will deform it slightly, resulting in a change in its resistance (Figure 3.5).

Example 3.1

A strain gauge is used to measure a force. Its resistance changes by 100 ohms/lb. What is the change in resistance when measuring 0.09 lb?

This problem is effectively asking for a conversion from pounds to ohms.*

$$100\,\text{ohms} = 1\,\text{lb}$$

$$0.09\,\cancel{\text{lb}}\left(\frac{100\,\text{ohms}}{1\,\cancel{\text{lb}}}\right) = 9\,\text{ohms}$$

3.2 MASS AND WEIGHT

The words *mass* and *weight* are sometimes used interchangeably, but in physics this is not correct.

- The *mass* of an object is a measure of its resistance to being accelerated.
- The *weight* of an object is a measure of the gravitational attraction the earth has for that object.

FIGURE 3.6 An astronaut has the same mass in outer space as on earth, but has very little weight because gravity is so small there.

Mass and weight are related to each other by a constant, g (free-fall acceleration). If an object is in outer space, it has hardly any weight, but it still has mass. The mass of an object is the same everywhere because the amount of matter contained in it stays the same, but weight is gravity-dependent, that is, an astronaut weighs less on the moon because gravity is smaller (Figure 3.6).

Just as we have units of length and time, we have units of mass and force. In the English system of units, force is measured in pounds and mass in a unit called slugs. In the SI system, mass is measured in kilograms, and force is measured in a unit called newton.

* We know, of course, that it does not make sense in terms of dimensional analysis to convert pounds into ohms, but we can think of them as effectively equal in the context of this problem.

	English System	SI System
Force	pounds (lb)	newton (N)
Mass	slugs (slugs)	kilograms (kg)
Conversions		
1 kg = 1,000 g	1 g is about the mass of a paper clip.	
1 lb = 4.45 N	1 newton is almost 1/4 lb.	
1 slug = 14.6 kg	1 slug has the equivalent weight of 32.2 lb.	

- The MKS system is part of the SI system. All units are recorded and calculated in meters, kilograms, and seconds (MKS).
- In the English system, calculations are done using pounds, slugs, feet, and seconds. This system is sometimes called the FPS system or the foot, pound, and second system.

3.3 GRAVITATIONAL FORCE

The force due to gravity explains why masses fall to earth, why the earth rotates around the sun, and why the moon rotates around the earth. The force of attraction of one mass on another is given by Newton's law of gravity:

$$F = \frac{G \cdot m_1 \cdot m_2}{r^2}$$

$$F = \frac{\left(6.67 \times 10^{-11} \frac{\text{N} \cdot \text{m}^2}{\text{kg}^2}\right) \cdot (\text{mass}_1(\text{kg})) \cdot (\text{mass}_2(\text{kg}))}{(\text{distance}(\text{m}))^2}$$

The force of gravity between two masses separated by a distance r equals the product of the masses times a constant G, divided by the distance between the masses squared.

Example 3.2

A man of mass 70 kg standing on the surface of the earth, of mass 5.98×10^{24} kg, at a distance of 6.37×10^6 m from the center of the earth will have a force of attraction:

$$F = \frac{G \cdot m_1 \cdot m_2}{r^2} = \frac{\left(6.67 \times 10^{-11} \frac{\text{N} \cdot \text{m}^2}{\text{kg}^2}\right)(5.98 \times 10^{24}\,\text{kg})(70\,\text{kg})}{(6.37 \times 10^6\,\text{m})^2}$$

$$F = 688.09\,\text{N}$$

The force of gravity can be simplified if we consider this force at the surface of the earth where the distance is fixed. The force of gravity equation then simplifies to:

$$F = m \cdot g$$

where g equals:

$$g = \frac{G \cdot m_{\text{earth}}}{r_{\text{center}}^2} = \frac{\left(6.67 \times 10^{-11} \frac{\text{N} \cdot \text{m}^2}{\text{kg}^2}\right)\left(5.98 \times 10^{24}\,\text{kg}\right)}{\left(6.37 \times 10^6\,\text{m}\right)^2}$$

$$g = 9.8 \frac{\text{m}}{\text{s}^2}$$

Example 3.3

What is the force of attraction of a man of mass 72 kg standing on the surface of the earth?

$$F = m \cdot g = (72\,\text{kg})\left(9.8 \frac{\text{m}}{\text{s}^2}\right)$$

$$F = 705.6\text{N}$$

Weight is the force of attraction of a mass with a planet.

The weight of the man in the previous example equals 705.6 N.

Because weight is a force, we shall write:

$$F_w = m \cdot g$$

Force due to weight = mass × acceleration due to gravity

where

$g = 9.8 \frac{\text{m}}{\text{s}^2}$ in the metric system

$g = 32.2 \frac{\text{ft}}{\text{s}^2}$ in the English system.

Example 3.4

An object has a mass of 10,000 g. What is its weight?

$$m = 10{,}000\,\text{g}$$

$$F_w = ?$$

We are working in the MKS system; therefore, the mass must be converted into kilograms.

$$10{,}000\,\text{g} = 10\,\text{kg}$$

$$F_w = m \cdot g$$

$$F_w = (10\,\text{kg}) \cdot \left(9.8 \frac{\text{m}}{\text{s}^2}\right) \leftarrow \text{Must use the metric acceleration due to gravity}$$

$$F_w = 98\,\text{N}$$

Example 3.5

An object has a mass of 10.0 slugs. What is its weight?

$$m = 10.0\,\text{slugs}$$

$$F_w = m \cdot g$$

$$F_w = (10\,\text{slugs}) \cdot \left(32.2\frac{\text{ft}}{\text{s}^2}\right) \leftarrow \text{Must use the English acceleration due to gravity}$$

$$F_w = 322\,\text{lb}$$

3.4 REDEFINING THE KILOGRAM

The kilogram, up until 2017, was defined to be the mass of a cylinder contained in a vault in France under a vacuum. This was a very inconvenient standard, because of the need to travel to France to compare a mass to the standard in the vault. To alleviate this problem, the kilogram has been redefined in terms of a constant called Plank's constant, a value from a branch of physics called Quantum Mechanics. Plank's constant never changes, and can be measured in labs around the world. The redefinition of the kilogram has joined other units such as the meter and the second that have been defined in terms of a physical constant (Figure 3.7).

FIGURE 3.7 The kilogram, up until 2017, was defined to be the mass of a cylinder contained in a vault in France under a vacuum.

3.5 NEWTON'S LAWS

Newton's first law says that if a body changes its motion from standing still to moving, or changes from one speed to another, or changes its direction of travel, it is because there are forces acting upon it that do not balance each other out. For example, in a game of tug-of-war, if each side pulls with equal force, then the flag does not move; however, if one side pulls with more force, there will be a change in the motion of the team members (Figure 3.8).

Newton's second law says that in order to change an object's motion, a net force has to be applied to it. How fast you change its motion depends on how big the net force is. If the object is very massive and at rest, you have to apply a lot of force to move it. If an object is moving, to slow it

FIGURE 3.8 If the forces do not balance each other, there will be an acceleration of team members.

down, speed it up, or change its direction, a force is needed. This statement can be summarized by the equation:

$$a = \frac{F}{m}$$

$$\text{acceleration} = \frac{\text{Force}}{\text{mass}}$$

This equation says that if something is being accelerated it is because there is a net force acting on it, and how much it accelerates depends on its mass and the applied force. The force used in this equation is the sum of all the forces acting on the mass.

We can rewrite the above force equation as

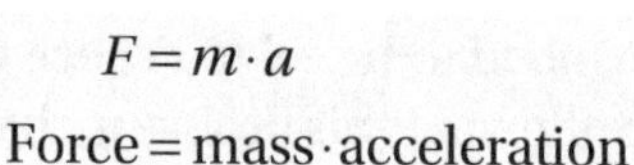

$$F = m \cdot a$$

$$\text{Force} = \text{mass} \cdot \text{acceleration}$$

FIGURE 3.9 A force causes the car to accelerate.

Example 3.6

If a car full of people has a mass of 1,600 kg and accelerates at $+2.0\frac{\text{m}}{\text{s}^2}$ to the right, what is the net force acting on the car (Figure 3.9)?

$$m = 1{,}600\,\text{kg}$$

$$a = +2.0\frac{\text{m}}{\text{s}^2}$$

$$F = m \cdot a$$

$$F = (1{,}600\,\text{kg})\left(2.0\frac{\text{m}}{\text{s}^2}\right)$$

$$F = 3{,}200\,\text{N to the right}$$

Example 3.7

A man pushes a crate on wheels that has a combined mass of 5.0 slugs with a force of 25 lb. What is the acceleration of the crate?

$$m = 5.0\,\text{slugs}$$

$$F = 25\,\text{lb}$$

$$a = ?$$

$$a = \frac{F}{m}$$

$$a = 25\,\text{lb}/(5.0\,\text{slugs})$$

$$a = 5.0\,\text{ft/s}^2$$

FIGURE 3.10 To represent force up, down, to the right or left, we will illustrate them this way.

3.6 MULTIPLE FORCES

When dealing with problems involving more than one force, a force diagram is used to help clarify the problem (Figure 3.10).

Example 3.8

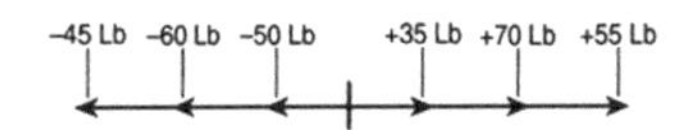

FIGURE 3.11 Forces graphically illustrated.

In a game of tug-of-war, two teams pull on a rope in the opposite directions, trying to pull the opposing team across a line. Suppose the members of the team on the left pull with forces 50, 60, and 45 lb, and members on the right team pull with forces of 35, 70, and 55 lb (Figure 3.11). What is the net force on the rope, and which way will it move?

$$
\begin{aligned}
F_{\text{left}} &= -50 - 60 - 45\,\text{lb} \\
&= -155\,\text{lb} \\
F_{\text{right}} &= 35 + 70 + 55\,\text{lb} \\
&= 160\,\text{lb} \\
F_{\text{total}} &= \sum_i F_i = F_{\text{right}} + F_{\text{left}} \\
&= 160 - 155\,\text{lb} \\
&= +5\,\text{lb} \leftarrow \text{5 lb to the right. The rope will move to the right.}
\end{aligned}
$$

Example 3.9

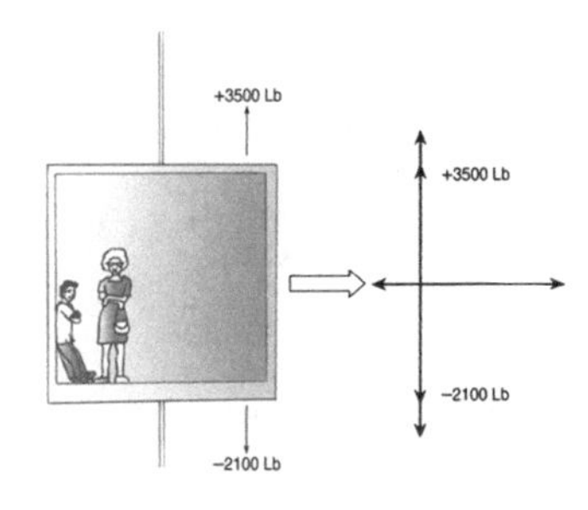

FIGURE 3.12 In our coordinate system, up is positive and down is negative.

An elevator weighing 2,100 lb is being pulled upward with a force of 3,500 lb (Figure 3.12). (a) What is the total force acting on the elevator? (b) What is its acceleration?
(a)

$$
\begin{aligned}
F_{\text{up}} &= +3{,}500\,\text{lb} \\
F_{\text{down}} &= -2{,}100\,\text{lb} \\
F_{\text{total}} &= ? \\
F_{\text{total}} &= F_{\text{up}} + F_{\text{down}} \\
F_{\text{total}} &= 3{,}500\,\text{lb} - 2{,}100\,\text{lb} \\
F_{\text{total}} &= +1{,}400\,\text{lb} \leftarrow \text{A positive force means is being accelerated upward.}
\end{aligned}
$$

(b)

$$
\begin{aligned}
F_w &= -2{,}100\,\text{lb} \\
F_{\text{total}} &= +1{,}400\,\text{lb} \\
a &= ? \\
m &= \frac{F_w}{g} \\
m &= \frac{-2{,}100\,\text{lb}}{-32.2\frac{\text{ft}}{\text{s}^2}} \\
m &= 65\,\text{slugs} \\
a &= \frac{\sum_i F_i}{m} \\
a &= \frac{+1{,}400\,\text{lb}}{65\,\text{slugs}} \\
a &= +22\frac{\text{ft}}{\text{s}^2}
\end{aligned}
$$

3.7 STATIC EQUILIBRIUM

Static equilibrium means all the forces are counterbalanced by other forces, and nothing is accelerating. In other words, static equilibrium means that:

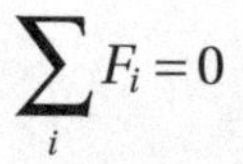

$$\sum_i F_i = 0$$

Sum of forces up - sum of forces down = 0.
Sum of forces to the right - sum of forces the left = 0.

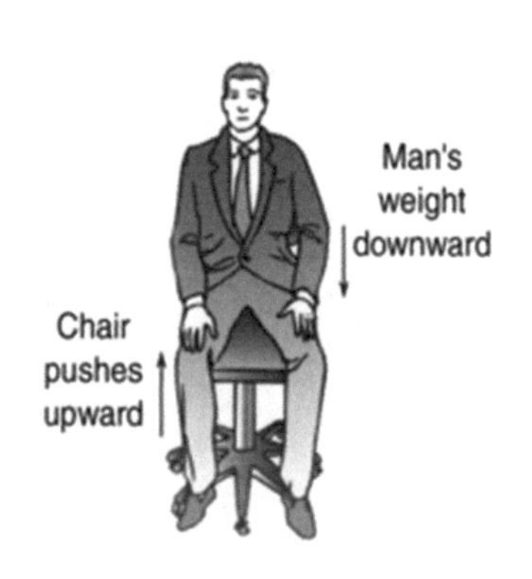

FIGURE 3.13 The chair and the man are in static equilibrium.

Example 3.10

A chair used to support a man weighing 165 lb must push upward with 165 lb, otherwise it will collapse under his weight (Figure 3.13).

Example 3.11

A bridge is designed to support dynamic loads, that is, loads that change in time (Figure 3.14). If a truck weighing 1 ton crosses the bridge, the bridge must exert a force of 1 ton upward, plus the force necessary to support itself, otherwise the truck will fall through the bridge.

FIGURE 3.14 A bridge is designed to support dynamic load.

3.8 FORCES ACTING IN TWO OR THREE DIMENSIONS

So far, all the forces we have encountered have been acting in one dimension, (i.e., all in a straight line, either up and down or left and right). When there is a combination of forces acting in two or three dimensions, life gets more complicated, and trigonometry will be needed to solve such problems. To solve these problems, the forces need to be broken up into *x*, *y*, and *z* components along the coordinate axes and then summed to determine the total force in each direction. This is demonstrated below.

Example 3.12

Suppose your car gets stuck in the mud, but that doesn't bother you because you know physics! You tie a cable to the bumper and to a tree and stretch it tight. You apply a force of 55 lb to the middle of the cable, creating a 5° angle with the horizontal (Figure 3.15). What tension is generated in the cable?

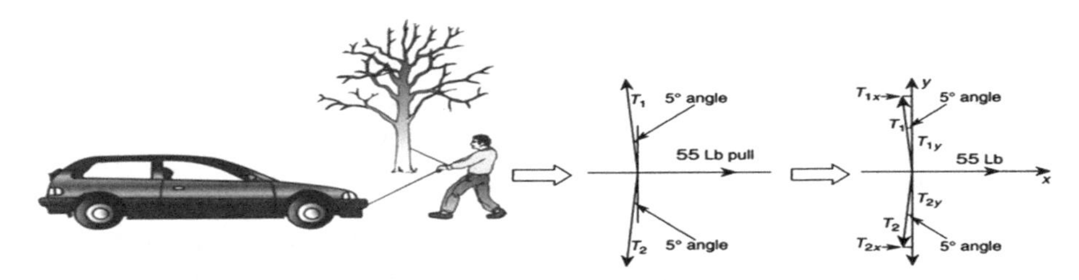

FIGURE 3.15 The problem can be simplified with a force diagram.

The problem is simplified using a force diagram (see Figure 3.15). To solve the problem, look at every force on the diagram and break it up into components along the x axis and y axis. If nothing is moving, we must have static equilibrium, that is,

Sum of the forces up = Sum of the forces down.

Sum of the forces to the left = Sum of the forces to the right.

The applied force is 55 lb.

$$T_3 = 55\text{lb}$$
$$\theta = 5°$$
$$T_1 = ?$$
$$T_2 = ?$$

By symmetry, $T_1 = T_2$ and

$T_{1y} = T_{2y}$ and $T_{1x} = T_{2x}$

Therefore, the sum of the forces to the left is:

$T_{1x} = T_{2x} = T_{1x} + T_{1x} = 2T_{1x}$

The sum of the forces to the left equals the forces to the right.

$$2T_{1x} = 55.0\text{lb}$$

$$T_{1x} = \frac{55.0\text{lb}}{2} = 27.5\text{lb} \leftarrow \text{Solving for } T_{1x}$$

$$\sin\theta = \frac{\text{opposite}}{\text{hypotenuse}}$$

Therefore

$$\sin\theta = \frac{T_{1x}}{T_1}$$

or

$$T_1 = \frac{T_{1x}}{\sin\theta}$$

Plugging in gives →

$$T_1 = \frac{27.5\text{lb}}{\sin(5°)} = 316\text{lb}$$

Therefore, since $T_1 = T_2$

$$T_2 = 316\text{lb}$$

Wow! You generated 316 lb with 55 lb. How is this possible? Why should the tension be higher than 55 lb in two cables? The reason is that the two cables are pulling not only to the right but also against each other, generating this higher tension.

FIGURE 3.16 A magnified view of two surfaces in contact.

3.9 FRICTION

The force of friction between two surfaces arises because the surfaces tend to dig into one another (Figure 3.16). At the points of contact between the two materials, molecular bonds form between the atoms of the two materials. These bonds produce an attractive force between the two materials which we call friction.

The small block will dig in deeper than the big block because the weight is focused on a smaller area. The big block, however, has more peaks and valleys to hold on with. These two results balance each other out, making the force of friction independent of the surface area.

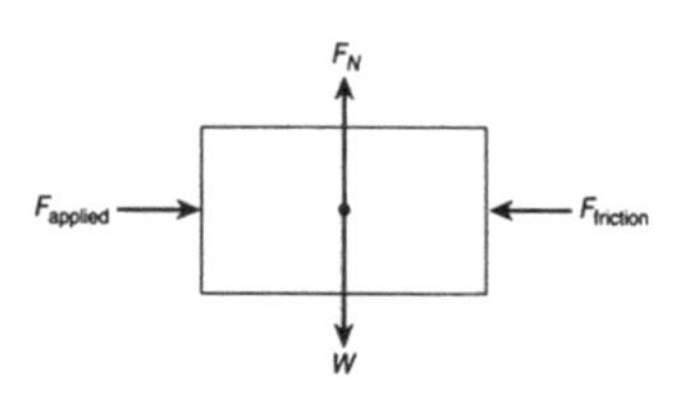

FIGURE 3.17 The normal force is the force perpendicular to the surface and the frictional force acts in an opposite direction to the applied force.

3.9.1 Key Points about Frictional Force

1. The force of friction between two surfaces is independent of the size of the two surfaces.
2. The force of friction between two surfaces is dependent on the quality of the surfaces.
3. The force of friction between two surfaces is dependent on the normal or perpendicular force to the surface. (If the surfaces are horizontal, the normal force is just the weight.)

These three statements can be summarized as:

$$F_f = \mu \cdot F_N$$

Frictional force = coefficient of friction × normal force

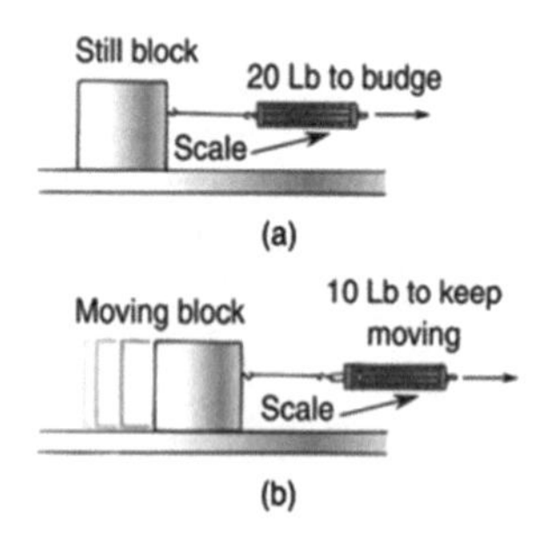

FIGURE 3.18 It takes more force to initially budge something than to keep it moving.

The normal force is the force perpendicular to the surface (Figure 3.17).

The coefficient of friction μ is a number without a unit. It is a measure of the slipperiness of two surfaces in contact. It takes more force to initially budge something than it does to sustain its motion (Figure 3.18). This is because once the object is moving, the surfaces don't dig into each other as much. The amount of friction depends on whether the two surfaces are sliding or rolling over one another, or whether the two surfaces are still.

- The coefficient friction for the case when the surfaces are sliding over one another is given by μ_k, where k stands for kinetic or moving.
- The coefficient friction for the case when the surfaces are rolling over one another is given by μ_r, where r stands for rolling.
- The coefficient friction for the case when the surfaces are still with respect to one another is given by μ_s, where s stands for static or still.

It takes more force to initially budge something than to sustain its motion. This is because once moving, the surfaces don't dig into each other as much. Therefore, the coefficient of stationary or static friction μ_s is greater than the coefficient of moving or kinetic friction μ_k.

$$\mu_s > \mu_k$$

The coefficient of starting or static friction (μ_s) is greater than the coefficient of sliding or kinetic friction (μ_k). See Table 3.1 for various coefficients.

TABLE 3.1
Coefficients of Friction (μ)

Material	μ_s Starting Friction	μ_k Sliding Friction
Steel on steel	0.58	0.20
Steel on steel (lubricated)	0.13	0.13
Glass on glass	0.95	0.40
Hardwood on hardwood	0.40	0.25
Steel on concrete		0.30
Aluminum on aluminum	1.9	
Rubber on dry concrete	2.0	1.0
Rubber on wet concrete	1.5	0.97
Aluminum on wet snow	0.4	0.02
Steel on Teflon	0.04	0.04

Example 3.13

How much force is needed to begin to slide a book across a table when the book weighs 9.5 lb and the coefficient of friction between the book and the table is $\mu = \mu_k = 0.2$? See Figure 3.19.

$$F_N = 9.5\text{lb}$$
$$\mu = 0.20$$
$$F_f = \mu F_N$$
$$F_f = (0.20)(9.5\text{lb})$$
$$F_f = 1.9\text{lb}$$

Values listed for the coefficient of friction are for ordinary pressures and temperatures, but may still vary somewhat from sample to sample.

FIGURE 3.19 In order to push the book, the frictional force between the book and the table needs to be overcome.

FIGURE 3.20 Improperly inflated tires add to rolling friction.

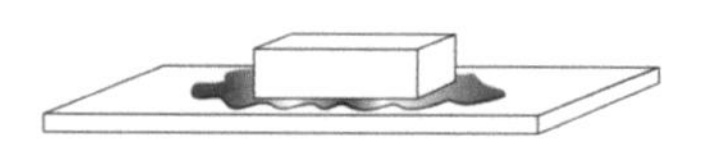

FIGURE 3.21 Lubricants act to fill in the voids in the surfaces in contact.

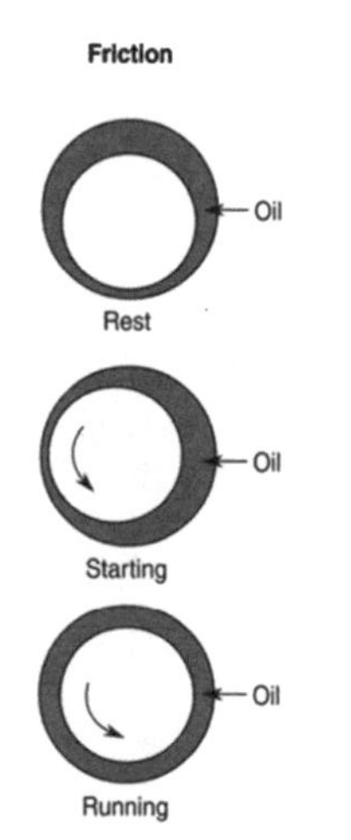

FIGURE 3.22 Lubrication between surfaces.

FIGURE 3.23 Bearings convert kinetic friction into rolling friction.

Example 3.14

How much force is required to budge a 100 lb wooden crate on a wooden floor compared with the force needed to sustain its movement?

$$F_N = 100\,\text{lb}$$
$$\mu_s = 0.4$$
$$\mu_k = 0.25$$

Static case:

$$F_f = \mu_s F_N$$
$$F_f = (0.40)(100\,\text{lb})$$
$$F_f = 40\,\text{lb} \leftarrow \text{Force needed to budge the crate}$$

Kinetic case:

$$F_f = \mu_k F_N$$
$$F_f = (0.25)(100\,\text{lb})$$
$$F_f = 25\,\text{lb} \leftarrow \text{Force needed to keep the crate moving}$$

3.9.2 Rolling Friction

A rolling tire has friction because it is deformed as it rolls. A lot of energy goes into deforming the tire. A tire that is almost flat has a much harder time rolling than a properly inflated tire (Figure 3.20). A steel wheel riding on steel, such as a train wheel, has almost no rolling friction because the wheel hardly deforms.

3.9.3 Reducing Friction

- Polishing surfaces or using lubricants can reduce friction. Too much polishing, however, can actually increase friction, such as in the case of two pieces of glass, which have difficulty sliding over one another.
- Lubricants act to fill in the voids in the surfaces that are in contact (Figure 3.21). Two surfaces with a lubricant between them will tend to float over one another.
- If the pressure between the two surfaces is great, a lubricant may have to be pumped between them (Figure 3.22). Examples of this are in an engine on the piston walls, valve stems, and on the crankshaft and cam.
- Ball and roller bearing can further reduce friction by turning sliding friction into rolling friction (Figure 3.23). The use of these bearings is sometimes necessary when there is no flow of lubricant possible, such as in a wheel.

Friction is normally thought of as something to be reduced, as in an engine. Friction is, however, a necessary thing. Think of how dangerous it can be driving on a wet road or how slippery a bathtub is when wet. In these cases, we want to increase friction for safety reasons (Figure 3.24). In fact, if it were not for fingerprints increasing friction, it would be difficult to pick up anything (Figure 3.25).

FIGURE 3.24 In the winter, some people place weight in the back of their pickup trucks to enhance traction on the slippery roads. The increase in weight will increase the friction between the tires and the road.

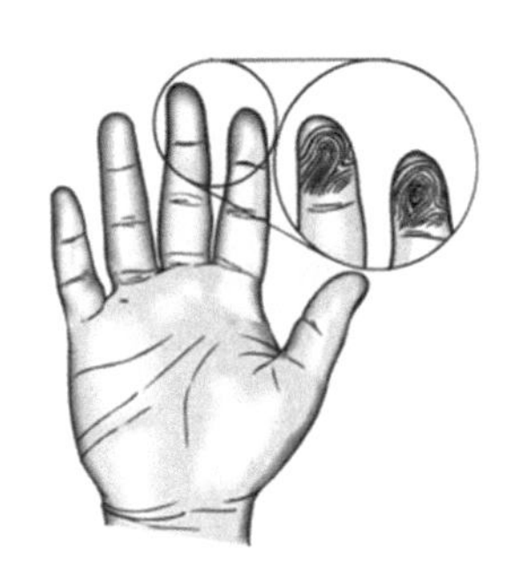

FIGURE 3.25 Fingerprints increase friction to aid in picking up objects.

3.10 MOMENTUM

If something is massive, it is difficult to move. Likewise, if something is massive and moving, it is difficult to stop. It is much more difficult to stop a semitrailer moving at 2 mph than a baby carriage moving at that speed. We say the semi has more momentum (Figure 3.26).

Mathematically, momentum is given by:

$$p = m \cdot v$$

$$\text{Momentum} = \text{mass} \times \text{velocity}$$

Something that is massive and moving fast has a lot of momentum. Momentum is a vector because it has both magnitude and direction. Since momentum is $m \times v$, it has units of:

$$\text{Momentum units: kg}\frac{\text{m}}{\text{s}} \text{ or slug}\frac{\text{ft}}{\text{s}}.$$

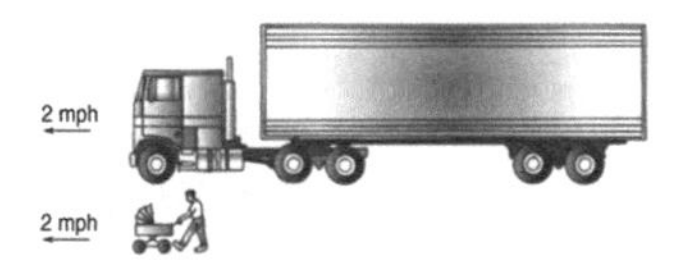

FIGURE 3.26 A semitrailer and a baby carriage are moving at the same speed. The semi is much more difficult to stop because it has a much greater momentum.

Example 3.15

A car with a mass of 890 kg is moving at +1.5 m/s (Figure 3.27). What is its momentum?

$$m = 890\text{kg}$$

$$v = +1.5\frac{\text{m}}{\text{s}}$$

$$p = m \cdot v$$

$$p = (890\text{kg}) \cdot \left(1.5\frac{\text{m}}{\text{s}}\right)$$

$$p = +1{,}335\text{kg}\frac{\text{m}}{\text{s}}$$

FIGURE 3.27 A car with a mass of 890 kg is moving at +1.5 m/s.

Momentum and force are related to each other by the equation:

$$F_{\text{average}} = \frac{\Delta p}{t} = \frac{p_f - p_i}{t}$$

$$\text{Force}_{\text{average}} = \frac{\text{momentum}_{\text{final}} - \text{momentum}_{\text{initial}}}{\text{time}}$$

This equation shows that if the momentum of something is changing in time, it is because there is a net nonzero force acting on it.

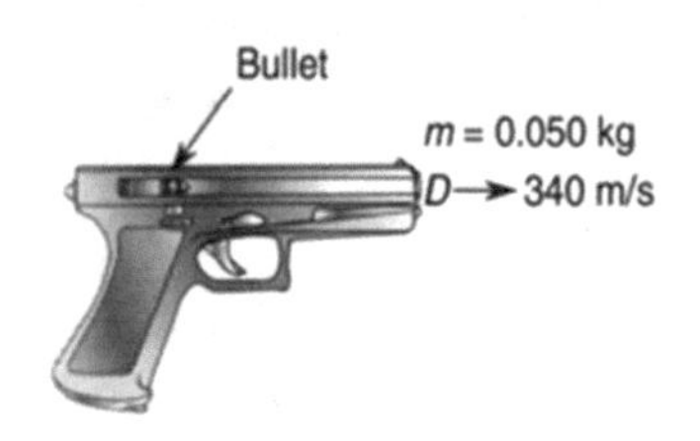

FIGURE 3.28 What was the average force applied to the bullet?

Example 3.16

A bullet of mass 0.050 kg is fired from a gun. It starts from rest and obtains a speed of 340 m/s in a time of 0.0010 s (Figure 3.28). What was the average force applied on the bullet?

$$p = m \cdot v$$

$$p_i = (0.050\,\text{kg})(0) = 0$$

$$p_f = (0.050\,\text{kg})(340\,\text{m/s}) = 17\,\text{kg m/s}$$

$$F_{\text{average}} = \frac{p_f - p_i}{t}$$

$$F_{\text{average}} = \frac{p_f - p_i}{t} = \frac{17\,\text{kg}\frac{\text{m}}{\text{s}} - 0\,\text{kg m/s}}{0.0010\,\text{s}}$$

$$F_{\text{average}} = 17{,}000\ \text{N}$$

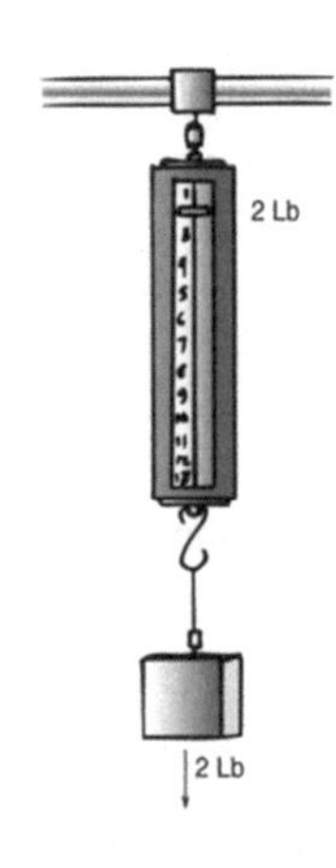

FIGURE 3.29 The block pulls down with 2 lb of force and the spring with 2 lb of force.

3.11 ACTION REACTION

Why does a gun recoil when fired? Why does a rocket move upward when the exhaust gas moves downward? When walking, why do you move forward when you push backward with each step? A law called *action reaction* is responsible.

Newton's third law of motion is sometimes called the law of action reaction. It states that for every force there is an equal and opposite force.

FIGURE 3.30 A man walking pushes back with 10 lb of force with each step, and he is therefore pushed forward with 10 lb of force with each step.

Example 3.17

A spring scale weighs a block. The block pulls down with 2 lb of force. The spring scale pulls up with 2 lb of force (Figure 3.29).

Example 3.18

A man walking pushes back with 10 lb of force with each step. He is therefore pushed forward with 10 lb of force with each step (Figure 3.30).

Example 3.19

A rocket exhaust gases downward with 5,000 lb of thrust (Figure 3.31). The rocket is therefore pushed upward with 5,000 lb of force.

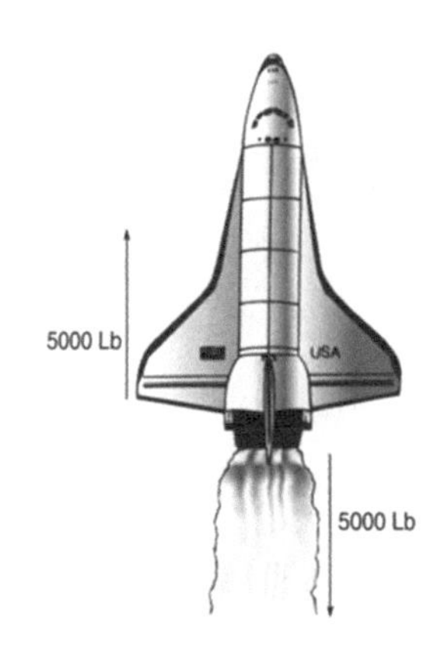

FIGURE 3.31 A rocket exhausts gases downward with 5,000 lb of thrust and it is therefore pushed upward with 5,000 lb of force.

Ion Engines are types of rocket engines that use an ionized gas as a propellant instead of a conventional chemical rocket. They are used on communication satellites to maintain their orbit because they are light and very efficient. Ion engines work by injecting a gas such as Xeon into a chamber. An electron beam is fired into this gas. A collision of the electrons with the Xeon atoms can result in electrons being knocked off the Xeon that is ionizing the atoms. A magnetic field keeps the electrons away from the walls of the chamber, increasing the chances of a collision with the Xeon. The ionized Xeon now has a net positive charge on it and is accelerated toward the high voltage grid, where it exits the engine. By Newton's third law, for every force there is an equal and opposite force, the engine will be propelled forward while expelling the gas backward. See Figure 3.32.

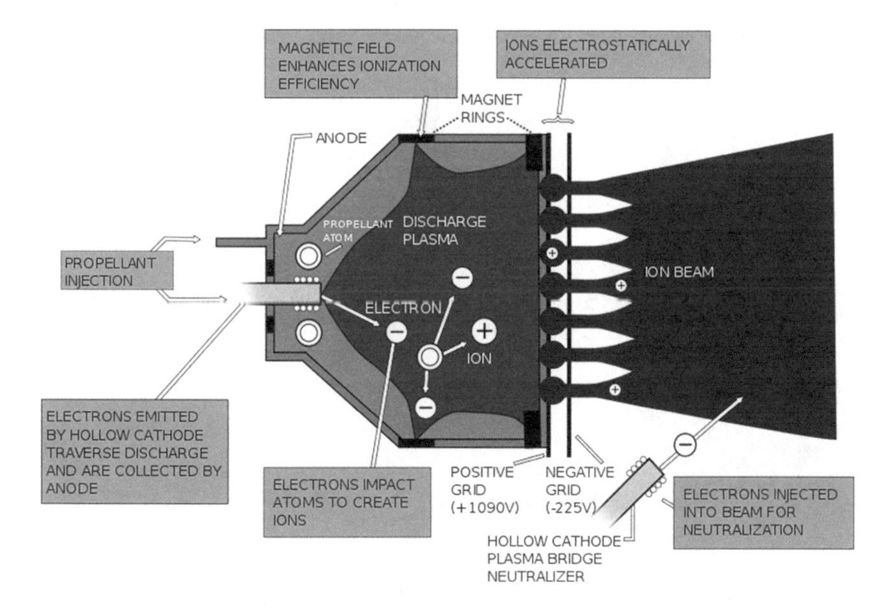

FIGURE 3.32 Ion engine.

3.12 CONSERVATION OF MOMENTUM

The principle of action reaction is related to a law called the *conservation of momentum*, which states that the total momentum of an isolated system always stays the same.

If we think of a system as a gun and a bullet or a rocket and the rocket's exhaust, the total momentum of these systems stays the same before and after firing. This is described by the equation:

$$p_{\text{total before}} = p_{\text{total after}}$$

$$\sum_{\text{before}} p_i = \sum_{\text{after}} p_i$$

This equation says, if you add up the momentum in a system before and after an event, the momentum is the same.

Conservation of momentum can tell us how the speed of an object will change before and after some event takes place. For instance, how fast does a gun recoil when fired?

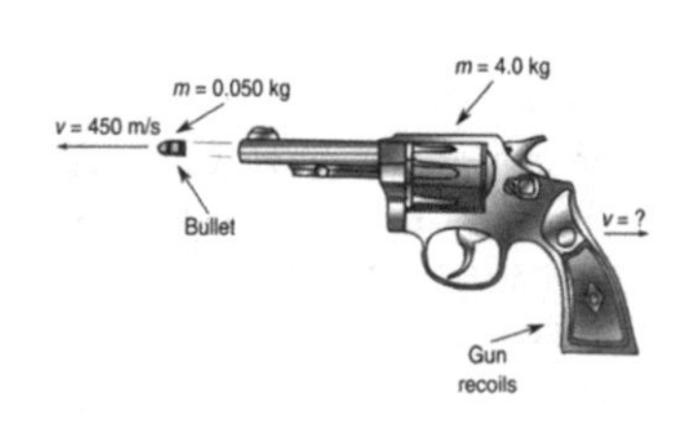

FIGURE 3.33 A gun with a mass of 4.0 kg fires a bullet of mass 0.050 kg with a velocity of 450 m/s.

Example 3.20

A gun with a mass of 4.0 kg fires a bullet of mass 0.050 kg with a velocity of 450 m/s (Figure 3.33). How fast does the gun recoil backward?

$$m_{\text{gun}} = 4.0\,\text{kg}$$
$$m_{\text{bullet}} = 0.050\,\text{kg}$$
$$v_{\text{bullet}} = 450\frac{\text{m}}{\text{s}}$$
$$v_{\text{gun}} = ?$$

Before the gun is fired, the gun and the bullet are still, therefore $p_{\text{total}} = 0$. By momentum conservation after the gun is fired, $p_{\text{total}} = 0$.

$$p_{\text{gun}} + p_{\text{bullet}} = 0$$
$$p_{\text{gun}} = -p_{\text{bullet}}$$
$$m_{\text{gun}} \times v_{\text{gun}} = -m_{\text{bullet}} \times v_{\text{bullet}}$$
$$\frac{\cancel{m}_{\text{gun}} \times v_{\text{gun}}}{\cancel{m}_{\text{gun}}} = \frac{m_{\text{bullet}} \times v_{\text{bullet}}}{m_{\text{gun}}}$$
$$v_{\text{gun}} = -\frac{m_{\text{bullet}} \times v_{\text{bullet}}}{m_{\text{gun}}}$$
$$v_{\text{gun}} = -\frac{.050\,\text{kg} \times 450\,\text{m/s}}{4.0\,\text{kg}}$$
$$v = -5.6\frac{\text{m}}{\text{s}}$$

The gun must move backward to cancel out the momentum of the bullet moving forward.

3.13 CHAPTER SUMMARY

Symbols used in this chapter:

Symbol	Unit
t—time	s
d—distance	m, ft
h—height	m, ft
v—velocity	m/s, ft/s
a—acceleration	m/s^2, ft/s^2
g—acceleration due to gravity	9.8 m/s^2, 32.2 ft/s^2
m—mass	kg, slug
G—gravitational constant	$6.67 \times 10^{-11} \frac{\text{N} \cdot \text{m}^2}{\text{kg}^2}$
F—force	newton (N), pound (lb)
F_w—force due to weight	newton (N), pound (lb)
F_f—force due to friction	newton (N), pound (lb)
μ—coefficient of friction	no unit
p—momentum	kg m/s, slug ft/s

3.13.1 Formulas

The force of attraction of one mass on another is given by Newton's law of gravity:

$$F = \frac{G \cdot m_1 \cdot m_2}{r^2}$$

$$F = \frac{\left(6.67 \times 10^{-11} \frac{\text{N} \cdot \text{m}^2}{\text{kg}^2}\right)(\text{mass}_1(\text{kg}))(\text{mass}_2(\text{kg}))}{\text{distance}(\text{m})^2}$$

Force due to weight = mass × acceleration due to gravity

$$F_w = m \cdot g$$

$$g = 9.8 \frac{\text{m}}{\text{s}^2} \text{ in the metric system}$$

$$g = 32.2 \frac{\text{ft}}{\text{s}^2} \text{ in the English system}$$

3.13.2 Conversions

1 kg = 1,000 g	1 g is about the mass of a paper clip.
1 lb = 4.45 N	1 newton is almost 1/4 lb.
1 slug = 14.6 kg	1 slug has the equivalent weight of 32.2 lb.

Newton's first law says that if a body changes its motion from standing still to moving, or from one speed to another, or changes its direction of travel, it is because there are forces acting upon it that do not balance each other out.

Newton's second law says that the sum of forces acting on a body equals the mass × acceleration.

$$F = m \cdot a$$

Newton's third law of motion is sometimes called the law of action reaction. For every force there is an equal and opposite force.

Static equilibrium means all forces are counterbalanced by other forces, and nothing is accelerating. In other words, the static equilibrium means that:

$$\sum_i F_i = 0$$

Sum of all the forces pointing upward = Sum of all the forces pointing downward.

Sum of all the forces pointing to the right = Sum of all the forces pointing to the left.

3.13.3 Key Points about Frictional Force

1. The force of friction between two surfaces is independent of the size of the two surfaces.
2. The force of friction between two surfaces is dependent on the quality of the surfaces.
3. The force of friction between two surfaces is dependent on the normal or perpendicular force to the surface. (If the surfaces are horizontal, the normal force is just the weight.)

These three statements can be summarized as:

$$F_f = \mu \cdot F_N$$

Frictional force = coefficient of friction × normal force

Something that is massive and moving fast has a lot of momentum. Momentum is a vector because it has both magnitude and direction.

$$\text{Momentum} = \text{mass} \times \text{velocity}$$

$$p = m \cdot v$$

$$\text{Momentum units: kg}\frac{\text{m}}{\text{s}} \text{ or slug}\frac{\text{ft}}{\text{s}}$$

Momentum and force are related to each other by the equation:

$$F_{\text{average}} = \frac{\Delta p}{t} = \frac{p_f - p_i}{t}$$

$$\text{Force}_{\text{average}} = \frac{\text{momentum}_{\text{final}} - \text{momentum}_{\text{initial}}}{\text{time}}$$

This equation shows that if the momentum of something is changing in time, it is because there is a net nonzero force acting on it.

Conservation of momentum says, if you add up the momentum in a system before and after an event, the momentum is the same.

$$p_{\text{total before}} = p_{\text{total after}}$$

$$\sum_{\text{before}} p_i = \sum_{\text{after}} p_i$$

PROBLEM SOLVING TIPS

1. Make a sketch of the problem, including a force diagram if relevant.
2. List the quantities you have and the quantity you need.
3. With your list of formulas in front of you, determine the relevant ones. If you have trouble, use the process of elimination.
4. Plug in numbers after you complete all the algebra; sometimes things cancel out.
5. Remember to pay attention to the units. Stay in the correct unit system, MKS (meter-kilogram-second) or FPS (foot-pound-second).

PROBLEMS

1. A force sensor emits a signal of 10 mV/lb. Suppose a weight of 100 lb is added to the scale. What is the voltage output from the sensor?
2. What is the mass of a 180 lb person in slugs and kilograms? What is the weight of this person in newton?

3. An automatic door with a mass of 12 kg opens by accelerating at 1 m/s^2. What force must a motor deliver to this door to cause this acceleration?
4. Four tires support a wagon weighing 22 lb. How much weight does each tire support?
5. Suppose the coefficient of kinetic friction between the rubber tires of a car and a concrete road is $\mu_k = 1.2$. Suppose the car weighs 3,500 lb. What is the deceleration of the car if it locks up the brakes?
6. What is the momentum of a hammer with mass 8 kg striking a nail at 5 m/s?
7. A nail stops the hammer in problem 6 in 0.08 s. What force is delivered to the nail?
8. The starship *Enterprise* fires a photon torpedo at a Klingon battle cruiser. Suppose the *Enterprise* has a mass of 1.2×10^{11} kg, and the effective mass of the photon torpedo is 500 kg traveling at 2.9×10^8 m/s. What is the recoil velocity of the *Enterprise*?
9. Suppose a bug hits your car windshield. Describe the force experienced by the bug relative to the force delivered to the windshield. Hint: Think about action reaction.
10. How can the conservation of momentum be used to determine what happened in an automobile accident?
11. What does Newton's third law say?
12. How would you design a device to measure acceleration using a resistive mass and a string?
13. A bridge must maintain static equilibrium in order for it to keep from collapsing. How could you monitor the bridge to see what kind of load it is under during typical use?
14. In electronics, there is a socket for an integrated circuit chip called a zero insertion force (ZIF) socket. This socket is designed to make inserting and removing chips easy. On a 16-pin chip, if 0.25 lb of force is applied to the chip when inserting, how much force does each pin "feel"?
15. A mass of 0.5 kg changes velocity from 2 to 3.5 m/s in 0.1 s. What was the average force on the mass?
16. A CD is ejected from a computer. The CD has a mass of 0.2 kg. What is the force on the CD if it is accelerated at 1.4 m/s^2?
17. An elevator is pulled upward with a force of 5,000 lb from a cable. The elevator weighs 3,000 lb. What is the total force on the elevator?
18. What is the acceleration of the elevator in the previous problem?

19. The ABS (Antilock Brake System) on a car is based on the principle that the coefficient of starting friction is greater than sliding friction. The coefficient of starting friction for rubber on wet concrete is 1.5, the coefficient of sliding friction for rubber on wet concrete is 0.97. How many times faster can a car stop using ABS versus locking up the brakes?
20. A robotic arm with a mass of 75 kg starts from rest and accelerates till it reaches a speed of 2.0 m/s in 3.0 s. What is the final momentum of the arm?
21. What was the force on the arm in the previous problem?
22. As a CD is ejected from a computer the computer recoils backward in order to conserve momentum. If the computer has a mass of 10 kg, the CD a mass of 0.2 kg and is ejected at a speed of 2.0 m/s, what is the recoil speed of the computer?
23. A weight scale is made using a force sensor that has an output of 10 mV/lb. If 25 lb is added to the scale, how many mV does the sensor put out?
24. A force is applied to the scanning arm inside a copy machine moving the arm across the page it is scanning. If the arm starts from rest, has a mass of 2.0 kg, and reaches a speed of 0.4 m/s in a time of 0.8 s, what was the net force on the arm?
25. An IC (integrated circuit) chip is inserted into a socket. The chip is pushed downward with a force of 0.1 lb. The socket has friction resisting the chip insertion with a force of 0.04 lb. If the chip weighs 1.0 Oz (16 Oz = 1 lb) what is its acceleration into the socket? Remember to convert weight into mass!
26. A computer weighing 12 lb is sitting on a table. If the table is to support the computer, what force must the table exert upward on the computer?
27. A motor weighs 2 lb. What is its mass?
28. A solenoid pulls in its plunger with a mass of 0.13 slugs with a force of 2.0 lb. What is the acceleration of the plunger?
29. When adding a lubricant to a surface the coefficient of sliding friction dropped from 0.4 to 0.3. If without the lubricant it took 10 lb of force to slide an object over the surface, what is the force required with the lubricant?
30. Another name for starting friction is static friction; it is the frictional force that is needed to budge a static or stationary object. If a power supply weighing 22 lb is to be slid across a table where the coefficient of starting friction is 0.5, how much force is needed to budge the supply?

31. An elevator weighing 2,000 lb is pulled upward with a force of 2,100 lb. What is the acceleration of the elevator?
32. A shopping cart weighing 55 lb gets away from a shopper heading for a stack of cans at a speed of 2.5 ft/s. What is the momentum of the cart?
33. A conveyer belt starting from rest transports a box of mass 15 kg at a speed of 0.4 m/s in 2.0 s. What force was applied on the box?
34. A doorbell consists of a solenoid with a plunger that moves in and out striking a bell. The plunger has a mass of 0.3 kg and is ejected at 2.1 m/s. The solenoid has a mass of 1.2 kg. What is the recoil velocity of the solenoid?
35. Two men are trying to push a piece of equipment weighing 100 lb. One man pushes with 50 lb and the other with 40 lb. The coefficient of starting friction is 0.35. How fast does the piece of equipment accelerate?
36. What is the gravitational force of attraction of an electron and a proton in a hydrogen atom? Use the following: $m_{\text{proton}} = 1.6727 \times 10^{-27}$ kg, $m_{\text{electron}} = 9.110 \times 10^{-31}$ kg, $r = 5.3 \times 10^{-11}$ m.
37. What is the gravitational force of attraction of the moon with the earth? Use $m_{\text{earth}} = 6.0 \times 10^{24}$ kg, $m_{\text{moon}} = 7.35 \times 10^{22}$ kg. The average distance from the earth's center to the moon is 384,400,000 m.

4

Energy, Work, and Power

Energy is used to make things move, light up, and heat up. It takes many forms, and it seems like we are always running out of it. But what exactly is energy? Does a gallon of gasoline have a lot of energy compared with a car battery or what can be obtained from a large solar cell or a windmill? This chapter will attempt to answer these questions, along with explaining how to calculate energy and power. When we put together these ideas of energy and power along with Newton's laws and the equations of motion, we will have a very powerful set of tools for analyzing almost any type of mechanical motion (Figure 4.1).

FIGURE 4.1 Electrical power plant.

So, what exactly is energy?

Energy is defined by what it can do. We can use energy to perform work.

Energy: *Energy* is the capacity to do work.

Work: Work in physics is different from work in everyday life. The definition of work is

$$W = F \cdot d \cdot \cos(\theta)$$

$$\text{Work} = \text{Force} \cdot \text{distance} \cdot \cos(\theta)$$

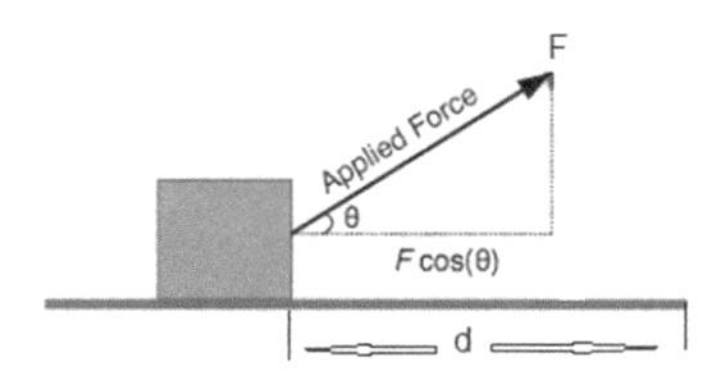

FIGURE 4.2 The work (W) done on an object equals product of the applied force (F) and the distance (d) the object moves through. $\cos(\theta)$ is the angle between the applied force and the direction of travel.

The work (W) done on an object equals product of the applied force (F) and the distance (d) the object moves through. $\cos(\theta)$ is the angle between the applied force and the motion. The term $F \cdot \cos(\theta)$ can be thought of as the component of force in the direction of motion (Figure 4.2).

Example 4.1

A car is pushed with 55 lb of force in the same direction in which the car moves (Figure 4.3). The car moves a distance of 6.0 ft. How much work is done?

$$F = 55\text{lb}$$

$$d = 6.0\text{ft}$$

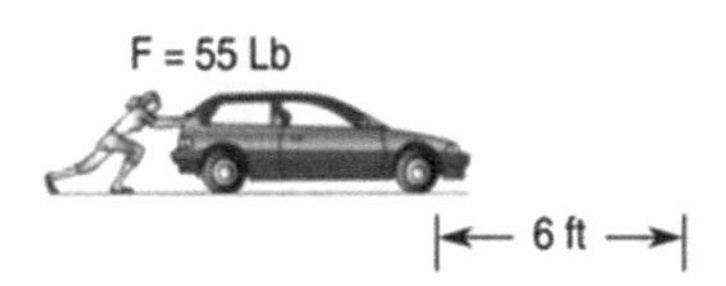

FIGURE 4.3 How much work was done?

$$\theta = 0° \leftarrow \text{The applied force is in the same direction as the motion.}$$

$$W = ?$$

$$W = F \cdot d \cdot \cos(\theta)$$

$$W = (55\text{lb})(6.0\text{ft}) \cdot \cos(0) = 330\text{ftlb}$$

4.1 ENERGY/WORK UNITS

In the English system, work or energy is measured in foot pounds (ft lb). In the SI system, it is measured in newton meters (Nm) or joules (J).

	English System	SI System
Energy/work	ft lb	Nm or joules (J)
Conversions $1\text{J} = 1\text{Nm}$ $1\text{J} = 0.74\,\text{ft} \cdot \text{lb}$		

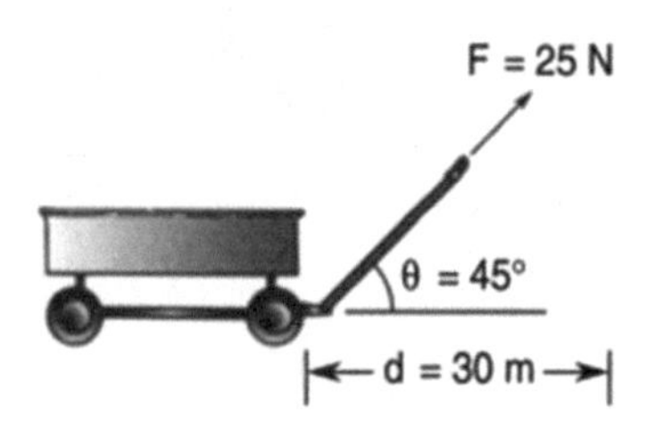

FIGURE 4.4 What is the work done?

Example 4.2

A cable is pulling a cart, making an angle of 45° with the direction of motion (Figure 4.4). The tension on the cable is 25 N and pulls the cart 3.0 m. What is the work done?

$$F = 25\text{N}$$

$$d = 3.0\text{m}$$

$$\theta = 45°$$

$$W = ?$$

$$W = F \cdot d \cos(\theta)$$

$$W = 25\text{N} \cdot 3.0\text{m} \cdot \cos(45°)$$

$$W = 53\text{Nm}$$

$$W = 53\text{J} \leftarrow \text{Because } 1\text{Nm} = 1\text{J}.$$

In a system without any energy loss such as friction, the energy needed to perform some amount of work is equal to the work.

Energy = Work performed ← For a system without any energy loss.

Example 4.3

How much energy is needed to perform 100J of work? To perform 100J of work, 100J of energy is needed (assuming no energy is lost). Energy and work have the same units.

4.2 EFFICIENCY

Efficiency is a measure of how much energy is converted into some useful work. For a system that is not friction free, the efficiency will always be less than 100%.

$$Efficiency = \left(\frac{work\ done}{energy\ consumed}\right) \times 100$$

FIGURE 4.5 An automobile engine.

Example 4.4

An automobile engine performs 9,500J of work in order to move a car. The car consumes 47,500J of energy by burning gasoline (Figure 4.5). What is the efficiency of this car?

$$W = 9{,}500\,\text{J}$$

$$E = 47{,}500\,\text{J}$$

$$\text{Efficiency} = ?$$

$$Efficiency = \left(\frac{work\ done}{energy\ consumed}\right) \times 100$$

$$Efficiency = \left(\frac{9{,}500\,J}{47{,}500\,J}\right) \times 100$$

$$Efficiency = 2\bar{0}\%$$

In a system with energy loss such as friction, the amount of work the system can perform is:

$$W = \text{Efficiency} \cdot E$$

$$\text{Work} = \text{Efficiency} \cdot \text{Energy Available}$$

Example 4.5

A motor has 2,000J of energy available to it, but it is only 80% efficient in converting this energy to work. How much work can it perform?

$$W = \text{Efficiency} \cdot E$$
$$W = 0.80 \cdot 2{,}000\,\text{J}$$
$$W = 1{,}600\,\text{J}$$

Energy can take two basic forms:

Potential energy is the energy stored in a system due to the arrangement of the position of its parts. A compressed spring, a weight raised to some height, and a battery with its charges separated all have potential energy (Figure 4.6). This energy is stored, waiting to be used. We will use the letter U to represent potential energy.

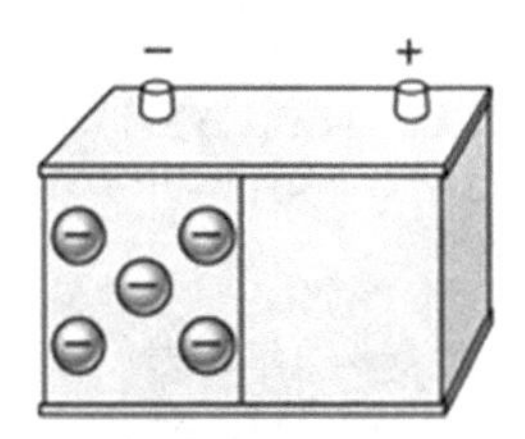

FIGURE 4.6 A weight raised to some height, a compressed spring, and a battery with its charges separated all have potential energies.

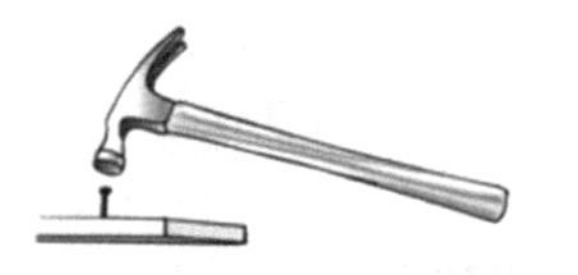

FIGURE 4.7 A hammer delivers its kinetic energy to a nail.

Kinetic energy means moving energy.

A hammer striking the head of a nail delivers its kinetic energy to the nail, forcing it into the wood some distance, performing some work. We will use the letter K to represent kinetic energy (Figure 4.7).

4.3 POTENTIAL ENERGY SOURCES

4.3.1 Gravitational

FIGURE 4.8 A pile driver. Potential energy is stored, then released, and turned into kinetic energy. The kinetic energy is delivered to the pile, performing some work.

Energy can be stored by lifting something above the ground. This is *gravitational potential energy.* The energy can then be released, allowing it to fall back to the ground, turning this potential into kinetic energy (Figure 4.8).

Gravitational potential energy is given by the following formula.

$$U = m \cdot g \cdot h$$

$$\text{Potential Energy (J)} = (\text{mass(kg)})\left(9.8\frac{\text{m}}{\text{s}^2}\right)(\text{height(m)})$$

Example 4.6

A water tower stores potential energy in the water by pumping it up into a tank, high above the ground. Suppose the tank is 10.0 m above the ground and contains 12,500 kg of water (Figure 4.9). What is the potential energy of this water?

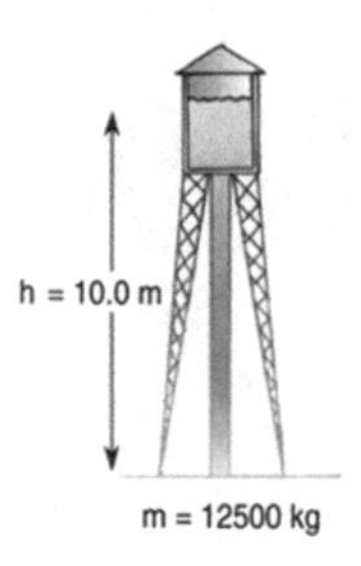

FIGURE 4.9 What is the potential energy of this water?

$$h = 10.0\text{m}$$

$$m = 12{,}500\text{kg}$$

$$U = ?$$

$$U = m \cdot g \cdot h$$

$$U = (12{,}500\text{kg})\left(9.8\frac{\text{m}}{\text{s}^2}\right)(10.0\text{m})$$

$$U = 1.225 \times 10^6 \text{ J}$$

4.3.2 Springs

FIGURE 4.10 A compressed or stretched spring stores potential energy.

In order to compress or elongate a spring, work must be done on the spring. The energy used to perform this work is stored in the spring, waiting to be released (Figure 4.10).

4.3.3 Batteries

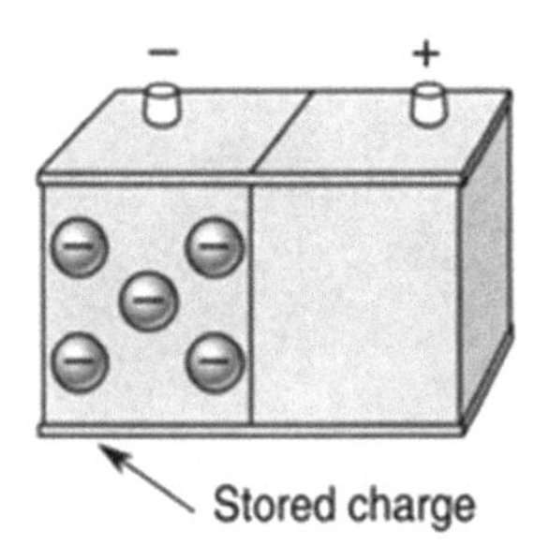

FIGURE 4.11 A battery separates charge to be used later to perform some useful work.

A *battery* is a source of potential energy. The amount of energy a battery can store is determined by how much charge it can store and its voltage (Figure 4.11).

Example 4.7

Suppose an electric car is powered by a 12 V battery containing 4.1 MJ of energy. If it takes 150 lb or 668 N of force to move the car at a constant rate, how far can the car travel?

Assume an efficiency of 70%.

$$E = 4.1\text{MJ}$$

$$F = 668\text{N}$$

$$\text{Efficiency} = 70\%$$

$$d = ?$$

Since there is only 70% efficiency, the total energy available to do work is

$$W = \text{Efficiency} \cdot E$$

$$W = 0.70 \times 4.1\text{MJ}$$

$$W = 2.87\,\text{MJ}$$

$$W = F \cdot d$$

$$d = \frac{W}{F}$$

$$d = \frac{2.87 \times 10^6\,\text{J}}{668\text{N}}$$

$$d = 4{,}296.4\text{m}$$

This is only about 3 miles! Better increase the number of batteries!

4.3.4 Fuels

Fuels store chemical potential energy. A molecule of gasoline can be thought of as a bunch of atoms separated by compressed springs (Figure 4.12). The springs are the intermolecular forces between atoms. With the addition of heat, the springs will be released from compression, hurling the atoms outward. This is an explosion of the gas molecule that turns the chemical potential into kinetic energy. Listed below in Table 4.1 are the energies released when these fuels are burned.

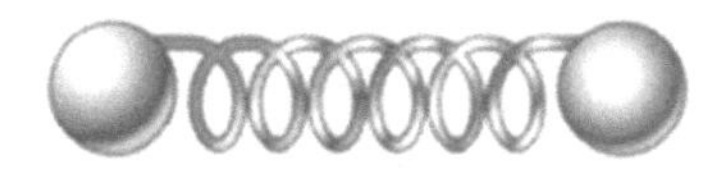

FIGURE 4.12 Chemical potential energy is analogous to the energy stored in compressed springs.

TABLE 4.1
Energy Contained in Fuels

	Heat of Combustion		
Fuel	**Btu/lb$_m$**	**MJ/kg**	**Air-Fuel Ratio for Complete Combustion (lb$_m$/lb$_m$ or g/g)**
Butane	20,000	46	15.4
Gasoline	19,000	44	14.8
Crude oil	18,000	42	14.2
Fuel oil	17,500	41	13.8
Coke	14,500	34	11.3
Coal, bituminous	13,500	31	10.4

Actual values will vary slightly depending on geographical origin. Values do not include latent heat in any water vapor formed as a product of combustion.

Note: 1 MJ/kg = 1,000 kJ/kg = 1,000 J/g

Example 4.8

Calculate the gas mileage a car would get if the force needed to propel it along at some speed is 215 lb and the efficiency of the car is 20%.

$$F = 215\,\text{lb}$$

$$\text{Efficiency} = 20\%$$

The energy needed to travel 1 mile, or 5,280 ft, is:

$$W = F \times d$$

$$W = 215\,\text{lb} \times 5,280\,\text{ft}$$

$$W = 1.14 \times 10^6\,\text{ft lb}$$

The car will expend more energy than this because it is only 20% efficient.

The energy consumed in traveling 1 mile can be determined from:

$$\text{Efficiency} = \left(\frac{\text{work done}}{\text{energy consumed}}\right) \times 100$$

$$\text{Energy consumed} = \left(\frac{\text{work done}}{\text{efficiency}}\right) \times 100$$

$$\text{Energy consumed} = \left(\frac{1.14 \times 10^6\,\text{ft lb}}{20}\right) \times 100$$

$$\text{Energy consumed} = 5.70 \times 10^6\,\text{ft lb}$$

Since the calculation was based on traveling 1 mile, the energy consumed per mile is:

$$\text{Energy consumed/mile} = 5.70 \times 10^6 \frac{\text{ft lb}}{\text{mile}}$$

Gasoline has an energy content of 9.7×10^7 ft lb/gal. The gas mileage is therefore:

$$\text{Gas mileage} = \frac{9.7 \times 10^7 \text{ ft lb/gal}}{5.70 \times 10^6 \text{ ft lb/mile}}$$

$$\text{Gas mileage} = 17 \frac{\text{mile}}{\text{gal}}$$

Can you estimate the gas mileage of your car from a calculation like this?

4.4 KINETIC ENERGY

Kinetic energy means moving energy. If something moving strikes something stationary, it will deliver a force to it, causing it to move by imparting its energy to the stationary object. The formula for kinetic energy is:

$$K = \frac{1}{2} m \cdot v^2$$

$$\text{Kinetic energy (J)} = \frac{1}{2} \text{mass (kg)} \cdot \left(\text{velocity} \left(\frac{\text{m}}{\text{s}} \right) \right)^2$$

Example 4.9

A rail gun consists of a U-shaped conductor with a sliding conducting rail lying across it. When a pulse of current is induced into the circuit, the rail will quickly move outward due to the magnetic fields generated in the loop (Figure 4.13).*

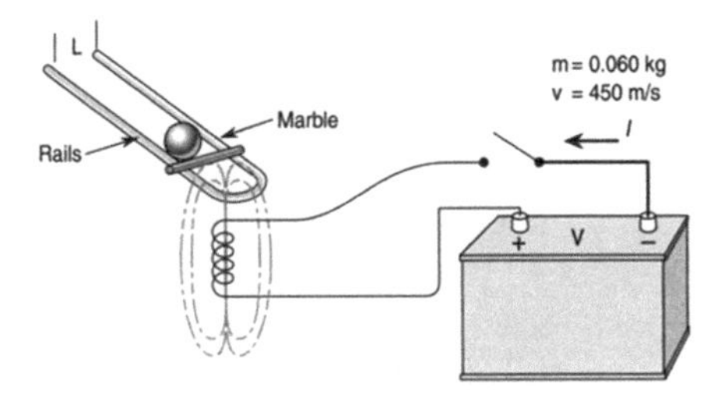

FIGURE 4.13 A rail gun.

* A magnetic field is generated around a current-carrying wire. When two current carrying wires are parallel to each other and the current in each wire is moving in the opposite direction, the wires will repel each other. This is similar to the repulsive force between two magnets, with like poles facing each other.

Suppose a marble with a mass of 0.060 kg is placed on the rail and launched at a speed of 450 m/s. What is the kinetic energy of the marble?

$$m = 0.060\,\text{kg}$$

$$v = 450\,\frac{\text{m}}{\text{s}}$$

$$K = \frac{1}{2} m \cdot v^2$$

$$K = \frac{1}{2}(0.060\,\text{kg}) \cdot \left(450\,\frac{\text{m}}{\text{s}}\right)^2$$

$$K = 6{,}075\,\text{J}$$

Example 4.10

How much work can be performed by an object that has 6,100 J of kinetic energy?

If there is a 100% efficient transfer of energy into work, then 6,100 J of energy can perform 6,100 J of work.

If the energy transfer into work is less than 100%, say 40%, the amount of work that can be done is:

$$40\%\text{ of } 6{,}100\,\text{J} = 0.4 \times 6{,}100\,\text{J} = 2{,}440\,\text{J}$$

Example 4.11

A batter hits a baseball with a mass of 2.1 kg, delivering 2,500 J of energy to it (Figure 4.14). At what speed is the ball traveling?

$$m = 2.1\,\text{kg}$$

$$K = 2{,}500\,\text{J}$$

$$K = \frac{1}{2} m \cdot v^2$$

$$v = \sqrt{\frac{2 \cdot K}{m}}$$

$$v = \sqrt{\frac{2 \cdot 2{,}500\,\text{J}}{2.1\,\text{kg}}}$$

$$v = 49\,\text{m/s}$$

FIGURE 4.14 At what speed the ball is traveling?

4.5 WORK AND KINETIC ENERGY

To perform work on a mass means to apply a force on the mass and move it. This results in its velocity changing. A change in velocity means its kinetic energy has changed, which is described by the **work-kinetic energy theorem:**

$$\Delta K = W$$
$$\text{Change in Kinetic Energy} = \text{Work}$$

Example 4.12

A mass of 2 kg at rest has a force of 3 N applied on it, moving it a distance of 4 m.

a. How much work was performed on the mass?
b. What is its change in kinetic energy?
c. What is its final velocity?

(a)

$$W = F \cdot d = (3\text{N})(4\text{m})$$

$$W = 12\text{J}$$

(b)

$$\Delta K = W = 12\text{J}$$

(c)

$$\Delta K = \frac{1}{2}mv_f^2 - \frac{1}{2}mv_i^2$$

The mass started from rest so $v_i = 0$.

$$\frac{1}{2}mv_f^2 = W$$

$$v_f = \sqrt{\frac{2W}{m}} = \sqrt{\frac{2(12\text{J})}{2\,\text{kg}}}$$

$$v_f = 3.46\text{m/s}$$

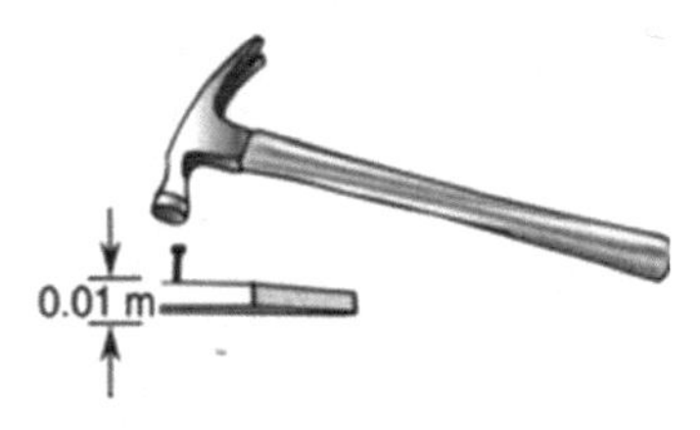

FIGURE 4.15 With what average force did the hammer strike the nail?

Example 4.13

A hammer with 500 J of kinetic energy strikes a nail, driving it 0.01 m into a piece of wood (Figure 4.15). With what average force did the hammer strike the nail?

$$K = 500\,\text{J}$$

$$d = 0.01\text{m}$$

$$F = ?$$

$$\text{Work} = \text{energy} \leftarrow \text{Assume 100\% of energy is converted into work.}$$

$$\text{Work} = 500\,\text{J}$$

$$W = F \cdot d$$

$$\frac{W}{d} = \frac{F \cdot \not{d}}{\not{d}} \leftarrow \textit{solving for } F$$

$$F = \frac{W}{d}$$

$$F = \frac{500\,\text{J}}{0.01\text{m}} = 50{,}000\,\text{N}$$

4.6 ENERGY CONSERVATION

The total energy of an isolated system is a constant. Energy is never created or destroyed; it just changes form. Even for a system that is less than 100% efficient, we can account for all of the energy used if we add up the work done, the energy wasted due to friction that went into heating up the system, any light or sound given off, or any other form the input energy may have turned into. Given this law, if something falls, losing some of its potential energy, the loss must be made up by gaining kinetic energy (neglecting air resistance).* Kinetic and potential energies are in balance. A pendulum illustrates this nicely (Figure 4.16).

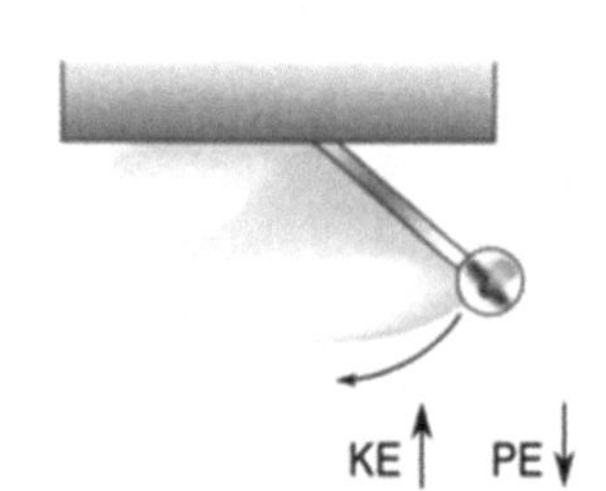

FIGURE 4.16 Consider a pendulum. If the potential energy decreases, the kinetic energy increases.

The total energy of a system is made up of the sum of the kinetic and potential energies.

$$E = K + U$$

The conservation of energy says the total energy of an isolated system always stays the same; it is a constant. This is described mathematically by the following equation:

$$\Delta E = \Delta K + \Delta U = 0$$

$$\text{Change in energy}_{\text{Isolated}} = \text{Change in energy}_{\text{Kinetic}} + \text{Change in energy}_{\text{Potential}} = 0$$

* We are, of course, neglecting air resistance, which will limit the object's velocity to a terminal velocity.

We can express the conservation of energy equation in another form:

$$\Delta K = -\Delta U$$

$$K_f - K_i = -\left(U_f - U_i\right)$$

$$\text{Kinetic}_{\text{final}} - \text{Kinetic}_{\text{initial}} = -(\text{Potential}_{\text{final}} - \text{Potential}_{\text{initial}})$$

FIGURE 4.17 The potential energy stored is turned into kinetic energy when the water flows downward.

Example 4.14

The water at the top of a waterfall has potential energy. If this water is allowed to fall back to the ground, its potential energy will be turned into kinetic energy (Figure 4.17). Suppose the potential energy of the water is 54,000J at a height of 10.0m. What is the kinetic energy of this water when it falls to the ground, and what is its velocity?

Using the conservation of energy:

$$K_f - K_i = -\left(U_f - U_i\right)$$

The initial potential energy equals: U_i = 54,000J.

All of this energy is turned into kinetic energy so:

$$K_f = 54{,}000\,\text{J}.$$

From the problem, $K_i = 0$ and $U_f = 0$.

$$K_f = U_i$$

$$\frac{1}{2}mv_f^2 = \text{mgh}$$

$$\frac{1}{2}v_f^2 = \text{gh} \leftarrow \text{m on both sides canceled out}$$

$$v_f = \sqrt{2\text{gh}} = \sqrt{2\left(9.8\frac{\text{m}}{\text{s}^2}\right)(10.0\text{m})}$$

$$v_f = 14\frac{\text{m}}{\text{s}}$$

4.7 WORK AND POTENTIAL ENERGY

Recall that the work done on a mass is the force applied on the mass in the direction of the motion of the mass. A conservative force is a force where the work done on a particle does not depend on the path the mass is pushed or pulled through by the force. An example of a conservative force is gravity, or electric forces, or the force associated with a spring. In the case of gravity or a spring, the change in potential energy and the work done on a particle depends only on the initial and final positions of the system, and not on the path taken. Conservative forces therefore have associated with them a potential energy. Work done by a frictional force is an example of a non-conservative force, because the work depends on the path and not just the end points.

Putting together the conservation of energy, the work-kinetic energy theorem, and the idea of a conservative force:

$$\Delta U + \Delta K = 0$$

$$\Delta U = -\Delta K$$

$$\Delta K = W$$

Finally, we get:

$$\Delta U = -W_C$$

$$\text{Change in Potential Energy} = -\text{Conservative Work Done}$$

Example 4.15

To illustrate $\Delta U = -W_C$ calculate the work done and change in the potential energy of a mass of 2 kg lifted from ground level to 3 m above the ground.

Calculating the change in potential energy:

$$\Delta U = m \cdot g \cdot h = (2\text{kg})\left(9.8\frac{\text{m}}{\text{s}^2}\right)(3\text{m})$$

$$\Delta U = 58.8\,\text{J}$$

Calculating the work done:

$$W = F \cdot d = m \cdot g \cdot d$$

$$W = (2\text{kg})\left(-9.8\frac{\text{m}}{\text{s}^2}\right)(3\text{m})$$

Here we use a = –g because the force points in the opposite direction as the motion.

$$W = -58.8\,\text{J}$$

Comparing W and ΔU we see:

$$\Delta U = -W.$$

4.8 POWER

Power is the rate at which energy is used.

$$P = \frac{E}{t}$$

$$\text{Power} = \frac{\text{Energy}}{\text{time}}$$

If a lot of energy is used quickly, then a lot of power is consumed.

Power is measured in watts, $\frac{\text{ft} \cdot \text{lb}}{\text{s}}$ or horsepower.

English Units	Metric Units
Power $\left(\frac{\text{ft} \cdot \text{lb}}{\text{s}}\right)$ or horsepower (hp)	watts (W)
Conversions	
$1 \text{ hp} = 550 \frac{\text{ft} \cdot \text{lb}}{\text{s}}$	
$1 \text{ hp} = 746 \text{ W}$	
$1 \text{ W} = 0.74 \frac{\text{ft} \cdot \text{lb}}{\text{s}}$	

FIGURE 4.18 How much power is being consumed?

Example 4.16

A light bulb consumes 100 J of energy every 2 s (Figure 4.18). How much power is being consumed?

$$E = 100\text{J}$$

$$T = 2\text{s}$$

$$P = \frac{E}{t}$$

$$P = \frac{100\text{J}}{2\text{s}}$$

$$P = 50 \text{ W}$$

Example 4.17

A crane lifts a crate with a force of 250 lb, 10 ft in the air in 5 s (Figure 4.19). How much work was done? How much energy was used and power consumed?

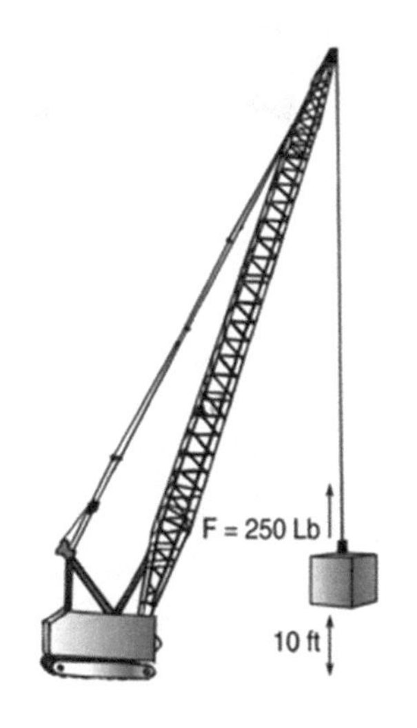

FIGURE 4.19 How much energy was used and power consumed?

$$F = 250\text{lb}$$

$$d = 10\text{ft}$$

$$t = 5\text{s}$$

$$W = F \cdot d$$

$$W = (250\text{lb})(1\bar{0}\text{ft}) = 2{,}500\text{ft} \cdot \text{lb}$$

$$\text{Energy} = \text{Work}$$

$$\text{Energy} = 2{,}500\,\text{ft}\cdot\text{lb}$$

$$P = \frac{E}{t}$$

$$P = \frac{2{,}500\,\text{ft}\cdot\text{lb}}{5\,\text{s}}$$

$$P = 500\frac{\text{ft}\cdot\text{lb}}{\text{s}}$$

FIGURE 4.20 What is this power in watts?

Example 4.18

A lawn mower is rated at 3 hp (Figure 4.20). What is this power in watts?

$$P = 3.0\,\text{hp}$$

$$1\,\text{hp} = 746\,\text{W}$$

$$3\,\cancel{\text{hp}}\left(\frac{746\,\text{W}}{1\,\cancel{\text{hp}}}\right) = 2{,}238\,\text{W}$$

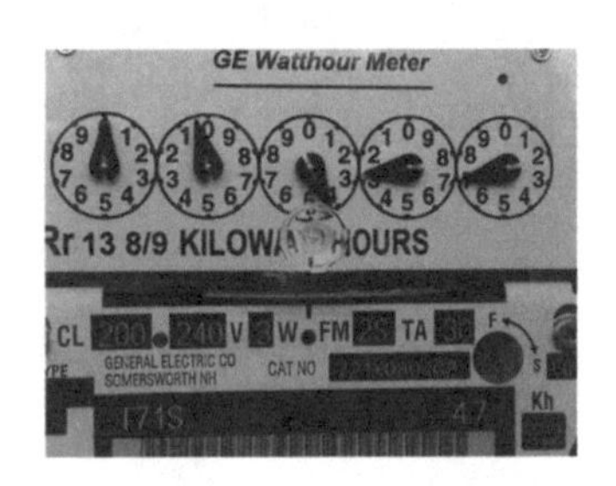

FIGURE 4.21 Electrical power meter.

4.9 KILOWATT-HOURS

The power meter in your house keeps track of how much energy you use (Figure 4.21). It does this by keeping track of how many kilowatts are used and the length of time those kilowatts are used for, remember (Energy = Power × Time). The meter reading is given in kilowatt-hours (kWh), because we pay for energy, not power.

$$\text{Energy}(\text{kWh}) = \text{kilowatts} \times \text{hours}$$

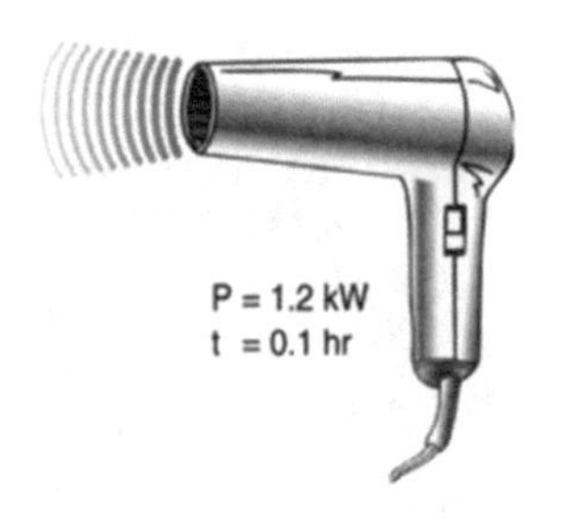

FIGURE 4.22 What is the energy consumed?

Example 4.19

A hair dryer consumes 1.2 kW of power and is run for 0.1 h (Figure 4.22). What is the energy consumed?

$$P = 1.2\,\text{kW}$$

$$t = 0.1\,\text{h}$$

$$E = (1.2\,\text{kW})(0.1\,\text{h}) = 0.12\,\text{kWh}$$

Example 4.20

The electrical power company charges you by the number of kilowatt-hours used. Suppose the power company charges 9 cents/kWh. How much does it cost you to leave a 100 W light on for 8 h? This problem is, effectively, a conversion from kilowatt-hours to cents.

$$P = 100\,\text{W}$$
$$P = 0.1\,\text{kW}$$
$$t = 8\,\text{h}$$
$$\text{kWh} = (0.1\,\text{kW})(8\,\text{h})$$
$$\text{kWh} = 0.8\,\text{kWh}$$
$$\text{Rate} = 9\,\text{cents/kWh}$$
$$\text{Cost} = \left(\frac{9\,\text{cents}}{\text{kWh}}\right)(0.8\,\text{kWh})$$
$$\text{Cost} = \left(\frac{9\,\text{cents}}{\cancel{\text{kWh}}}\right)(0.8\,\cancel{\text{kWh}})$$
$$\text{Cost} = 7.2\,\text{cents}$$

4.10 POWER, FORCE, AND VELOCITY

Power, force, and velocity are related through the formula:

$$P = F \cdot v$$

$$\text{Power} = \text{force} \times \text{velocity}$$

Example 4.21

A girl pedals her bike, pushing down on the pedals while generating 55 lb of force between the wheel and the road (Figure 4.23). She is traveling at 10 ft/s. How much power is the girl transmitting to the bike?

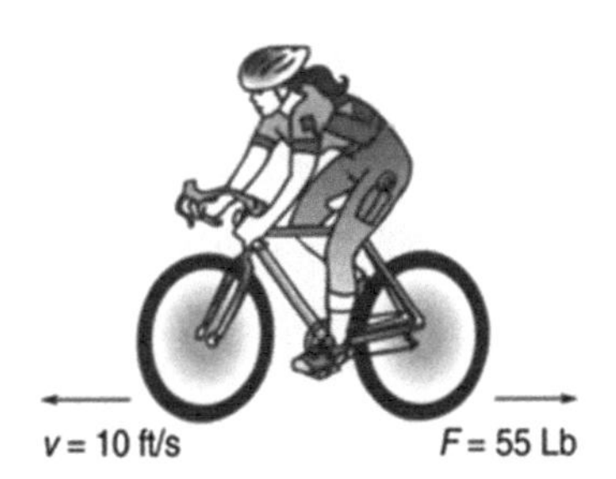

FIGURE 4.23 How much power the girl is transmitting to the bike?

$$F = 55\,\text{lb}$$

$$v = 10\,\text{ft/s}$$

$$P = ?$$

$$P = Fv$$

$$P = (55\,\text{lb}) \cdot (10\,\text{ft/s})$$

$$P = 550\,\text{ft lb/s}$$

$$P = 1\,\text{hp}$$

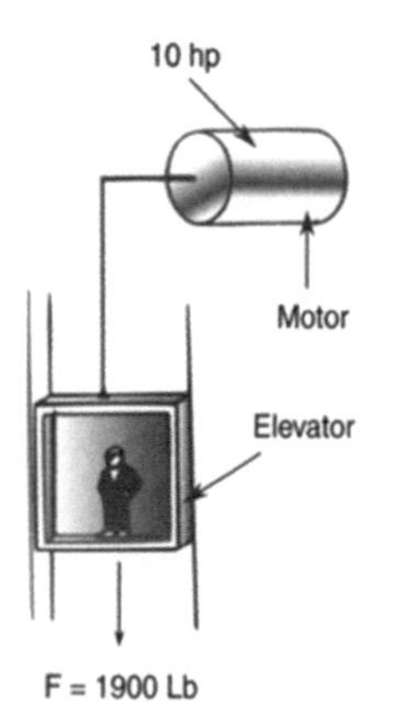

FIGURE 4.24 How fast will the elevator car move?

Example 4.22

You are asked to design an elevator to transport 1,900 lb of freight at a constant speed using a 10 hp motor (Figure 4.24). How fast will the elevator car move?

$$F = 1{,}900\,\text{lb}$$

$$P = 10\,\cancel{\text{hp}}\left(\frac{550\,\text{ft lb/s}}{1\,\cancel{\text{hp}}}\right) = 5{,}500\,\text{ft lb/s}$$

$$v = ?$$

$$P = F \cdot v$$

$$\frac{P}{F} = \frac{\cancel{F} \cdot v}{\cancel{F}} \Leftarrow \text{Solving for } v.$$

$$v = \frac{P}{F}$$

$$v = \frac{5{,}500\,\text{ft}\,\cancel{\text{lb}}/\text{s}}{1{,}900\,\cancel{\text{lb}}}$$

$$v = 2.9\,\text{ft/s}$$

4.11 OTHER SOURCES OF ENERGY

Comparing Energy Sources: What is the difference between gas, coal, and nuclear power? The electrical energy generated from gas, coal, and nuclear fuel is not a direct conversion. The heat generated from these three fuel sources is used to boil water, to make steam. This steam is then directed at a turbine to turn an electrical generator (Figure 4.25).

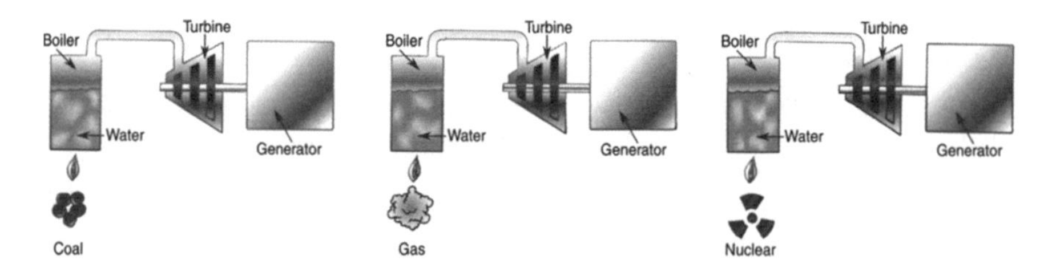

FIGURE 4.25 Gas, coal, and nuclear power are all used to heat water to generate steam, to turn a turbine connected to an electrical generator. The only difference between them is the fuel used to heat the water.

4.11.1 Fuel Cells

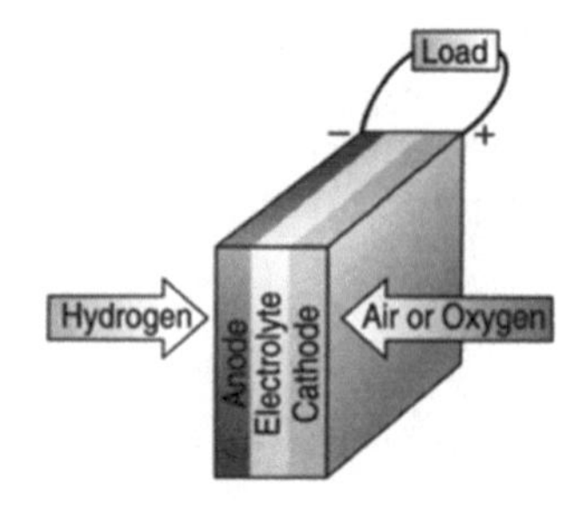

FIGURE 4.26 A fuel cell.

Fuel cells function like a battery in that they derive their electrical energy from a chemical reaction (Figure 4.26). Unlike a battery, they never need to be recharged. They do require fuel, however, in the form of hydrogen and oxygen. A fuel cell consists of two electrodes contained in an electrolyte.

Hydrogen flows over one electrode and oxygen flows over the other. A chemical reaction takes place at the electrodes, removing an electron from the hydrogen, attaching itself to one electrode (the cathode) and the positive proton to the other electrode (the anode). This buildup of opposite charges on two electrodes makes for a device that functions like a battery.

4.11.2 Wind

A *windmill* can take the kinetic energy of the wind and use it to pump water, mill corn, and generate electricity (Figure 4.27). How much energy does wind have? The energy contained in the wind depends on its velocity, density, and moisture content.

FIGURE 4.27 A windmill converts the wind's kinetic energy into rotational energy to be used for some purpose, such as generating electricity or pumping water.

4.11.3 Hydroelectric Energy

Hydroelectric energy is the energy captured from falling water (Figure 4.28).

FIGURE 4.28 Hydroelectric power is power generated from falling water.

4.11.4 Solar Energy

Solar cells convert the energy contained in sunlight into electrical energy (Figure 4.29). On a sunny day, about 1,000 W/m^2 of energy is contained in the sunlight that reaches earth. Currently, solar cells can at best convert about 20% of this energy into electricity. A solar cell is basically made of two materials called P and N materials. The P and N are placed next to one another, and this combination is called a PN junction. As light makes its way to the junction, it knocks free electrons on the P side, and they are transported to the N side because of the electric field present in every PN junction. A grid is placed over the solar cell to collect these electrons on the front, and a metal film on the back acts as the positive terminal. The cell now acts like a battery, deriving its energy from light instead of some chemical reaction (Figure 4.30). Currently, at about 12.2 cents/kWh, solar power is becoming comparable to electricity from the utility companies (about 12 cents/kWh).

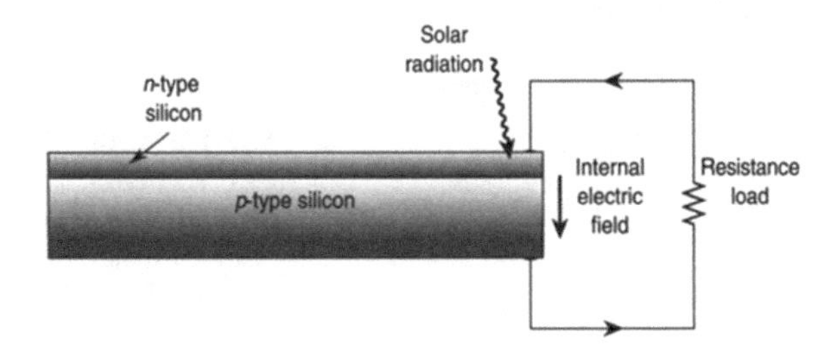

FIGURE 4.29 A solar cell.

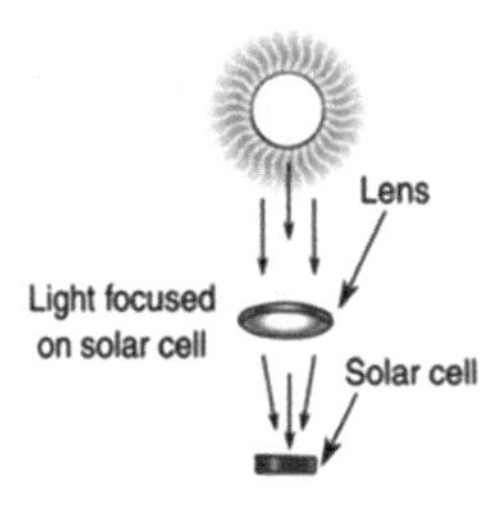

FIGURE 4.30 Lenses help collect more light to focus on the solar cell.

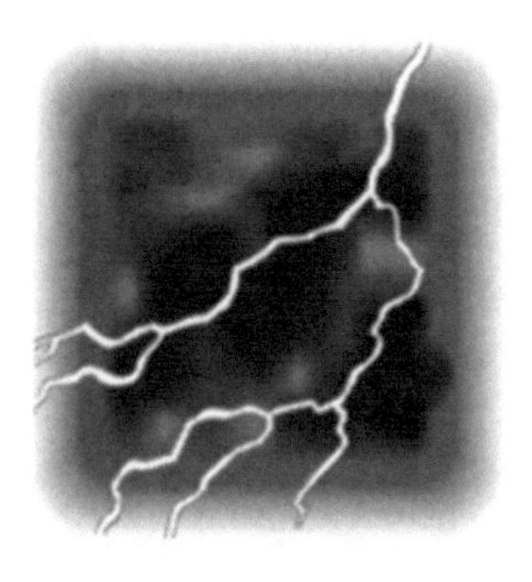

FIGURE 4.31 A lightning bolt contains millions of joules of energy.

4.11.5 Lightning

A bolt of *lightning* contains millions of joules of energy (Figure 4.31). It would be great if we could capture and use it. Research is currently seeking ways to capture this energy. One of the biggest problems to overcome in this field is finding a way to collect the energy without destroying the equipment used to collect it!

4.11.6 Fusion

Fusion Power, ITER: International Thermonuclear Experimental Reactor is the world's largest nuclear fusion reactor located in southern France. It consists of a magnetic "bottle", shaped like a doughnut surrounded by a magnetic field. This design is called a Tokomak, Figure 4.32. A gas of isotopes of hydrogen, tritium, and deuterium are contained in the Tokomak. The gas is heated to temperatures of 100,000,000°C by running a current through the gas. The magnetic field keeps the plasma away from the walls of the reactor to prevent it from melting. At such high temperatures, the tritium and deuterium are moving very quickly inside the vessel and they occasionally slam into one another with such force that they stick together forming a new atom, helium, Figure 4.33. Heat energy is given off in the process. This process is called fusion and is the source of the sun's energy. The heat released in this process is used to boil water to create steam to power an electrical generator.

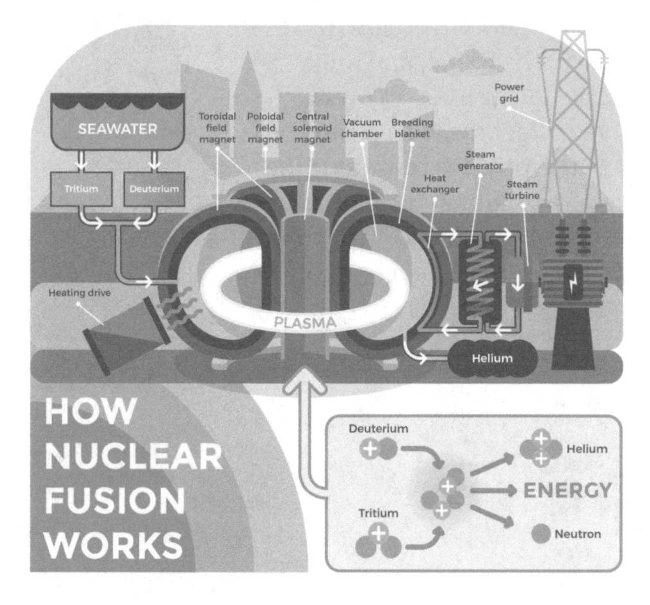

FIGURE 4.32 A Tokomak fusion reactor consists of a magnetic "bottle", shaped like a doughnut surrounded by a magnetic field to contain the hot plasma undergoing fusion.

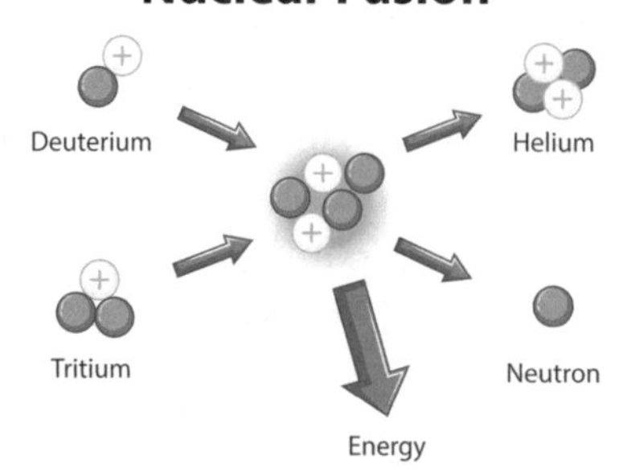

FIGURE 4.33 At high temperatures, the tritium and deuterium are moving very quickly inside the vessel and they occasionally slam into one another with such force that they stick together forming a new atom, helium. Heat energy is given off in the process.

4.12 CHAPTER SUMMARY

Symbols used in this chapter:

Symbol	Unit
t—time	s
d—distance	m
h—height	m
v—velocity	m/s, ft/s
a—acceleration	m/s^2, ft/s^2
g—acceleration due to gravity	9.8 m/s^2, 32.2 ft/s^2
F—force	N, lb
m—mass	kg, slug
Efficiency	%
W—work	J, ft lb
U—potential energy	J, ft lb
K—kinetic energy	J, ft lb
P—power	W, ft lb/s, hp
kWh—kilowatt-hours	kWh

4.12.1 Conversions

1 J = 1 Nm

1 J = 0.74 ft lb

$$1\,\text{hp} = 550\frac{\text{ft}\cdot\text{lb}}{\text{s}}$$

1 hp = 746 W

$$1\,\text{W} = 0.74\frac{\text{ft}\cdot\text{lb}}{\text{s}}$$

Energy is the capacity to do work.

$$W = F \cdot d\cos(\theta)$$

$$\text{Work} = \text{Force}\cdot\text{distance}\cdot\cos(\theta)$$

$$\text{Efficiency} = \left(\frac{\text{work done}}{\text{energy consumed}}\right)\times 100$$

$$U = m \cdot g \cdot h$$

Potential energy = mass (kg) × gravity (m/s^2) × height (m)

$$K = \frac{1}{2}m \cdot v^2$$

$$\text{Kinetic energy} = \frac{1}{2}\text{mass}(\text{kg})(\text{velocity}(\text{m/s}))^2$$

Conservation of energy says: $\Delta E = \Delta K + \Delta U = 0$

$$\text{Change in energy}_{\text{Isolated}} = \text{Change in energy}_{\text{Kinetic}} + \text{Change in energy}_{\text{Potential}}$$

Power is the rate at which energy is used.

$$P = \frac{E}{t}$$

$$\text{Power} = \frac{\text{Energy}}{\text{time}}$$

Power is measured in watts, $\frac{\text{ft} \cdot \text{lb}}{\text{s}}$ or horsepower.

English and Metric units

Power: $\frac{\text{ft} \cdot \text{lb}}{\text{s}}$, hp, W

$$\text{Energy}(\text{kWh}) = \text{kilowatts} \times \text{hours}$$

Power, force, and velocity are related through the formula:

$$P = F \cdot v$$

$$\text{Power} = \text{Force} \times \text{velocity}$$

PROBLEM SOLVING TIPS

- All problems dealing with electricity and energy should be solved using the MKS system of units.
- There are a lot of formulas in this chapter, so keep a formula sheet in front of you while solving the problems.

PROBLEMS

1. A computer's CD drive pushes out a CD holder with 0.5 lb of force, through a distance of 0.4 ft and takes only 1 s to eject. How much work was done?
2. Convert 25 J into ft lb.
3. An elevator motor consumes 81,000 J of energy in lifting 3,500 N of weight at a constant velocity through a distance of 4 m. What is the efficiency of this motor?
4. How much energy is required to pick a 1 kg book off the ground and raise it to a height of 1.2 m?

5. A battery contains 500,000 J of energy and is connected to a 100 W light bulb. How long will the light remain lit?
6. A kick boxer delivers a punch to his opponent. Suppose his fist weighs 5 lb and is moving at 10 ft/s. What is the energy contained in his fist (remember to convert the weight to mass first)? If all this energy is delivered to his opponent, pushing his jaw back by 1 in., what is the average force contained in his punch (remember to convert inches to feet)?
7. You build an electric generator by connecting your 3.5 hp lawn mower engine to a car alternator. Assume the efficiency of this system is 40%. How many watts can you generate with this generator?
8. What is the minimum power required by a conveyer belt to lift a 150 kg package to a height of 2.0 m in 3.0 s?
9. In a hydroelectric power plant, suppose that water falls for a distance of 110 m before reaching the turbine/generator. What is the speed of the water when it reaches the turbine?
10. Suppose the water in problem 9 strikes the turbine with 1.2×10^6 N of force. What power is delivered to the turbine? Suppose the efficiency of this system is 21%. How many watts does the system put out?
11. How much work is done in pulling a wagon with a force of 15 lb for a distance of 10.0 ft when the handle makes an angle of 45° with the horizontal?
12. Use energy conservation to calculate how fast an apple will hit Isaac Newton's head if he is sitting below the tree and the apple falls 5.0 ft onto his head.
13. What is the approximate amount of energy a bat must deliver to a baseball if it is to knock it out of the park for a distance of 450 ft?
14. An electric car is driven by a 65 hp electric motor that is 80% efficient. If it is traveling at 55 mile/h, for a distance of 56 miles, how much energy did the batteries consume?
15. A mass of 0.5 kg starting at rest then has 1,100 J of work performed on it. What is its final velocity?
16. A motor is used to lift an elevator upward with 5,000 N of force for a distance of 3 m. How much work was performed?
17. In the previous problem, if the motor consumes 20,000 J of energy, what is the efficiency of the system?
18. If a motor consumes 20,000 J of energy in 1.2 s, what is the power of the motor?
19. A "Killer Satellite" uses a rail gun to shoot another satellite. The gun delivers 6,100 J of energy to a projectile with a mass of 0.3 kg. What is the velocity of the projectile?

20. The rail gun in problem 19 delivers its energy to the projectile over a distance of 1.0 m. What force was applied on the projectile?
21. A 300 W TV set is left on for 8 h. How many kWh of energy was consumed?
22. Your mom yells at you for leaving a 300 W TV on for 8 h. If electricity costs 9 cents/kWh, how much do you owe her?
23. A computer with a mass of 12 kg falls off a table 1.2 m high. As it fell the potential energy that it had was turned into kinetic energy. What was the initial potential energy?
24. In the previous problem, what velocity did the computer hit the ground with?
25. If the energy consumption/mile of a car is 6.0×10^6 ft lb/mile and gas has an energy content/gallon of 9.7×10^7 ft lb/gal, what is the gas mileage of car in miles per gallon (mile/gal)?
26. 5.0 lb of force is required to move a part on a conveyer belt 12 ft in 6 s. How much work was done?
27. How much power was used in moving the part in the previous problem?
28. If the drive system on an elevator consumes 6,000 J of energy, lifting the elevator 2.0 m with 1,000 N of force, what is the efficiency of the system?
29. A spring is compressed from its uncompressed state, changing it potential energy by 100 J. What was the work done on the spring?
30. CD drives are hermetically sealed. However, suppose the seal is broken and a piece of dust of mass 0.001 g lands on the spinning disk. It then flies off at a velocity of 3.1 m/s. What is the kinetic energy of the dust in joules?
31. An electric hoist lifts a 10 kg mass to a height of 1.0 m. What is the change in the potential energy of the mass?
32. A dart gun consists of a spring under compression and a projectile, the dart. When the trigger is pulled at, the dart is launched. Assume the dart has a mass of 0.01 kg and the dart leaves the gun at 5 m/s. What was the change in the potential energy stored in the spring?
33. If we add up all the sources of energy in the universe, the sum seems to equal zero. How much energy did the universe have during the "Big Bang"?
34. If a 0.5 kg object falls from rest from a height of 1.0 m and lands on a nail, driving it 1 cm into a piece of wood, what velocity did the object strike the nail with?
35. What power did the object deliver to the nail in the previous problem?

5

Rotational Motion

In this chapter, we will discuss rotational motion. We will apply our equations to wheels, motors, and anything else that rotates (Figure 5.1).

FIGURE 5.1 A top.

Any point on a circle can be described by its radius r from the center and an angle θ. The angle can be measured in degrees, revolutions, or radians (Figure 5.2).

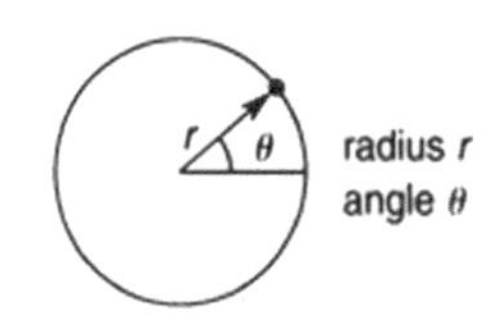

FIGURE 5.2 A point on a circle can be described by its radius r from the center and an angle θ measured in degrees, revolutions, or radians.

Comparing the three units:

$$1\,\text{rev} = 360° = 2\pi\,\text{rad}$$

See Figure 5.3 for a comparison of angular units.

The unit *radians* may be unfamiliar. It has its origin in nature. Take any circle and place a string around it to measure its circumference. Then measure its diameter. Divide the circumference by the diameter, and the number 3.14159..., which is written as π, will result.

Converting between angular units,

$$1\,\text{rev} = 360° = 2\pi\,\text{rad}$$

Example 5.1

How many revolutions are there in 180°?

$$\theta = 180°$$

$$1\,\text{rev} = 360°$$

$$180\cancel{°}\left(\frac{1\,\text{rev}}{360\cancel{°}}\right) = 0.50\,\text{rev}$$

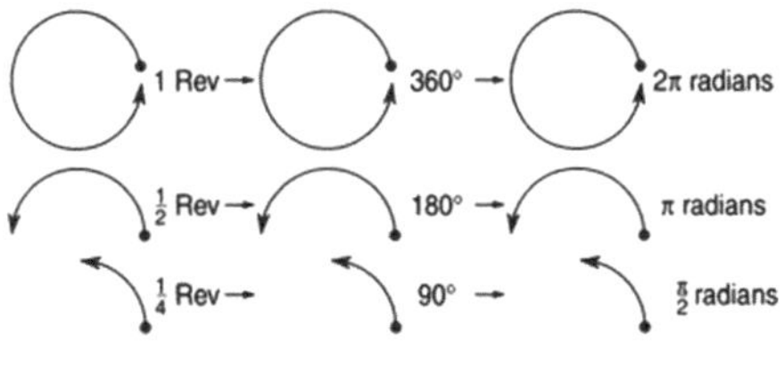

FIGURE 5.3 A comparison of angular units.

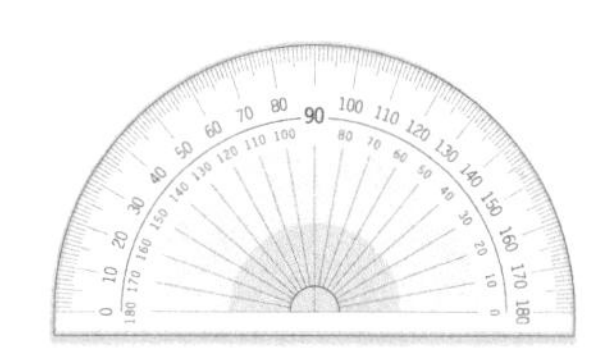

FIGURE 5.4 A protractor.

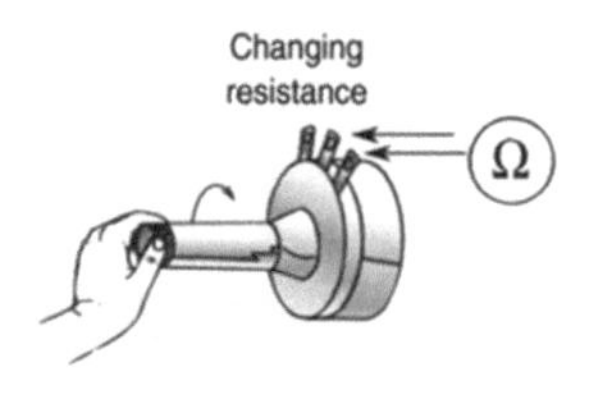

FIGURE 5.5 A potentiometer used to measure angular orientation.

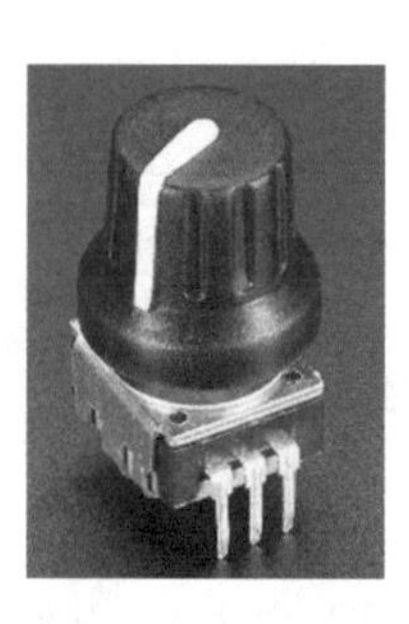

FIGURE 5.6 A rotary encoder, courtesy of adafruit.com.

FIGURE 5.7 A computer mouse converts the revolutions of the ball using an optical encoder in the mouse to a position on the computer screen.

Example 5.2

How many radians are there in 320°?

$$\theta = 320^\circ$$

$$2\pi \text{ rad} = 360^\circ$$

$$320^{\not\circ} \frac{2\pi \text{ rad}}{360^{\not\circ}} = 5.6 \text{ rad}$$

Example 5.3

How many radians are there in 200 rev?

$$\theta = 20\bar{0} \text{ rev}$$

$$1 \text{ rev} = 2\pi \text{ rad}$$

$$20\bar{0} \cancel{\text{rev}} \frac{2\pi \text{ rad}}{1 \cancel{\text{rev}}} = 1,256 \text{ rad}$$

5.1 MEASURING ANGLES

Aside from using a protractor, there are a number of electronic ways of measuring an angle (Figure 5.4).

5.1.1 Potentiometer

A potentiometer connected to the center of rotation can be used to determine the angular position of something undergoing a rotation (Figure 5.5).

5.1.2 Mechanical Rotary Encoder

A mechanical rotary encoder consists of a shaft that opens and closes a switch during the rotation of the shaft. Counting the number of switch activations determines the angle that the shaft rotates through (Figure 5.6).

5.1.3 Optical Rotary Encoder

An old type of computer mouse contains two small wheels with spokes that are driven by the ball at the bottom of the mouse. A light source and detector are placed on either side of the wheel. The light detector measures how many times the light beam is broken as the wheel turns. Simply counting the number of interruptions of the light beam determines the angle through which the ball rotates or the distance the mouse has moved (Figure 5.7).

5.1.4 Gyroscope

A mechanical gyroscope consists of a spinning wheel mounted sideways on a frame. Because the wheel is spinning, its momentum resists having its orientation changed. As the instrument is rotated, the spinning wheel keeps pointing in the same direction. The angle measured is the angle between the instrument and the spinning wheel with the fixed orientation (Figure 5.8).

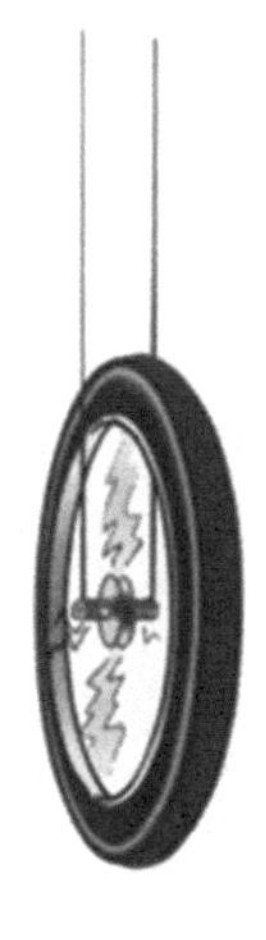

FIGURE 5.8 A gyroscope demonstrated using a spinning bicycle wheel hanging by a cable.

5.2 DISTANCE MEASUREMENTS ALONG A CIRCLE

Any motion that takes place in the circle can be described with two parameters:

1. The distance (r) by which the object is away from the center of rotation.
2. The angle (θ) through which it moves.

Suppose a point travels along the arc of a circle. How far has this point traveled?

$$d = r \cdot \theta \leftarrow \theta \text{ must be in radians}$$

$$\text{Distance along an arc} = \text{radius} \times \text{angle}(\text{in radians})$$

If a wheel of radius r rolls along a surface through an angle θ, the wheel has traveled a distance d (Figure 5.9).

$$d = r \cdot \theta \leftarrow \theta \text{ must be in radians}$$

$$\text{Distance a wheel rolls} = \text{radius} \times \text{angle}(\text{in radians})$$

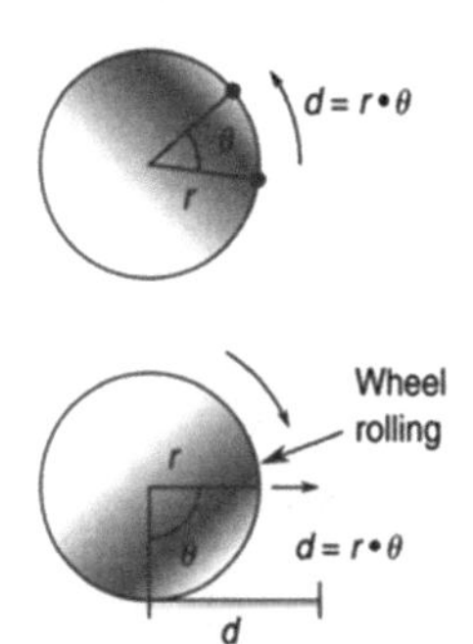

FIGURE 5.9 If a wheel of radius r rolls along a surface through an angle θ, the wheel has traveled a distance d, $d = r \cdot \theta$, θ must be in radians.

Example 5.4

A point on the rim of a wheel with a radius of 2.0 ft travels through an angle of 1.5 rad (Figure 5.10). (a) How far did the point travel on the wheel? (b) If the wheel was rolling on the ground, how far did it roll?

(a)

$$r = 2.0\,\text{ft}$$

$$\theta = 1.5\,\text{rad}$$

$$d = ?$$

$$d = r \times \theta$$

$$d = (2.0\,\text{ft})(1.5\,\text{rad})$$

$$d = 3.0\,\text{ft} \leftarrow \text{We don't retain the unit rad.}$$

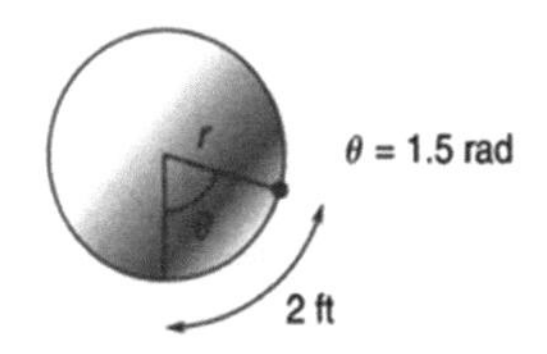

FIGURE 5.10 A point on the rim of a wheel with a radius of 2.0 ft travels through an angle of 1.5 rad.

(b)

The distance the wheel rolls for is the same as the distance the point travels through on the rim, $d = 3.0\,\text{ft}$.

Note that the unit radian drops out in the calculation of $d = r{\cdot}\theta$ because the unit radian is dimensionless.

When using the formula $d = r{\cdot}\theta$, if θ was in any unit other than radians, the answer would be wrong. Why are there so many different units for angles, and when do you use which one? The answer depends on the context of the problem.

As a general rule, however, always use radians when a calculation involves an angle times another quantity.

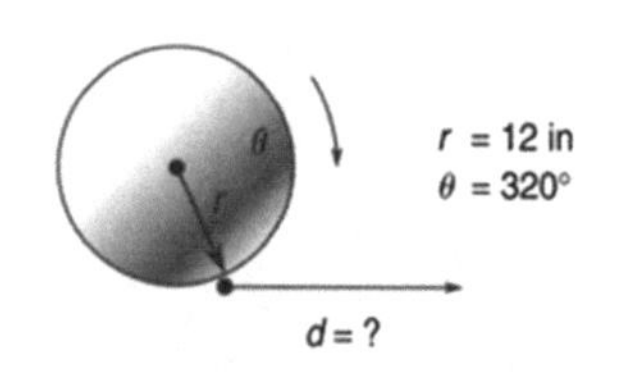

FIGURE 5.11 How far has this wheel rolled?

Example 5.5

A wheel with a radius of 12 in. rotates through an angle of 320°. How far has this wheel traveled?

In order to use this formula, θ must be in radians (Figure 5.11).

$$r = 12\text{in.}$$

$$\theta = 320°$$

$$d = ?$$

$$d = r \cdot \theta \Leftarrow \text{In order to use this fomula } \theta \text{ must be in radians.}$$

$$(320\not{°})\left(\frac{2\pi}{360\not{°}}\right) = 5.6\text{rad}$$

$$d = (12\text{in.})(5.6\text{rad})$$

$$d = 67\,\text{in.}$$

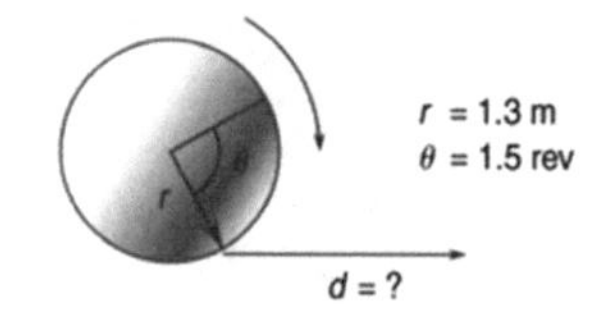

FIGURE 5.12 How far has the wheel traveled?

Example 5.6

Suppose a wheel with a radius of 1.3 m rolls along the ground through an angle of 1.5 revolutions (Figure 5.12). How far has the wheel traveled?

$$r = 1.3\text{m}$$

$$\theta = 1.5\,\text{Rev}$$

$$d = ?$$

$$d = r \cdot \theta \Leftarrow \text{In order to use this fomula } \theta \text{ must be in radians.}$$

$$(1.5\,\not{\text{rev}})\left(\frac{2\pi}{1\,\not{\text{rev}}}\right) = 9.4\text{rad}$$

$$d = (1.3\text{m})(9.4\text{rad})$$

$$d = 12.25\text{m}$$

5.3 ANGULAR VELOCITY

Angular velocity is the measure of how fast something is spinning (Figure 5.13). Everyone is familiar with the unit revolutions/min (rpm), which is a measure of how fast something is spinning. It describes how many revolutions a shaft or a wheel turns in 1 min.

FIGURE 5.13 Angular velocity is a measure of how fast something is spinning.

Example 5.7

A shaft on a motor turns 6,000 rev in 1 min. What is its angular velocity?

$$6{,}000\,\text{rev}/1\,\text{min} = 6{,}000\,\text{rpm}$$

$$(1\,\text{rpm} = 1\,\text{rev/min})$$

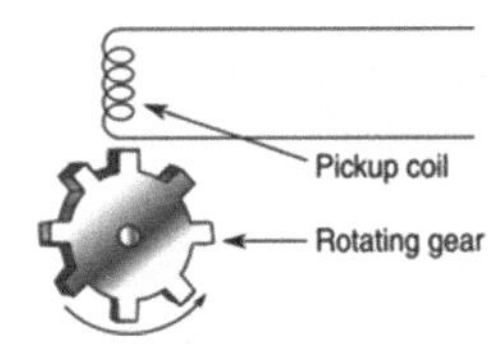

FIGURE 5.14 A coil used as a proximity sensor functioning as a tachometer.

5.3.1 Measuring Angular Velocity

If a shaft or wheel is moving slowly enough, simply counting the number of revolutions in a given amount of time is enough to determine the angular velocity. However, it's more probable that the wheel is turning too fast to perform such a measurement, and therefore another means is necessary.

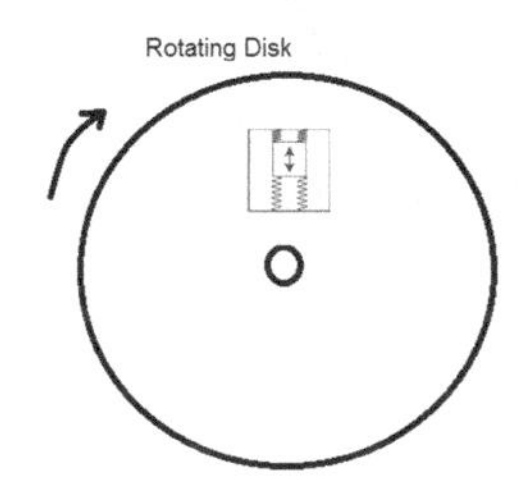

FIGURE 5.15 A *MEMS gyroscope* is micro machined silicon wafer mass suspended by polysilicon springs. During a rotation the mass is displaced, the distance between the plates changes, producing a small electrical signal.

5.3.2 Tachometer

A tachometer is used to determine angular velocity. There are a number of types of tachometers. Typically, they are simply proximity sensors such as the one described below.

5.3.3 Magnetic Pickup Coil Tachometer

In Figure 5.14, a coil is placed near the rotating gear. The inductance of the coil will change as a tooth on the gear positions itself directly below the coil. This change in inductance will be monitored and converted into a pulse and counted for some interval of time. If one knows the number of teeth on the gear and the number of pulses occurring in some interval of time, the angular velocity can be found.

MEMS Gyroscope: A MEMS (microelectromechanical systems) gyroscope is micro machined, about the size of a human hair, silicon wafer mass suspended by polysilicon springs. The wafer has a plate attached to it that acts as a plate of a capacitor, while the other plate is fixed, Figure 5.15. There are three of these wafers, one along each axis, Figure 5.16. During a rotation along a particular axis, the mass is displaced and the capacitance

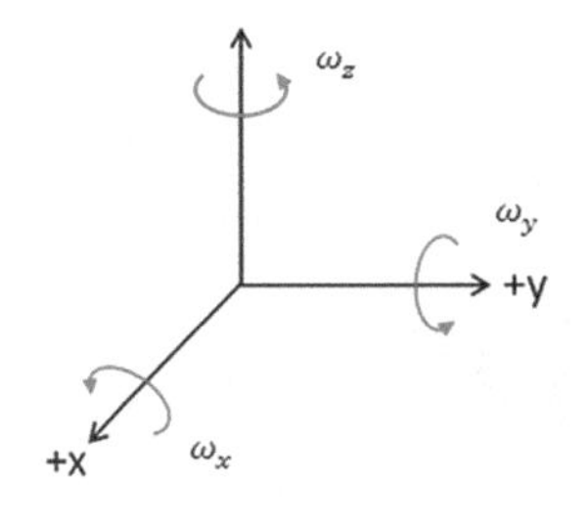

FIGURE 5.16 Rotations along x, y, and z axes.

between the plates changes. This change in capacitance produces a change in voltage that is proportional to the angular velocity (Figure 5.17).

FIGURE 5.17 A three-axis gyroscope capable of measuring rotation in the *x*, *y*, and *z* axes from adafruit.com.

5.4 ANGULAR VELOCITY CALCULATIONS

The average angular velocity is a rate at which an angle changes and is given by:

$$\omega = \frac{\Delta\theta}{\Delta t}$$

$$\text{Angular velocity}_{\text{average}} = \frac{\text{Change in angle}}{\text{Change in time}}$$

ω has units of (degrees or revolutions or radians)/(time unit).

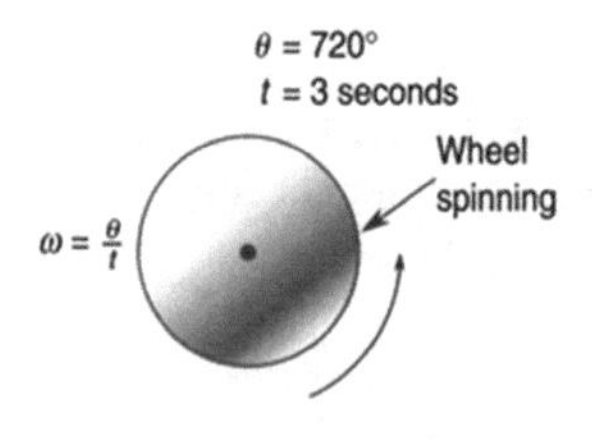

FIGURE 5.18 A wheel turns 720° in 3 s.

Example 5.8

A wheel turns 720° in 3 s (Figure 5.18). What is its angular velocity?

$$\omega = \frac{\Delta\theta}{\Delta t}$$

$$\omega = \frac{720^\circ}{3\,\text{s}}$$

$$\omega = 240^\circ/\text{s}$$

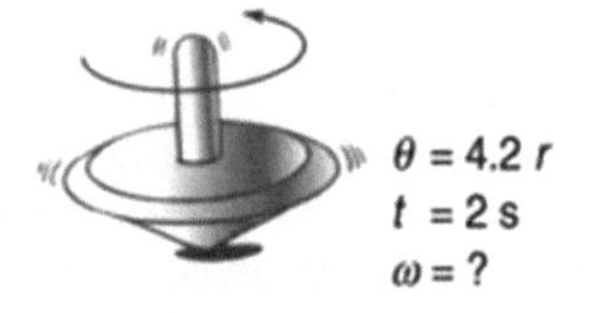

FIGURE 5.19 A top revolves 4.2 radians in 2 s.

Example 5.9

A top revolves 4.2 rad in 2 s (Figure 5.19). What is its angular velocity?

$$\omega = \frac{\Delta\theta}{\Delta t}$$

$$\omega = \frac{4.2\text{rad}}{2\text{s}}$$

$$\omega = 2.1\frac{\text{rad}}{\text{s}}$$

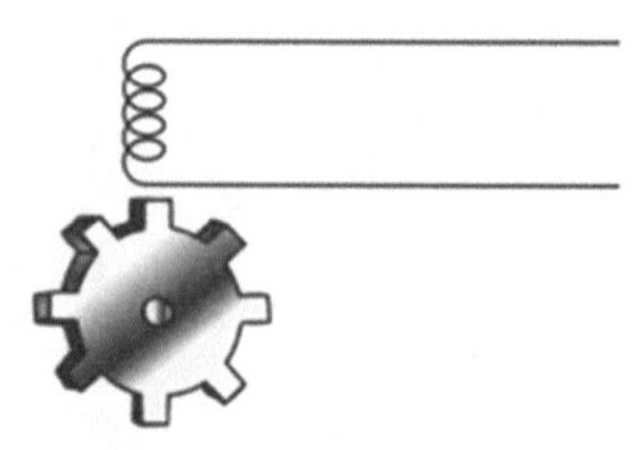

FIGURE 5.20 What is the angular velocity of the gear?

Example 5.10

Suppose a gear is rotating and being monitored by the tachometer described above. If the gear has ten teeth and produces a count of 500 counts in 1 min, what is the angular velocity of the gear?

Since the gear has ten teeth, in one complete revolution there will be ten pulses out of the tachometer. Therefore, we want to convert 500 counts into revolutions (Figure 5.20).

$$500\,\cancel{\text{counts}}\ \frac{1\text{rev}}{10\,\cancel{\text{counts}}} = 50\text{rev}$$

$$\omega = \frac{\theta}{t}$$

$$\omega = \frac{50\text{rev}}{1\text{min}} = 50\text{rpm}$$

5.5 CIRCULAR VELOCITY

We can now calculate the velocity of an object moving in a circular path. Recall that velocity $v = \frac{d}{t}$.

For a circle:

$$d = r \cdot \theta$$

$$\omega = \frac{\Delta\theta}{\Delta t}$$

We can write $\omega = \frac{\theta}{t}$ instead of $\omega = \frac{\Delta\theta}{\Delta t}$, when ω and t start at 0.

Therefore,

$$v = \frac{d}{t}$$

$$v = \frac{r \cdot \theta}{t}$$

$$v = r \cdot \frac{\theta}{t}$$

$$v = r \cdot \omega$$

$$\text{velocity}(\text{m/s}) = \text{radius}(\text{m}) \times \text{angular velocity}(\text{rad/s})$$

In the English system,

$$\text{Velocity}(\text{ft/s}) = \text{radius}(\text{ft}) \times \text{angular velocity}(\text{rad/s})$$

$$v = r \cdot \omega$$

This equation is the velocity of an object moving in a circle or it can describe the velocity of a wheel rotating at ω rolling along a surface (Figure 5.21).

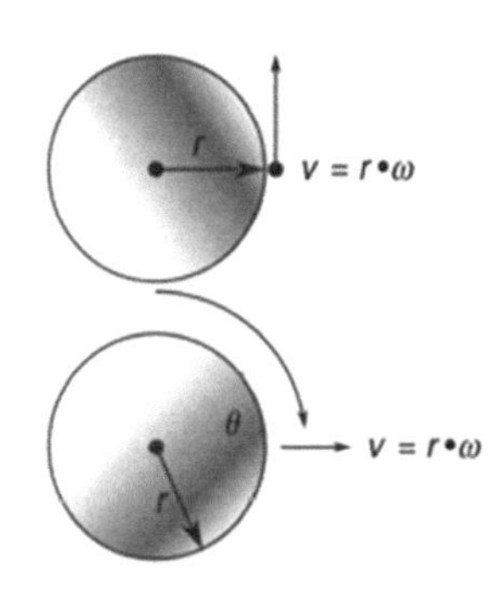

FIGURE 5.21 Velocity of an object moving in a circle (ω in radians/time) or the velocity of a wheel rolling along a surface can be determined by the equation $v = r \cdot \omega$.

FIGURE 5.22 The speed of the car depends on how fast the wheels are spinning and their radius.

Example 5.11

A wheel with a radius of 1.25 ft is rotating at 4.2 rad/s. How fast is the car moving? (See Figure 5.22.)

$$r = 1.25\,\text{ft}$$
$$\theta = 4.2\frac{\text{rad}}{\text{s}}$$
$$v = ?$$
$$v = r \cdot \omega$$
$$r = 1.25\,\text{ft}$$
$$\theta = 4.2\frac{\text{rad}}{\text{s}}$$
$$v = ?$$
$$v = r \cdot \omega$$
$$v = (1.25\,\text{ft})\left(4.2\frac{\text{rad}}{\text{s}}\right)$$
$$v = 5.3\frac{\text{ft}}{\text{s}} \Leftarrow$$ Note that radians does not appear in the answer because it is dimensionless.

5.6 CENTRIPETAL ACCELERATION

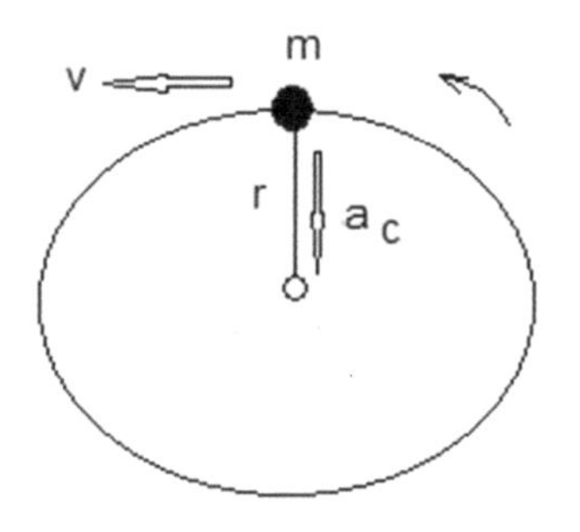

FIGURE 5.23 A mass accelerating in a circle.

An object moving in a circle has a velocity vector that is constantly changing, because its direction is constantly changing. The object may be moving at a constant speed as it moves around the circle, but because it is changing direction it is accelerating. Recall, a change in velocity is an acceleration. This change in velocity is directed toward the center and is called centripetal acceleration. Centripetal acceleration is given by (Figure 5.23):

$$a_c = r \cdot \omega^2$$

$$\text{acceleration}_{\text{centripetal}}\left(\frac{\text{m}}{\text{s}^2}\right) = \text{radius(m)} \times \text{angular velocity}\left(\frac{\text{rad}}{\text{s}}\right)^2$$

The centripetal acceleration can be rewritten in another form if we use the formula for circular velocity.

$$v = r \cdot \omega$$

$$\omega = v/r$$

$$a_c = r \cdot \omega^2 \leftarrow \text{substitute in for } \omega, \ {}^{v}\!/_{r}$$

$$a_c = r \cdot \left({}^{v}\!/_{r}\right)^2$$

$$a_c = \frac{v^2}{r}$$

Example 5.12

A wheel of radius 0.4 m in a car is spinning at an angular velocity of 100 rad/s. What is the centripetal acceleration of the wheel?

$$a_c = r \cdot \omega^2 = (0.4\,\text{m})(100\,\text{rad/ s})^2$$

$$a_c = 4{,}000 \frac{\text{m}}{\text{s}^2}$$

Example 5.13

A shoe in a dryer is spinning at 1.5 m/s in a drum of radius 0.5 m. What is the centripetal acceleration of the shoe?

$$a_c = \frac{v^2}{r}$$

$$a_c = \frac{\left(1.5\frac{\text{m}}{\text{s}}\right)^2}{0.5\text{m}}$$

$$a_c = 4.5\text{m/s}^2$$

5.7 CENTRIPETAL FORCE

Changing the motion of something means acceleration, and the force on a body that is accelerating is given by $F = m \times a$. If a body is being accelerated in a circular path, it is accelerating toward the center of the circle and its acceleration is $a_c = r{\cdot}\omega^2$. Therefore, the centripetal force is directed toward the center of rotation and is given by:

$$F_c = m \cdot a_c$$

$$\text{Force}_{\text{centripetal}} = \text{mass}(\text{kg}) \cdot \text{centripetal acceleration}(\text{m/s}^2)$$

Example 5.14

What is the centripetal force on a shoe in a washing machine if the shoe has a mass of 0.35 kg and the inside of the machine is rotating at 120 rpm and has a radius of 0.30 m (Figure 5.24)?

Converting ω to $\frac{\text{radians}}{\text{s}}$

$$\omega = \left(120 \frac{\cancel{\text{rev}}}{\cancel{\text{min}}}\right)\left(\frac{2\pi\text{rad}}{1\cancel{\text{rev}}}\right)\left(\frac{1\cancel{\text{min}}}{60\,\text{s}}\right) = 13\text{rad/s}$$

$$F_c = m \cdot a_c$$

$$F_c = m \cdot r \cdot \omega^2$$

$$F_c = 0.35\ \text{kg} \cdot 0.30\text{m} \cdot \left(13 \frac{\text{rad}}{\text{s}}\right)^2$$

$$F_c = 18\text{N}$$

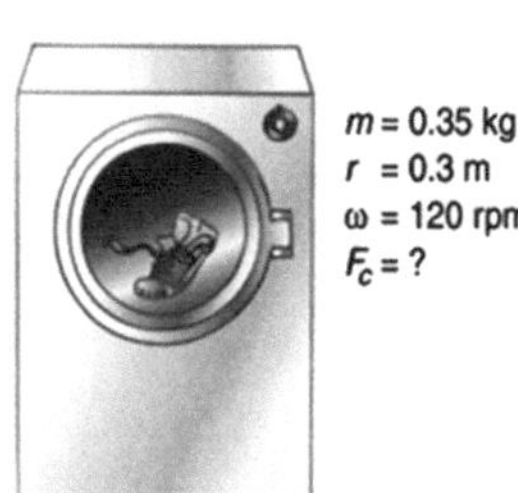

FIGURE 5.24 What is the centripetal force on the shoe in a washing machine?

5.8 CENTRIFUGAL FORCE

When driving in a car turning a corner, you feel yourself being pushed in the opposite direction to that you are turning. The faster or the tighter you make the turn, the greater the force. This force is called centrifugal force. The origin of this force can be explained as follows. When driving a car straight, your body has momentum straight ahead. When making a turn, in order for your body to move with the car, you must hold on to the wheel tightly to "force" your body into this new direction, otherwise you would move in a straight path. This force is called a fictitious force because there is no real force pushing you. It is a result of your body's inertia resisting the change in direction. Although fictitious, we can still use this centrifugal force to calculate correctly the behavior of rotating objects. The centrifugal force has the same equation as the centripetal force; only it is directed in the opposite direction (Figures 5.25 and 5.26).

FIGURE 5.25 The driver experiences a centrifugal force.

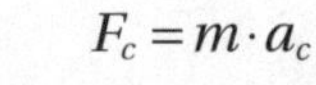

$$F_c = m \cdot a_c$$

$$\text{Force}_{\text{centrifugal}} = \text{mass}(\text{kg}) \cdot \text{centripetal acceleration}(\text{m/s}^2)$$

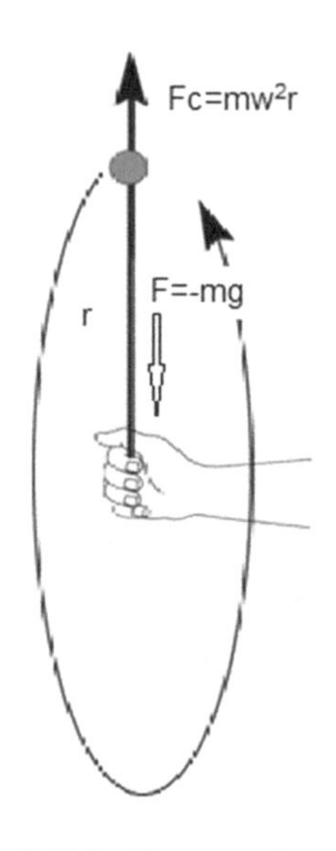

FIGURE 5.27 How fast must the rock be rotated so that it does not fall?

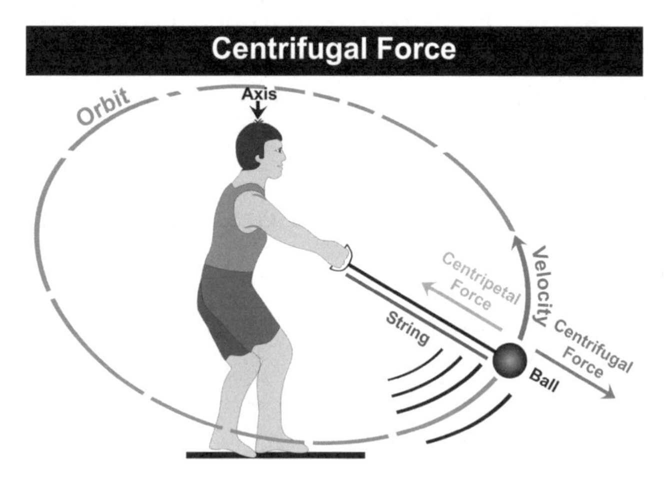

FIGURE 5.26 The mass experiences a centrifugal force.

Example 5.15

A rock is tied to the end of a 1.0 m long string. How fast must it be rotated so that it does not fall? The rock must be rotated fast enough so that the force downward due to gravity equals the centrifugal force upward. Setting the two forces equal to each other (Figure 5.27):

$$F_{\text{centrifugal}} = F_{\text{gravity}}$$
$$m \cdot r \cdot \omega^2 = m \cdot g$$
$$\omega^2 = g/r$$
$$\omega = \sqrt{g/r}$$
$$\omega = \sqrt{\frac{9.8\frac{\text{m}}{\text{s}^2}}{1.0\text{m}}}$$
$$\omega = 3.13\text{rad/s}$$

5.8.1 Geosynchronous Orbits

FIGURE 5.28 A satellite in a geosynchronous orbit.

A satellite can be placed in an orbit around the earth so that it remains fixed at one point over the earth. This orbit is called a geosynchronous orbit. A geosynchronous satellite has the same angular velocity as the earth and remains at a fixed distance from the earth (Figure 5.28).

To find the geosynchronous orbit distance, consider the forces on the satellite.

$$F = ma = F_{\text{centrifugal}} - F_{\text{gravity}} = m \cdot \omega^2 \cdot r - G\frac{m \cdot M_{\text{Earth}}}{r^2}$$

$$a = \frac{F}{m} = \omega^2 \cdot r - G\frac{M_{\text{Earth}}}{r^2}$$

For a geosynchronous orbit, the distance to the satellite is fixed and therefore the acceleration in the radial direction r equals zero.

$$a = 0 \rightarrow \omega^2 \cdot r = G\frac{M_{\text{Earth}}}{r^2}$$

$$\text{Solving for } r \rightarrow r = \left(G\frac{M_{\text{Earth}}}{\omega^2}\right)^{\frac{1}{3}}$$

$$r = \left(G\frac{M_{\text{Earth}}}{\omega^2}\right)^{\frac{1}{3}} = \left(6.67 \times 10^{-11}\frac{\text{m}^3}{\text{kg}\,\text{s}^2} \times \frac{5.97 \times 10^{24}\,\text{kg}}{\left(7.3 \times 10^{-5}\frac{\text{rad}}{\text{s}}\right)^2}\right)^{\frac{1}{3}}$$

$$r = 42.1 \times 10^6\ \text{m with respect to the center of the earth}$$

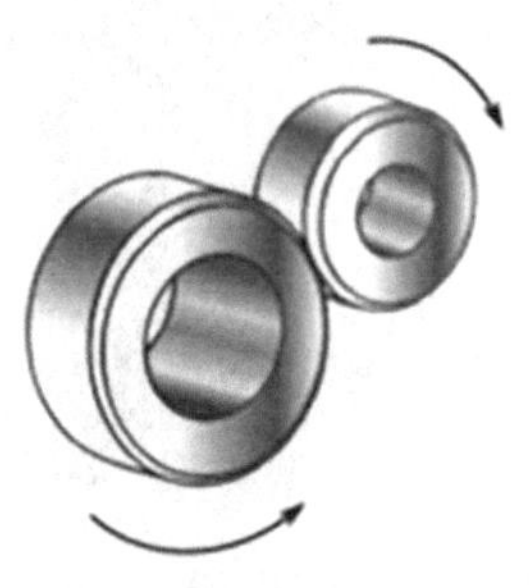

FIGURE 5.29 Couple gears and wheels together to change speed, direction of rotation, and torque capability.

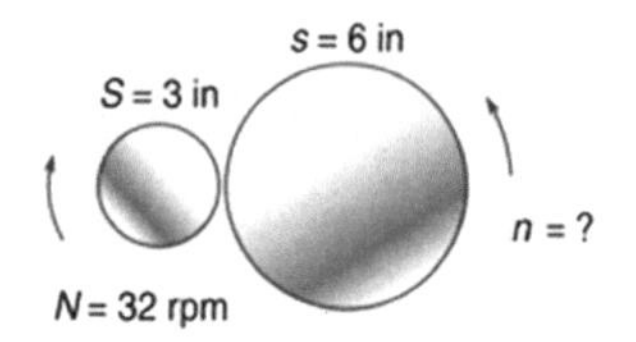

FIGURE 5.30 How fast is the driven wheel spinning?

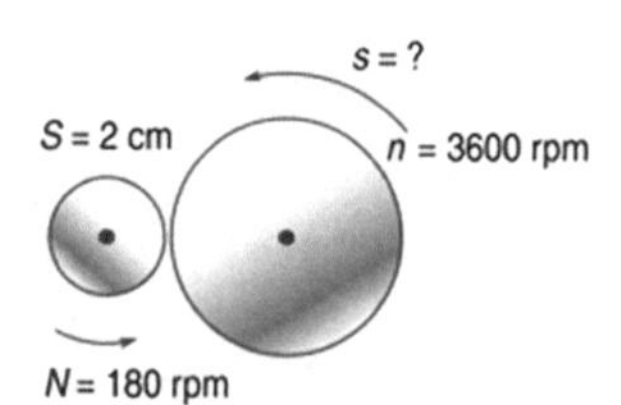

FIGURE 5.31 What diameter does the driven wheel have?

The earth has a mean radius of 6.4×10^6 m, therefore the distance to the satellite from the earth's surface is:

$$r = 42.1 \times 10^6 \text{ m} - 6.4 \times 10^6 \text{ m}$$

$$r = 35.7 \times 10^6 \text{ m or approximately 22,000 miles above the surface of the earth}$$

5.9 TRANSMISSIONS

Gears and wheels can be coupled in such a way as to change their speed, direction of rotation, and torque capability (Figure 5.29).

To determine the speed of rotation of a wheel or gear being driven by a driving wheel or gear, we use the following formulas:

$$N \cdot S = n \cdot s$$

N = angle or angular speed of the driver

S = size of driving diameter or number of teeth

n = angle or angular speed of the driven

s = size of driven diameter or number of teeth

Example 5.16

A wheel of diameter 3 in. spinning at 32 rpm is driving a wheel of diameter 6 in. (Figure 5.30). How fast is the driven wheel spinning?

$$S = 3\text{in.}$$

$$s = 6\text{in.}$$

$$N = 32\text{rpm}$$

$$n = ?$$

$$NS = ns$$

$$n = \frac{NS}{s}$$

$$n = \frac{(32\text{rpm})(3\,\cancel{\text{in.}})}{6\,\cancel{\text{in.}}}$$

$$n = 16\text{rpm}$$

Example 5.17

If a wheel that must spin at 3,600 rpm is being driven by a wheel of diameter 2 cm spinning at 180 rpm, what diameter should the driven wheel have? (Figure 5.31)

$$S = 2\text{cm}$$
$$N = 180\text{rpm}$$
$$n = 3{,}600\text{rpm}$$
$$s = ?$$
$$NS = ns$$
$$\frac{NS}{n} = \frac{\cancel{n}s}{\cancel{n}}$$
$$s = \frac{NS}{n}$$
$$s = \frac{(180\ \cancel{\text{rpm}})(2\text{cm})}{3{,}600\ \cancel{\text{rpm}}}$$
$$s = 0.1\text{cm}$$

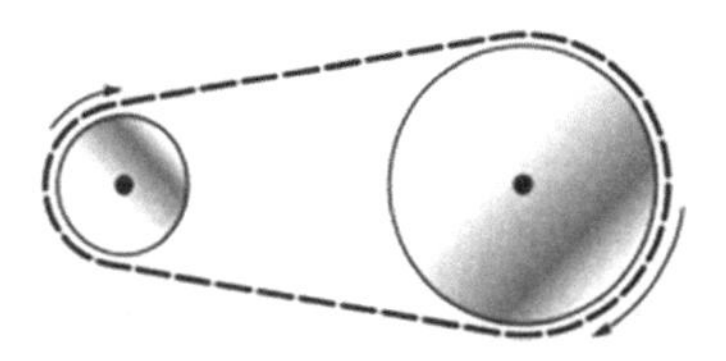

FIGURE 5.32 A distance can separate the driving and driven wheels or gears by using a belt or chain, and the two wheels or gears now rotate in the same direction.

A distance can separate the driving and driven wheels or gears by using a belt or chain, and the two wheels or gears now rotate in the same direction (Figure 5.32).

Two wheels can be made to rotate in the opposite directions if the belt is twisted (Figure 5.33).

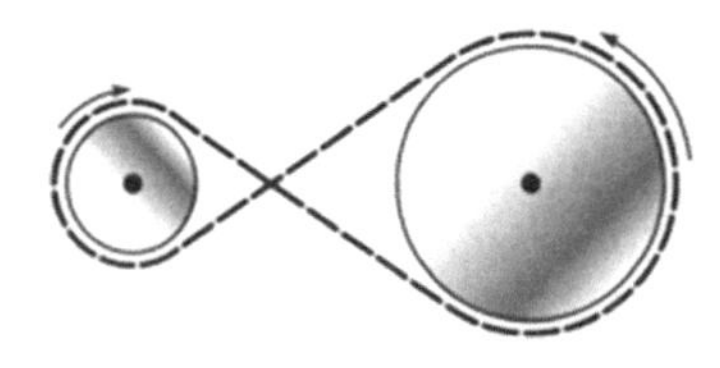

FIGURE 5.33 Two wheels can be made to rotate in opposite directions if the belt is twisted.

Example 5.18

A gear with 80 teeth spinning at 100 rpm is driving a gear with 20 teeth (Figure 5.34). How fast is the driven gear spinning?

$$S = 80\text{teeth}$$
$$s = 20\text{teeth}$$
$$N = 100\text{rpm}$$
$$n = ?$$
$$NS = ns$$
$$\frac{NS}{s} = \frac{n\cancel{s}}{\cancel{s}} \leftarrow \text{solving for } n$$
$$n = \frac{NS}{s}$$
$$n = \frac{(100\text{rpm})(80\ \cancel{\text{teeth}})}{20\ \cancel{\text{teeth}}}$$
$$n = 400\text{rpm}$$

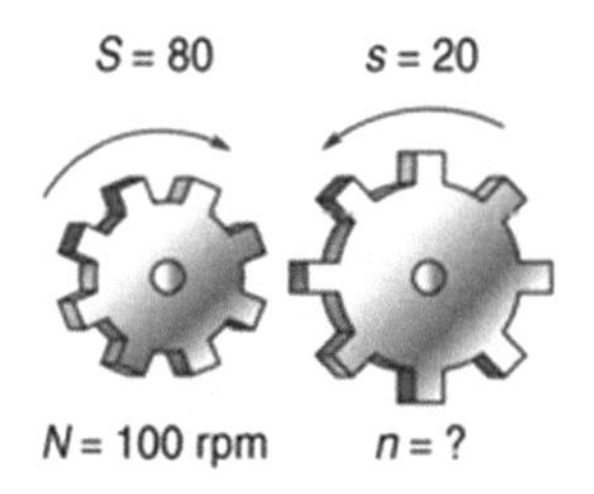

FIGURE 5.34 How fast is the driven gear spinning?

Multiple wheels and gears can be connected to create multiple speeds and rotations (Figure 5.35). To solve problems such as these, apply these equations to two wheels or gears at a time.

FIGURE 5.35 Multiple wheels and gears can be connected to create multiple speeds and rotations.

5.10 TORQUE

FIGURE 5.36 A longer wrench has more torque than a shorter one.

Everyone who has ever used a wrench knows that it is easier to turn a wrench with a long handle than a short one. The longer wrench has more torque (Figure 5.36).

Torque is defined mathematically as:

$$T = F \cdot r$$

$$\text{Torque(N/m)} = \text{Force(N)} \times \text{radius arm(m)}$$

In the English system,

$$\text{Torque}(\text{ft lb}) = \text{Force}(\text{lb}) \times \text{radius arm}(\text{ft})$$

Torque has units of foot pounds or newton meters. The force is applied perpendicular to the radius arm as shown in Figure 5.37.

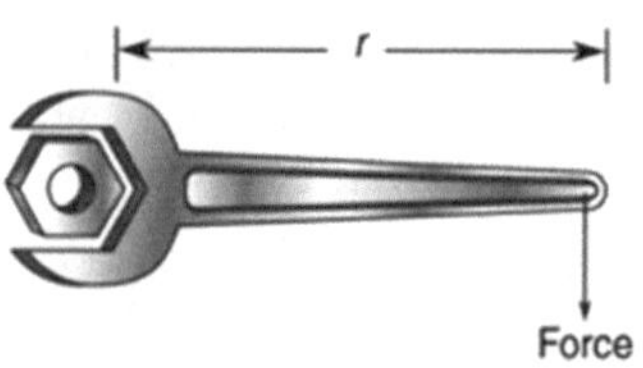

FIGURE 5.37 Force is applied perpendicular to the radius arm to generate the torque.

Example 5.19

A torque wrench measures the torque applied to a bolt (see Figure 5.38). Suppose a torque wrench handle is 1 ft long and that 100 lb is applied to the handle. What does the torque wrench read?

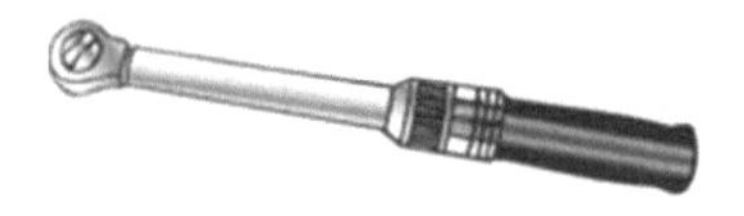

FIGURE 5.38 A mechanic's torque wrench.

$$F = 100\text{lb}$$

$$r = 1\text{ft}$$

$$T = ?$$

$$T = F \times r$$

$$T = (100\text{lb})(1\text{ft})$$

$$T = 100 \text{ ft lb}$$

5.11 TORQUE AND POWER

Every motor has horsepower and torque as two of its specifications. *Torque* is a measure of a motor's strength to rotate something connected to its shaft some distance from the center of rotation. Torque and power are related by:

$$P = T \cdot \omega$$

$$\text{Power}(\text{W}) = \text{Torque}(\text{Nm}) \times \text{angular velocity}(\text{rad/s})$$

In the English system,

$$\text{Power}(\text{ft lb / s}) = \text{Torque}(\text{ft lb}) \times \text{angular velocity}(\text{rad/s})$$

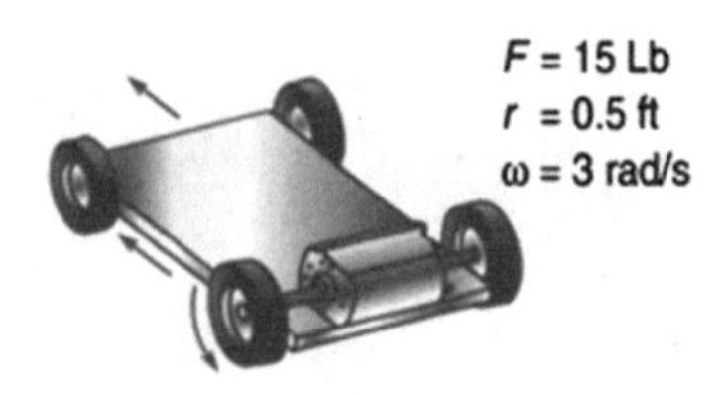

FIGURE 5.39 A motor connected to a wheel drives a cart.

Example 5.20

A motor connected to a wheel drives a cart. The cart requires 15 lb of force to move it. The wheel has a radius of 0.5 ft and is rotating at 3.0 rad/s (Figure 5.39). (a) What torque is the motor putting out? (b) What power is the motor generating? (c) What is this power in horsepower and watts?

(a)

$$F = 15\text{lb}$$

$$r = 0.5\text{ft}$$

$$\omega = 3.0\frac{\text{rad}}{\text{s}}$$

$$1\text{hp} = 550\frac{\text{ft lb}}{\text{s}}$$

$$1\text{hp} = 746\text{W}$$

$$T = ?$$

$$T = F \cdot r$$

$$T = (15\text{lb})(0.5\text{ft}) = 7.5\text{ft lb}$$

(b)

$$P = ?$$

$$P = T \cdot \omega$$

$$P = (7.5\text{ft lb})\left(3\frac{\text{rad}}{\text{s}}\right) = 23\frac{\text{ft lb}}{\text{s}}$$

(c) Converting foot pounds into horsepower,

$$23\frac{\text{ft lb}}{\text{s}}\left(\frac{1\text{hp}}{550\frac{\text{ft lb}}{\text{s}}}\right) = 0.042\text{hp}$$

Converting horsepower into watts,

$$0.042\text{hp}\left(\frac{746\text{W}}{1\text{hp}}\right) = 31\text{W}$$

5.12 CHANGING TORQUE

Sometimes it is necessary to increase the torque delivered to a wheel, for example, when driving up a hill. The torque can be changed in multiple wheel or gear systems by changing the relative size of the driving and driven gears or wheels (Figure 5.40) according to:

$$\frac{T_d}{T_D} = \frac{s}{S}$$

T_D = driving torque

T_d = driven torque

S = size of driving diameter or number of teeth

s = size of driven diameter or number of teeth

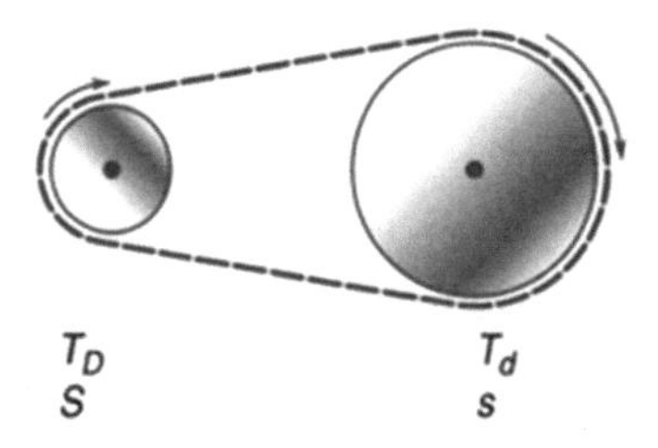

FIGURE 5.40 The torque can be changed in multiple wheel or gear systems by changing the relative size of the driving and driven gears or wheels.

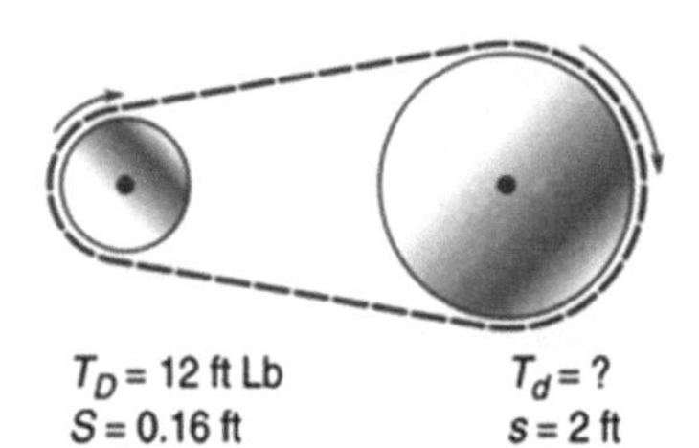

FIGURE 5.41 Centripetal force generated by changing the motion of a body into a circular path.

Example 5.21

What is the increase in torque when the driving torque is 12 ft lb, the driving wheel diameter is 0.16 ft, and the driven wheel is 2 ft in diameter? (Figure 5.41)

$$T_D = 12\,\text{ft lb}$$

$$S = 0.16\,\text{ft}$$

$$s = 2.0\,\text{ft}$$

$$T_d = ?$$

$$\frac{T_d}{T_D} = \frac{s}{S}$$

$$T_d = T_D \frac{s}{S}$$

$$T_d = 12\,\cancel{\text{ft}}\,\text{lb}\ \frac{2.0\,\text{ft}}{0.16\,\cancel{\text{ft}}}$$

$$T_d = 150\,\text{ft lb}$$

5.13 CHAPTER SUMMARY

Symbols used in this chapter:

Symbol	Unit
t—time	s
d—distance	m
r—radius	m
v—velocity	m/s
a—acceleration	m/s²
θ—angle	rad, deg, °, rev
ω—angular velocity	rad/s, deg/s, rev/s
N—driver angle or speed	rad, rad/s
n—driven angle or speed	rad, rad/s
S—driver diameter or teeth	m or (number of teeth)
s—driven diameter or teeth	m or (number of teeth)
T—torque	Nm, ft lb
F_c—force$_{\text{centripetal}}$	N, lb
F_c—force$_{\text{centrifugal}}$	N, lb

$$1\,\text{rev} = 360° = 2\,\text{pi rad}$$

$$d = r \cdot \theta \Leftarrow \theta \text{ must be in radians.}$$

$$\text{distance} = \text{radius} \times \text{angle}(\text{in radians})$$

If a wheel of radius r rolls along a surface through an angle θ, the wheel has traveled a distance d, where $d = r \cdot \theta \Leftarrow \theta$ must be in radians.

$$\omega = \frac{\Delta\theta}{\Delta t}$$

$$\text{Angular velocity}_{\text{average}} = \frac{\text{Change in angle}}{\text{Change in time}}$$

$$v = r \cdot \omega$$

$$\text{velocity} = \text{radius} \times \text{angular velocity}$$

$$a_c = r \times \omega^2$$

$$\text{acceleration}_{\text{centripetal}}\left(\frac{\text{m}}{\text{s}^2}\right) = \text{radius}(\text{m}) \times \text{angular velocity}\left(\frac{\text{rad}}{\text{s}}\right)^2$$

Alternatively:

$$a_c = \frac{v^2}{r}$$

$$F_c = m \times a_c$$

$$\text{Force}_{\text{centripetal}}(\text{N}) = \text{mass}(\text{kg}) \cdot a_c\left(\frac{\text{m}}{\text{s}^2}\right) \leftarrow \text{Force directed inward}$$

$$\text{Force}_{\text{centrifugal}}(\text{N}) = \text{mass}(\text{kg}) \cdot a_c\left(\frac{\text{m}}{\text{s}^2}\right) \leftarrow \text{Force directed outward}$$

$$N \cdot S = n \cdot s$$

N = Angle or angular speed of the driver.
S = Size of driving diameter or number of teeth.
n = Angle or angular speed of the driven.
s = Size of driven diameter or number of teeth.

$$T = F \cdot r$$

Torque = Force × radius arm
Torque has units of ft lb or N m

$$P = T \cdot \omega$$

Power = torque × angular velocity (in radians/time)

$$\frac{T_d}{T_D} = \frac{s}{S}$$

T_D = Driving torque.

T_d = Driven torque.

S = Size of driving diameter or number of teeth

s = Size of driven diameter or number of teeth

PROBLEM SOLVING TIPS

- Always use angular units in radians for ω and θ when calculating.

$$\text{Distance: } d = r \cdot \theta \quad \Leftarrow \theta \text{ must be in radians.}$$

$$\text{Velocity: } v = r \cdot \omega \quad \leftarrow \omega \text{ in radians/time}$$

$$\text{Power: } P = T \cdot \omega \quad \leftarrow \omega \text{ in radians/time}$$

$$\text{Force: } F_c = mr\omega^2 \quad \leftarrow \omega \text{ in radians/time}$$

- When solving transmission problems, capital letters refer to the drivers, lowercase to the driven.

PROBLEMS

1. A point on a CD rotates through an angle of 1 rev. The radius of the CD is 2.25 in. How far has this point traveled? Remember that θ must be in radians.
2. Suppose you are fixing a broken CD player because it is not spinning fast enough. You measure that the CD completes 5 rev in 1 s. What is the angular velocity in rpm?
3. You are asked to design a robot that moves at a speed of 120 in./min. The drive wheel is connected to a motor spinning at 25 rpm. What should the radius of the wheel be?
4. A wheel of diameter 3 in. is being driven by another wheel of diameter 6 in. How many times faster is the driven wheel spinning compared with the driving wheel?
5. A gear with 100 teeth is being driven by another gear. The driven gear is spinning at 10 rpm, and the driving gear is spinning at 5 rpm. How many teeth does the driving gear have?
6. A wrench with an 8 in. handle is used to tighten a bolt. Twenty-two pounds of force is applied to the handle. What is the torque in ft lb applied to the bolt?

7. A motor drives a conveyer belt. If the motor's angular speed is 1,870 rpm and its torque is 35 ft lb, what is its horsepower? Remember that 1 hp = 550 ft lb/s.
8. Using two gears, how can you double the torque of a motor?
9. A motor of torque 12 ft lb is driving a wheel of 2 in. in diameter. This wheel is connected by a belt to another driven wheel of 1 in. diameter. What is the torque of the 1 in. wheel?
10. What is the angular velocity of the earth on its own axis?
11. A floppy disk in a computer rotates at 31.4 rad/s. How many revolutions does it make in a 30 s reading of the disk? (Be careful of the units!)
12. Suppose you wanted to build an elevator to outer space by attaching a 5.98×10^{24} kg mass in space to a cable 22,000 miles long and drop the other end of the cable to the earth and anchor it to the ground. The centripetal force would keep the cable taut, and you could ride up the cable to space. What would the centripetal force be on the cable? Why doesn't NASA do this?
13. What is the centripetal acceleration of the tip of a fan blade moving at 1.1 m/s at a distance of 0.5 m from the axis of rotation?
14. To prevent a car from skidding on a turn, the frictional force of the tire on the road must be greater than or equal to the centripetal force. How fast can the car make a turn if the coefficient of the tire with the road is $\mu = 0.9$, the car weighs 9,345 N, and the radius of the turn is 25 m. Hint: Maximum speed occurs when $F_c = F_f$.
15. If you build an electric cart and the drive motor that rotates at 1,750 rpm is connected to a drive wheel of radius 6.0 in., how fast will this cart move?
16. A motor rotates at 3,600 rpm. Convert this number into rad/s.
17. A go-kart is made by mounting a washing machine motor onto a cart. The motor drives a wheel of radius 6 in. If the motor is spinning at 2.0 rad/s, how fast is the cart moving in in./s?
18. A wheel of diameter 10 cm is spinning at 200 rpm and is driving another wheel of diameter 5 cm. How fast is the driven wheel spinning (in rpm)?
19. The head bolts on a certain car require that they be tightened with a torque of 110 ft lb. You are able to push on the wrench with 50 lb of force. How long should the wrench be?
20. The specification plate on a motor states that the motor puts out 452.4 W of power and runs at 3,600 rpm. What is the torque of this motor? (Remember to convert rpm into rad/s!)
21. The transmission on a car is designed to change the rpm and torque of an engine to the proper levels at the wheels to drive the

car effectively. If the engine has 130 ft lb of torque and the torque at the wheels is 260 ft lb, how many times bigger is the driven wheel compared to the driving wheel?

22. What is the centripetal force on a 75 kg man turning a corner of radius 100 m in a car at an angular velocity of 1.1 rad/s?
23. A 2,000 lb car exits off the expressway onto an exit ramp at 45 mph or 66 ft/s. The ramp had a turning radius of 300 ft. What is the centripetal force on the car? (Remember to convert weight to mass!)
24. A robot moves across the floor on wheels of radius 2 in. rotating at 25 rpm. How fast is the robot moving in in./s?
25. A conveyer belt requires 22 lb of force to move. Mounted on the shaft of the motor that drives the belt is a wheel of radius 6 in. What torque must the motor have in ft lb?
26. A motor is spinning at 1,750 rpm. What is this in rad/s?
27. What is the angular velocity of the earth on its own axis in rpm?
28. The carriage on a copy machine is translated back and forth on wheels of radius 1.0 cm. If the wheels are spinning at 5 rpm, how fast is the carriage moving in cm/min?
29. A motor rotating at 50 rpm is connected to a wheel of radius 12 cm. This wheel is driving another wheel of radius 4 cm. How fast is the 4 cm wheel spinning?
30. A cart on wheels requires 8 N of force to keep it moving steadily. If the cart is driven by a motor connected to a wheel of radius 0.03 m, what torque is the motor generating?
31. If the motor in the previous problem is spinning at 25 rpm, how much power is the motor generating?
32. A motor has a torque of 1.1 Nm with a gear on its shaft with 20 teeth. Another gear is being driven by this gear with 60 teeth. What is the torque of the driven gear?
33. A mass of 0.4 kg is placed on a turntable at a distance of 0.10 m from the center of rotation. If the turntable is spinning at 45 rpm, what is the centripetal force on the mass?
34. The wheels on a cart of radius 2.0 in. turn through 50 rev. How far has the cart moved?
35. A wheel of diameter 10 cm spinning at 200 rpm drives another wheel of diameter 5 cm. How fast is the driven wheel spinning?
36. A wheel starts from rest and reaches an angular velocity of 20 rad/s in 4 s. What was its angular acceleration?
37. What is the centrifugal force on the driver of a car if she has a mass of 72 kg and is turning a corner of radius 10 m at a velocity of 10 m/s?

6

Machines

This chapter describes the basic types of machines. *Machines* are devices that help us perform work that could not otherwise be performed. Typically, they generate more force than we are capable of generating and are found in many types of electromechanical equipment (Figure 6.1).

FIGURE 6.1 Machines are devices that help us to perform work that could not otherwise be performed.

Theoretically machines work on the principle: Work In = Work Out.

The amount of work you put into the machine equals the amount you get out.

Note that work in = work out is only an idealization because there will always be friction that just wastes energy.

Rewriting this equation using the equation for work, $W = F \cdot d$,

$$F_{in} \cdot d_{in} = F_{out} \cdot d_{out}$$

$$\text{Force in} \times \text{distance in} = \text{Force out} \times \text{distance out}$$

This equation says that when you apply a force to a part of the machine and move it some distance (d), another part of the machine will generate a force out and move it a distance out. All machines work on this principle. Figure 6.2 illustrates the six basic machines.

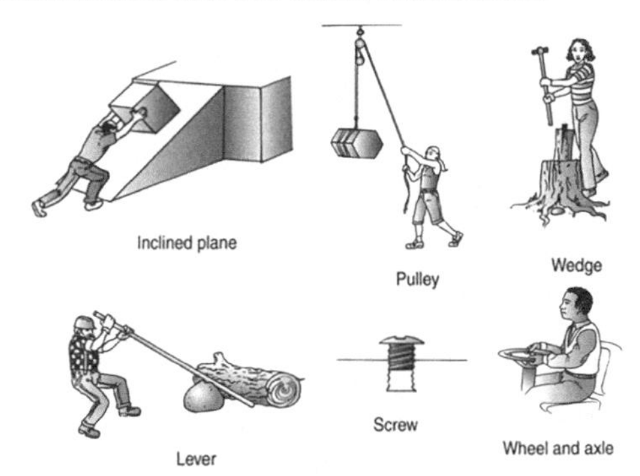

FIGURE 6.2 The six basic machines.

Most machines you are familiar with are *compound machines*, that is, they consist of one or more of these basic machines (Figure 6.3).

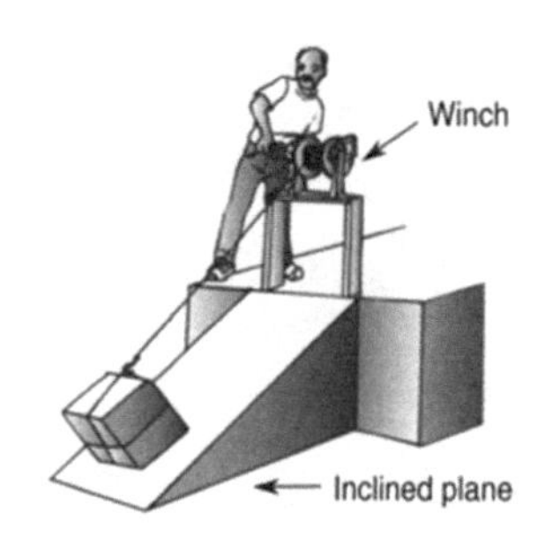

FIGURE 6.3 A compound machine consists of several basic machines.

6.1 MECHANICAL ADVANTAGE

Mechanical advantage (MA) is a measure of how many times your strength is multiplied by using a machine. Because a machine is not 100% efficient, some of the work going in is wasted because of friction. How much force a machine can actually generate is defined by:

$$\text{Mechanical advantage} \rightarrow \text{MA} = \frac{F_o}{F_i} = \frac{\text{force out}}{\text{force in}}$$

If the force generated by the machine is large as compared with the input force, the MA is large.

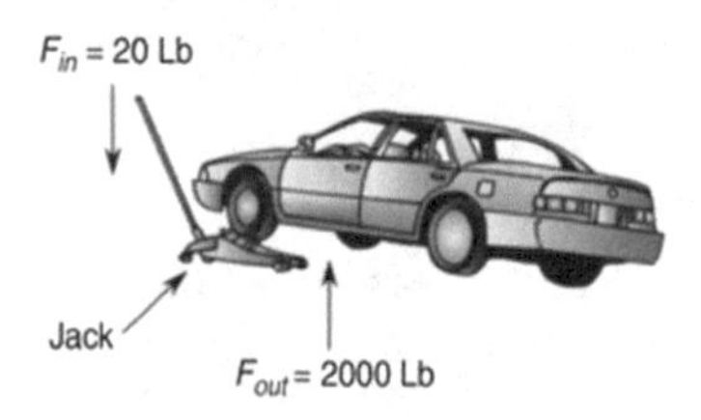

FIGURE 6.4 A machine generates 2,000 lb with 20 lb of force, what is the MA?

Example 6.1

Using a machine to generate 2,000 lb with 20 lb of force, what is the MA? (Figure 6.4)

$$F_{\text{out}} = 2{,}000\,\text{lb}$$

$$F_{\text{in}} = 20\,\text{lb}$$

$$\text{MA} = ?$$

$$\text{MA} = \frac{F_o}{F_i}$$

$$\text{MA} = \frac{2{,}000\,\text{lb}}{20\,\text{lb}} = 100 \leftarrow \text{This machine can increase your strength 100 times.}$$

Example 6.2

A machine has a MA of 20. What weight can be moved with an applied force of 5.0 lb?

$$\text{MA} = 20$$

$$F_i = 5.0\,\text{lb}$$

$$F_o = ?$$

$$\text{MA} = \frac{F_o}{F_i}$$

$$F_o = \text{MA} \cdot F_i$$

$$F_o = 20 \cdot 5\,\text{lb}$$

$$F_o = 100\,\text{lb}$$

6.2 EFFICIENCY

The *efficiency* of a machine is a measure of how much work it puts out versus how much work goes in. It will always be less than one because of friction.

$$\text{Efficiency} = \frac{\text{Work Out}}{\text{Work In}}$$

Example 6.3

A machine puts out 89 J of work for an input of 98 J. What is its efficiency?

$$\text{Work In} = 98\,\text{J}$$
$$\text{Work Out} = 89\,\text{J}$$
$$\text{Efficiency} = \frac{\text{Work Out}}{\text{Work In}}$$
$$\text{Efficiency} = \frac{89\,\text{J}}{98\,\text{J}}$$
$$\text{Efficiency} = 0.91 \leftarrow \text{The machine is 91\% efficient.}$$

Applying input force to a machine using a crank or a handle and moving that implement by a distance larger than that of the output of the machine will give you a very large MA. Think about jacking up a car. You have to pump the jack handle through a large distance to make the car move just a little (Figure 6.5), but with that little force you are able to lift a car!

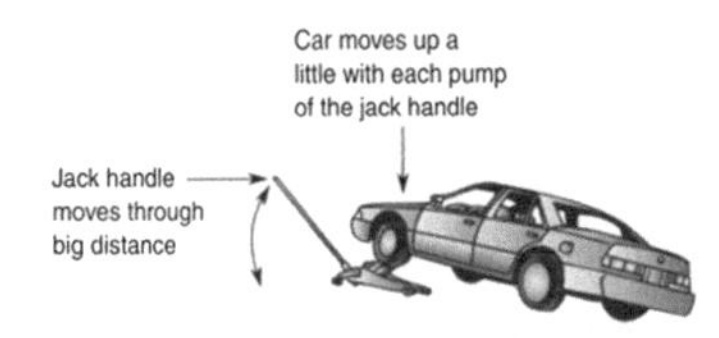

FIGURE 6.5 When jacking a car, you must pump the jack handle a big distance to move the car up just a little.

From the definition of efficiency, the MA can also be expressed in another way:

$$\text{MA} = \text{efficiency} \times \frac{d_i}{d_o}$$

$$\text{Mechanical advantage} = \text{efficiency} \times \frac{\text{effort distance in}}{\text{load distance out}}$$

Example 6.4

When jacking up a car, suppose the jack handle moves through a distance of 36 in. and the car moves up 0.5 in. What is the MA? Assume a 40% efficiency (Figure 6.6).

$$d_i = 36\,\text{in.}$$
$$d_o = 0.50\,\text{in.}$$
$$\text{Efficiency} = 0.4$$
$$\text{MA} = ?$$
$$\text{MA} = \text{efficiency} \times \frac{d_i}{d_o}$$
$$\text{MA} = 0.40 \times \frac{36\,\text{in.}}{0.50\,\text{in.}}$$
$$\text{MA} = 29 \leftarrow \text{Your strength has increased 29 times!}$$

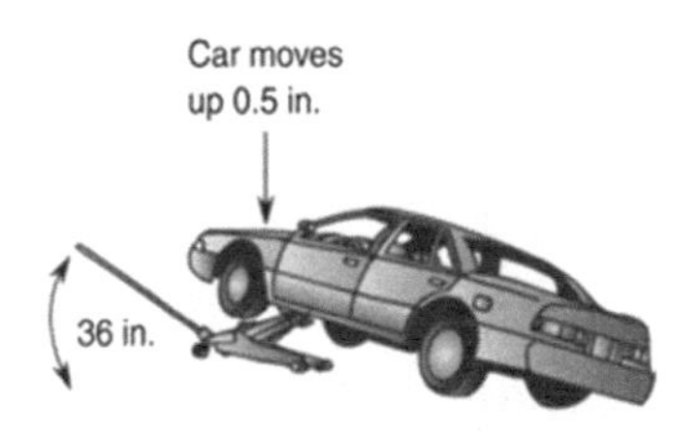

FIGURE 6.6 What is the mechanical advantage of the jack?

6.3 THE LEVER

There are three types of levers: class 1, class 2, and class 3 (Figures 6.7–6.9).

The lever has little friction. Therefore, the efficiency is 100% and

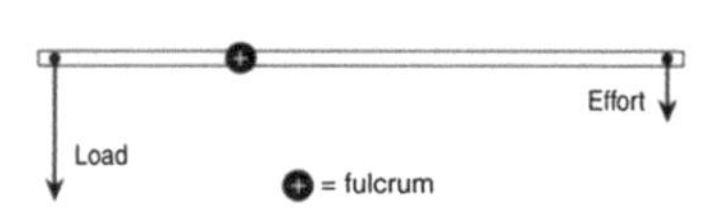

FIGURE 6.7 A class 1 lever.

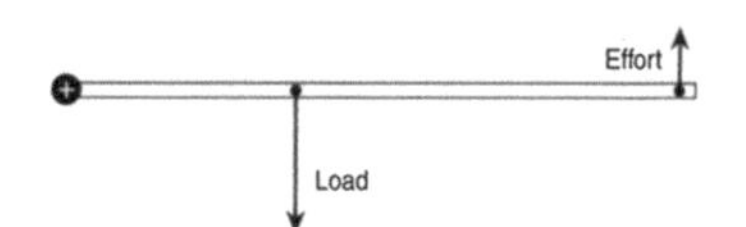

FIGURE 6.8 A class 2 lever.

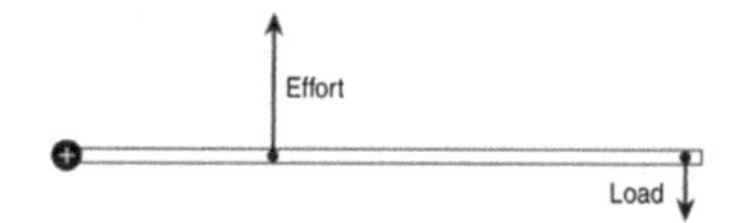

FIGURE 6.9 A class 3 lever.

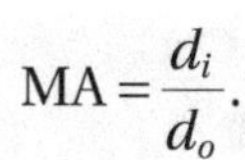

$$MA = \frac{d_i}{d_o}.$$

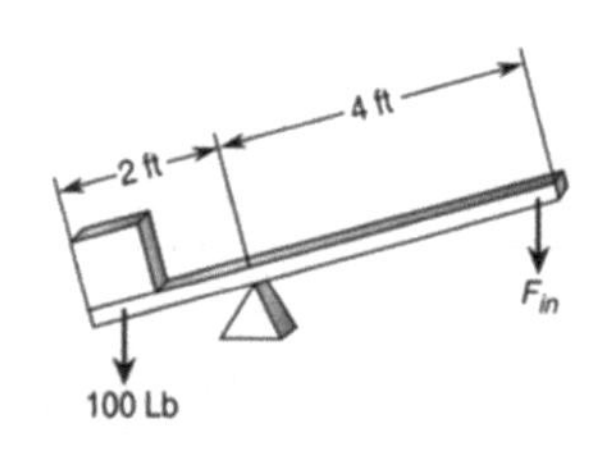

FIGURE 6.10 What is the MA of this lever, and how much input force is needed to raise the load?

Example 6.5

A lever of class 1 is used to lift a 100 lb weight placed 2 ft from the fulcrum. The input force is placed 4 ft from the fulcrum (Figure 6.10). What is the MA of this lever, and how much input force is needed to raise the load?

$$d_i = 4\,\text{ft}$$

$$d_o = 2\,\text{ft}$$

$$F_{\text{out}} = 100\,\text{lb}$$

$$F_i = ?$$

$$MA = ?$$

$$MA = \frac{d_i}{d_o}$$

$$MA = \frac{4\,\text{ft}}{2\,\text{ft}} = 2$$

$$MA = \frac{F_o}{F_i}$$

$$F_o = MA \cdot F_i$$

$$\frac{F_o}{MA} = \frac{\cancel{MA} \cdot F_i}{\cancel{MA}}$$

$$F_i = \frac{F_o}{MA}$$

$$F_i = \frac{100\,\text{lb}}{2} = 50\,\text{lb}$$

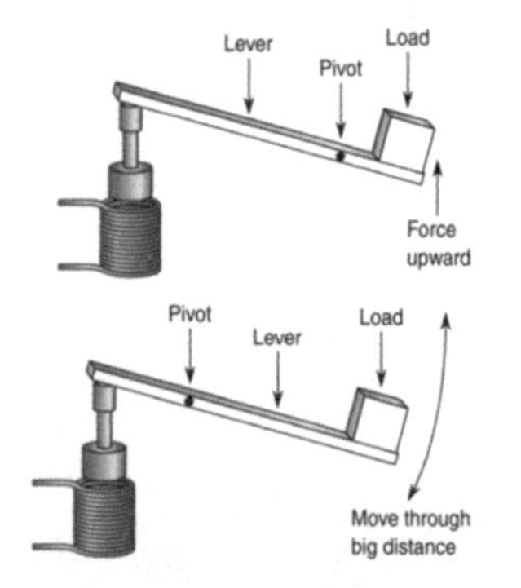

FIGURE 6.11 A class 1 lever is sometimes used with a solenoid either to generate more force or to move something through a greater distance.

Electromechanical devices use levers. See Figure 6.11.

Example 6.6

A wheelbarrow is a class 2 lever and is used to lift a 500 lb weight placed 1 ft from the fulcrum. The input force is placed 10 ft from the fulcrum (Figure 6.12). What is the MA of this lever, and how much input force is needed to raise the load?

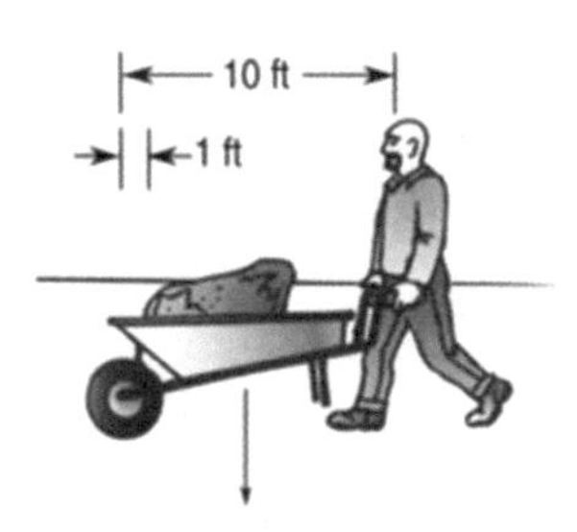

FIGURE 6.12 What is the MA of this wheelbarrow and how much input force is needed to raise the load?

$$d_i = 10\,\text{ft}$$
$$d_o = 1\,\text{ft}$$
$$F_o = 500\,\text{lb}$$
$$F_i = ?$$
$$\text{MA} = ?$$
$$\text{MA} = \frac{10\,\cancel{\text{ft}}}{1\,\cancel{\text{ft}}} = 10$$
$$\text{MA} = \frac{F_o}{F_i}$$
$$\text{MA} \cdot F_i = \left(\frac{F_o}{\cancel{F_i}}\right)\cancel{F_i} \leftarrow \text{solving for } F_i$$
$$\frac{\cancel{\text{MA}} \cdot F_i}{\cancel{\text{MA}}} = \frac{F_o}{\text{MA}}$$
$$F_i = \frac{F_o}{\text{MA}}$$
$$F_i = \frac{500\,\text{lb}}{10} = 50\,\text{lb}$$

Example 6.7

A class 3 lever is used to lift a 50.0 lb weight placed 6.0 in. from the fulcrum. The input force is placed 2.0 in. from the fulcrum (Figure 6.13). What is the MA of this lever, and how much input force is needed to raise the load?

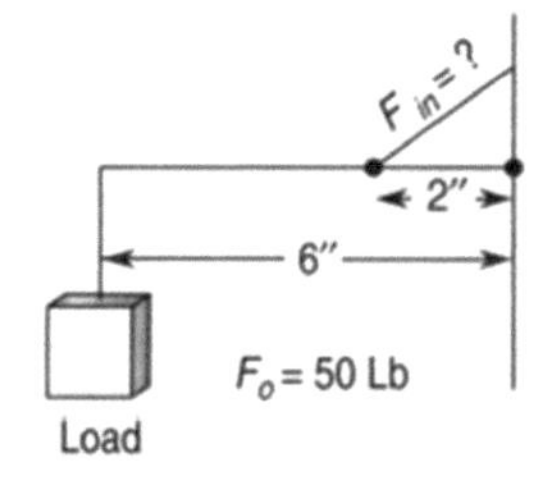

FIGURE 6.13 What is the MA of this class 3 lever, and how much input force is needed to raise the load?

$$d_i = 2\,\text{in.}$$
$$d_o = 6\,\text{in.}$$
$$F_o = 50\,\text{lb}$$
$$F_i = ?$$
$$\text{MA} = ?$$
$$\text{MA} = \frac{d_i}{d_o}$$
$$\text{MA} = \frac{2\,\cancel{\text{in.}}}{6\,\cancel{\text{in.}}} = \frac{1}{3}$$
$$\text{MA} = \frac{F_o}{F_i}$$
$$\text{MA} \cdot F_i = \left(\frac{F_o}{\cancel{F_i}}\right)\cancel{F_i} \leftarrow \text{solving for } F_i$$
$$F_o = \text{MA} \cdot F_i$$

$$\frac{F_o}{\text{MA}} = \frac{\cancel{\text{MA}} \cdot F_i}{\cancel{\text{MA}}}$$

$$F_i = \frac{F_o}{\text{MA}}$$

$$F_i = \frac{50\,\text{lb}}{1/3} = 150\,\text{lb}$$

6.4 THE PULLEY

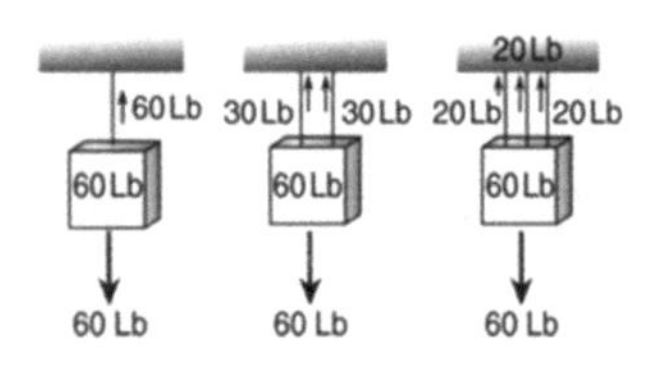

FIGURE 6.14 Ropes suspending weight.

Ropes suspending weight will have the following tensions in them (see Figure 6.14).

If we make the ropes movable by using pulleys, we can lift the load and still have the reduced tension in the strands of rope (Figure 6.15).

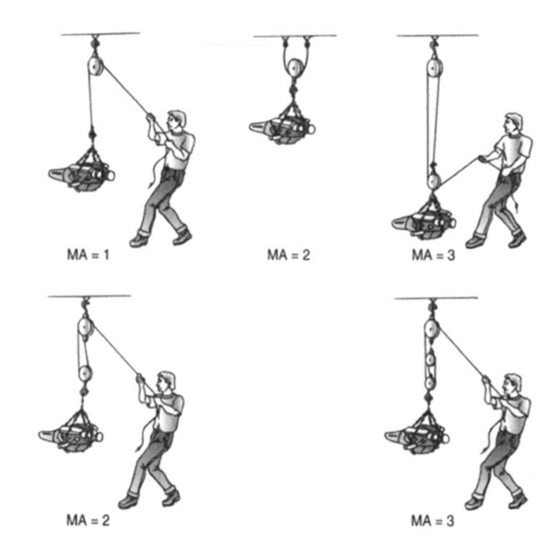

FIGURE 6.15 A block and tackle.

The bearings will have some friction in them. Therefore, if the input force moves in the direction of the load, then:

$$\text{MA} = \text{bearing efficiency} \times \text{number of strands}.$$

If the input force moves in the opposite direction of the load, then:

$$\text{MA} = \text{bearing efficiency} \times (\text{number of strands} - 1)$$

These machines are sometimes called a *block and tackle*. To build one, wind the rope from outside in, always winding either clockwise or counterclockwise (Figure 6.16).

Clockwise Counterclockwise

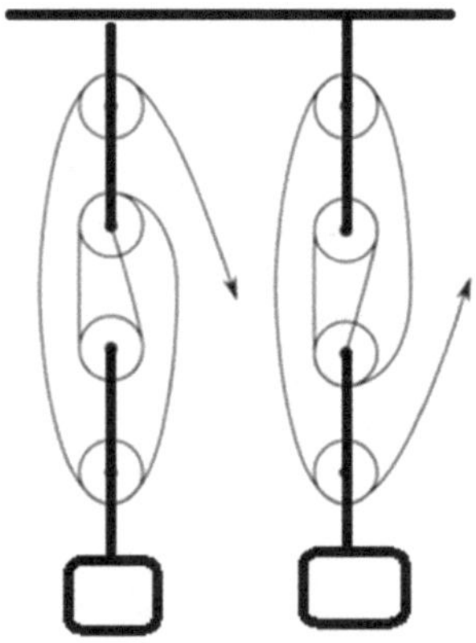

FIGURE 6.16 To a wind a block and tackle, wind the rope from outside in, always winding either clockwise or counterclockwise.

Example 6.8

A 300 lb engine is being lifted out of a car with a block winch, shown in Figure 6.17. How hard does the mechanic have to pull to lift the engine? Assume an efficiency of 90%.

There are four strands, with the last one being pulled in the opposite direction of the engine, therefore,

$$MA = \text{efficiency} \times (\text{number of strands} - 1)$$

$$MA = 0.90 \times (4 - 1)$$

$$MA = 0.90 \times 3$$

$$MA = 2.7$$

$$MA = \frac{F_o}{F_i}$$

$$F_i = \frac{F_o}{MA}$$

$$F_i = \frac{300\,\text{lb}}{2.7}$$

$$F_i = 111\,\text{lb}$$

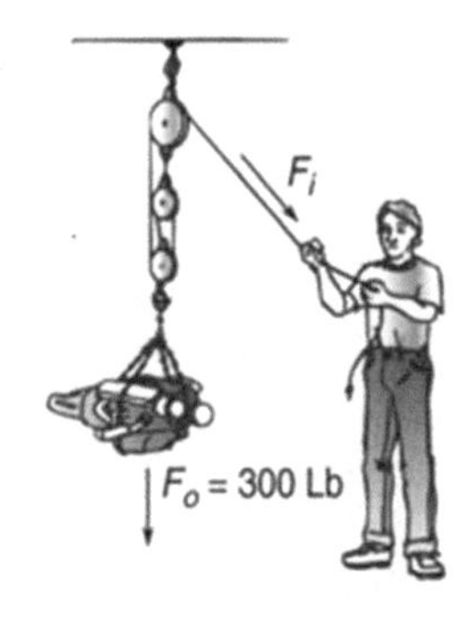

FIGURE 6.17 How hard does the mechanic have to pull to lift the engine?

6.5 THE WHEEL AND AXLE

A *wheel and axle* machine converts a large rotary motion into a small one, but it increases the output force (Figure 6.18).

The distance in (d_i) and the distance out (d_o) are the radii or diameters of the wheel and axle.

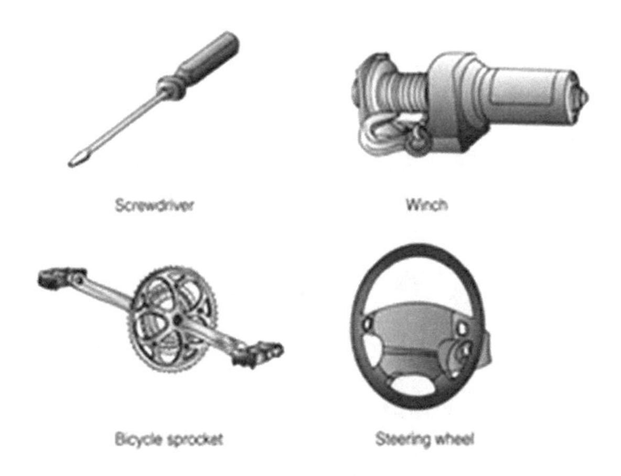

FIGURE 6.18 A steering wheel, a bicycle sprocket and pedals, a screwdriver, and a winch are examples of a wheel and axle machine.

$$MA = \text{bearing efficiency} \times \frac{d_i}{d_o}$$

$$MA = \text{bearing efficiency} \times \frac{\text{wheel diameter or radius}}{\text{axle diameter or radius}}$$

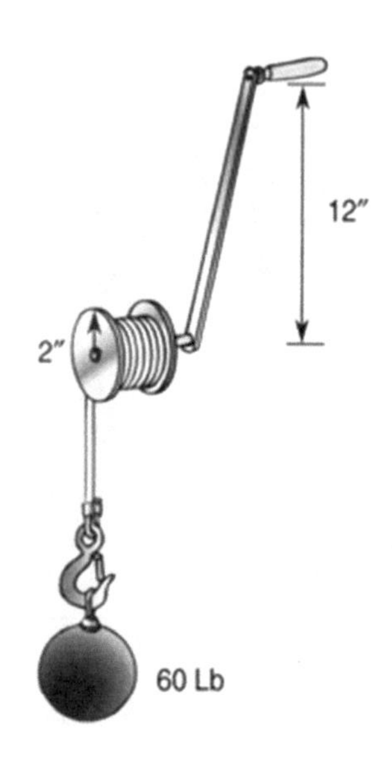

FIGURE 6.19 What is the mechanical advantage?

Example 6.9

A winch handle is 12 in. long and rotates a drum, which is 2 in. in radius. The load is 60 lb. The efficiency of the bearings is 92% (Figure 6.19). What is the MA, and how much input force is needed to lift the weight?

$$d_i = 12\text{in.}$$

$$d_o = 2\text{in.}$$

$$F_o = 60\text{lb}$$

$$\text{Efficiency} = 92\%$$

$$F_i = ?$$

$$\text{MA} = ?$$

$$\text{MA} = \text{efficiency} \times \frac{d_i}{d_o}$$

$$\text{MA} = 0.92 \times \frac{12\,\text{in.}}{2\,\text{in.}} = 5.5$$

$$\text{MA} = \frac{F_o}{F_i}$$

$$\text{MA} \cdot F_i = \left(\frac{F_o}{F_i}\right) F_i \leftarrow \text{solving for } F_i$$

$$F_o = \text{MA} \cdot F_i$$

$$\frac{F_o}{\text{MA}} = \frac{\text{MA} \cdot F_i}{\text{MA}}$$

$$F_i = \frac{F_o}{\text{MA}}$$

$$F_i = \frac{60\text{lb}}{5.5} = 11\text{lb}$$

6.6 THE INCLINED PLANE

Sliding a heavy object up an inclined plane takes less force than actually picking it up and raising it to that height. See Figure 6.20. An example of an inclined plane is a cam shaft in an engine. See Figure 6.21.

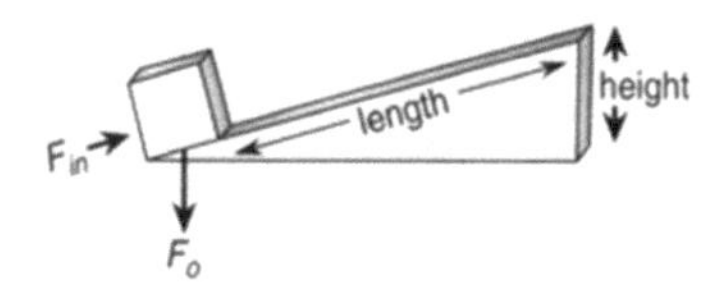

FIGURE 6.20 An inclined plane is one of the basic machines.

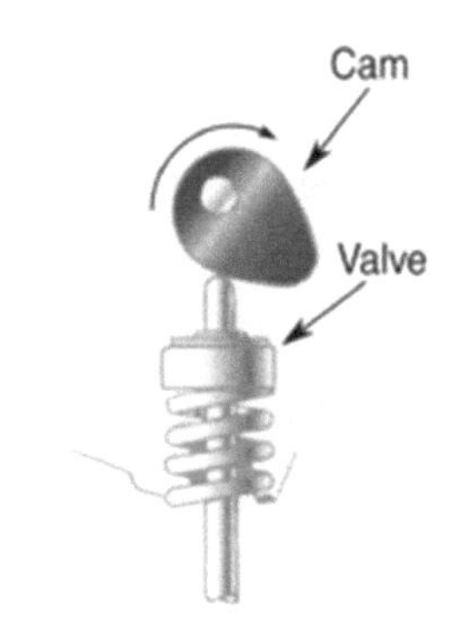

FIGURE 6.21 A cam is an example of an inclined plane.

$$\text{For an inclined plane, MA is: MA} = \text{efficiency} \times \frac{\text{length}}{\text{height}}$$

Example 6.10

A 75 lb crate is slid up a greased inclined plane that is 12.0 ft long and 3.0 ft high. What is the MA, and how much input force is needed? Assume an efficiency of 65% (Figure 6.22).

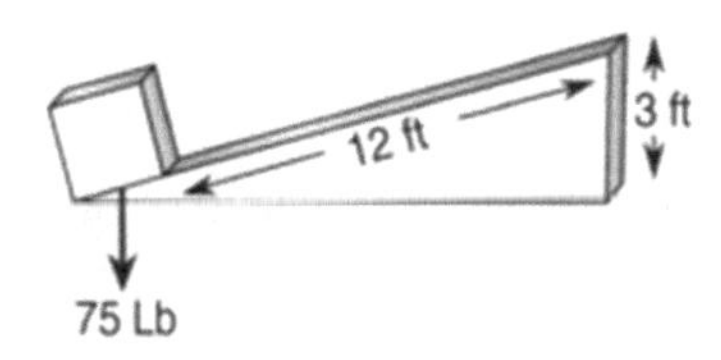

FIGURE 6.22 An inclined plane is used to raise a crate with less force then the weight of the crate.

$$\text{Length} = 12.0\,\text{ft}$$

$$\text{Height} = 3.0\,\text{ft}$$

$$F_o = 75\,\text{lb}$$

$$\text{Efficiency} = 0.65$$

$$F_i = ?$$

$$\text{MA} = ?$$

$$\text{MA} = \text{efficiency} \times \frac{\text{length}}{\text{height}}$$

$$\text{MA} = 0.65 \times \frac{12.0\,\text{ft}}{3.0\,\text{ft}}$$

$$\text{MA} = 2.6$$

$$\text{MA} = \frac{F_o}{F_i}$$

$$\text{MA} \cdot F_i = \left(\frac{F_o}{\cancel{F_i}}\right)\cancel{F_i} \leftarrow \text{solving for } F_i$$

$$F_o = \text{MA} \cdot F_i$$

$$\frac{F_o}{\text{MA}} = \frac{\cancel{\text{MA}} \cdot F_i}{\cancel{\text{MA}}}$$

$$F_i = \frac{F_o}{\text{MA}}$$

$$F_i = \frac{75\,\text{lb}}{2.6} = 29\,\text{lb}$$

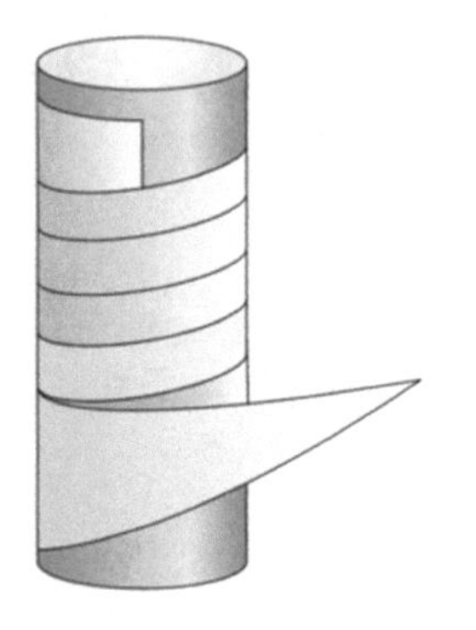

FIGURE 6.23 A screw is an inclined plane wrapped around a cylinder.

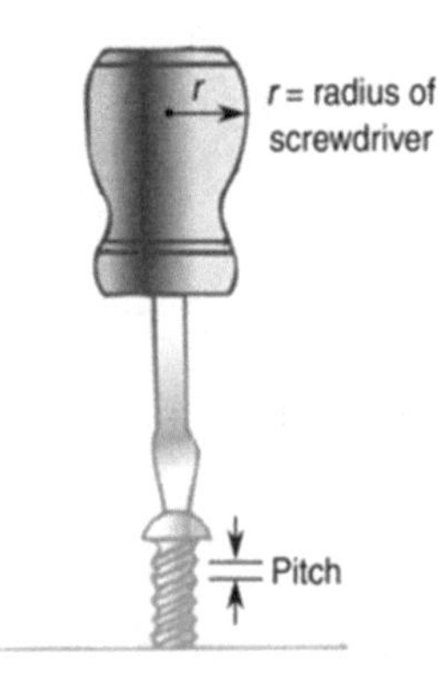

FIGURE 6.24 The pitch of a screw is the distance between adjacent threads.

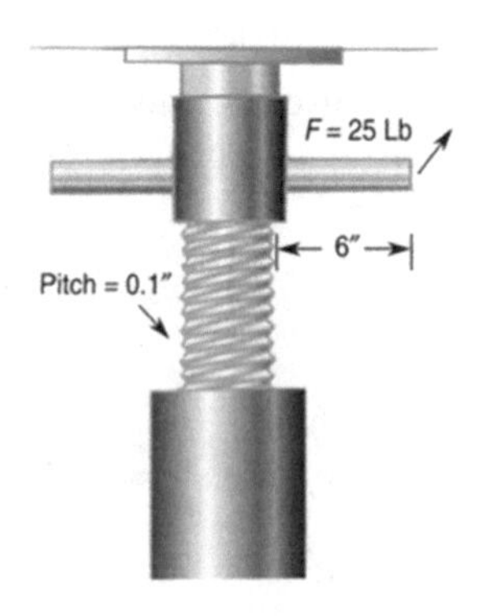

FIGURE 6.25 An screw is used to generate more force than is applied.

6.7 THE SCREW

A *screw* is essentially an inclined plane wrapped around a cylinder (Figure 6.23). The pitch of a screw is the distance between adjacent threads (Figure 6.24).

When the screw is rotated through one complete revolution, it rises up one pitch distance. A handle on a screwdriver or a handle mounted on the screw allows the screw to be turned with ease. Lubrication will maximize efficiency. The MA is:

$$\text{MA} = \text{efficiency} \times \frac{2\pi r}{\text{pitch}}$$

$$\text{MA} = \text{efficiency} \times \frac{2\pi \cdot \text{handle radius}}{\text{pitch}}$$

Example 6.11

A jackscrew has a pitch of 0.10 in. and a handle that is 6.0 in. in length (Figure 6.25). What is the MA, and how much weight can it lift with 25 lb of input force? Assume 85% efficiency.

$$r = 6.0\,\text{in.}$$
$$\text{pitch} = 0.10\,\text{in.}$$
$$F_i = 25\,\text{lb}$$
$$\text{Efficiency} = 0.85$$
$$F_o = ?$$
$$\text{MA} = ?$$
$$\text{MA} = \text{efficiency} \times \frac{2\pi r}{\text{pitch}}$$
$$\text{MA} = 0.85 \times \frac{2\pi(6.0\,\cancel{\text{in.}})}{0.10\,\cancel{\text{in.}}}$$
$$\text{MA} = 320$$
$$\text{MA} = \frac{F_o}{F_i}$$
$$\text{MA} \cdot F_i = \left(\frac{F_o}{\cancel{F_i}}\right)\cancel{F_i} \leftarrow \text{solving for } F_o$$
$$F_o = \text{MA} \cdot F_i$$
$$F_o = 320 \cdot 25\,\text{lb}$$
$$F_o = 8{,}\bar{0}00\,\text{lb}$$

Screws have an enormous MA!

6.8 WEDGE

Wedges are generally used as aids in splitting things apart. A wedge typically has a lot of friction, and therefore its efficiency is very low (Figure 6.26). We can approximate the MA of the wedge by:

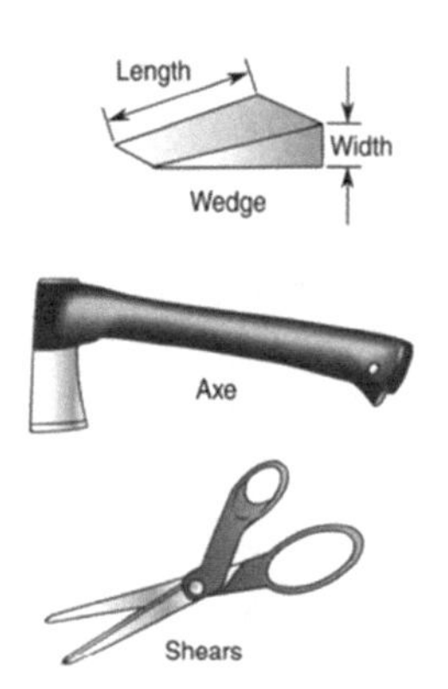

FIGURE 6.26 A wedge typically has a lot of friction and very low efficiency.

$$\text{MA} = \text{efficiency} \times \frac{\text{length}}{\text{width}}$$

Example 6.12

What is the MA of a wedge with a length of 9 in., a width of 1 in., and an efficiency of 30%? See Figure 6.27.

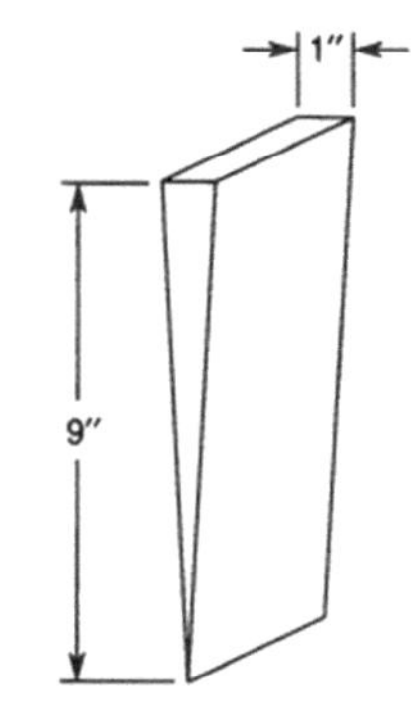

FIGURE 6.27 What is the MA, and how much input force is needed?

$$\text{Length} = 9\,\text{in.}$$

$$\text{Width} = 1\,\text{in.}$$

$$\text{Efficiency} = 30\%$$

$$\text{MA} = 0.30 \times \frac{9\,\cancel{\text{in.}}}{1\,\cancel{\text{in.}}}$$

$$\text{MA} = 3$$

6.9 COMPOUND MACHINES

Using two or more basic machines at the same time results in a much higher MA. The MA of a compound machine is: $\text{MA}_{\text{total}} = \text{MA}_1 \times \text{MA}_2 \times \text{MA}_3 \times \ldots$

Example 6.13

A compound machine consists of a six-strand pulley system with an efficiency of 88%, a wheel and axle with the wheel radius of 4.0 in. and the axle radius of 2.0 in., and an efficiency of 98% (Figure 6.28). What is the MA of this compound machine?

For the pulley:
Six strands

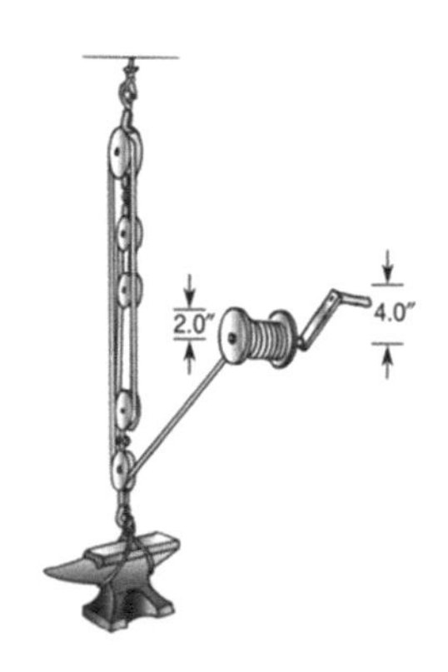

FIGURE 6.28 Compound machines.

$$\text{Efficiency} = 0.88$$

$$\text{MA} = \text{efficiency} \times \text{number of strands}$$

$$\text{MA} = 0.88 \times 6$$

$$\text{MA} = 5.3$$

For the wheel and axle:

$$d_i = 4.0\,\text{in.}$$

$$d_o = 2.0\,\text{in.}$$

$$\text{Efficiency} = 0.98$$

$$\text{MA} = \text{efficiency} \times \frac{d_i}{d_o}$$

$$\text{MA} = 0.98\,\frac{4.0\,\text{in.}}{2.0}$$

$$\text{MA} = 2.0$$

The total MA is:

$$\text{MA}_{\text{total}} = \text{MA}_1 \times \text{MA}_2$$

$$\text{MA}_{\text{total}} = 5.3 \times 2.0$$

$$\text{MA}_{\text{total}} = 11$$

6.10 CHAPTER SUMMARY

Symbols used in this chapter:

Symbol	Unit
d—distance	m
d—diameter	m
Length	m
r—radius	m
F—force	N
MA—mechanical advantage	no unit
Efficiency	no unit or %

The basic law of machines assumes no friction: work in = work out
Written another way:

$$F_{\text{in}} \cdot d_{\text{in}} = F_{\text{out}} \cdot d_{\text{out}}$$

Force in × distance in = force out × distance out

$$\text{Efficiency} = \frac{\text{work out}}{\text{work in}}$$

Mechanical advantage:

$$MA = \frac{F_o}{F_i} = \frac{\text{force out}}{\text{force in}}$$

or

$$MA = \text{efficiency} \times \frac{d_i}{d_o}$$

$$\text{Mechanical advantage} = \text{efficiency} \times \frac{\text{effort distance in}}{\text{load distance out}}$$

There are six basic machines: the lever, pulley, wheel and axle, inclined plane, screw, and the wedge.

1. The lever: $MA = \text{efficiency} \times \frac{d_i}{d_o}$
2. The pulley: If the input force moves in the direction of the load, then MA = efficiency × number of strands. If it moves in the opposite direction of the load, then MA = efficiency × (number of strands – 1).
3. The wheel and axle: $MA = \text{efficiency} \times \frac{d_i}{d_o}$
4. The inclined plane: $MA = \text{efficiency} \times \frac{\text{length}}{\text{height}}$
5. The screw: $MA = \text{efficiency} \times \frac{2\pi r}{\text{pitch}}$
6. The wedge: $MA = \text{efficiency} \times \frac{\text{length}}{\text{width}}$

For compound machines: $MA_{\text{total}} = MA_1 \times MA_2 \times MA_3 \times \cdots$

PROBLEM SOLVING TIPS

- If you have trouble determining which formula to use, remember that MA is a measure of how many times your strength is increased by a machine. It is determined by

$$MA = \text{efficiency} \times \frac{\text{The distance you move a handle, lever, etc. through.}}{\text{The distance by which the machines moves what it's working on.}}$$

PROBLEMS

1. A machine is able to lift 500 lb with 5 lb of applied force. What is the MA of this machine?
2. With an applied force of 25 lb a machine lifts 100 lb. Suppose the applied force is moved through a distance of 10 in. and the machine lifts a load of 2 in. What are the MA and the efficiency of this machine?
3. A wheelbarrow with a 6 ft handle lifts 300 lb of weight placed 1 ft from the wheel. What is the MA of this machine assuming a 98% efficiency? How much force is needed to lift this load?
4. A car stuck in the mud can be lifted up with a lever. The car is 2 ft from the fulcrum, and the applied force is 10 ft from the fulcrum. If a person stands on the lever with 180 lb of body weight, how hard is the car pushed upward? Assume 100% efficiency.
5. A body builder is working her arms. Her bicep is attached 2.0 in. from the elbow joint. She holds a 55 lb weight in her hands at a distance of 12 in. from the elbow. What is the MA of her bicep, and how much force does the bicep has to generate to lift this weight? Assume 100% efficiency.
6. A block and tackle with five strands is used to lift a 200 lb motor onto a roof. The motor rises upward, while the rope is pulled downward. What force is needed to lift this motor? Assume an efficiency of 92%.
7. A bicycle pedal is 10 in. from the center of rotation and is connected to a gear 6 in. in radius. What is the MA of this machine? Assume an efficiency of 96%.
8. When building the pyramids in Egypt, workers pulled heavy stone blocks up the side of the pyramids by using skids. If the blocks weighed 4,500 lb and the pyramid was sloped at 40°, how hard did the workers have to push? Assume an efficiency of 62%. How could a pulley system be incorporated to increase the MA?
9. How much force can a nutcracker generate when the handle is 10 in. long, the thread pitch is 0.25 in., and the applied force is 4 lb? Assume an efficiency of 84%.
10. A compound machine consists of a lever with an efficiency of 100%, a block and tackle with an efficiency of 94%, and a wheel and axle with an efficiency of 94% (having MAs of 3, 4, and 5, respectively). What is the total MA for this compound machine? If 20 N of force is applied, what is the output force?
11. A knob on an oscilloscope has a radius of 0.50 in. and turns a rotary switch it is connected to with a radius of 0.125 in. What is the MA of this knob? Assume an efficiency of 99%.

12. Suppose an elevator has to exert a force of 435 N to move a crate straight up a distance of 3.0 m. If an inclined plane was used, less force would have been needed to raise it by the same distance, but it should have been pulled farther along the inclined plane. How long would this plane need to be if it is to be pulled with only 50.0 N of force? Assume an efficiency of 74%.
13. Design a compound machine that could have aided the ancient Egyptians in building a pyramid. The machine should incorporate five of the basic machines and have a MA of 10,000. Show the dimensions of each machine and their respective MAs.
14. An axe head has a length of 20.0 cm and a width of 5.0 cm. If a hammer delivers a blow of 150 N to the head, how much force is delivered to the wood? Assume an efficiency of 34%.
15. How could you redesign your bicep muscle so that you could win arm wrestling contests more easily?
16. An elevator consists of a motor and pulley system to lift the carriage up and down. If the work into the system is 2,000 ft lb and the system performs 1,500 ft lb of work, what is the efficiency of the system?
17. A car jack delivers 1,200 lb of force to a bumper of a car. A force of 10 lb is delivered to the jack handle. What is the MA of the jack?
18. Inside a copy machine is a compound machine consisting of a screw with a MA of 1,200 and a pulley system with a MA of 4. What is the total MA?
19. An automated soldering station consists of a soldering iron that is translated back and forth on a threaded shaft. The pitch of the threads is 0.05 in. and at the end of the shaft is a handle of radius 1.0 in. and is driven by a motor. If it takes 1.5 lb of force to translate the iron back and forth, how much force does the motor deliver to the shaft? Assume an efficiency of 0.9.
20. A knob on an oscilloscope has a diameter of 1.5 in. and is connected to a shaft of 0.25 in. If the knob is turned with 1.1 lb of force, what force is the shaft turned with if the efficiency equals 1?
21. A vending machine dispenses some candy by pushing the candy up an inclined plane of length 9 in. and a height 3 in. The candy then falls down a shoot to be removed. If the candy weighs 0.15 lb, how much force is the candy pushed with? Assume an efficiency of 0.8.
22. A robotic arm raises a 25 N weight. The weight is 50 cm from the elbow joint and a cable is connected 20 cm from the joint and is used to raise and lower the arm. How much force is needed to raise the weight? Assume the efficiency is 1.

23. The efficiency of a log-splitting wedge is very low. Assume it is 0.2. The wedge is 12 in. long and 1 in. wide and a hammer strikes the wedge with a force of 100 lb. What force is delivered to the log?
24. A robot contains a compound machine consisting of three basic machines with MAs of 4, 5, and 6. If the drive system to the machines can generate a force of 10 lb, what weight can the robot lift?
25. If a compound machine consists of two machines, one with an efficiency of 0.9 and the other an efficiency of 0.8, what is the total efficiency?
26. What is the MA of a machine, if with 10 lb of input force the machine can lift 120 lb?
27. A machine has 100 ft lb of work put into it but only performs 80 ft lb of work. What is the percent efficiency of the machine?
28. A machine not well lubricated will have a low efficiency because of friction. If the efficiency of the machine is only 0.4 and 20 Nm of work is put into the machine, what is the output of the machine?
29. The handle of a car jack moves through a distance of 2.5 ft and the car moves up by 0.005 ft with each pump of the jack handle. If the efficiency of the jack is 0.6, what is the MA?
30. A read/write head on a hard drive is mounted on a lever. The head is mounted 10 cm from a pivot and the actuator of the lever arm is mounted 2 cm from the pivot. Assume 100% efficiency, what is the MA of this lever?
31. A pulley system mounted inside a robotic arm consists of four strands. The applied force to the last strand moves in the same direction as the load put on the arm. If the input force to the pulleys is 20 lb, how much can the arm lift? (Assume an efficiency of 0.75.)
32. A compound machine consists of two basic machines with efficiencies of 0.9 and 0.8. What is the total efficiency of this compound machine?
33. A steering wheel on a car of diameter 50 cm is connected to a shaft 1 cm. If 12 N of force is applied to the wheel, what force is output to the 1 cm shaft? (Assume an efficiency of 1.)
34. A ¼ 20 screw is used to level the legs on a table. The table weighs 100 lb so each leg supports 25 lb. If the screws are turned with a wrench that is 6 in. long, how much force is needed to raise the table? (Assume an efficiency of 0.75.)
35. A compound machine consists of three simple machines with MAs of 12, 14, and 50. What is the total MA of this machine?

7

Strength of Materials

This chapter describes the mechanical properties of materials. Some properties of solid materials are strength, elasticity, malleability, and ductility. The ever-growing field of materials engineering is dedicated to studying these properties as well as others.

7.1 STRENGTH

What determines the strength of a material? When a material is stretched, its atoms are pulled apart. When it is compressed, its atoms are pushed together. The electrical forces between the atoms resist this change in length. When the applied force stretching or compressing the material is greater than the electric forces between atoms, then a permanent deformation will occur. A material's *elastic limit* is the maximum amount by which it can be stretched or compressed without it becoming permanently deformed (Figure 7.1). A spring is designed to operate within its elastic limit. A spring when stretched or compressed returns to its original shape after the force is removed.

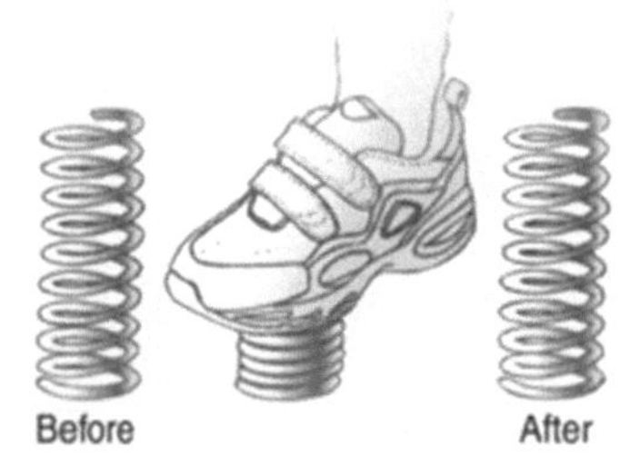

FIGURE 7.1 A material's elastic limit is the maximum amount by which it can be stretched or compressed without it becoming permanently deformed.

7.2 SPRINGS

Springs operate below the elastic limit. They return to their original shape after being stretched or compressed, if not stretched or compressed too far. The force needed to change the length of a spring is shown in *Hooke's law.*

$$F = k \cdot \Delta x$$

$$\text{Force}(\text{N}) = \text{spring constant}(\text{N/m}) \times \text{length change}(\text{m})$$

In the English system:

$$\text{Force}(\text{lb}) = \text{spring constant}(\text{lb/in.}) \times \text{length change}(\text{in.})$$

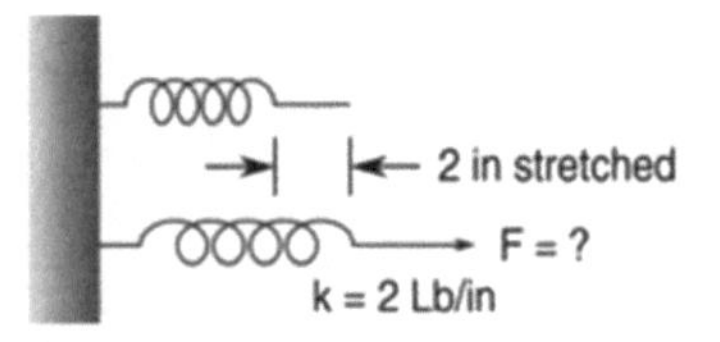

FIGURE 7.2 What is the force stretching the spring?

This equation says that the force needed to stretch or compress a spring is equal to how stiff the spring is and how much its length is changed. The spring constant (k) has units of force/length.

Example 7.1

A spring has a spring constant of 5 lb/in. and is stretched by 2 in. (Figure 7.2). What is the force stretching the spring?

$$k = 5\frac{\text{lb}}{\text{in.}}$$

$$\Delta x = 2\text{in.}$$

$$F = k \cdot \Delta x$$

$$F = \left(5\frac{\text{lb}}{\text{in.}}\right)2\text{in.}$$

$$F = 10\text{lb}$$

7.3 STRESS

To change the shape of a material, a stress must be applied to it. *Stress* is the ratio of the applied force to the area it is applied to.

$$\text{Stress} = \frac{F}{A}$$

$$\text{Stress}\left(\frac{\text{N}}{\text{m}^2}\right) = \frac{F(\text{N})}{A(\text{m}^2)}$$

In the English system of units:

$$\text{Stress}\left(\frac{\text{lb}}{\text{in.}^2}\right) = \frac{F(\text{lb})}{A(\text{in.}^2)}$$

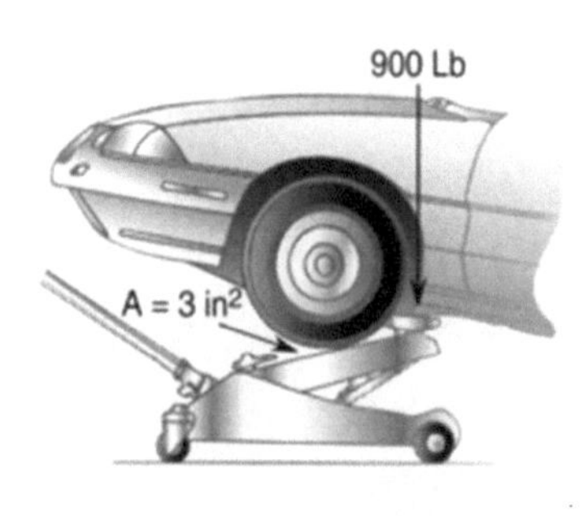

FIGURE 7.3 What is the stress on the head of the jack?

Example 7.2

A car jack is supporting 900 lb of weight applied to an area of 3 in.^{2} (Figure 7.3). What is the stress on the head of the jack?

$$F = 900\text{lb}$$

$$A = 3\text{in.}^2$$

$$\text{Stress} = \frac{F}{A}$$

$$\text{Stress} = \frac{900\,\text{lb}}{3\,\text{in.}^2}$$

$$\text{Stress} = 300\,\frac{\text{lb}}{\text{in.}^2}$$

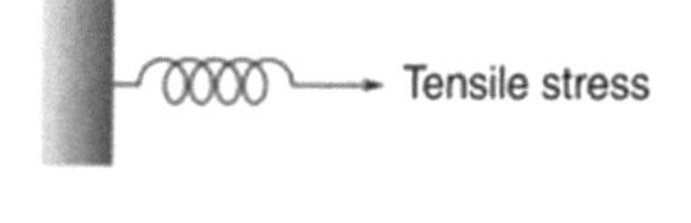

FIGURE 7.4 Tensile stress is associated with stretching.

There are three types of stresses: tensile stress, compressive stress, and shear stress (Figures 7.4–7.6).

Bending consists of both compressive and tensile stresses. In Figure 7.7, compression is taking place in the top layers of the materials and a tensile stress in the bottom layers.

FIGURE 7.5 Compressive stress is associated with squashing or compression.

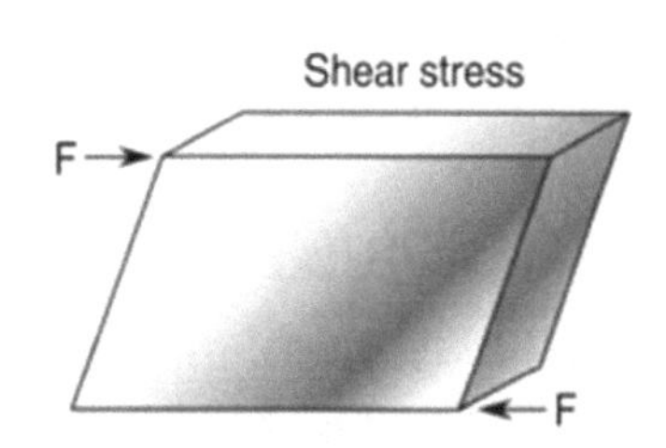

FIGURE 7.6 Shear stress is associated with two forces separated from each other, applied in opposite directions. Shear stresses are usually applied to crack something or cut or tear it apart.

FIGURE 7.7 Bending consists of both compressive and tensile stresses.

7.4 STRAIN

A stress applied to a material results in a strain on the material. The longitudinal strain is the ratio of the change in length of a material to its original length. Strain has no units.

$$\text{Strain} = \frac{\Delta L}{L}$$

$$\text{Strain} = \frac{\text{Length}_{\text{change}}\,(\text{m})}{\text{Length}_{\text{original}}\,(\text{m})}$$

In the English system of units:

$$\text{Strain} = \frac{\text{Length}_{\text{change}}\,(\text{in.})}{\text{Length}_{\text{original}}\,(\text{in.})}$$

Example 7.3

A bar under tension is stretched from 2.10 in. to 2.16 in. (Figure 7.8). What is the strain on the bar?

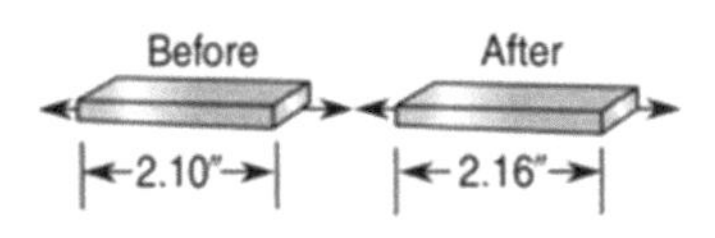

FIGURE 7.8 What is the strain on the bar?

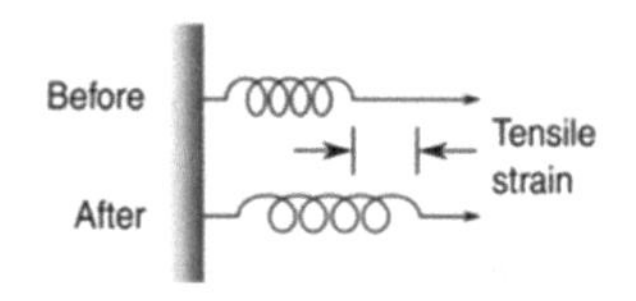

FIGURE 7.9 Tensile strain is associated with stretching.

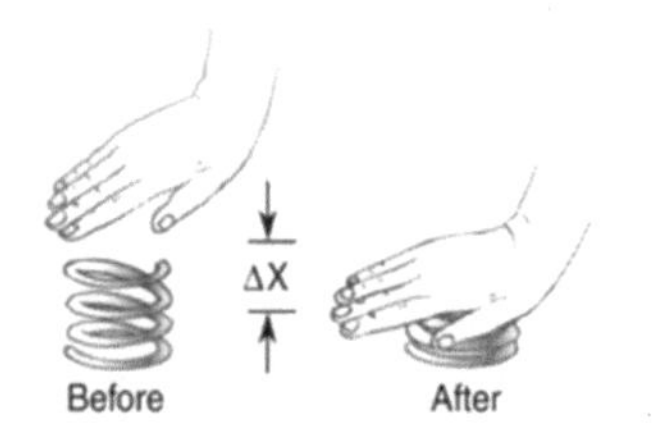

FIGURE 7.10 Compressive strain is associated with squashing or compression.

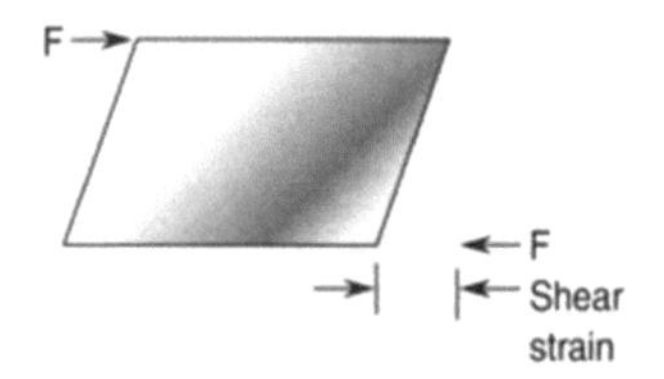

FIGURE 7.11 Shear strain is associated with two forces separated from each other and applied in opposite directions.

$$\text{Original length} = 2.10\,\text{in.}$$

$$\text{Final length} = 2.16\,\text{in.}$$

$$\Delta L = 2.16\,\text{in.} - 2.10\,\text{in.} = 0.060\,\text{in.}$$

$$\text{Strain} = \frac{\Delta L}{L}$$

$$\text{Strain} = \frac{0.060\,\text{in.}}{2.10\,\text{in.}}$$

$$\text{Strain} = 2.9 \times 10^{-2} \leftarrow \text{no unit}$$

There are three types of strains: tensile strain, compressive strain, and shear strain (Figures 7.9–7.11).

7.5 STRESS/STRAIN BEHAVIOR

Various regions describing how a material behaves for a particular stress are illustrated in Figure 7.12.

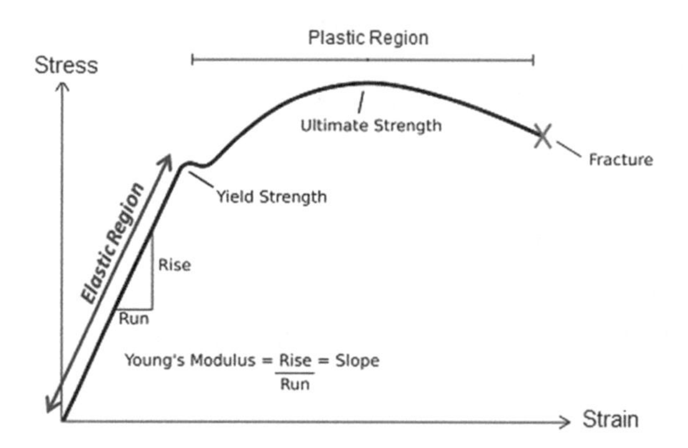

FIGURE 7.12 Various regions describing how a material behaves for a particular stress.

Elastic range is the material's elastic limit or the maximum amount by which it can be stretched or compressed without it becoming permanently deformed. Springs operate in the elastic range.

Yield strength is the end point of the elastic range where anymore stress will result in a permanent deformation of the material (Table 7.1).

Plastic range is just beyond the elastic limit. The plastic range is where stretching or compressing results in a permanent change in shape. When making objects such as coins or horseshoes a metal is worked within its

TABLE 7.1
Yield Strengths and Ultimate Strengths of Materials

Material	Yield Strength (MPa) 1 MPa = 10^6 N/m^2	Ultimate Strength (MPa)	Density (g/cm^3)
Structural steel ASTM A36 steel	250	400	7.8
Steel, API 5L X65 (Fikret Mert Veral)	448	531	7.8
Steel, high strength alloy ASTM A514	690	760	7.8
Steel, high tensile	1,650	1,860	7.8
Steel wire			7.8
Steel, piano wire	2,000		7.8
High density polyethylene (HDPE)	26–33	37	0.95
Polypropylene	12–43	19.7–80	0.91
Stainless steel AISI 302 – Cold-rolled	520	860	8.03
Cast iron 4.5% C, ASTM A-48	130	200	7.3
Titanium alloy (6% Al, 4% V)	830	900	4.51
Aluminum alloy 2014-T6	400	455	2.7
Copper 99.9% Cu	70	220	8.92
Cupronickel 10% Ni, 1.6% Fe, 1% Mn, balance Cu	130	350	8.94
Brass		250	
Tungsten		1,510	19.25
Glass (St Gobain "R")	4,400 (3,600 in. composite)		2.53
Bamboo	142	265	0.4
Marble	N/A	15	
Concrete	N/A	3	
Carbon fiber	N/A	5,650	1.75
Spider silk	1,150	1,200	
Silkworm silk	500		
Kevlar	3,620		1.44
Vectran		2,850–3,340	
Pinewood (parallel to grain)		40	
Bone (limb)		130	
Nylon, type 6/6	45	75	1.15
Rubber	–	15	
Boron	N/A	3,100	2.46
Silicon, monocrystalline (m-Si)	N/A	7,000	2.33
Sapphire (Al_2O_3)	N/A	1,900	3.9–4.1
Carbon nanotube (see note below)	N/A	62,000	1.34

Note: Multiwall carbon nanotubes have the highest tensile strength of any material yet measured.

FIGURE 7.13 To permanently form a metal into some shape, you must work in the plastic range of the metal.

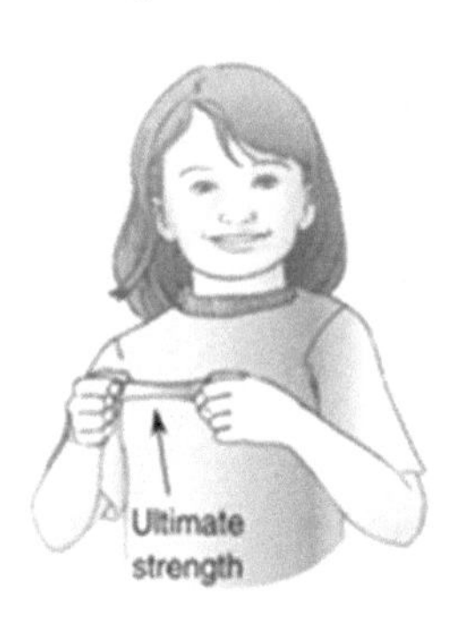

FIGURE 7.14 In the ultimate strength region of a material, stretching the material becomes increasingly easy.

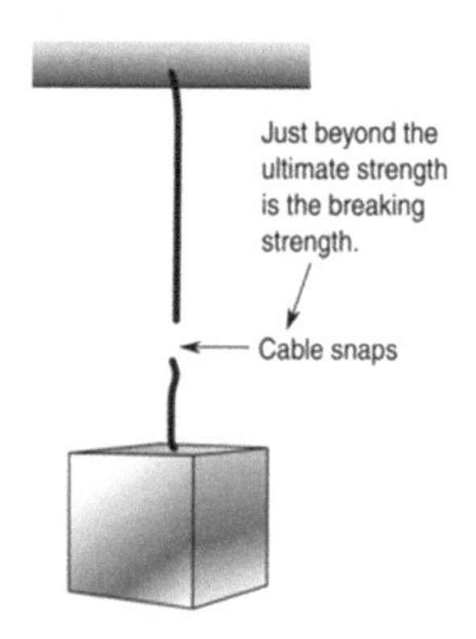

FIGURE 7.15 The breaking strength region occurs just before rupture.

plastic range, Figure 7.13. At the end of the plastic range, if the material is stretched any further, the molecules will begin to flow over one another. Stretching the material becomes increasingly easy (Figure 7.14). The material becomes taffy-like in this state. This is called the ultimate strength region.

Ultimate strength is the maximum tensile stress that a material can withstand.

Breaking strength is the point on the stress strain curve at the point of rupture, Figure 7.15.

Tensile strength is a measure of the maximum force pulling on a material before breaking it.

7.5.1 Space Elevator

A space elevator is a theoretical method of transportation for traveling from earth to space. It consists of a cable anchored to the earth at the equator, and stretching to space with a large mass attached to its end. The large mass would be placed at a distance from the earth to obtain a geosynchronous orbit so as to remain above the anchoring point on the earth. Centrifugal force will produce a tension in the cable to keep it taught, so an elevator car can ride up along it to space. For a geosynchronous orbit, the satellite will be at a distance $r = 42.1 \times 10^6$ m with respect to the center of the earth or 35,786 km above mean sea level. The problem with realizing this design has been finding a cable with a high enough tensile strength to endure the extreme tension in the cable needed to function. The cable must be able to support its own weight plus that of the elevator. Each length of cable must support the section below it. Typical designs have a tapper where the cable has a larger diameter in space than the anchor point on the earth. The *breaking length*, also known as the self-supporting length, is defined to be the maximum length of a material of a fixed diameter that can support its weight hanging vertically. Carbon nanotubes, having an ultimate strength of 62,000 MPa, seem to be the leading candidate for the cable, but can be manufactured in lengths of only several centimeters. Clearly, we will not be riding a space elevator to space anytime soon (Figure 7.16).

7.5.2 Young's Modulus of Elasticity

Modulus of elasticity is the ratio of stress over strain. The result of stress on a material is strain. If we are working within the elastic limit, the ratio of stress over strain is given by the modulus of elasticity.

Young's modulus: Young's modulus applies to tension and compression stresses.

$$Y = \frac{F/A}{\Delta L/L}$$

$$\text{Young's modulus} = \frac{\text{stress}}{\text{strain}}$$

The *modulus of elasticity* is the measure of how much a material changes its shape due to an applied force and returns to its original shape after the force is released. Table 7.2 lists the elastic modulus for various materials.

FIGURE 7.16 Space elevator.

TABLE 7.2
Elastic Constants for Various Materials in USCS Units

Material	Young's Modulus Y (lb/in.^{2})	Shear Modulus S (lb/in.^{2})	Bulk Modulus B (lb/in.^{2})	Elastic Limit, (lb/in.^{2})	Ultimate Strength, (lb/in.^{2})
Aluminum	10×10^6	3.44×10^6	10×10^6	19,000	21,000
Brass	13×10^6	5.12×10^6	8.5×10^6	55,000	66,000
Copper	17×10^6	6.14×10^6	17×10^6	23,000	49,000
Iron	13×10^6	10×10^6	14×10^6	24,000	47,000
Steel	30×10^6	12×10^6	23×10^6	36,000	71,000

Example 7.4

Expansion joints in a bridge allow concrete to thermally expand during the summer. Suppose that a 100 ft section of road expands by 1 in. (Figure 7.17). How much pressure will the road exert on an object stuck between the joint?

$$L = 100.0\,\text{ft}$$

$$\Delta L = 1.0\,\text{in.} = 0.083\,\text{ft}$$

$$Y_{\text{concrete}} = 10.6 \times 10^6 \frac{\text{lb}}{\text{in.}^2}$$

$$Y = \frac{F/A}{\Delta L/L}$$

$$F/A = Y \cdot \Delta L/L$$

$$F/A = 10.6 \times 10^6 \frac{\text{lb}}{\text{in.}^2} \cdot \frac{0.083\,\cancel{\text{ft}}}{100.0\,\cancel{\text{ft}}}$$

$$F/A = 8{,}798 \frac{\text{lb}}{\text{in.}^2}$$

A road under this kind of pressure would buckle if an expansion joint is not used.

FIGURE 7.17 How much pressure will the road exert on an object stuck between the joint?

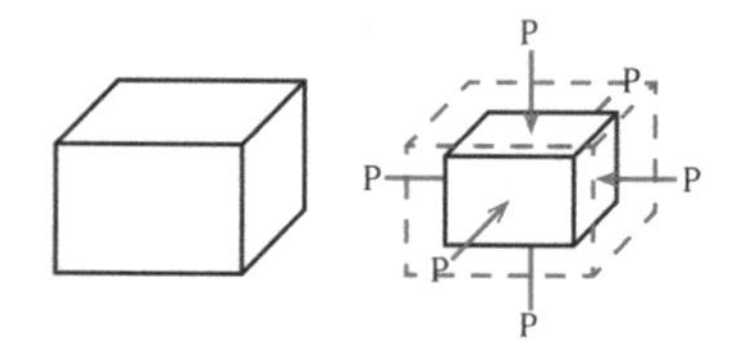

FIGURE 7.18 Bulk modulus.

7.6 BULK MODULUS

The *bulk modulus* refers to compression stresses applied to a volume. If a compressive stress is uniformly applied to the volume, the compressive force will cause the volume to shrink (Figure 7.18).

$$B = -\frac{F/A}{\Delta V/V}$$

$$\text{Bulk modulus} = -\frac{\text{volume stress}}{\text{volume strain}}$$

The minus sign in the equation is there to cancel out the minus sign from the volume change, which is negative, because it shrinks.

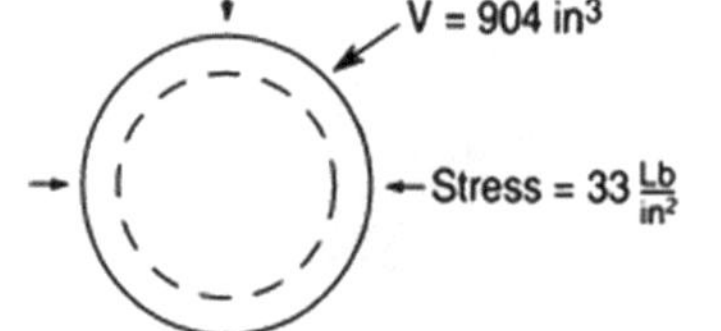

FIGURE 7.19 What is its change in volume under water?

Example 7.5

Suppose an inflatable ball of original volume 904 in.^{3} is placed under water with an outside pressure applied to it of 33 lb/in.^{2}. If the bulk modulus of the ball is 10^4 lb/in.^{2} (Figure 7.19), what is its change in volume under water?

$$V = 904\,\text{in.}^3$$

$$\frac{F}{A} = 33\,\text{lb/in.}^2$$

$$B = 10^4\,\text{lb/in.}^2$$

$$B = \frac{F/A}{\Delta V/V}$$

$$B = \frac{\frac{F}{A}\cdot V}{\Delta V} \leftarrow \text{Simplifying the equation}$$

$$\Delta V = -\frac{\frac{F}{A}\cdot V}{B} \leftarrow \text{Solving for } \Delta V \text{ gives}$$

$$\Delta V = -\frac{33\frac{\cancel{\text{lb}}}{\text{in.}^2}\cdot 904\,\text{in.}^3}{10^4\frac{\cancel{\text{lb}}}{\text{in.}^2}}$$

$$\Delta V = \frac{33\cdot 904}{10^4}\,\text{in.}^3$$

$$\Delta V = 3\,\text{in.}^3$$

7.7 SHEAR MODULUS

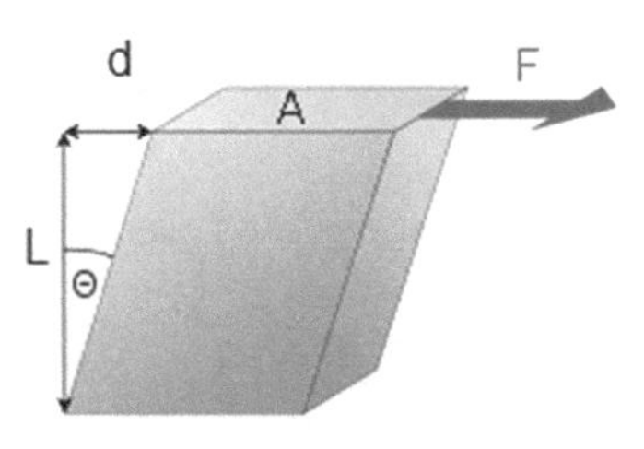

FIGURE 7.20 Shear modulus.

The *shear modulus* applies to a solid under a shearing force. The volume of the object does not change; only its shape changes (Figure 7.20). The shear modulus is given by:

$$S = \frac{F/A}{d/l}$$

$$\text{Shear modulus} = \frac{\text{shearing stress}}{\text{shearing strain}}$$

7.8 HARDNESS

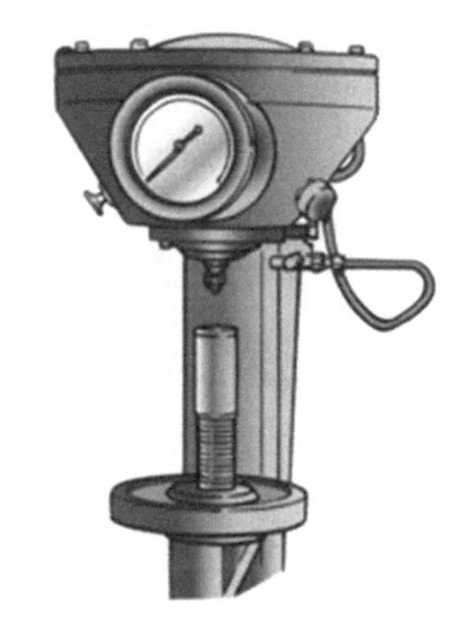

FIGURE 7.21 A material being tested for hardness.

Hardness of a material is determined by the interatomic forces between the atoms resisting a change in separation-distance with an applied force (Table 7.3). The hardness of a material can be determined by how easily it scratches. The Brinell method is a test for determining the hardness of a material. It is performed by applying 29,400 N of force on a material with a 10 mm hardened chrome-steel ball. The size of the imprint on the material being tested determines how hard it is (Figure 7.21).

Some extremely hard materials can be very strong under compression, but very weak with a bending or shear force or a tension. For example,

TABLE 7.3
Relative Hardness of Materials (Rated 1–10, with 10 being the hardest)

Aluminum	2–2.9
Brass	3–4
Copper	2.5–3
Diamond	10
Gold	2.5–3
Iron	4–5
Lead	1.5
Magnesium	2.0
Marble	3–4
Mica	2.8
Phosphorbronze	4
Platinum	4.3
Quartz	7
Silver	2.5–4
Steel	5–8.5
Tin	1.5–1.8
Zinc	2.5

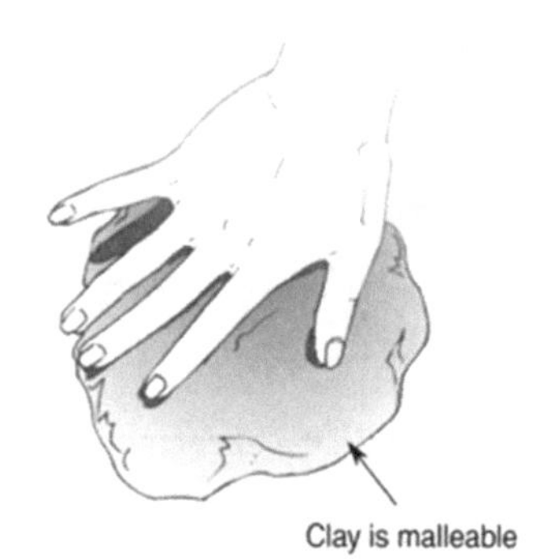

FIGURE 7.22 The malleability of a material is a measure of how easily it can change shape.

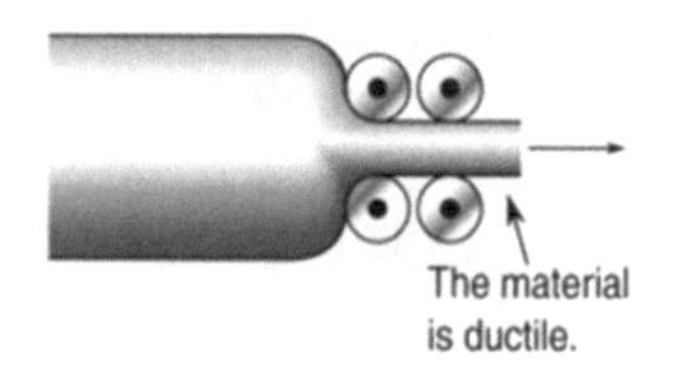

FIGURE 7.23 Materials that can easily be extruded through a die are ductile.

brick and tungsten carbide are very strong in the compression mode but not with tensile or shear forces.

7.9 MALLEABILITY

The *malleability* of a material is a measure of how easily it can change shape (Figure 7.22). For example, lead or gold can easily be sculpted because they are very malleable. Malleable materials have a large plastic range when under compression. When these materials are shaped, the atoms tend to slide or roll over one another.

7.10 DUCTILITY

Ductile materials are easily deformed when under tension, such as when metal is extruded through a die to make a wire (Figure 7.23). Ductile materials have a large plastic range under tension. Steel is both malleable and ductile when heated above 1,400°F.

7.11 CHAPTER SUMMARY

Symbols used in this chapter:

Symbol	Unit
t—time	s
x—length	m
Δx—change in length	m
d—length	m
L—length	m
k—spring constant	N/m or lb/in
ΔL—change in length	m
A—area	m^2
V—volume	m^3 or $in.^3$
ΔV—change in volume	m^3 or $in.^3$
Y—Young's modulus	N/m^2
B—bulk modulus	N/m^2
S—shear modulus	N/m^2
F—force	N

Hooke's law:

$$F = k \cdot \Delta x$$

$$\text{Force(N)} = \text{spring constant(N/m)} \times \text{length change(m)}$$

In the English system:

$$\text{Force(lb)} = \text{spring constant(lb/in.)} \times \text{length change(in.)}$$

$$\text{Stress} = \frac{F}{A}$$

$$\text{Stress}\left(\frac{\text{N}}{\text{m}^2}\right) = \frac{F(\text{N})}{A(\text{m}^2)}$$

In the English system of units:

$$\text{Stress}\left(\frac{\text{lb}}{\text{in.}^2}\right) = \frac{F(\text{lb})}{A(\text{in.}^2)}$$

$$\text{Strain} = \frac{\Delta L}{L}$$

$$\text{Strain} = \frac{\text{Length}_{\text{change}}(\text{m})}{\text{Length}_{\text{original}}(\text{m})}$$

In the English system of units:

$$\text{Strain} = \frac{\text{Length}_{\text{change}}(\text{in.})}{\text{Length}_{\text{original}}(\text{in.})}$$

Young's modulus applies to tension and compression stresses.

$$Y = \frac{F/A}{\Delta L/L}$$

$$\text{Young's modulus} = \frac{\text{stress}}{\text{strain}}$$

The *bulk modulus* refers to compression stresses applied to a volume.

$$B = -\frac{F/A}{\Delta V/V}$$

$$\text{Bulk modulus} = -\frac{\text{volume stress}}{\text{volume strain}}$$

The *shear modulus* applies to a solid under a shearing force.

$$S = \frac{F/A}{d/l}$$

$$\text{Shear modulus} = \frac{\text{shearing stress}}{\text{shearing strain}}$$

PROBLEM SOLVING TIP

- When solving problems dealing with a modulus and its units, be careful to use similar units in the other variables in the equation.

PROBLEMS

1. Suppose a man who weighs 165 lb gets in his car and the car lowers by 0.10 in. What is the spring constant of the coils in the suspension?
2. If the same car in problem 1 is loaded with 552 lb, how far down will the car lower?
3. A brick building stands 130 ft tall. Each brick weighs 10 lb and has dimensions of 8 in. × 4 in. × 2 in. What is the stress on the bottom layer of bricks? How tall could a building be before the elastic limit is reached?
4. Using Table 7.1, find out which material has the highest tensile strength.
5. What does the elastic limit mean?
6. What is the difference between ductility and malleability?
7. If a 560 lb weight is hanging from a steel cable with a cross-sectional area of 0.5 in.^{2} and a length of 25 ft, how much will it stretch?
8. If a spring has a spring constant $k = 34$ N/m, how far will it stretch under 2.0 N of force?
9. What material would you choose to build something under a high compressive stress and keep it relatively light?
10. What is the difference between stress and strain?
11. What is the stress on your body from atmospheric pressure?
12. What is the change in the volume of a steel submarine at the bottom of the sea if the pressure at the bottom of the sea is 1,400 psi, and the volume of the sub at the surface is 12×10^4 ft^3?
13. If a steel bolt of 0.50 in. diameter is poking out of a machine and has a 500.0 lb load on it, how much does it bend?

14. How could you use a spring to measure weight? Discuss how you would calibrate it.
15. How much will a 0.25 in. diameter steel cable stretch if it is holding up 250 lb?
16. A computer disk is inserted into a computer. The disk presses against a spring with a spring constant of 0.5 lb/in. If the spring compresses 0.25 in., what force was applied on the disk?
17. A 0.5 in. radius steel cable is pulled with 550 lb of force. What is the stress on the cable?
18. If the cable in the previous problem was 100 ft before the stress was applied, what is the change in length?
19. A weight scale consists of a pan hanging from the end of a spring. The spring has a spring constant of 25 lb/in. If the spring stretches by 0.5 in. after a weight is added to the pan, how much weight was added?
20. The bulk modulus of a certain balloon is 5,000 lb/in.^{2}. If a balloon of volume 240 in.^{3} has 2 lb/in.^{2} of air added to it, what is the change in volume?
21. A jack to raise a very heavy load by a very small distance can be made by supporting the load with a cylinder of metal and heating it until it expands by the desired length. Suppose a steel rod of length 8 in. and area 4 in.^{2} expands 0.0001 in. after heating, how heavy a load can it raise?
22. An iron rod of length 6.0 in. and cross-sectional area 0.5 in.^{2} protrudes out of a wall with a 50 lb weight on the end of it. How much does it bend?
23. An electronic weight scale consists of springs and a potentiometer that changes resistance when a load is added. If the spring constant of the scale is 0.1 lb/in., the resistive sensor changes its resistance as 10.0 ohms/in. and if a 115 lb woman steps on the scale, what is the change in resistance?
24. The elasticity of one's lungs can be measured by applying air pressure to the mouth and measuring a change in the volume of the lungs. If an applied pressure of 5.0 psi results in a 2% change in volume, what is the bulk modulus of the lungs?
25. A pogo stick has a spring with a spring constant of 12 lb/in. If the spring is compressed by 6.0 in., what force does it push its rider up with?
26. If a material has a large Young's modulus, is it easily stretched?
27. If a material has a small bulk modulus, is it easily compressed?
28. What are the strongest and weakest materials listed in Table 7.1?

29. Looking at Table 7.2, find out which material can withstand the highest shearing force.
30. A brick contained in an archway experiences what types of stresses? Choose from shear, compressive, and tensile.
31. True or false: A material "worked" in its plastic range will return to its original shape.
32. What does it mean for a material to be stressed to its ultimate strength?
33. Which material is most promising to build a "space elevator" with?
34. Which are the hardest and the softest materials listed in Table 7.3?
35. What is the unit of strain?

8 Fluids

This chapter describes the properties of fluids. Anything that flows can be considered a fluid. Even gases, such as air, can be treated as fluids. The principles developed in this chapter and the next will help you understand cooling electronics, working with hydraulics, aeronautics, buoyancy, pumps, compressors, and heating and ventilation systems (Figure 8.1).

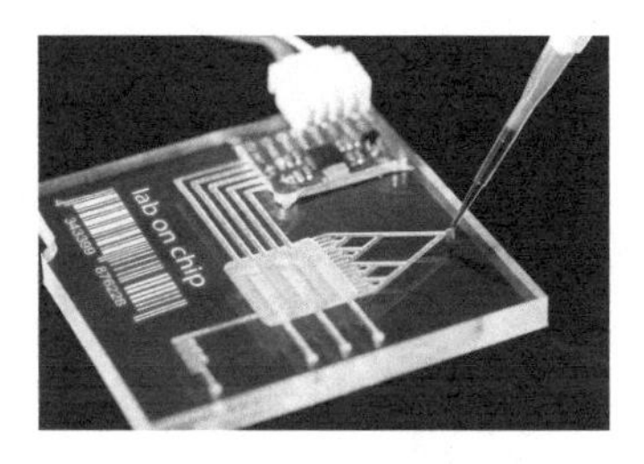

FIGURE 8.1 Microfluidics is a technology of precisely controlling fluids for applications such as the "lab on a chip", where medical diagnostic tests can be performed in the field.

8.1 PROPERTIES OF FLUIDS

A fluid is a material without a fixed shape. Place it in a container, however, and it will take that shape. If you squeeze some fluid out of an eyedropper, it takes a spherical shape. A droplet takes this shape because of *cohesion*, which is the attractive force between like molecules (Figure 8.2).

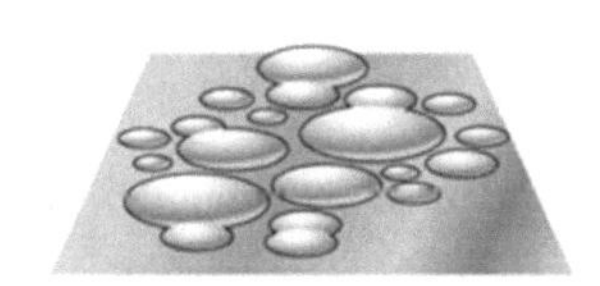

FIGURE 8.2 Cohesion pulls the fluid into a spherical shape.

Surface tension in a fluid is responsible for supporting small objects, like a water bug, on the surface of water (Figure 8.3).

FIGURE 8.3 Surface tension in a fluid supports small objects.

Surface tension results from cohesion. It reduces water's ability to flow into small places, such as fabrics. Surface tension goes down when temperature goes up. Detergents also reduce the surface tension of water. Washing your clothes in warm water with detergent reduces this tension, facilitating the water's ability to carry away the dirt.

Adhesion is the attractive force between unlike molecules. Glue and tape are considered as adhesives. Whether a fluid beads up on a surface or disperses depends on the relative strength of the cohesion between the molecules of the fluid compared with the adhesion between the fluid and the surface (Figures 8.4 and 8.5).

8.2 CAPILLARY ACTION

Capillary action is the action of fluids rising up through narrow tubes and is the reason why some paper towels are so absorbent. Capillary action arises from the adhesion between the fluid and the wall of the tube. Fluid

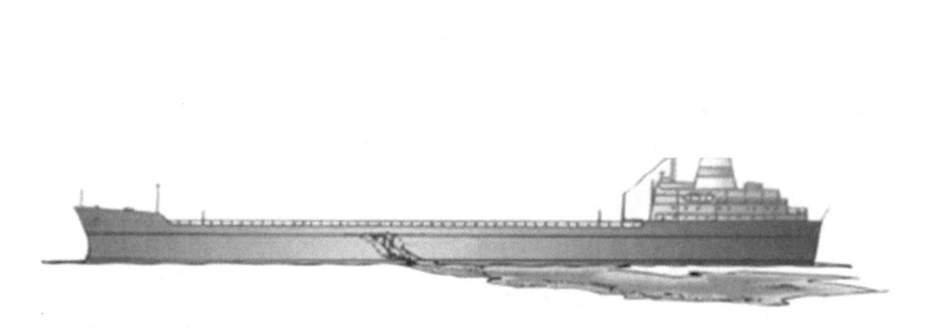

FIGURE 8.4 Oil does not mix with water because the cohesion of the oil molecules is greater than the adhesion between the oil and water as illustrated with this spill from an oil tanker.

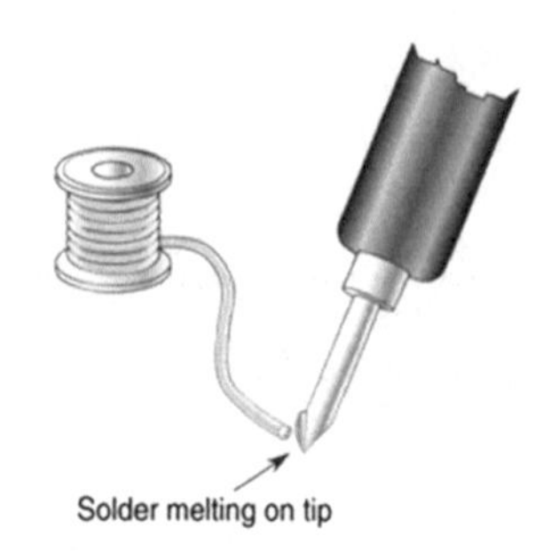

FIGURE 8.5 "Wetting" a soldering iron tip. A hot iron reduces the cohesion of the molten solder until it flows over the tip.

will rise upward against the force of gravity until a balance is struck between adhesion, cohesion, and gravity (Figure 8.6).

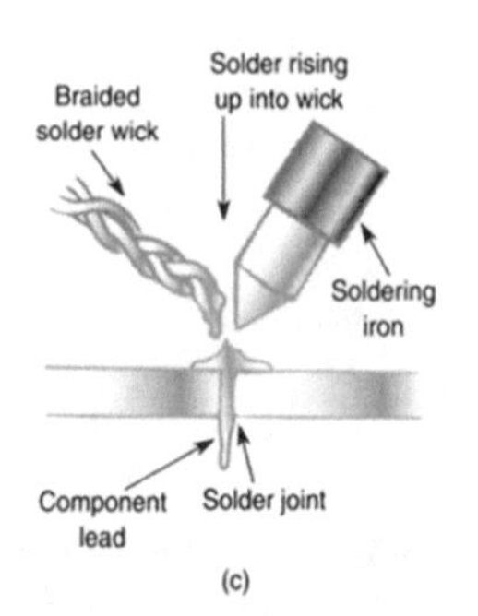

FIGURE 8.6 Capillary action.

8.3 VISCOSITY

The *viscosity* of a fluid is a measure of how easily it flows under its own weight (Figure 8.7).

The viscosity of a fluid can be reduced with increasing temperature. Motor oils have a viscosity rating as shown in the SAE (Society of Automotive Engineers) number. Oil with a low SAE rating of flows more easily than that with higher SAE rating. In cold weather, oils with low SAE numbers are used, otherwise the engine would not turn over. In warm weather, high-viscosity oils are used to prevent the oil from becoming too thin and leaking from the seals in the engine.

8.4 SPECIFIC GRAVITY

The *specific gravity* of a substance is its density divided by the density of water.

$$\text{Specific gravity} = \frac{\text{density of substance}}{\text{density of water}}$$

A substance will sink in water if its specific gravity is greater than that of water.
A substance will rise in water if its specific gravity is less than that of water.
A substance will remain still in water if its specific gravity is equal to that of water.

Specific gravity has no unit; it is just a number. The density of water is 1 g/cm³. The density of lead is 11.4 g/cm³. Therefore, the specific gravity of lead is 11.4. This is much higher than that of water, and therefore lead will sink in water.

FIGURE 8.7 Syrup is a very viscous fluid compared with water.

8.5 DETERMINING CONCENTRATIONS OF SOLUTIONS

How can you tell if you have enough antifreeze in your radiator? If the concentration is not high enough, the water may freeze! We can use the specific gravity of substances to determine it.

8.6 HYDROMETER

A hydrometer measures the specific gravity of a substance (Figure 8.8). The hydrometer contains a vial sealed with air and a weight, and the combination has the same density as water. As the solution is drawn into the hydrometer, the vial will float to some level where the buoyancy force balances out the weight of the vial. This level is dependent on the density of the fluid. The specific gravity is determined from a scale on the side of the hydrometer.

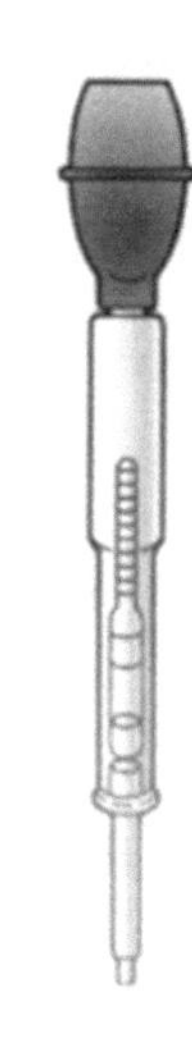

FIGURE 8.8 A hydrometer.

Antifreeze is made from ethylene glycol and can prevent the water in your car radiator from freezing, depending on the concentration in your radiator. The hydrometer can read the specific gravity of the radiator fluid and therefore determine the concentration of the antifreeze. The concentration determines the fluid's freezing temperature (see Table 8.1).

A car battery contains a solution of sulfuric acid and water. If the battery is fully charged, the solution is about 38% sulfuric acid. When the battery is dead, the solution is about 26% sulfuric acid. The specific gravity

TABLE 8.1
Freezing Points for Various Concentrations of Antifreeze

Specific Gravity	Percent of Ethylene Glycol, by Volume	Freezing Point (°C)
1.000	0	0
1.013	9.2	–3.6
1.026	18.3	–7.9
1.040	28.0	–14.0
1.053	37.8	22.3
1.067	47.8	–33.8
1.079	58.1	–49.3
1.109	100	17.4

changes from 1.28 fully charged to about 1.18 discharged. By measuring the specific gravity of the battery's solution, the condition of the battery can be determined.

FIGURE 8.9 Archimedes' principle describes why objects float in air and water.

8.7 BUOYANCY

What makes a helium balloon float in the air or a piece of wood float in the water (Figure 8.9)?

Consider a bottle floating in water. The part of the bottle under water is displacing some volume of water. Before the bottle displaced the water, that volume of water was supported by a force equal to its weight. Otherwise, it would have sunk. This fact leads to Archimedes' principle (Figure 8.10).

8.8 ARCHIMEDES' PRINCIPLE

The Archimedes' principle states that the upward force on a body is equal to the weight of the volume of fluid displaced by the object.

The upward force is called **buoyant force**.

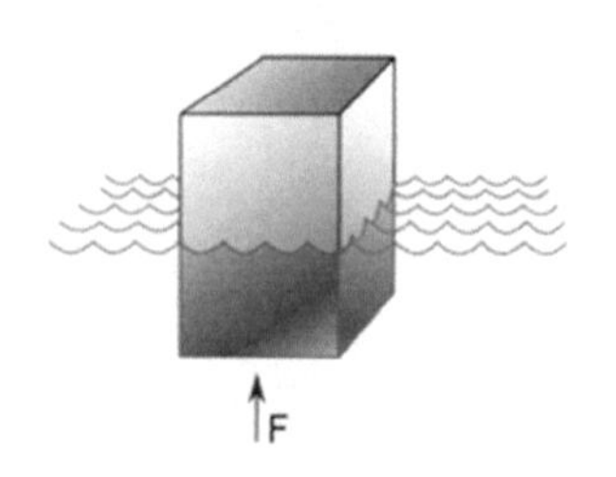

FIGURE 8.10 A buoyant force F on a partially submerged object.

$$F_b = V \cdot g \cdot \rho$$

$$\begin{matrix}\text{Buoyant} \\ \text{force(N)}\end{matrix} = \begin{matrix}\text{volume} \\ \text{displaced}(\text{m}^3)\end{matrix} \times \text{gravity}\left(9.8\ \text{m/s}^2\right) \times \text{density}\left(\text{kg/m}^3\right)$$

In the English system of units, density is often given in terms of lb/ft^3, in this case use:

$$F_b = V \cdot \rho_w$$

$$\text{Buoyant force(lb)} = \text{volume displaced}\left(\text{ft}^3\right) \times \text{weight density}\left(\text{lb/ft}^3\right)$$

If you lift someone in a swimming pool, they feel much lighter because of Archimedes' principle. (Figure 8.11)

FIGURE 8.11 A person feels lighter in water because of the buoyant force.

Example 8.1

Suppose a 180 lb person is partially submerged in a swimming pool. Suppose the volume of the submerged part equals $1.5\ \text{ft}^3$. What is the buoyant force, and how much does the person now weigh? (Figure 8.12)

Weight = 180 lb

$V = 1.5\ \text{ft}^3$

$\rho_{\text{water}} = 62.4\ \text{lb/ft}^3$

$$F_b = V \cdot \rho$$

$$F_b = \left(1.5\,\cancel{ft^3}\right)\left(62.4\frac{lb}{\cancel{ft^3}}\right) = 94\,lb$$

The person now weighs,

$$\text{New weight} = 180\,lb - 94\,lb = 86\,lb$$

No wonder we can pick up someone much bigger than us in a swimming pool!

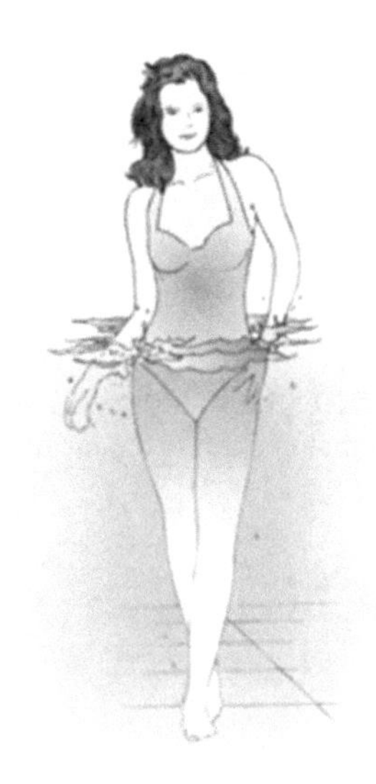

FIGURE 8.12 What is the buoyant force, and how much does the person now weigh?

8.9 PRESSURE

Pressure is a measure of how much force is applied on a given area. Consider placing a 5 lb weight on your arm. That might not feel like much, but what if you apply 5 lb on top of a needle onto your arm. Ouch! That same force, applied on a small area, creates a lot of pressure. Mathematically, pressure is determined by the following equation:

$$P = \frac{F}{A}$$

$$\text{Presure (Pa)} = \frac{\text{Force (N)}}{\text{Area}(\text{m}^2)}$$

The unit Pa is pronounced as pascal. 1 Pa = 1 N/m^2.

In the English system of units, pressure is often given in terms of psi or lb/in.^{2}.

$$\text{Presure (psi)} = \frac{\text{Force (lb)}}{\text{Area}(\text{in.}^2)}$$

English System	SI System
Pressure lb/in.^{2} (psi)	N/m^2 (pascals, Pa)
Conversions	
Meteorologists generally measure pressures in hectopascals (hPa) unit. Smartphones display pressure in hPa. Some other commonly used units are:	
1 atm = 14.7 lb / in.^{2}	
1 atm = 101.3 kPa	
1 atm = 1013.25 hPa	
1 atm = 1013.25 mbar	
1 hPa = 100 Pa	
1 atm = 30 in. of Hg	

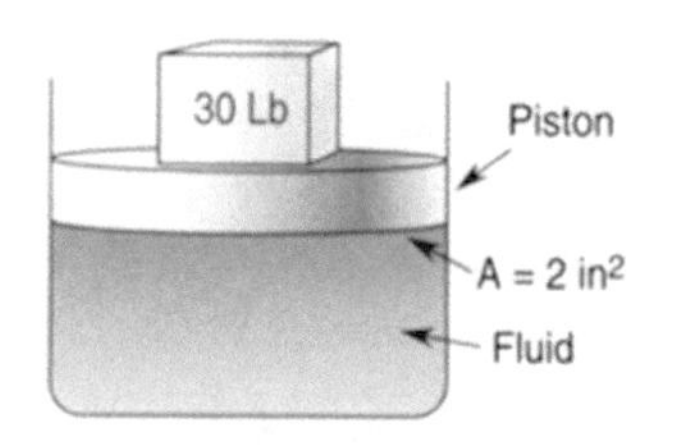

FIGURE 8.13 What pressure is the piston applying to the fluid?

Example 8.2

A 30.0 lb weight pushes down on a fluid through a piston of area 2.0 in.^{2}. What pressure is the piston applying on the fluid? (Figure 8.13)

$$A = 2.0\,\text{in.}^2$$

$$F = 30.0\,\text{lb}$$

$$P = ?$$

$$P = F/A$$

$$P = 30.0\,\text{lb}/2.0\,\text{in.}^2$$

$$P = 15\,\text{lb/in.}^2$$

FIGURE 8.14 You cannot localize a pressure on part of a fluid.

One of the differences between a fluid and a solid is that you cannot localize a pressure on part of a fluid (Figure 8.14). The liquid squishes out of the way. Pressure applied on a volume of fluid at the same depth has the same pressure throughout (Figure 8.15).

8.10 MEASURING PRESSURE

When you measure your tire pressure, what is it exactly that the pressure gauge is measuring?

Inside a tire there is a lot of air, which is confined to a small place. The air is composed of molecules moving around, with some striking the inside of the inner tube. The tube bulges out because it is being struck by the air molecules from the inside, forcing it outward (Figure 8.16). If you step on the tube, the rest of the tube bulges out more because the same amount of air is confined to a smaller place, resulting in more molecules pushing or applying force on a smaller area. A pressure gauge measures how much force is applied on a given area (Figure 8.17).

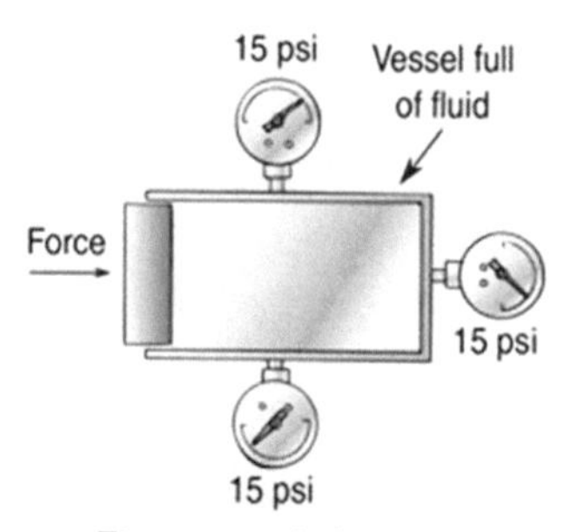

FIGURE 8.15 Pressure applied on a volume of fluid at the same depth has the same pressure throughout.

An electronic pressure gauge typically uses a *strain gauge*, a device that, when strained or made to change shape, will change its electrical properties (Figure 8.18). In a pressure gauge, the strain gauge is mounted in some type of chamber that can be attached to the liquid or gas volume being measured. As the liquid or gas molecules strike the gauge, they deliver a force to the resistive pattern the sensor is made of. This will cause it to deform, resulting in a change in resistance.

A *differential pressure gauge* measures the difference in pressure between a fixed pressure and the atmosphere. Most pressure gauges measure relative to the atmospheric pressure. These relative pressures are called *gauge pressures* and are designated as psig, which stands for pounds per square inch relative. psia means pounds per square inch absolute. This is a scale relative to a perfect vacuum. See Figure 8.19.

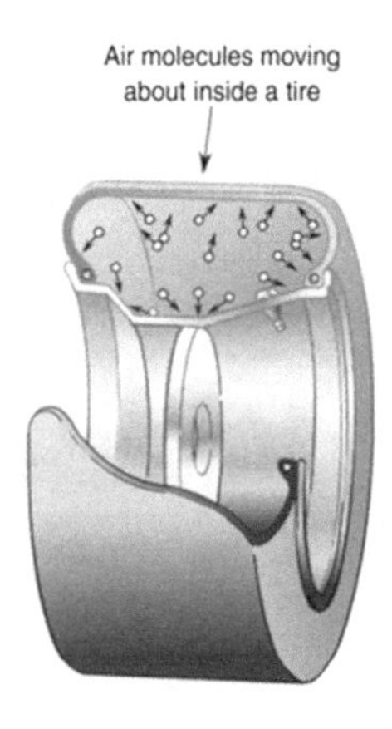

FIGURE 8.16 Air moving around inside a tire pushes the walls of the tire outward.

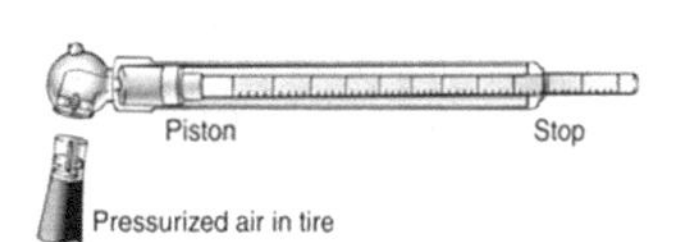

FIGURE 8.17 A typical tire gauge consists of a piston contained in a cylinder constrained by a spring. The piston is connected to a calibrated rod from which the pressure is read.

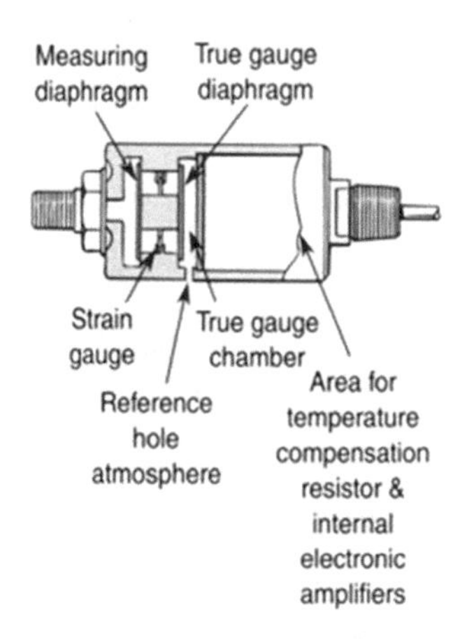

FIGURE 8.18 A strain gauge used to measure pressure.

If a tire has a pressure of 32 psi, this refers to psig or gauge pressure.

To convert to absolute pressure, you must add the atmospheric pressure to this number.

$$\text{psia} = \text{psig} + 14.7\,\text{lb/in.}^2$$

At the bottom of a swimming pool, the pressure is higher than near the top. This is because at the bottom there is a lot of water above you that is pushing down on you. The atmosphere itself exerts a pressure on us. There are miles of atmosphere above us, weighing us down (Figure 8.20). Air is very light, but enough of it can weigh a substantial amount.

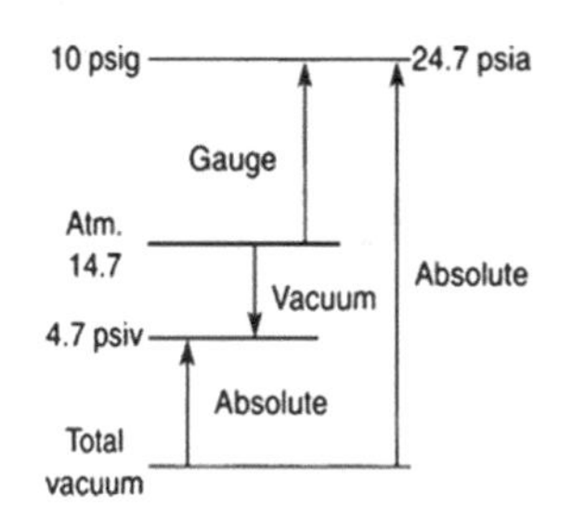

FIGURE 8.19 Chart comparing psia and psig.

8.11 BAROMETER

A *barometer* measures the pressure of the atmosphere. In a simple barometer, the height of the liquid depends on the weight of the liquid versus the force on the liquid from the atmosphere (Figure 8.21). An electronic barometer is shown in Figure 8.22.

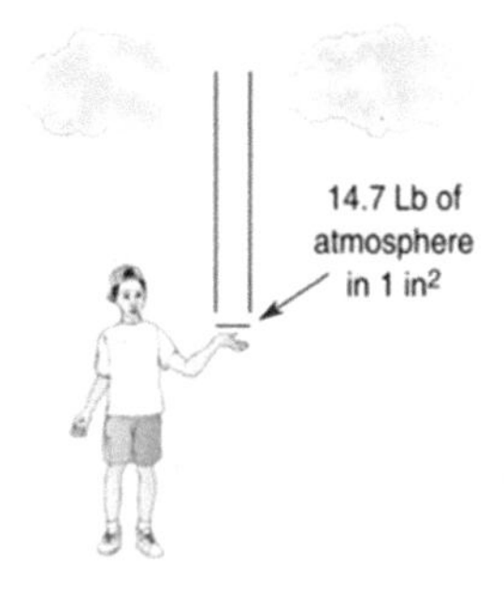

FIGURE 8.20 You are essentially supporting the atmosphere above you. If you hold out your hand, 1 in.^{2} on your palm is supporting 14.7 lb of air above you.

8.12 PRESSURE VERSUS DEPTH

Pressure in a fluid increases with depth. The pressure at the bottom of a swimming pool is higher than near the surface. The pressure does not depend on the shape of the container, but only on the depth and weight density and is determined by:

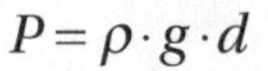

$$P = \rho \cdot g \cdot d$$

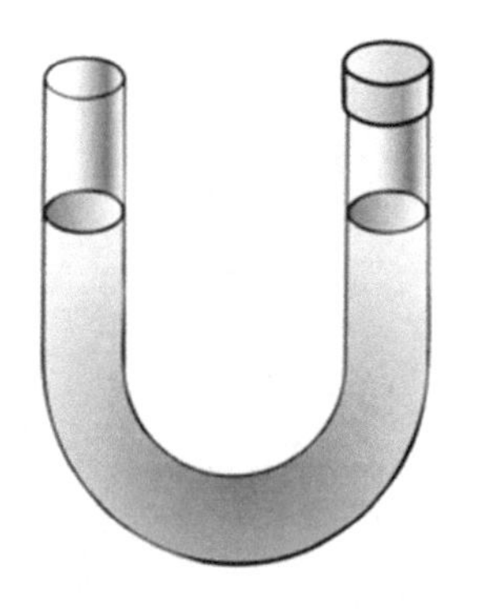

FIGURE 8.21 A simple barometer.

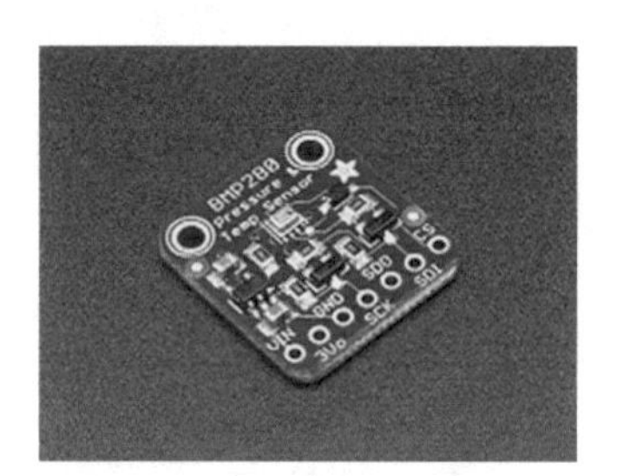

FIGURE 8.22 An electronic barometer, courtesy of adafruit.com.

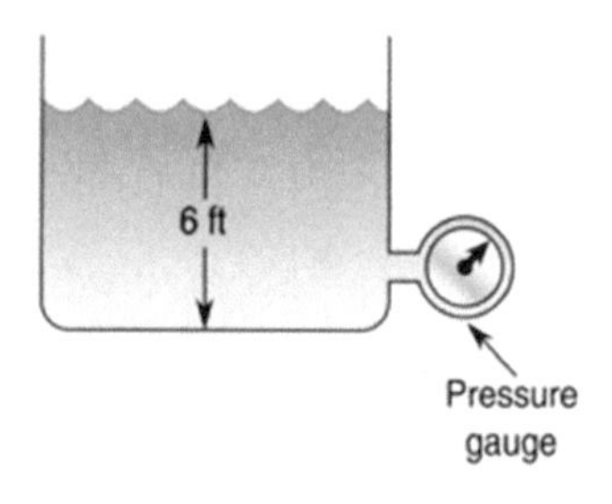

FIGURE 8.23 The pressure at the bottom of a pool is a combination of the pressure due to the water plus atmospheric pressure.

$$\text{Pressure(Pa)} = \text{density}\left(\text{kg/m}^3\right) \times \text{gravity}\left(9.8\,\text{m/s}^2\right) \times \text{depth(m)}$$

In the English system of units, density is often given in terms of lb/ft^3, in this case use:

$$P = \rho_w \cdot d$$

$$\text{Pressure}\left(\text{lb/ft}^2\right) = \text{weight density}\left(\text{lb/ft}^3\right) \times \text{depth(ft)}$$

This equation says that the pressure in a fluid increases with the depth below the surface.

Example 8.3

A swimming pool is 6 ft deep. What is the pressure at the bottom in psia and psig?

The pressure at the bottom of a pool is a combination of the pressure due to the water plus atmospheric pressure (Figure 8.23).

$$d = 6.00\,\text{ft}$$

$$\rho = 62.4\,\text{lb/ft}^3 \Leftarrow \text{Weight density of water}$$

$$P = \rho_w \cdot d$$

$$P = \left(6.00\,\cancel{\text{ft}}\right)\left(62.4\frac{\text{lb}}{\text{ft}^{\cancel{3}\,2}}\right)$$

$$P = 374\frac{\text{lb}}{\text{ft}^2}$$

Converting from $\frac{\text{lb}}{\text{ft}^2}$ to $\frac{\text{lb}}{\text{in.}^2}$ gives

$$374\frac{\text{lb}}{\text{ft}^2}\left(\frac{1\,\text{ft}}{12\,\text{in.}}\right)^2 = 374\frac{\text{lb}}{\cancel{\text{ft}^2}}\left(\frac{\cancel{\text{ft}^2}}{12^2\,\text{in.}^2}\right) = 2.60\frac{\text{lb}}{\text{in.}^2}$$

$$P = 2.6\frac{\text{lb}}{\text{in.}^2}\ \text{psig gauge pressure}$$

The pressure at the bottom of the pool is $2.6\frac{\text{lb}}{\text{in.}^2}$ greater than the surface.

$$\text{Absolute pressure} = 2.60\frac{\text{lb}}{\text{in.}^2} + 14.7\frac{\text{lb}}{\text{in.}^2}$$

$$\text{Absolute pressure} = 17.3\frac{\text{lb}}{\text{in.}^2}$$

Dams are built with a much wider base than a top because of this increasing water pressure with depth (see Figure 8.24). A device to measure depth (Figure 8.25).

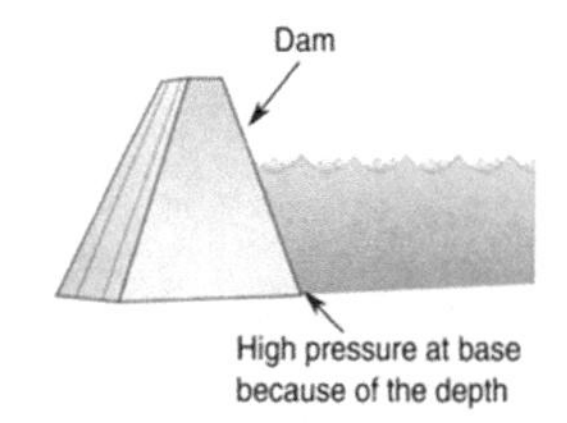

FIGURE 8.24 Pressure at the bottom of a dam is much greater than at the top. Therefore, the dam is built much wider near the bottom to support the extra pressure.

8.13 HYDRAULICS

Hydraulics is about transferring pressure from one region to another to obtain some mechanical advantage. A hydraulic jack can generate a force much greater than an applied force on the jack handle (Figure 8.26). Inside the jack there are two chambers with pistons. The jack handle is connected to a small piston in the small chamber that pressurizes the fluid. When the jack handle is pushed down with some force, a pressure in the fluid builds up and is transferred to the big piston, creating a much greater force upward than the applied force. One-way valves prevent the fluid from moving in the wrong direction.

In a hydraulic jack, the pressures inside the chambers are equal. The pressure is generated by the small piston and transferred through the fluid to the big piston. Because the pressures are equal:

$$P_{in} = P_{out}$$

$$\frac{F_{in}}{A_{in}} = \frac{F_{out}}{A_{out}}$$

Therefore,

$$F_{out} = \left(\frac{A_{out}}{A_{in}}\right) \cdot F_{in}$$

This equation says that the force going in can be increased if the area of the out-piston is bigger than that of the in-piston.

A hydraulic jack is a machine and therefore obeys the law of machines:

$$F_{in} \cdot d_{in} = F_{out} \cdot d_{out}$$

It therefore follows that

$$\frac{F_{out}}{F_{in}} = \frac{A_{out}}{A_{in}} = \frac{d_{in}}{d_{out}}$$

This equation says that the small piston will travel through a greater distance than the big piston if the output force is larger than the input force. In other words, you have to pump the jack handle up and down many times to lift a car just a little.

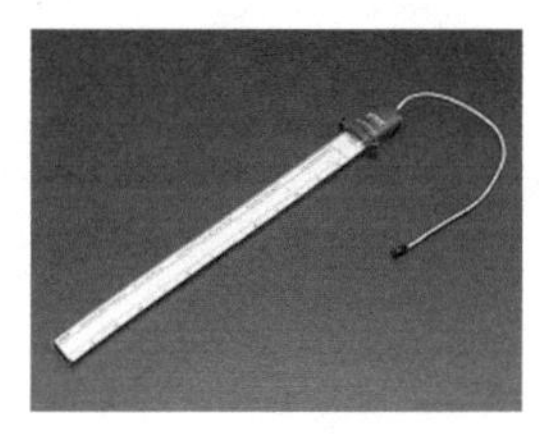

FIGURE 8.25 An eTape Liquid Level Sensor has a resistive sensor that varies with the level of the fluid, similar to a resistive strain gauge. The pressure of the fluid in which it is immersed causes a change in resistance related to the distance from the top of the sensor to the surface of the fluid, courtesy of adafruit.com.

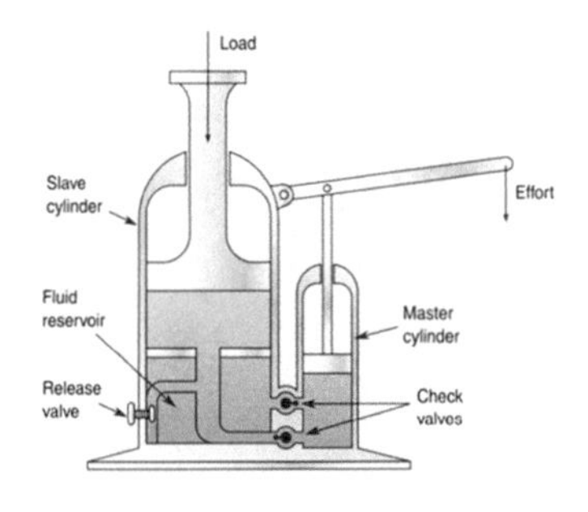

FIGURE 8.26 A hydraulic jack.

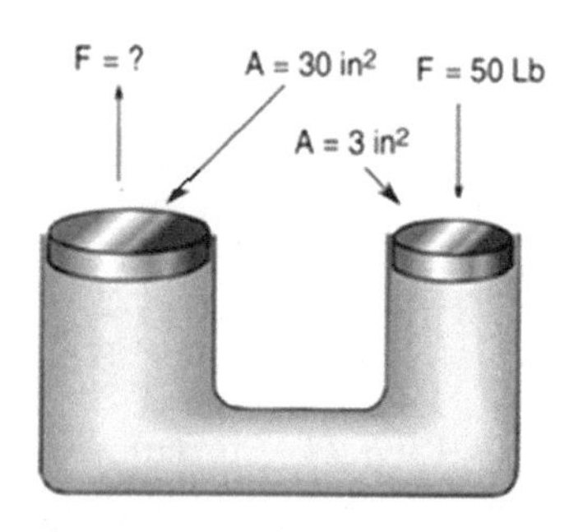

FIGURE 8.27 How much force is generated?

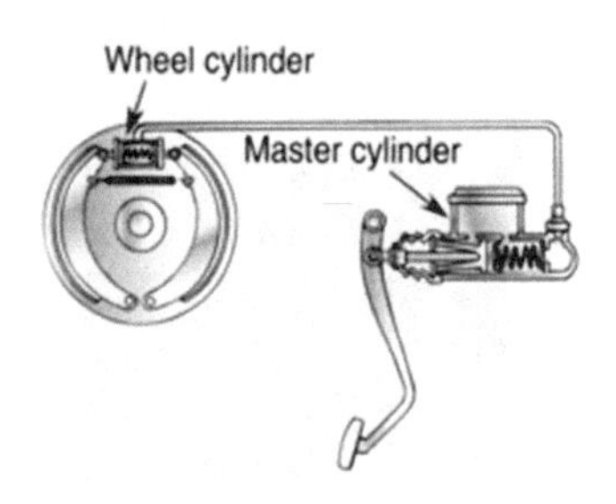

FIGURE 8.28 A hydraulic brake system.

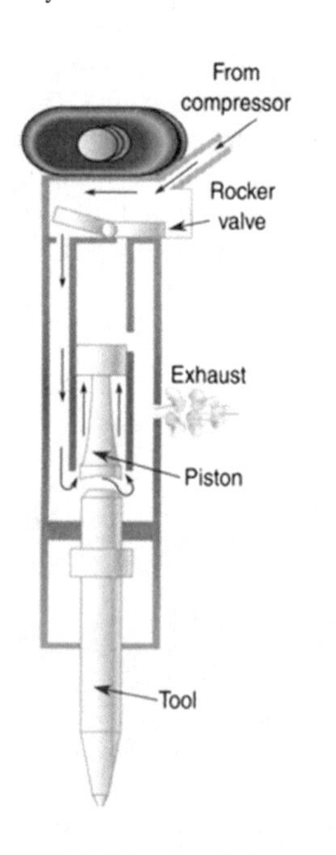

FIGURE 8.29 An air hammer.

Example 8.4

A hydraulic press contains a small and a large piston of areas 3 and 30 in.^{2}, respectively. The handle of the jack to the small piston delivers 50 lb of force. How much force is generated? (Figure 8.27)

$$A_{out} = 30\,\text{in.}^2$$

$$A_{in} = 3\,\text{in.}^2$$

$$F_{in} = 50\,\text{lb}$$

$$F_{out} = \left(A_{out}/A_{in}\right)F_{in}$$

$$F_{out} = \left(\frac{30\,\cancel{\text{in.}^2}}{3\,\cancel{\text{in.}^2}}\right)50\,\text{lb}$$

$$F_{out} = 500\,\text{lb}$$

8.14 BRAKING SYSTEMS

Automotive brakes work on a hydraulic system (Figure 8.28). They consist of a master cylinder, which is a reservoir of fluid, and a brake cylinder at each wheel. The master cylinder contains two pistons, one for the front and one for the back brakes. When the brake pedal is pressed, the pressure generated by the master cylinder is transmitted to each wheel cylinder, causing the brake shoes or the calipers to apply a mechanical pressure on the drums or calipers to slow the vehicle down.

Hydraulic systems are also used in heavy equipment. For this use, a pump generates the pressure in the master cylinder.

8.15 PNEUMATICS

Pneumatics is the fluid mechanics of air. Air is considered a low-viscosity fluid. One application of pneumatics is the air hammer, which consists of a heavy piston that oscillates up and down with air pressure controlled by a rocker valve (Figure 8.29).

8.16 CHAPTER SUMMARY

Symbols used in the chapter:

Symbol	Unit
t—time	s
d—depth	m
d—distance	m
m—mass	kg
ρ—density	kg/m^3
ρ_w—weight density	lb/ft^3
g—acceleration due to gravity	$9.8\,m/s^2$
F—force	N, lb
A—area	m^2 or in^2
P—pressure	Pa, $lb/in.^2$

Cohesion is the attractive force between like molecules.

Adhesion is the attractive force between unlike molecules.

Capillary action is the behavior of fluids as they rise up through narrow tubes and the reason why some paper towels are so absorbent.

Viscosity of a fluid is a measure of how easily it flows under its own weight.

The *specific gravity* of a substance is its density divided by the density of water.

$$\text{Specific gravity} = \frac{\text{density of substance}}{\text{density of water}}$$

Archimedes' Principle

The Archimedes' principle states that the upward force on a body is equal to the weight of the volume of fluid displaced by the object.

The upward force is called **buoyant force**.

$$F_b = V \cdot g \cdot \rho$$

$$\text{Buoyant force}(\text{N}) = \text{volume displaced }(\text{m}^3) \times \text{gravity}(9.8\text{ m/s}^2) \times \text{density}(\text{kg/m}^3)$$

In the English system of units, density is often given in terms of lb/ft^3, in this case use:

$$F_b = V \cdot \rho_w$$

$$\text{Buoyant force}(\text{lb}) = \text{volume displaced}(\text{ft}^3) \times \text{weight density}(\text{lb / ft}^3)$$

Pressure

Pressure is a measure of how much force is applied on a given area.

$$P = \frac{F}{A}$$

$$\text{Presure (Pa)} = \frac{\text{Force (N)}}{\text{Area}(\text{m}^2)}$$

The unit Pa is pronounced as pascal. 1 Pa = 1 N/m^2.

In the English system of units, pressure is often given in terms of psi or lb/in.^{2}.

$$\text{Presure (psi)} = \frac{\text{Force (lb)}}{\text{Area}(\text{in.}^2)}$$

To convert to absolute pressure, you must add the atmospheric pressure to this number.

$$\text{psia} = \text{psig} + 14.7\,\text{lb/in.}^2$$

Pressure in a fluid increases with depth and is determined by:

$$P = \rho \cdot g \cdot d$$

$$\text{Pressure}(\text{Pa}) = \text{density}\left(\text{kg/m}^3\right) \times \text{gravity}\left(9.8\,\text{m/s}^2\right) \times \text{depth}(\text{m})$$

In the English system of units, density is often given in terms of lb/ft^3, in this case use:

$$P = \rho_w \cdot d$$

$$\text{Pressure}\left(\text{lb/ft}^2\right) = \text{weight density}\left(\text{lb/ft}^3\right) \times \text{depth}(\text{ft})$$

For a hydraulic jack:

$$\frac{F_{\text{out}}}{F_{\text{in}}} = \frac{A_{\text{out}}}{A_{\text{in}}} = \frac{d_{\text{in}}}{d_{\text{out}}}$$

PROBLEM SOLVING TIPS

- Be careful to monitor your units, making sure you are in the same system of units throughout each problem.

PROBLEMS

1. What is the difference between cohesion and adhesion?
2. Explain capillary action.
3. Why are some motor oils multi grades?
4. If pressure is responsible for moving fluids, can you explain why a low barometric pressure reading indicates that a storm is approaching?
5. What is the difference between psig and psia?
6. What is the pressure relative to the atmosphere at the bottom of a swimming pool 6 ft deep?
7. If a ship displaces 25,000 ft^3 of water, what is the buoyant force?
8. Given an air bag and an air compressor, how would you raise a sunken ship?
9. Design a hydraulic jack that is able to lift 1,500 lb with 50 lb of force. What does the ratio of the pistons have to be?
10. What is the purpose of the oil in a hydraulic jack?
11. Why does a hot air balloon float?
12. If a rectangular piece of wood measuring 2.5″ deep × 3.5″ wide × 6.0″ long is thrown into water, how much of the wood will be submerged?
13. If the specific gravity of your radiator fluid is 1.053, what temperature will it freeze at?
14. How does a solder wick remove solder?
15. How does a detergent aid in cleaning your clothes?
16. Suppose a pressure gauge uses a strain gauge that changes resistance with pressure as 125 Ω/psi, what is the pressure change if the resistance of the gauge changes by 290 Ω?
17. If a flow sensor monitors the flow through a pipe and the viscosity of the fluid decreases, what happens to the flow rate?
18. A pressure sensor outputs 0–25 mV for a change in pressure of 0–60 psig. What is its output when it is measuring 30 psig?
19. A chamber contains a gas under a pressure of 43 psi. Inside the chamber is a port of area 4.5 $in.^2$. What is the force on the port?
20. An automobile tire is filled with a pressure of 32 psi. What is the absolute pressure inside the tire?

21. An air shock consists of a piston that fits into a cylinder containing air. If the piston has an area of 12.2 in.^{2} and a force of 200 lb is placed onto the piston, what is the pressure inside the cylinder?
22. A barometer contains a column of mercury 76 cm high. The density of mercury is 13.6 g/cm^3. What is the pressure at the bottom of the column?
23. A sensor measuring the pressure at the bottom of an oil liquid filled tank records a pressure of 122 psi. If the oil has a density of 54.2 lb/ft^3, what is the depth of the oil? Be careful of the units!
24. A man jumps into a swimming pool displacing 1.1 ft^3 of water. What is the buoyant force exerted on the man?
25. A spring scale is measuring the weight of a volume of 2.5 cm^3. It is then submerged into a liquid and its weight is re-measured. If the difference in weight in and out of the liquid is 5.0 N, what is the density of the liquid?
26. A hydraulic jack contains pistons of areas 2.0 and 4.0 in.^{2}. If 25 lb is applied on the little piston, what force does the jack generate?
27. If the output force is three times as big as the input force on a hydraulic jack, how many times farther does the input piston move relative to the output piston?
28. Why is the base of a dam wider than the top?
29. What is the specific gravity of a substance that has a density of 4.6 g/cm^3?
30. If the piston in the master cylinder in a car has an area of 2.1 in.^{2} and the piston in the wheel cylinder has an area of 4.2 in.^{2}, and 25 lb of force is applied on the master cylinder, what is the force generated at the wheel cylinder?
31. The metric unit of pressure is the pascal (Pa), 1 Pa = 1 N/m^2. If 10.0 N of force is acting upon an area of 2.0 m^2, what is the pressure in Pa?
32. If a barometer is measuring 30.1 in. of mercury, what is the atmospheric pressure in lb/in.^{2}?
33. What is the cause of surface tension in a fluid?
34. What is Archimedes' principle?
35. Does a tire pressure gauge measure psia or psig?
36. What is a hydrometer?
37. A plant uses capillary action to draw water up its stem. In terms of adhesion and cohesion how does it do this?

9 Fluid Flow

This chapter describes fluid flow and its applications. Understanding fluid flow is important in the electronics field. Air can be considered a fluid, and its flow around electrical equipment is vital for cooling. Knowledge of this subject should also help you understand plumbing, pneumatics, aerodynamics, and ventilation (Figure 9.1).

FIGURE 9.1 Atmospheric rivers are narrow channels of concentrated moisture transported in the atmosphere. The "Pineapple Express" is the name given to such an atmospheric river originating over the Hawaiian tropics that follow a path toward California. Its flow can exceed that of the world's largest river, the Amazon River.

9.1 TYPES OF FLUID FLOW

Basically, there are two types of fluid flows: laminar and turbulent flows (Figure 9.2). *Laminar flow* is smooth, streamlined flow. *Turbulent flow* is nonlaminar flow, characterized by whirlpools and eddies.

The reduction of turbulent flow is of major concern to automobile and aeronautical engineers (Figure 9.3). Fluid flow is impeded by friction. *Friction* in a flow is determined by the viscosity of the fluid and the type of flow, that is, whether it is streamlined or turbulent. Creating a smooth surface for the fluid to flow over can reduce friction (Figure 9.4). The drag on a moving vehicle is high when there is turbulent flow. Turbulent flow wastes energy, and minimizing it is almost always desired.

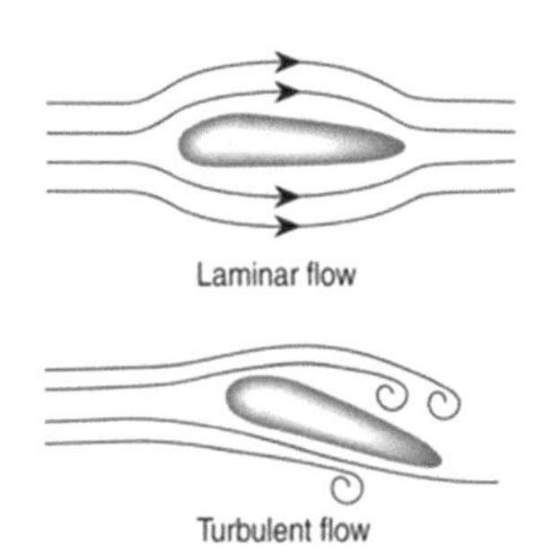

FIGURE 9.2 Laminar and turbulent flows.

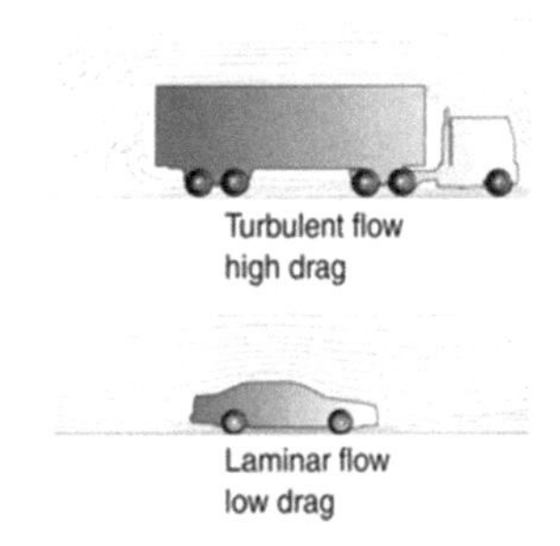

FIGURE 9.3 Turbulent flow can be minimized by changing the shape of a vehicle.

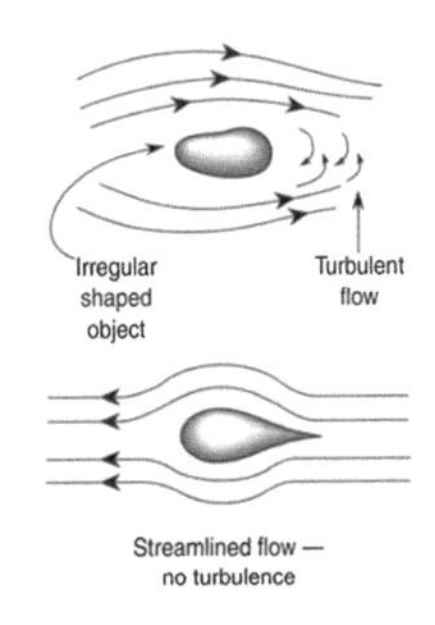

FIGURE 9.4 A teardrop has the lowest drag of any shape for speeds below the speed of sound.

Aerodynamic drag on an automobile is small at low velocities, but is responsible for most of the reduction in fuel economy at speeds over 40 mph. The drag force in the metric system is given by:

$$F_{Drag} = \frac{1}{2}\rho \cdot v^2 \cdot C_d \cdot A$$

$$Force_{Drag}(N) = \frac{1}{2}\, density\left(kg/m^3\right) \times \left(velocity(m/s)\right)^2 \times coefficient \times Area\left(m^2\right)$$

TABLE 9.1
Table of Some Aerodynamic Drag Coefficients

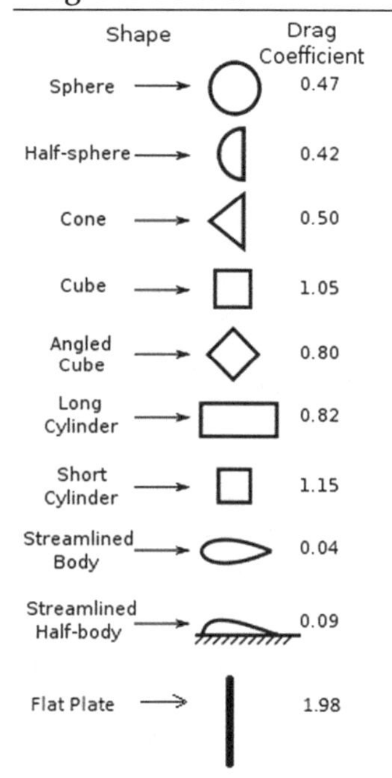

Shape	Drag Coefficient
Sphere	0.47
Half-sphere	0.42
Cone	0.50
Cube	1.05
Angled Cube	0.80
Long Cylinder	0.82
Short Cylinder	1.15
Streamlined Body	0.04
Streamlined Half-body	0.09
Flat Plate	1.98

The drag force in the English system is given by:

$$F_{Drag} = \frac{\rho \cdot v^2 \cdot C_d \cdot A}{29.91}$$

$$Force_{Drag}(lb) = \frac{density\left(lb/ft^3\right) \times \left(velocity(mph)\right)^2 \times coefficient \times Area\left(ft^2\right)}{29.91}$$

The drag coefficient is a measure of how streamlined an object is, and has no unit. The drag force is a function of the density of the fluid. Air is considered a fluid of low viscosity. The density of air varies with temperature, humidity, and pressure. At 20°C and 101.3 kPa, dry air has a density of 1.2041 kg/m³. At 70°F and 14.696 psi, dry air has a density of 0.075 lb/ft³. The drag coefficients of cars include the total frontal area of a car including the area facing forward including mirrors, luggage rack, etc. (Table 9.1).

Example 9.1

Suppose you put your hand, with an area of 0.15 ft², out of a car window moving at 65 mph. What is the force on your hand? (Figure 9.5)

$$v = 65\,\text{mph}$$
$$A_f = 0.15\,\text{ft}^2$$

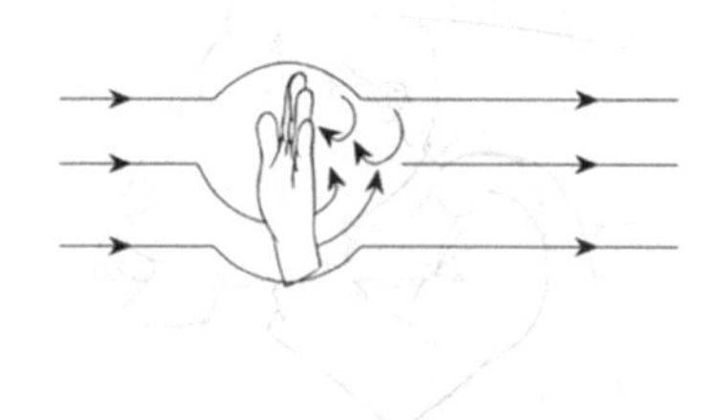

FIGURE 9.5 What is the force on your hand?

Let us approximate your hand as a square plate having a $C_d = 1.98$, and the air having a density of 0.075 lb/ft³.

$$F_{Drag} = \frac{\rho \cdot v^2 \cdot C_d \cdot A}{29.91}$$

$$F_{Drag} = \frac{\left(0.075\,\text{lb}/\text{ft}^3\right)\left(65\,\text{mph}\right)^2\left(1.98\right)\left(0.15\,\text{ft}^2\right)}{29.91}$$

$$F_{Drag} = 3.1\ \text{lb}$$

Turbulent flow can be used to your advantage when keeping a car firmly on the road at high speeds. A spoiler changes the laminar flow over the car to turbulent flow to keep the car from lifting off the road like an airplane (Figure 9.6).

FIGURE 9.6 A spoiler.

9.2 MEASURING THE VOLUME OF A FLUID IN FLOW

In order to charge you properly, the gas pump must measure how much gasoline you use. Likewise, the utility company must measure how much water you use. To measure the volume of a fluid, a volumetric meter is used (Figure 9.7). As the water enters the meter, it fills the chambers that have a known volume. As the chambers fill, they rotate, emptying their contents into the output. A recording mechanism records how many revolutions the chambers have made.

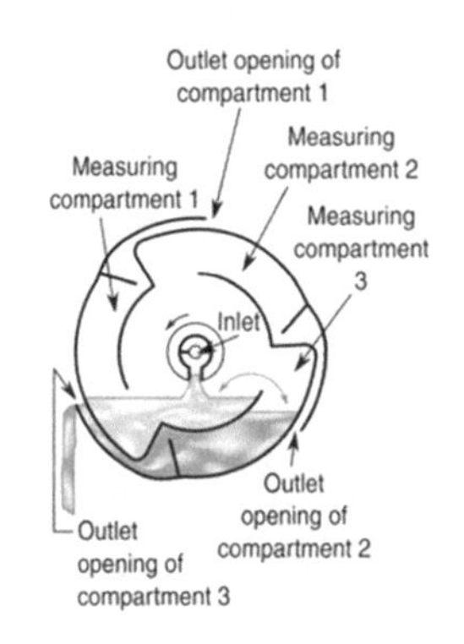

FIGURE 9.7 Volumetric flow meter.

Another type of volumetric flow meter commonly used in metering household water consumption uses a rotating impeller (Figure 9.8). As the fluid flows through the meter, the impeller rotates. The number of rotations is proportional to the amount of fluid passing through the meter. A counting mechanism is mounted on the rotating shaft, indicating the volume passed.

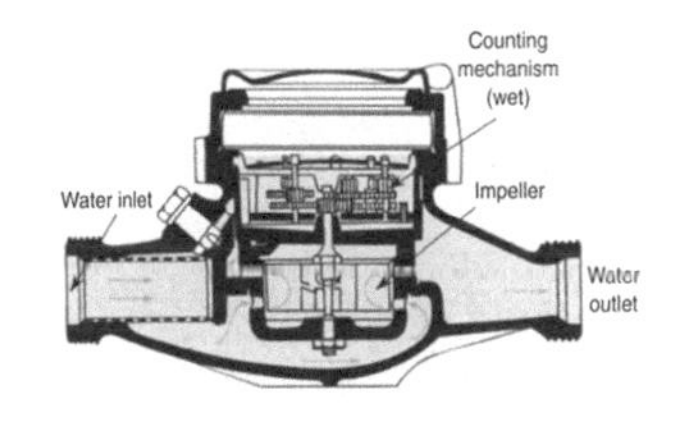

FIGURE 9.8 Impeller-type volumetric flow meter.

9.3 MEASURING FLOW RATE

Fans are used to keep electronics cool by sustaining a proper flow rate. Flow rates are measured in cubic feet per minute (cfm) or liters per minute (lpm) or some other volume per time unit. Flow rate can be measured using a tachometer attached to a propeller turning in the flow (Figure 9.9) or a thermal sensor (Figure 9.10).

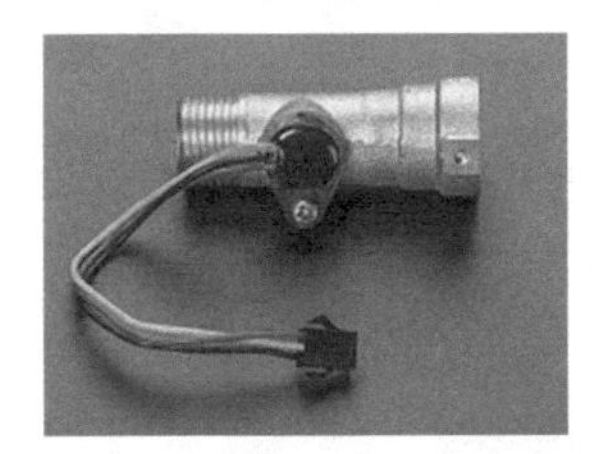

FIGURE 9.9 A liquid flow sensor measures fluid flow by measuring the rate of a pinwheel, spinning in line with the flowing fluid. Each rotation of the wheel corresponds to a certain volume of fluid flowing through the line. This volume multiplied by the rate of spin of the wheel gives the flow rate. Courtesy of adafruit.com.

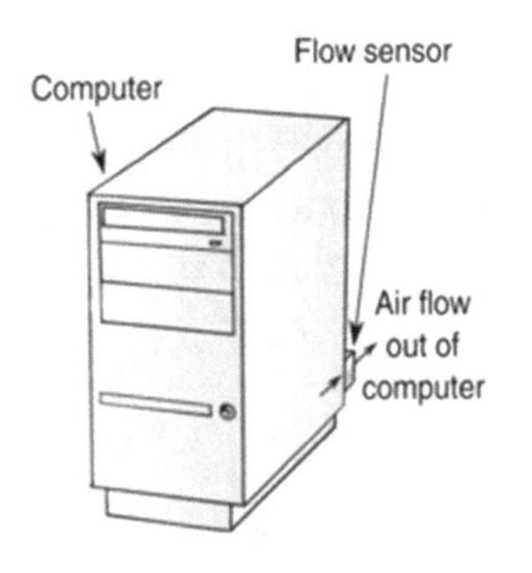

FIGURE 9.10 A thermal flow sensor measuring the air flow through a computer.

A thermal flow sensor contains a heater and an electronic thermometer placed next to one another. As the fluid flows over the heater, the heater cools. The greater the flow, the cooler the heater will be. The temperature is therefore a measure of the flow rate.

9.4 CALCULATING FLOW RATE

The rate at which a volume fluid flows through a pipe is given in terms of

$\frac{\text{gal}}{\text{min}}, \frac{\text{cm}^3}{\text{min}}$, or any other $\frac{\text{volume}}{\text{time}}$ unit

The rate of flow is given by:

$$Q = A \cdot v$$

$$\text{Rate of flow}\left(\text{cm}^3/\text{s}\right) = \text{Area of pipe}\left(\text{cm}^2\right) \cdot \text{velocity of fluid}\left(\text{cm/s}\right)$$

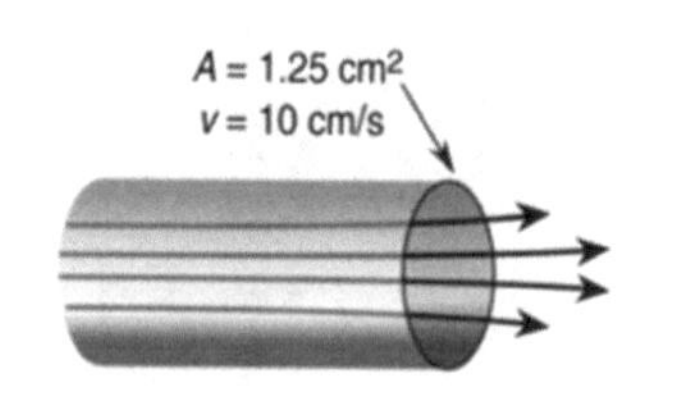

FIGURE 9.11 What is the flow rate?

This equation says that a large pipe with fluid flowing quickly through it will have a high flow rate.

Example 9.2

A pipe with a cross-sectional area of 1.25 cm² has water flowing through it at 10.0 cm/s. What is the flow rate? (Figure 9.11)

$$Q = A \cdot v = \left(1.25\,\text{cm}^2\right)\left(10.0\,\text{cm/s}\right)$$

$$Q = 12.5\,\frac{\text{cm}^3}{\text{s}}$$

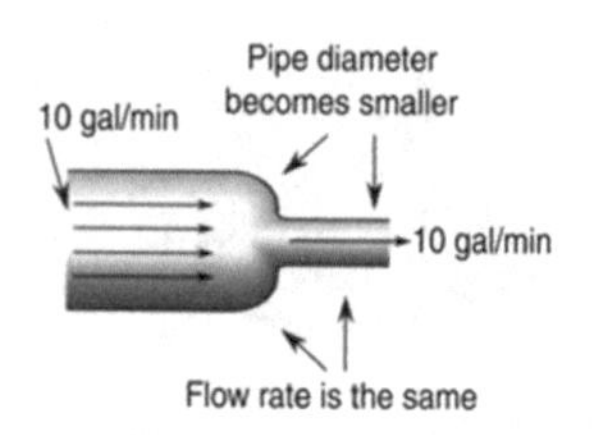

FIGURE 9.12 The rate at which an amount of fluid flows through a pipe does not change even if the pipe size changes.

Suppose a fluid flows through a pipe and the diameter of the pipe decreases. If the fluid is incompressible (not compressible under pressure) then the rate of fluid that flows into the big pipe must equal the rate of flow out of the small pipe (Figure 9.12).

As a fluid flows through an area, if the area gets smaller, the fluid speeds up. If the area becomes larger, the fluid moves more slowly. This is similar to a deep, slow-moving river turning into shallow, quick-moving rapids. This is summarized in what is called the *continuity equation:*

$$A_i \cdot v_1 = A_2 \cdot v_2$$

$$\text{Area 1} \times \text{velocity 1} = \text{Area 2} \times \text{velocity 2}$$

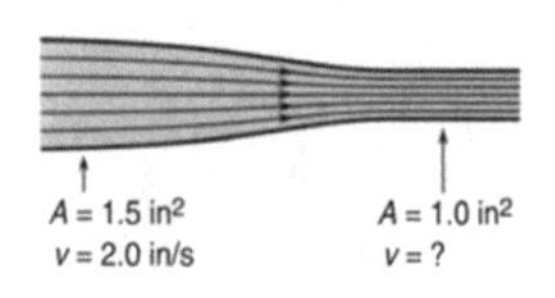

FIGURE 9.13 What is the velocity of the water in the 1.0 in.² section?

Example 9.3

Suppose water is flowing through a pipe with a cross-sectional area of 1.5 in.² and a velocity of 2.0 in./s. If the pipe area decreases to 1.0 in.², what is the new velocity of water? (Figure 9.13)

$$A_1 = 1.5\,\text{in.}^2$$
$$v_1 = 2.0\,\text{in./s}$$
$$A_2 = 1.0\,\text{in.}^2$$
$$v_2 = ?$$
$$A_1 \cdot v_1 = A_2 \cdot v_2$$
$$v_2 = \frac{A_1 \cdot v_1}{A_2}$$
$$v_2 = \frac{(1.5\,\cancel{\text{in.}^2})(2.0\,\text{in./s})}{1.0\,\cancel{\text{in.}^2}}$$
$$v_2 = 3.0\,\text{in./s}$$

9.5 FLUID PUMPS

Fluid flows when there is a pressure difference, and pumps are used to generate this pressure. Pumps are rated in terms of their flow rate, volume/time, and the pressure they generate.

9.5.1 Piston Pumps

On the first stroke, the piston moves inward, creating a partial vacuum, which allows the atmospheric pressure to force the fluid into the inlet of the pump. On the second stroke, the piston pushes the fluid out (Figure 9.14). Inlet and outlet valves ensure that the fluid moves only in the desired direction (Figure 9.15).

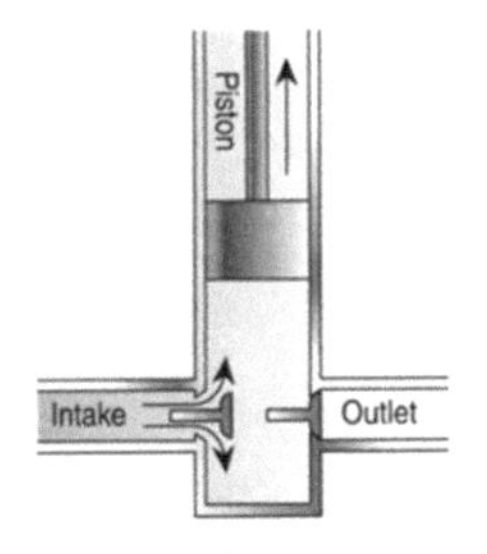

FIGURE 9.14 A piston pump alternately draws fluid in, then out, with each stroke.

9.5.2 Centrifugal Pump

In a centrifugal pump, a rotating impeller deflects the incoming fluid along the axis of the impeller to the outlet port (Figure 9.16). A squirrel cage blower is a type of centrifugal pump (Figure 9.17). The flow outward is continuous, unlike with the piston pump.

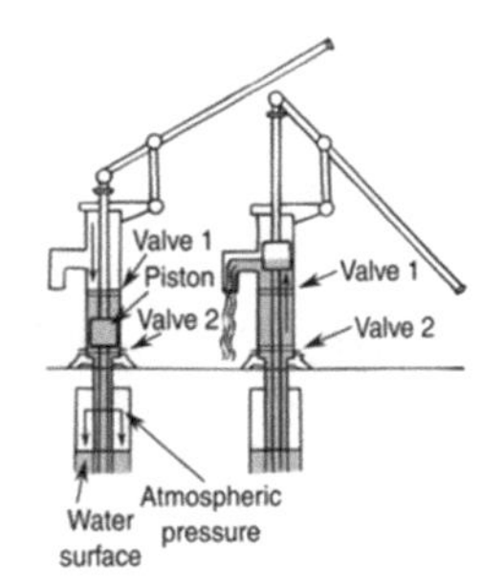

FIGURE 9.15 A well pump is a type of piston pump.

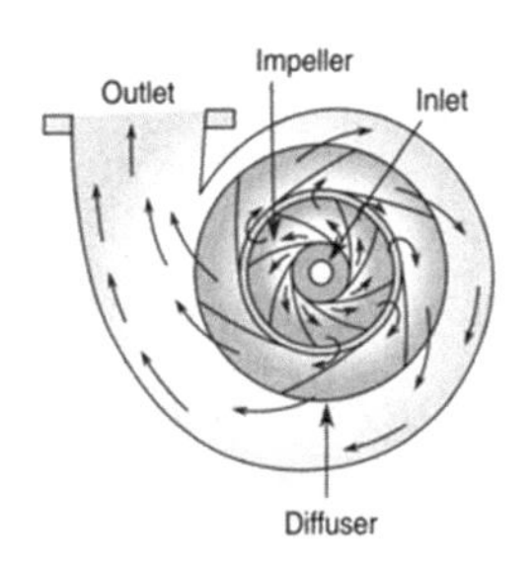

FIGURE 9.16 A centrifugal pump.

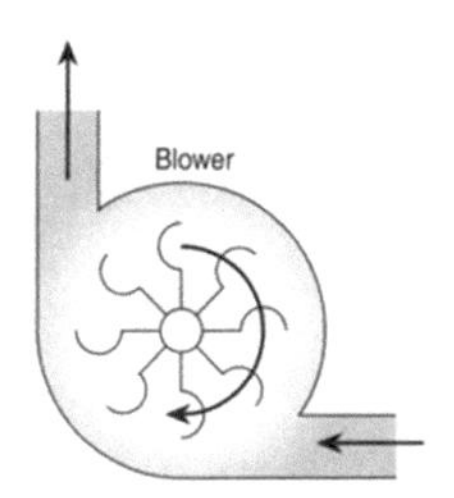

FIGURE 9.17 A squirrel cage blower for air is an example of a centrifugal pump.

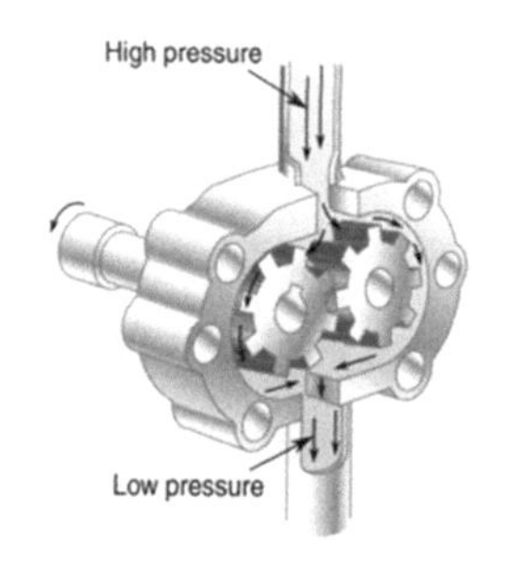

FIGURE 9.18 A gear pump.

9.5.3 Gear Pump

A gear pump drives the fluid around the outside of the gears toward the outlet in a continuous manner (Figure 9.18).

9.6 PUMP POWER

The power of a pump needed to deliver a fluid at a given flow rate Q and pressure P can be calculated by,

$$\text{Power} = P \cdot Q$$

$$\text{Power(W)} = \text{Pressure(Pa)} \times \text{Flow rate}(\text{m}^3/\text{s})$$

The power of a pump in the English system is given by:

$$\text{Power} = \frac{P \cdot Q}{1{,}714}$$

$$\text{Power(hp)} = \frac{\text{Pressure(psi)} \times \text{Flow rate(gal/min)}}{1{,}714}$$

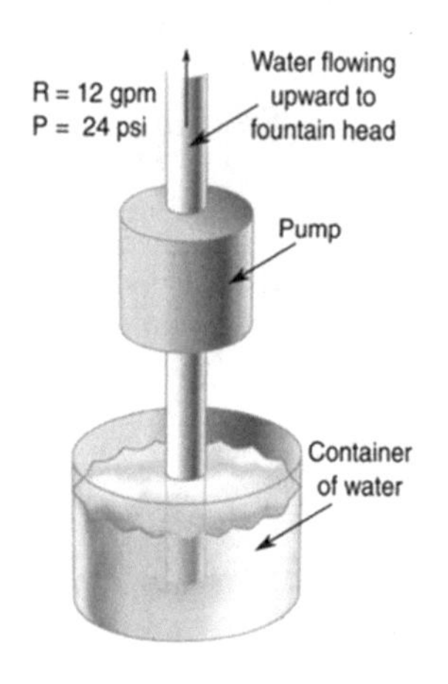

FIGURE 9.19 What horsepower does the pump have?

Example 9.4

A fountain pump is needed to pump 12 gal/min or 12 gpm of water at a pressure of 24 psi. What horsepower does the pump need to have? (Figure 9.19)

$$Q = 12\ \text{gpm}$$
$$P = 24\ \text{psi}$$
$$P = ?$$
$$\text{Power} = \frac{P \cdot Q}{1{,}714}$$
$$\text{Power} = \frac{24\ \text{psi} \cdot 12\ \text{gpm}}{1{,}714}$$
$$\text{Power} = 0.17\ \text{hp}$$

9.6.1 Pressure in a Moving Fluid and Bernoulli's Principle

Bernoulli's principle states that an increase in the velocity of a fluid is accompanied by a decrease in pressure or a decrease in the fluid's potential energy (Figure 9.20).

Bernoulli's principle is a consequence of the conservation of energy. In the absence of friction, the energy of the fluid is constant throughout the

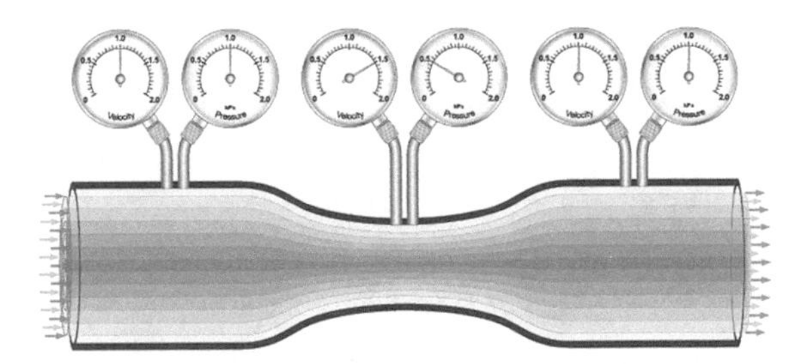

FIGURE 9.20 *Bernoulli's principle* states that an increase in the velocity of a fluid is accompanied by a decrease in pressure or a decrease in the fluid's potential energy.

pipe. The energy of a fluid flowing through a pipe is made up from three sources, the kinetic energy of the moving fluid, energy from pressure, and the gravitational potential energy of the fluid.

The total energy of the fluid is:

$$\text{Total Energy} = \text{Kinetic energy} + \text{Energy from pressure} + \text{Gravitational potential energy}$$

$$E = \frac{1}{2}mv^2 + P \cdot V + mgh$$

Taking this energy equation and dividing by the volume V and using $\rho = \frac{m}{V}$ gives:

$$\frac{E}{V} = \frac{1}{2}\rho v^2 + P + \rho gh$$

The energy/volume is also a constant throughout the pipe. This means the energy/volume in one part of a system is the same as the energy/volume in another part of the system (Figure 9.21):

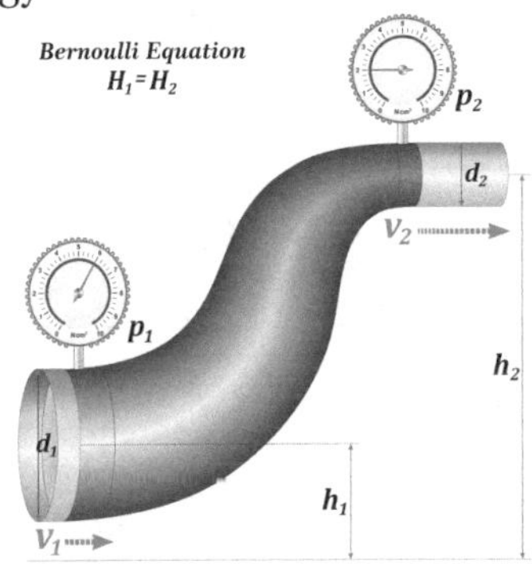

FIGURE 9.21 Bernoulli's principle with a changing level.

$$\frac{1}{2}\rho \cdot v_1^2 + P_1 + \rho \cdot g \cdot h_1 = \frac{1}{2}\rho \cdot v_2^2 + P_2 + \rho \cdot g \cdot h_2$$

$$\rho = \text{density (kg/m}^3), v = \text{velocity (m/s)}, P = \text{Pressure(Pa)},$$

$$h = \text{height(m)}, g = 9.8\,\text{m/s}^2$$

Example 9.5

Water flows through a pipe at a rate of 0.02 m³/s. The pipe has an area of 0.04 m² in section one. It then enters section two of the pipe of area 0.03 m² at a pressure of 101.299 kPa. Both sections are at the same level. What is the pressure in section one? (Figure 9.22)

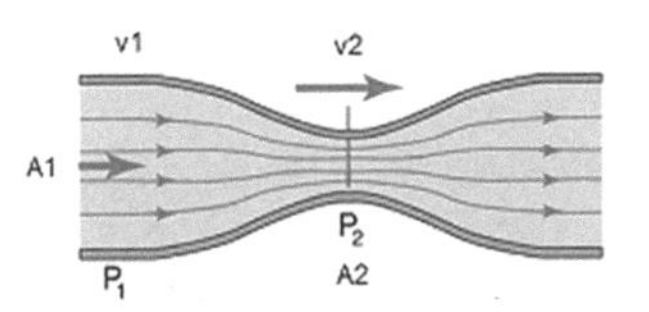

FIGURE 9.22 What is the pressure in section one?

The rate of flow Q of the water is the same throughout the pipe; the speed can change however. We solve for the velocities in each section using:

$$Q = A_1 \cdot v_1 = A_2 \cdot v_2$$

$$v_1 = \frac{Q}{A_1} = \frac{0.02\text{m}^3/\text{s}}{0.04\text{m}^2}$$

$$v_1 = 0.5\text{m / s}$$

$$v_2 = \frac{Q}{A_2} = \frac{0.02\text{m}^3/\text{s}}{0.03\text{m}^2}$$

$$v_2 = 0.67\text{m/s}$$

Both sections are at the same height so the term $\rho \cdot g \cdot h$ cancels out on both sides of the equation:

$$\frac{1}{2}\rho \cdot v_1^2 + P_1 + \rho \cdot g \cdot h_1 = \frac{1}{2}\rho \cdot v_2^2 + P_2 + \rho \cdot g \cdot h_2$$

Solving for P_1 gives:

$$P_1 = \frac{1}{2}\rho \cdot (v_2^2 - v_1^2) + P_2$$

$$P_1 = \frac{1}{2} \cdot 1{,}000\frac{\text{kg}}{\text{m}^3}\left(\left(0.67\frac{\text{m}}{\text{s}}\right)^2 - \left(0.5\frac{\text{m}}{\text{s}}\right)^2\right) + 101.299 \times 10^3\,\text{Pa}$$

$$P_1 = 101.398 \times 10^3\,\text{Pa}$$

So why would the pressure be lowest where the fluid velocity is highest? As the fluid enters the smaller section of the pipe, it speeds up. Therefore, its kinetic energy goes up. If the total energy is constant, one of the other energy terms must decrease, that is, the pressure term. The fluid speeds up at an energy cost to the energy in the pressure term. You may have experienced this in the shower. The water flow reduces the pressure in the shower (Figure 9.23).

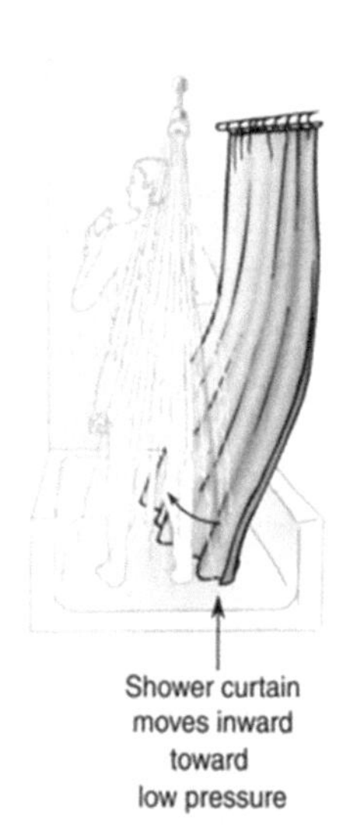

FIGURE 9.23 Low pressure created by a rapidly moving fluid.

The effect of low pressure created by a rapidly moving fluid is partially responsible for how an airplane gets its lift. The plane's wing is shaped so that the air moving over the top of the wing has to move faster to keep up with the air moving below it. As a result, the pressure is higher below the wing than above it, and the difference in air pressure pushes the plane upward (Figures 9.24 and 9.25).

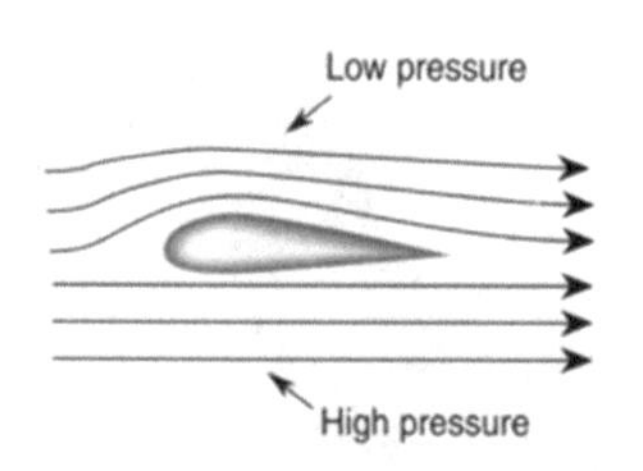

FIGURE 9.24 When the wing is horizontal, air moving over the top has to move faster to keep up with the air moving below, resulting in a pressure difference.

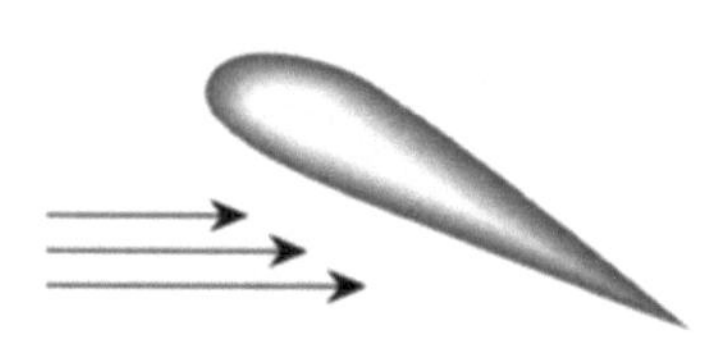

FIGURE 9.25 The plane also receives a lift when the wing is tilted slightly upward. In this case, air is deflected off the bottom of the wing, pushing it upward.

9.7 CARBURETOR

The engine in a lawn mower has a carburetor; it mixes air and fuel to burn in the combustion chamber. The carburetor contains a venturi, or a narrowing in the throat of the carburetor, where the gas and air mix

(Figure 9.26). In this region, air flows quickly, reducing the pressure here. As a result of this lower pressure, gas flows into the venturi from a higher-pressure bowl of gas. A needle valve controls the amount of gas in the bowl. The lower pressure is responsible not only for the gas flow into the venturi, but also for mixing the gas and air more thoroughly because the gas evaporates more easily at lower pressures.

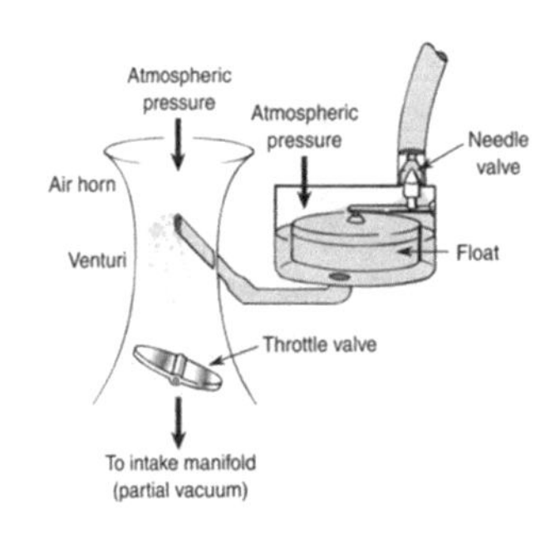

FIGURE 9.26 The carburetor.

9.8 POWER GENERATION

Kinetic energy captured from the wind and flowing water can be turned into electrical energy. The power that can be captured from a moving fluid obeys the following:

$$P=\frac{1}{2}\rho\cdot A\cdot v^3\cdot eff$$

$$\text{Power}(\text{W})=\frac{1}{2}\text{density}\left(\text{kg}/\text{m}^3\right)\times\text{Area}\left(\text{m}^2\right)\times\left(\text{velocity}(\text{m/s})\right)^3\times\text{efficiency}$$

The efficiency has no unit and depends on how well the machine can capture the energy of the moving fluid.

9.8.1 Wind Power

A windmill can use kinetic energy from wind for pumping water, milling corn, generating electricity, etc. How much energy does wind have? The energy contained in the wind depends on the density and moisture content of the wind. Wind turbines are based on wind conditions at sea level at a temperature of 15°C (59°F). Air has a density of approximately 1.225 kg/m^3 under this condition.

$$P=\frac{1}{2}\rho\cdot A\cdot v^3\cdot eff$$

$$P=\frac{1}{2}\left(1.225\,\text{kg}/\text{m}^3\right)\cdot A\cdot v^3\cdot eff$$

$$P=0.61\cdot A\cdot v^3\cdot eff$$

$$P_{\text{wind}}=0.61\cdot A\cdot v^3\cdot eff$$

$$\text{Power}_{\text{wind}}(\text{W})=0.61\times\text{Area}\left(\text{m}^2\right)\times\left(\text{velocity}(\text{m}/\text{s})\right)^3\times\text{efficiency}$$

FIGURE 9.27 A windmill.

The equivalent wind power in the English system is given by:

$$P_{\text{wind}} = 5.1 \times 10^{-6} \cdot A \cdot v^3 \cdot eff$$

$$\text{Power}_{\text{wind}}(\text{hp}) = 5.1 \times 10^{-6} \times \text{Area}(\text{ft}^2) \times (\text{velocity}(\text{mph}))^3 \times \text{efficiency}$$

Example 9.6

Suppose the blades of a windmill can effectively capture an area of wind 8.5 m² in size and the wind is blowing at 4.0 m/s, with an efficiency of 0.35. How much power can be captured? (Figure 9.27).

$$P = 0.61 \cdot A \cdot v^3 \cdot eff$$

$$P = 0.61(8.5\text{m}^2)(4.0\text{m/s})^3(0.35)$$

$$P = 116.1\text{W}$$

9.8.2 Hydroelectric Power

Hydroelectric power is power generated from moving water. How much energy is contained in the moving water of a horizontal stream? We can use the power formula above and use the density of water $\rho = 1{,}000\,\text{kg/m}^3$ to get:

$$P_{\text{water}} = 500 \cdot A \cdot v^3 \cdot eff$$

$$\text{Power}_{\text{water}}(\text{W}) = 500 \times \text{Area}(\text{m}^2) \times (\text{velocity}(\text{m/s}))^3 \times \text{efficiency}$$

FIGURE 9.28 Hydroelectric power.

The equivalent power in the English system is given by (Figure 9.28):

$$P_{\text{water}} = 5.56 \times 10^{-3} A \cdot v^3 \cdot eff$$

$$\text{Power}_{\text{water}}(\text{hp}) = 5.56 \times 10^{-3} \times \text{Area}(\text{ft}^2) \times (\text{velocity}(\text{mph}))^3 \times \text{efficiency}$$

Example 9.7

A hydroelectric power plant consists of a dam with water turbines spinning electrical generators. Suppose the water is flowing at 15 m/s and is spinning a turbine with an effective area of 12 m². Assume the efficiency is 1. How much power is being captured?

$$v = 15\text{m}/\text{s}$$

$$A = 12\text{m}^2$$

$$P = 500 \cdot v^3 \cdot A$$

$$P = 500 \cdot (15\text{m}/\text{s})^3 \cdot 12\text{m}^2$$

$$P = 2\bar{0}\,\text{MW}$$

9.9 CHAPTER SUMMARY

Symbols used in this chapter:

Symbol	Unit
t—time	s
h—height	m
g—gravity	$9.8\,m/s^2$
A—area	m^2 or ft^2
V—volume	m^3 or ft^3
v—velocity	m/s or mph
C_d—drag coefficient	no unit
eff—efficiency	no unit
Q—rate of flow	cm^3/s or m^3/s or gpm (gal/min)
ρ—density	kg/m^3 or lb/ft^3
F—force	newton (N)
E—energy	joule (J)
P—power	watts (W) or horsepower (hp)
P—pressure	pascal (Pa)

Types of fluid flow: There are two basic types of fluid flows, laminar and turbulent flows.

Laminar flow: Laminar flow means smooth, streamlined flow.

Turbulent flow: Turbulent flow is nonlaminar flow, characterized by whirlpools and eddies.

The drag force is given by:

$$F_{Drag} = \frac{1}{2}\rho \cdot v^2 \cdot C_d \cdot A$$

$$Force_{Drag}(N) = \frac{1}{2}\ density\left(kg/m^3\right) \times \left(velocity(m/s)\right)^2 \times coeffiecient \times Area\left(m^2\right)$$

The drag force in the English system is given by:

$$F_{Drag} = \frac{\rho \cdot v^2 \cdot C_d \cdot A}{29.91}$$

$$Force_{Drag}(lb) = \frac{density\left(lb/ft^3\right) \times \left(velocity(mph)\right)^2 \times coefficient \times Area\left(ft^2\right)}{29.91}$$

The rate of flow is given by:

$$Q = A \cdot v$$

$$\text{Rate of flow}\left(\text{cm}^3/\text{s}\right) = \text{Area of pipe}\left(\text{cm}^2\right) \cdot \text{velocity of fluid}(\text{cm/s})$$

The continuity equation:

$$A_i \cdot v_1 = A_2 \cdot v_2$$

$$\text{Area 1} \times \text{velocity 1} = \text{Area 2} \times \text{velocity 2}$$

The power of a pump needed to deliver a fluid at a given flow rate *Q* and pressure *P* can be calculated by:

$$\text{Power} = P \cdot Q$$

$$\text{Power}(\text{W}) = \text{Pressure}(\text{Pa}) \times \text{Flow rate}(\text{m}^3/\text{s})$$

The power of a pump in the English system is given by:

$$\text{Power} = \frac{P \cdot Q}{1{,}714}$$

$$\text{Power(hp)} = \frac{\text{Pressure}(\text{psi}) \times \text{Flow rate}(\text{gal/min})}{1{,}714}$$

Bernoulli's principle states that an increase in the velocity of a fluid is accompanied by a decrease in pressure or a decrease in the fluid's potential energy. Bernoulli's principle is a consequence of the conservation of energy. In the absence of friction, the energy of the fluid is constant throughout the pipe.

The total energy of the fluid is:

$$\begin{matrix}\text{Total} \\ \text{Energy}\end{matrix} = \text{Kinetic energy} + \text{Energy from pressure} + \text{Gravitational potential energy}$$

$$E = \frac{1}{2} m \cdot v^2 + P \cdot V + m \cdot g \cdot h$$

The energy/volume in one part of a system is the same as the energy/volume in another part of the system:

$$\frac{1}{2}\rho \cdot v_1^2 + P_1 + \rho \cdot g \cdot h_1 = \frac{1}{2}\rho \cdot v_2^2 + P_2 + \rho \cdot g \cdot h_2$$

$\rho = \text{density}\,(\text{kg}/\text{m}^3), v = \text{velocity}\,(\text{m/s}), P = \text{Pressure}(\text{Pa}), h = \text{height}(\text{m}),$

$$g = 9.8\,\text{m/s}^2$$

Power captured from a moving fluid:

$$P = \frac{1}{2}\rho \cdot A \cdot v^3 \cdot eff$$

$$\text{Power}(\text{W}) = \frac{1}{2}\text{density}\left(\text{kg}/\text{m}^3\right) \times \text{Area}\left(\text{m}^2\right) \times \left(\text{velocity}(\text{m}/\text{s})\right)^3 \times \text{efficiency}$$

Power captured by a windmill:

$$P_{\text{wind}} = 0.61 \cdot A \cdot v^3 \cdot eff$$

$$\text{Power}_{\text{Wind}}(\text{W}) = 0.61 \times \text{Area}\left(\text{m}^2\right) \times \left(\text{velocity}(\text{m}/\text{s})\right)^3 \times \text{efficiency}$$

The equivalent of wind power in the English system is given by:

$$P_{\text{wind}} = 5.1 \times 10^{-6}\; A \cdot v^3 \cdot eff$$

$$\text{Power}_{\text{wind}}(\text{hp}) = 5.1 \times 10^{-6} \times \text{Area}\left(\text{ft}^2\right) \times \left(\text{velocity}(\text{mph})\right)^3 \times \text{efficiency}$$

Power captured by a water wheel:

$$P_{\text{water}} = 500 \cdot A \cdot v^3 \cdot eff$$

$$\text{Power}_{\text{water}}(\text{W}) = 500 \times \text{Area}\left(\text{m}^2\right) \times \left(\text{velocity}(\text{m}/\text{s})\right)^3 \times \text{efficiency}$$

The equivalent power in the English system is given by:

$$P_{\text{water}} = 5.56 \times 10^{-3} A \cdot v^3 \cdot eff$$

$$\text{Power}_{\text{water}}(\text{hp}) = 5.56 \times 10^{-3} \times \text{Area}\left(\text{ft}^2\right) \times \left(\text{velocity}(\text{mph})\right)^3 \times \text{efficiency}$$

PROBLEM SOLVING TIPS

- In this chapter, there are a number of formulas that work only with a specific set of units, such as the FPS system. Be mindful of your units!
- There are three formulas for power in this chapter. Choose the correct formula according to the context of the problem.

PROBLEMS

1. Water is flowing through a pipe of radius 0.25 in. at a velocity of 2.0 ft/s. What is the rate of flow?
2. A fluid is flowing through a pipe of area 0.375 in.^{2} at a velocity of 3.0 ft/s. It enters the pipe with a cross section of 0.185 in.^{2}. What is the velocity of the fluid in the smaller section of pipe? What are the rates of flow in both sides?
3. Suppose you build a windmill and connect it to a car alternator to power your home. If the wind is blowing at 12.5 m/s and the area of the blades takes up 3.5 m^2 and the efficiency of the whole system is 21%, how much power can be generated?
4. Without using a pump, how can you increase the speed of a moving fluid?
5. If the power given to a hydroelectric generator by falling water is 12 MW and the turbines have an effective area of 5.5 m^2, how fast is the water moving?
6. What is the difference between laminar and turbulent flows?
7. How many more times is the air drag on a car moving at 25 mph versus 50 mph?
8. A pump is to deliver a flow rate of 1,200 gpm at 33 psi, what is the horsepower of the pump?
9. How much power can a windmill generate with an effective gathering area of 4.5 m^2, and an efficiency of 35% for a wind moving at 95 m/s?
10. How would you measure the flow rate and volume of a leaky faucet?
11. What shape should the auto industry make cars to minimize air drag?
12. If you ride with your foot out of the window traveling at 55 mph, estimate the force on your foot.
13. If a fan is mounted over a CPU and blows at a rate of 10 cfm through a cross-sectional area of 2 ft^3, how fast is the air moving?

14. How could you measure the flow rate through a piece of electronics in order to ensure it is being cooled properly?
15. What are the two ways an airplane gets its lift?
16. You are driving down the highway at 55 mph with your foot out of the window. If your foot has a cross-sectional area of 0.3 ft^2 with a drag coefficient of 1.17, what force is the air pushing on your foot with?
17. How many times more drag force is there on a car traveling at 70 mph compared to 50 mph?
18. An electronic flow sensor has a sensitivity of 100 mV/gpm. What is its output when measuring a flow of 25 gpm?
19. Some electronic equipment is cooled by passing air through it with a fan. If air is flowing through an opening with an area of 3.0 cm^2 at a velocity of 15 cm/s, what is the flow rate?
20. Water is flowing through a large section of pipe at a rate of 15 in.^{3}/s. The pipe then narrows to a cross-sectional area of 5 in.^{2}. What is the velocity of the water in the narrow section of the pipe?
21. You are playing in the backyard, trying to squirt your brother with the water hose. You place your thumb over the end of the hose, resulting in the water squirting farther. By placing your thumb over the hose, you have decreased the area through which the water flows by four times. How many times faster does the water flow out the end of the hose than in it?
22. Very high-power electrical equipment is sometimes cooled with water. Water is pumped through a water jacket thermally connected to the equipment transferring the heat out. If the flow rate is 50 gpm at a pressure of 30 psi, what horsepower pump is needed?
23. If a 3 hp fluid pump is operating at 60 psi, what is the flow rate out of it?
24. In the absence of friction, if a fluid is flowing through a pipe and the pipe narrows and the speed of the fluid increases so that its kinetic energy doubles, how many times does the fluid's energy from pressure change?
25. A windmill used to create electrical power can be made from a slot car motor connected to some type of device to catch the wind. As the motor spins it will generate a voltage, i.e. it will act as a generator. If the effective area of air the windmill catches is 0.4 m^2, and the wind is blowing at 5.0 m/s, what power can the windmill generate? Assume the generator is 100% efficient.

26. A hydroelectric generator can be made from a slot car motor connected to a turbine placed in flowing water. As the motor spins it will generate a voltage, i.e. it will act as a generator. If the effective area of the turbine is $0.2\,m^2$, and the water is flowing at 4.0 m/s, what power can the windmill generate? Assume the generator is 100% efficient.
27. Using Table 9.1 what shape has the lowest drag coefficient?
28. If 10 gpm is flowing into a volumetric flow meter, what volume does it measure after 20 min?
29. How does a thermal flow sensor work?
30. In a series electrical circuit, the current is the same everywhere. Likewise, the flow of a fluid through a set of pipes in series will be the same regardless of any changes in diameter of the pipes. If the flow at one point in a set of pipes in series is 4 gpm, what is the flow at any other point?
31. A squirrel cage blower mounted on the back of a piece of electronic equipment is blowing air at a rate of 22 cfm, that is, $22\,ft^3/min$. What is this rate in cm^3/min?
32. If a 5 hp pump delivers a pressure of 100 psi, how many gallons per minute is it pumping?
33. Why does the shower current move toward you while taking a shower?
34. How much more power is there in a wind traveling at 20 m/s as compared to 10 m/s?
35. A toilet has the unit l pf written on top of it. What does this stand for?
36. Water flows through a pipe at a rate of $0.01\,m^3/s$. The pipe has an area of $0.05\,m^2$ in section one. It then enters section two of the pipe of area $0.02\,m^2$ at a pressure of 101.31 kPa. Both sections are at the same level. What is the pressure in section one?
37. Water is sitting still in a pipe. One end of the pipe is 2.0 m higher than the other end, and the pipe is filled with water. What is the pressure difference in the water between the ends of the pipe?

10

Temperature and Heat

This chapter will focus on temperature, heat, and heat transfer. We will describe how to measure these quantities and how to control them (Figure 10.1).

Exactly what is a thermometer measuring when it measures temperature? When water is heated, its molecules or atoms begin vibrating or moving about faster, increasing their kinetic energies (Figure 10.2). This energy is transferred to the thermometer. The thermometer is therefore measuring the average kinetic energy of the water.

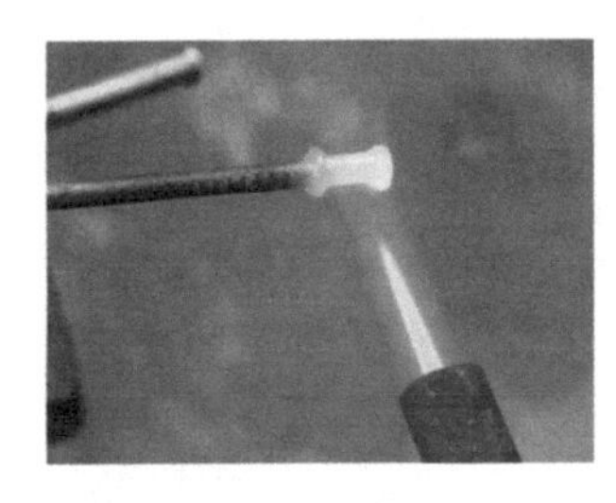

FIGURE 10.1 Heat delivered to a nail.

10.1 TEMPERATURE SCALES

There are many temperature scales. We will discuss three of them: Celsius, Fahrenheit, and kelvin. The Celsius and Fahrenheit scales are defined in an arbitrary way. For example, in the Celsius scale, 0° and 100° are arbitrarily placed at the freezing and boiling points of water, respectively.

If temperature is a measure of moving energy, then a substance that is completely still should have a temperature of zero. The kelvin temperature scale is based on this idea, with 0° set to be the lowest possible temperature (Figure 10.3).

FIGURE 10.2 When a pot of water is boiled, heat is applied to the pot. The heat causes the water molecules in the pot to move faster.

	Fahrenheit (°F)	Celsius (°C)	Kelvin (K)
Freezing water	32	0	273
Boiling water	212	100	373

Let's compare a few scales.

To convert between these scales, apply the following formulas:

$$T_C = \frac{5}{9}(T_F - 32)$$
$$T_F = \frac{9}{5}T_C + 32$$
$$T_C = T_K - 273$$
$$T_K = T_C + 273$$

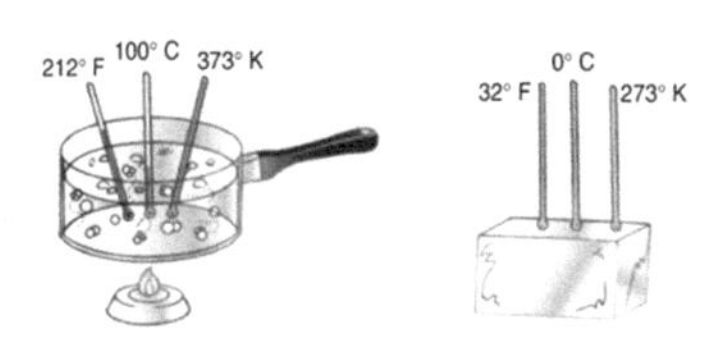

FIGURE 10.3 A comparison of temperature scales.

FIGURE 10.4 Liquid in a thermometer rises because the heat causes the liquid molecules to move faster and therefore expand.

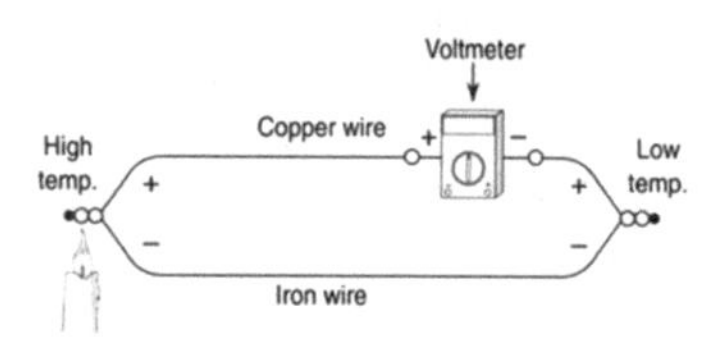

FIGURE 10.5 A thermocouple consists of two dissimilar pieces of wire connected to each other, producing a voltage proportional to the temperature. A reference junction scales the output to a known reference temperature.

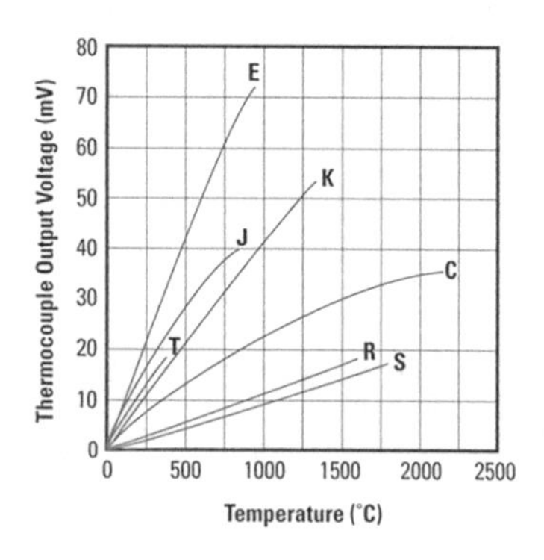

FIGURE 10.6 Thermocouple's output voltage versus temperature.

Example 10.1

Convert 72°F into Celsius.

$$T_F = 72°\text{F}$$

$$T_C = \frac{5}{9}(T_F - 32)$$

$$T_C = \frac{5}{9}(72°\text{F} - 32)$$

$$T_C = 22°\text{C}$$

Example 10.2

Convert 39°C into Fahrenheit.

$$T_C = 39°\text{F}$$

$$T_F = \frac{9}{5}T_C + 32$$

$$T_F = \frac{9}{5}39°\text{F} + 32$$

$$T_F = 102°\text{F}$$

Example 10.3

Convert 34°C into kelvin.

$$T_C = 34°\text{C}$$

$$T_K = T_C + 273$$

$$T_K = 34°\text{C} + 273 = 307\text{K}$$

10.2 MEASURING TEMPERATURE

10.2.1 Glass Thermometer

A *glass thermometer* has a liquid inside it that expands in a definite way when heated. When a glass thermometer is placed into a pot of boiling water, the molecules of the liquid inside the thermometer will begin to move faster, causing the liquid to expand and rise up in the thermometer (Figure 10.4). How much it rises determines the temperature.

10.2.2 Thermocouple

A *thermocouple* consists of two types of wires that are connected at one end. A voltage is developed across the junction and is proportional to the temperature of the junction. A reference junction is used to obtain measurements with respect to a known reference temperature (Figures 10.5 and 10.6; Table 10.1).

TABLE 10.1
Thermocouple Types and Sensitivities

Type	Couples	Seebeck Coefficient (μV/K)
E	Chromel-constantan	60
J	Iron-constantan	51
T	Copper-constantan	40
K	Chromel-alumel	40
N	Nicrosil-nisil	38
S	Pt (10% Rh)-Pt	11
B	Pt (30% Rh)-Pt (6% Rh)	8
R	Pt (13% Rh)-Pt	12

Thermocouples can be found in hot water heaters. They are used as a safety device to monitor the pilot light. If for some reason the pilot light goes out, the gas should be shut off to keep the house from filling up with gas. If the pilot light goes out, the thermocouple will alert the gas valve to shut off.

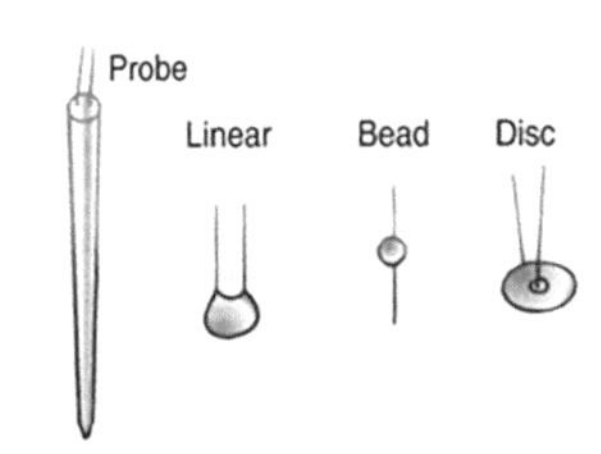

FIGURE 10.7 A thermistor is a variable resistor that changes resistance with temperature.

10.2.3 Thermistor

A thermistor is a semiconductor device that changes its resistance with temperature (Figure 10.7). As the temperature goes up, its resistance goes down (Figure 10.8).

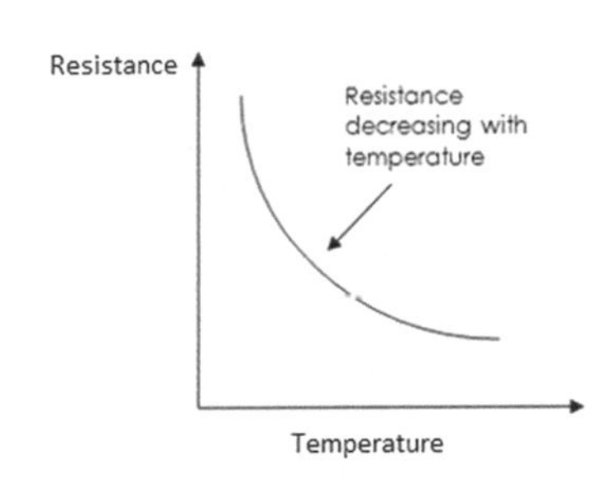

FIGURE 10.8 Thermistor's resistance decreases with increasing temperature.

10.2.4 Bimetal Strip

Some home thermostats contain a coil consisting of two types of metals layered one on top of the other (Figure 10.9).

As the temperature changes, the two metals will expand by different amounts, causing the strip to bend. The strip is used as a switch to open or close a circuit, telling the furnace or air conditioner to turn on or off.

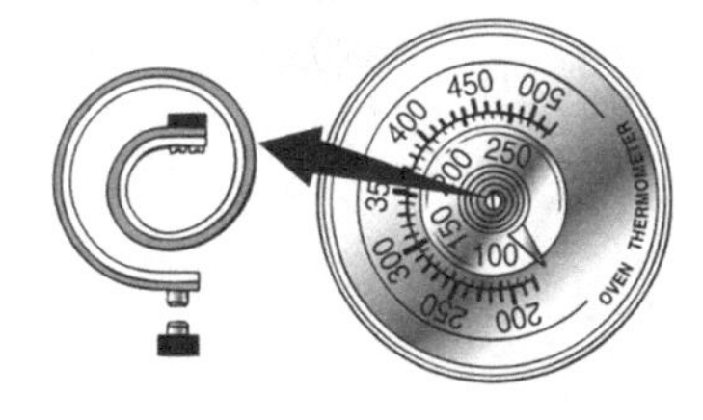

FIGURE 10.9 A bimetal strip thermostat.

10.2.5 Optical Pyrometer

Optical pyrometers determine temperature by measuring the light emitted by a hot object (Figure 10.10). For instance, as a metal is heated, it changes color from red to white. Each shade of color corresponds to a particular temperature. The pyrometer converts the color it sees into a temperature.

An optical pyrometer determines the temperature by comparing the color of an internal filament to the color of the radiation emitted by an object whose temperature is being measured.

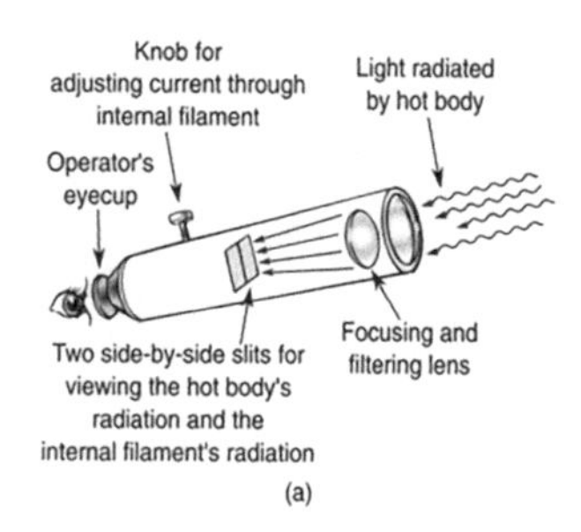

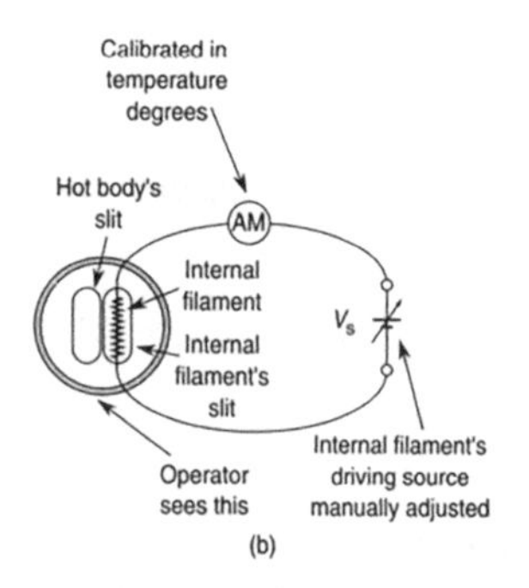

FIGURE 10.10 An optical pyrometer.

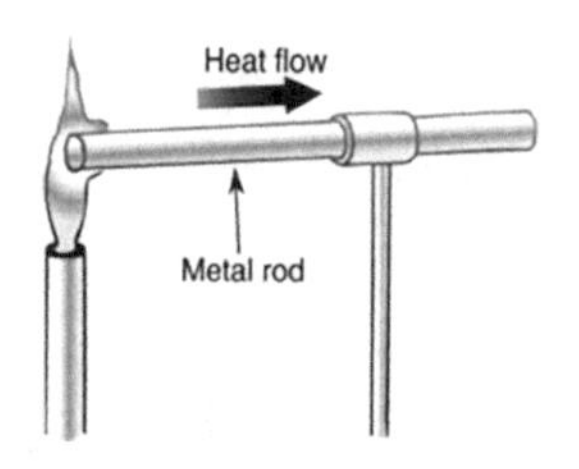

FIGURE 10.11 Heat flows from hot to cold objects.

FIGURE 10.12 A hot cup of coffee emits heat to the room, which is at a lower temperature.

10.3 HEAT

In everyday language, people use the words *heat* and *temperature* to mean the same thing, but in science they are not the same. So exactly what is heat?

Heat is the energy transferred between two systems due to a temperature difference. The temperature differences between two things determine which direction, and at what rate, heat flows.

If something gets hot, it is because heat flows into it.

If something gets cold, it is because heat flows out of it.

Heat flows from hot to cold objects (Figure 10.11).

If something feels hot to the touch, that means heat is flowing into your hand. If something feels cold, it is because heat is leaving your body.

How does a thermos bottle know to keep hot coffee hot and cold water cold? The thermos bottle doesn't know. It just prevents the flow of heat (Figure 10.12).

Heat has the same units as energy, such as joules or foot pounds. A common unit used is the *British thermal unit* (Btu). It takes 1 Btu of heat to raise the temperature of 1 lb of water by 1°F.

Heat units and conversions:

$$1\,\text{Btu} = 1{,}055\,\text{J}$$

$$1\,\text{Btu} = 778\,\text{ft lb}$$

$$1\,\text{ft lb} = 1.36\,\text{J}$$

$$1\,\text{cal} = 4.186\,\text{J}$$

$$1\,\text{food calorie} = 1\,\text{Cal} < \text{—Note the capital C}$$

$$1\,\text{Cal} = 4{,}186\,\text{J}$$

Example 10.4

We can build a heat source by connecting a battery to a resistor (Figure 10.13). Suppose the power dissipated by the circuit equals 100.0 W, and we let the circuit run for 50.0 s. How much heat does the resistor emit?

$$P = 100\,\text{W}$$

$$t = 50\,\text{s}$$

$$P = \frac{E}{t} \leftarrow \text{formula for power}$$

$$P \cdot t = \frac{E}{\cancel{t}} \cdot \cancel{t} \text{ solving for E}$$

$$E = P \cdot t$$

$$E = (100\,\text{W})(50\,\text{s})$$

$E = 5{,}000$ J of heat coming out of the resistor

Converting to Btu:

$$5{,}000\,\cancel{\text{J}}\ \frac{1\,\text{Btu}}{1{,}055\,\cancel{\text{J}}} = 4.7\,\text{Btu}$$

This amount of energy could raise the temperature of 1 lb of water by 4.7°F.

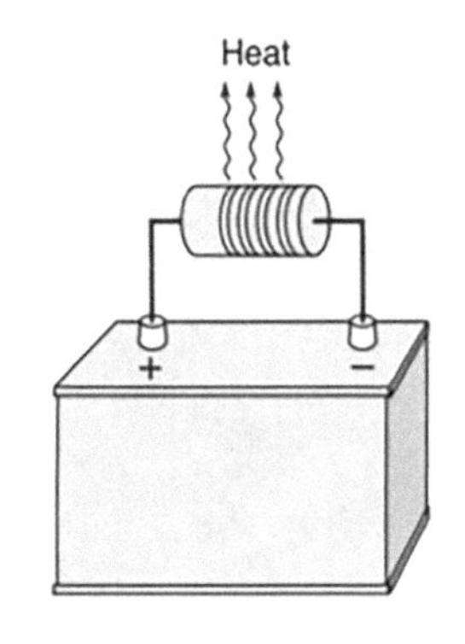

FIGURE 10.13 How much heat does the resistor emit?

10.4 HEAT TRANSFER

There are three ways to transfer heat from one place to another: conduction, convection, and radiation.

10.4.1 Conduction

Conduction is heat transferred between two materials in contact (Figure 10.14).

How easily heat flows through a material depends on the following:

- Heat flows because of a temperature difference across a material. The bigger the temperature differences, the larger the heat flow.
- Some materials, such as metals, allow heat to flow easily. Other materials, such as Styrofoam, prevent the flow of heat. How well they allow heat to flow is determined by their thermal conductivities (k) (Table 10.2). Thermal conductivity is similar to electrical conductivity, but it is applied to heat, not electricity.
- A material that is short and has a large cross-sectional area can pass a lot more heat in a given time than a long, skinny piece of material.

FIGURE 10.14 Heat transfer by conduction between two materials that are physically touching.

These points can be summarized by the following formula:

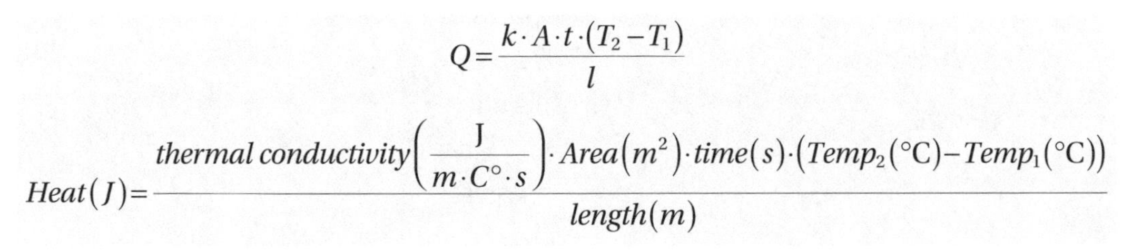

$$Q = \frac{k \cdot A \cdot t \cdot (T_2 - T_1)}{l}$$

$$Heat(J) = \frac{thermal\ conductivity\left(\frac{\text{J}}{m \cdot C^{\circ} \cdot s}\right) \cdot Area(m^2) \cdot time(s) \cdot (Temp_2(°\text{C}) - Temp_1(°\text{C}))}{length(m)}$$

TABLE 10.2
Thermal Conductivities

Substance	J/(s m °C)	Btu/(ft °F h)
Air	0.025	0.015
Aluminum	230	140
Brass	120	68
Brick/concrete	0.84	0.48
Cellulose fiber (loose fill)	0.039	0.023
Copper	380	220
Corkboard	0.042	0.024
Glass	0.75	0.50
Gypsum board (sheetrock)	0.16	0.092
Mineral wool	0.045	0.026
Plaster	0.14	0.083
Polystyrene foam	0.035	0.020
Polyurethane (expanded)	0.024	0.014
Steel	45	26

Or in the English system:

$$Heat(Btu) = \frac{thermal\ conductivity\left(\frac{Btu}{ft \cdot F^\circ \cdot h}\right) \cdot Area\left(ft^2\right) \cdot time(h) \cdot \left(Temp_2(^\circ\mathrm{F}) - Temp_1(^\circ\mathrm{F})\right)}{length(ft)}$$

Example 10.5

The handle of a cast-iron skillet has a length of 0.5 ft, a cross-sectional area of 0.0075 ft², a thermal conductivity of $k = 26\frac{\mathrm{Btu}}{\mathrm{ft\,^\circ F h}}$, and a temperature of 450°F at one end and 72°F at the other end. If we wait 0.3 h for the heat to flow, how much heat has passed through the handle? (Figure 10.15 and Table 10.2)

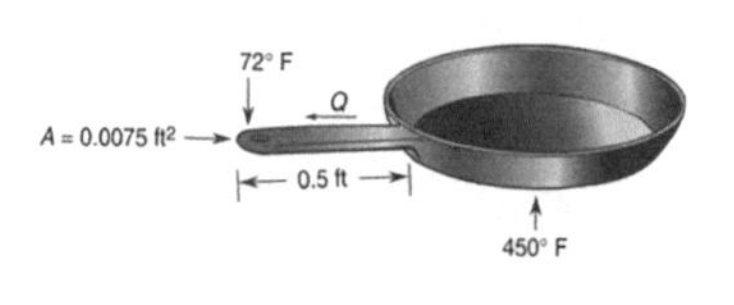

FIGURE 10.15

$$k = 26\frac{\mathrm{Btu}}{\mathrm{ft\,^\circ Fh}}$$

$$A = 0.0075\,\mathrm{ft}^2$$

$$l = 0.5\,\mathrm{ft}$$

$$t = 0.3\,\mathrm{h}$$

$$T_2 = 450^\circ\mathrm{F}$$

$$T_1 = 72^\circ\mathrm{F}$$

$$Q = ?$$

$$Q = \frac{k \cdot A \cdot t \cdot (T_2 = T_1)}{l}$$

$$Q = \frac{26 \frac{\text{Btu}}{\cancel{\text{ft}} \cdot \cancel{°\text{F}} \cdot \cancel{\text{h}}} \cdot 0.0075\,\cancel{\text{ft}}^2 \cdot 0.3\,\cancel{\text{h}} \cdot (450°\text{F} - 72°\cancel{\text{F}})}{0.5\,\cancel{\text{ft}}} \leftarrow \text{All units cancel but Btu.}$$

$$Q = 44\ \text{Btu}$$

10.5 THERMAL INSULATION

The thermal insulation you buy from a store has a value attached to it, which determines its insulating properties. This is called the R-value. A high R-value means it's a good insulator (Figure 10.16 and Table 10.3).

FIGURE 10.16 Insulation that is very thick and has a small thermal conductivity has a high R-value. Insulation with higher R-values insulate better.

$$R = \frac{l}{k}$$

$$R = \frac{\text{thickness}}{\text{thermal conductivity}} \left(\frac{\text{m}^2 \cdot °\text{C} \cdot \text{s}}{\text{J}} \right) \text{or} \left(\frac{\text{ft}^2 \cdot °\text{F} \cdot \text{h}}{\text{Btu}} \right)$$

For multiple layers of insulation, the total R-value is just the sum of the individual R-values:

TABLE 10.3
R-Values for Some Common Materials

Material	R-value ($\text{ft}^2 \cdot °\text{F} \cdot \text{h/Btu}$)
Hardwood siding (1.0 in. thick)	0.91
Wood shingles (lapped)	0.87
Brick (4.0 in. thick)	4.00
Concrete block (filled cores)	1.93
Styrofoam (1.0 in. thick)	5.0
Fiber-glass batting (3.5 in. thick)	10.90
Fiber-glass batting (6.0 in. thick)	18.80
Fiber-glass board (1.0 in. thick)	4.35
Cellulose fiber (1.0 in. thick)	3.70
Flat glass (0.125 in. thick)	0.89
Insulating glass (0.25 in. space)	1.54
Vertical air space (3.5 in. thick)	1.01
Air film	0.17
Dry wall (0.50 in. thick)	0.45
Sheathing (0.50 in. thick)	1.32

$$R_{\text{Total}} = \sum_i \frac{l_i}{k_i}$$

$$R_{\text{Total}} = \sum_i R_i$$

Therefore, heat transferred by conduction can be rewritten as:

$$Q = \frac{A \cdot t \cdot \Delta T}{R_{\text{Total}}}$$

Example 10.6

You want to insulate your home, so you make a trip to the hardware store and consider buying insulation. Comparing two types, you see an expensive one with a *R*-value of 30 (*R*-30) and a cheaper one labeled *R*-15. How much difference is there between the two?

R-30 is twice as big as *R*-15; therefore, *R*-30 will have half as much heat flow as *R*-15. In other words, *R*-30 is two times better at insulating than *R*-15.

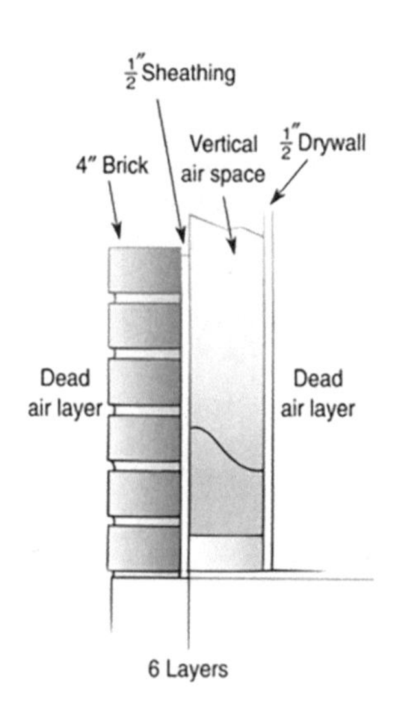

FIGURE 10.17

Example 10.7

(a) Calculate the total *R*-value for a multilayered wall in your home if the layers consist of 4.0 in. thick brick, 0.50 in. of sheathing, a 3.5 in. thick vertical air space, 0.5 in thick of drywall, and a dead-air layer, or air film, inside and outside your home. (b) Calculate the heat flow through a wall measuring 10.0 ft by 8.00 ft over a 24 h period if the average temperature is 32.0°F outside and 72.0°F inside (Figure 10.17).

(a)

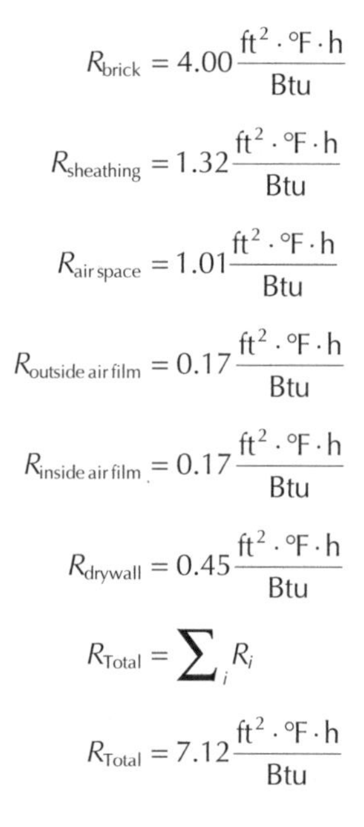

$$R_{\text{brick}} = 4.00 \frac{\text{ft}^2 \cdot {}^\circ\text{F} \cdot \text{h}}{\text{Btu}}$$

$$R_{\text{sheathing}} = 1.32 \frac{\text{ft}^2 \cdot {}^\circ\text{F} \cdot \text{h}}{\text{Btu}}$$

$$R_{\text{air space}} = 1.01 \frac{\text{ft}^2 \cdot {}^\circ\text{F} \cdot \text{h}}{\text{Btu}}$$

$$R_{\text{outside air film}} = 0.17 \frac{\text{ft}^2 \cdot {}^\circ\text{F} \cdot \text{h}}{\text{Btu}}$$

$$R_{\text{inside air film}} = 0.17 \frac{\text{ft}^2 \cdot {}^\circ\text{F} \cdot \text{h}}{\text{Btu}}$$

$$R_{\text{drywall}} = 0.45 \frac{\text{ft}^2 \cdot {}^\circ\text{F} \cdot \text{h}}{\text{Btu}}$$

$$R_{\text{Total}} = \sum_i R_i$$

$$R_{\text{Total}} = 7.12 \frac{\text{ft}^2 \cdot {}^\circ\text{F} \cdot \text{h}}{\text{Btu}}$$

(b)

$$t = 24.0\,\text{h}$$
$$T_{\text{outside}} = 32.0°\text{F}$$
$$T_{\text{inside}} = 72.0°\text{F}$$
$$A = 10.0\,\text{ft} \times 8.00\,\text{ft}$$
$$A = 80.0\,\text{ft}^2$$

$$Q = \frac{A \cdot t \cdot \Delta T}{R_{\text{Total}}}$$

$$Q = \frac{80.0\,\cancel{\text{ft}^2} \cdot 24.0\,\cancel{\text{h}} \cdot (72.0°\cancel{\text{F}} - 32.0°\cancel{\text{F}})}{7.12\dfrac{\cancel{\text{ft}^2} \cdot °\cancel{\text{F}} \cdot \cancel{\text{h}}}{\text{Btu}}}$$

$$Q = 10.8 \times 10^3\,\text{Btu} \leftarrow \text{Heat flow through one wall}$$

For the house to remain at the same temperature, any heat loss must be replaced by heat generated from the furnace.

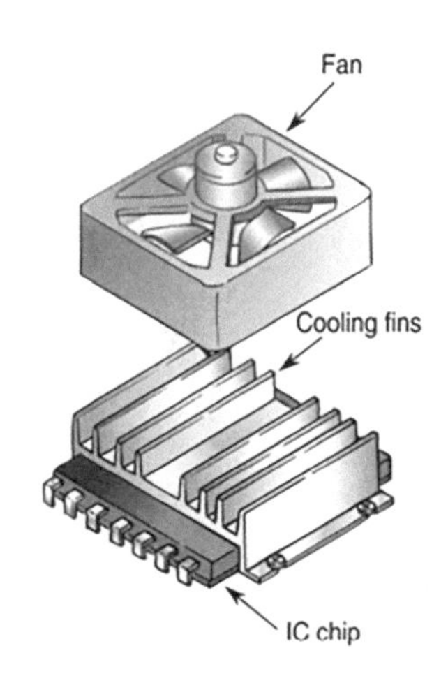

FIGURE 10.18 Convection transfers heat through a gas or liquid. In this sketch, an IC chip is cooled by convection.

10.6 CONVECTION

Convection is heat transferred by the motion of a fluid or gas (Figure 10.18). A gas or liquid can carry heat to or from an object.

10.6.1 Automobile Engine Cooling

One example of convection is the cooling system in a car (Figure 10.19). Water circulates through the engine block, carrying heat away to the radiator. At the radiator, convection takes place with air blowing through the radiator fins, removing the heat to the atmosphere.

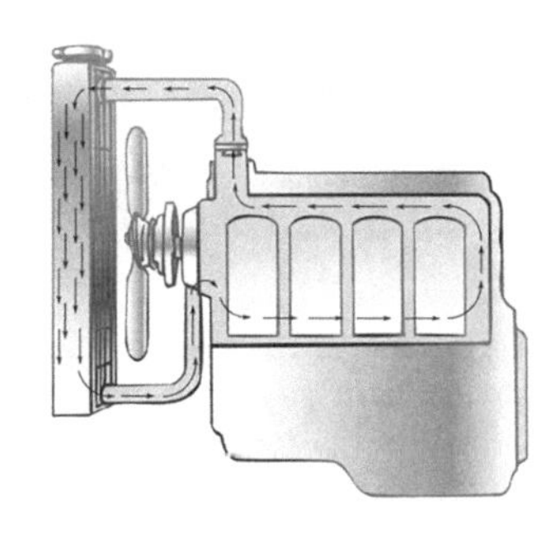

FIGURE 10.19 Convection cooling in an automobile.

10.6.2 Home Heating

Many homes are heated with convection heating (Figure 10.20). A fan blows air over a furnace core, transferring its heat throughout the home using heating ducts. A cold air return is provided to allow for circulation.

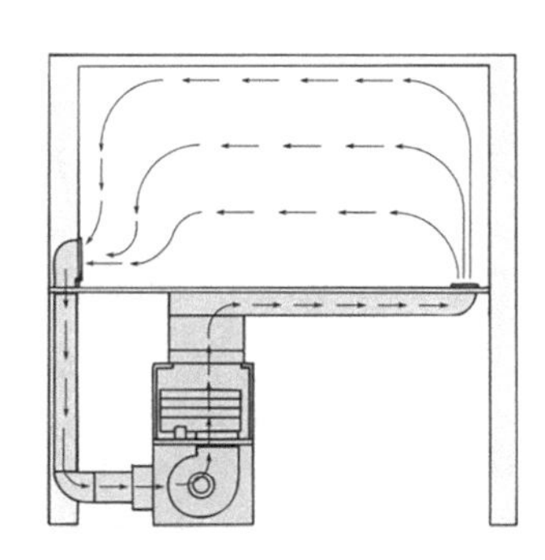

FIGURE 10.20 Convection heating in a home.

10.6.3 Solar Water Heater

A passive solar water heater does not use a mechanical pump to circulate the water (Figure 10.21). Because hot water rises, it will automatically circulate.

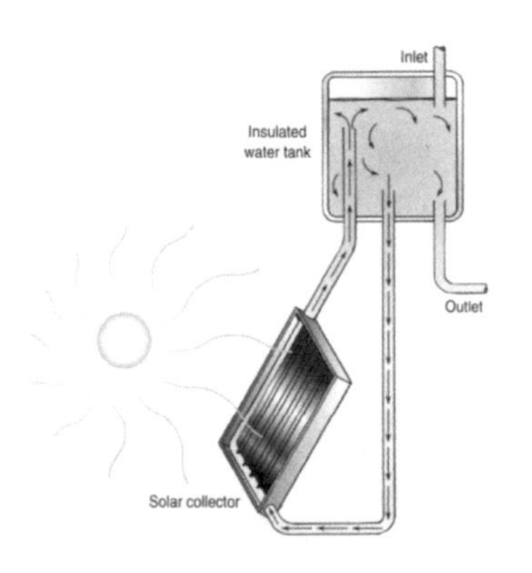

FIGURE 10.21 A solar water heater.

Convection depends on a number of factors:

- The faster the fluid or air flows, the quicker the heat will be transferred.
- Increasing the surface area will increase the rate of transfer, that is, big cooling fins mean faster cooling.
- Fluids or gases with high thermal conductivities transfer heat better.
- The bigger the temperature difference between the fluid or gas and the object transferring the heat, the faster the heat will get transferred, that is, cold water will cool an object faster than warm water.

10.7 RADIATION

Objects can heat up or cool down by either absorbing or emitting electromagnetic radiation (Figure 10.22).

The sun emits light at all wavelengths. It emits radio waves, infrared, visible, and even x-rays. It needs nothing to travel on or in. Like the sun, any object that is hotter than its environment emits some infrared radiation. Radiation is one of the ways a hot object cools down or a cool object warms up (Figures 10.23–10.25).

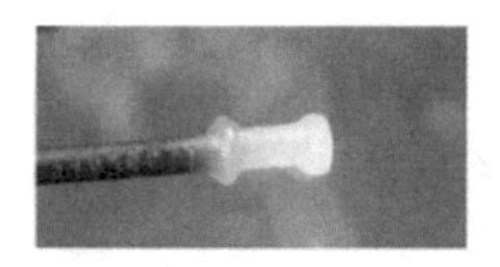

FIGURE 10.22 Hot objects transfer heat by giving off electromagnetic radiation; cool objects warm up by absorbing electromagnetic radiation.

10.7.1 Stephan–Boltzmann Law

The Stephan-Boltzmann law relates the temperature of an object to the electromagnetic energy given off by or absorbed by an object. An object of surface area A at a temperature T, sitting in an environment of temperature T_0, gives off heat Q at a rate of:

FIGURE 10.23 Having lots of south-facing windows in a house can reduce a heating bill because the windows allow in radiation from the sun.

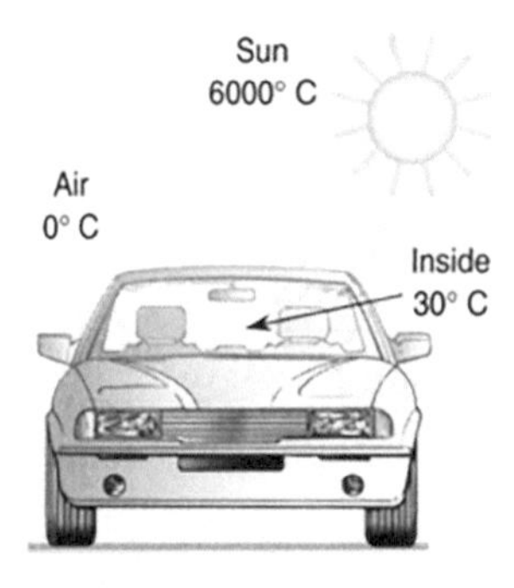

FIGURE 10.24 Even on a cold winter day, the inside of a car can become very warm from the sun's radiation.

FIGURE 10.25 A microwave oven heats food using electromagnetic radiation.

$$\frac{Q}{\Delta t} = e \cdot \sigma \cdot A \cdot \left(T^4 - T_0^4\right)$$

$$\frac{\text{Heat(J)}}{\text{time(s)}} = \text{emissivity} \times 5.67 \times 10^{-8}\left(\frac{\text{W}}{\text{m}^2\text{K}^4}\right) \times \text{Area}\left(\text{m}^2\right) \times \left(\text{Temp(K)}^4 - \text{Temp(K)}_{\text{Ambient}}^4\right)$$

Here the constant σ is called the Stephan-Boltzmann constant = $5.67 \times 10^{-8} \frac{\text{W}}{\text{m}^2\text{K}^4}$.

The emissivity e has no unit, varies from 0 to 1, and is related to how well an object gives off or absorbs heat. Black objects have an emissivity near 1 because they absorb and give off heat better than silver colored objects that have an emissivity near 0, that reflect heat. Objects that absorb or radiate heat perfectly have an emissivity of 1 and are called "black bodies".

Recall that heat is a form of energy and that power=energy/time or power = heat/time. So,

$$P = \frac{Q}{\Delta t} \text{ has units of power or Watts.}$$

Example 10.8

A steel sphere of radius 0.01 m, with an emissivity of 0.5, sitting in a room of temperature 298 K, is heated till it glows red. What is the power it is radiating?

A quick Google search shows that steel heated till it starts to glow red is at a temperature of 460°C. This is at a kelvin temperature of:

$$T(\text{K}) = 460°\text{C} + 273\text{K} = 733\text{K}$$

The area of the sphere is:

$$A = 4\pi r^2 = 4\pi\left(0.01\text{m}\right)^2$$

$$A = 0.00126\text{m}^2$$

$$P = \frac{Q}{\Delta t} = e \cdot \sigma \cdot A \cdot \left(T^4 - T_0^4\right)$$

$$P = (0.5)\left(5.67 \times 10^{-8}\left(\frac{\text{W}}{\text{m}^2\text{K}^4}\right)\right)\left(0.00126\text{m}^2\right)\left(\left(733\text{K}\right)^4 - \left(298\text{k}\right)^4\right)$$

$$P = 10.0\text{W}$$

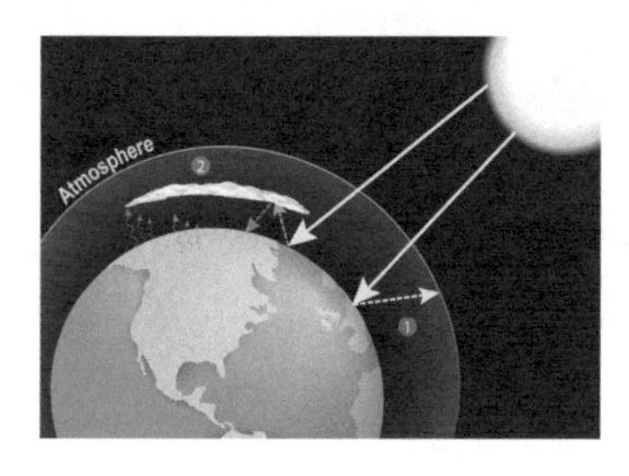

FIGURE 10.26 Global warming is a result of an energy imbalance. More energy from the sun is entering the atmosphere than leaving due to the "blanket" of greenhouse gases building up in the atmosphere.

10.7.2 Global Warming

The earth is heated by the sun. The earth will maintain its temperature if the absorbed heat is then reradiated out into space. We can write in terms of power, $P_{in}=P_{out}$. Clouds and other gasses in the atmosphere act like a blanket keeping in the heat absorbed by the sun. This is necessary if the earth is to maintain a temperature. Imagine, however, if you put on too many blankets at night when you are sleeping. You will begin to heat up! This is exactly what is happening to the earth now with all the CO_2 and methane that man is putting into the atmosphere. We are blanketing the earth too much, and the planet is heating up! (Figure 10.26)

10.8 CONTROLLING TEMPERATURE

Controlling the temperature of something like an oven requires knowing how much heat to add. Adding too much or too little heat will result in the oven being too hot or too cold. How much heat to add to reach a certain temperature will depend on how big the oven is and what it is made of. Not all materials change temperature by the same amount for the same heat input and weight. For example:

- 1 Btu of heat added to 1 lb of steel raises the temperature by 8.7°F.
- 1 Btu of heat added to 1 lb of water raises the temperature by 1°F.
- 4,184 J of heat added to 1 kg of water raises the temperature by 1°C.

How much a material changes temperature for a given weight and heat input is determined by its specific heat capacity.

10.9 SPECIFIC HEAT CAPACITY

Specific heat capacity is the amount of heat needed to raise the temperature of a given amount of a substance by 1°C. The specific heat capacity is given by the letter *c*.

It takes 4,184 J of heat added to water to raise the temperature of 1 kg of water by 1°C.

$$c_{\text{water}} = 4{,}184 \frac{\text{J}}{\text{kg °C}}$$

It takes 1 Btu of heat added to water to raise the temperature of 1 lb of water by 1°F.

$$c_{\text{water}} = 1 \frac{\text{Btu}}{\text{lb °F}}$$

Mathematically, the change in temperature of a material of a given weight w or mass m, with specific heat capacity c, and heat exchanged Q is given by:

$$\Delta T = \frac{Q}{c \cdot m}$$

$$\text{Temperature change}\,(^\circ\text{C}) = \frac{\text{heat}\,(\text{J})}{\text{specific heat capacity}\left(\frac{\text{J}}{\text{kg}^\circ\text{C}}\right) \cdot \text{mass}\,(\text{kg})}$$

OR

$$\text{Temperature change}\,(^\circ\text{F}) = \frac{\text{heat}\,(\text{Btu})}{\text{specific heat capacity}\left(\frac{\text{Btu}}{\text{lb}^\circ\text{F}}\right) \cdot \text{weight}\,(\text{lb})}$$

Which equation you choose depends on the units of the heat capacity. This equation says how much a material's temperature changes depends on the amount of heat exchanged and is inversely proportional to its specific heat capacity and its weight or mass. If the specific heat and weight (or mass) are very large, a lot of heat is needed to change the temperature (Table 10.4).

TABLE 10.4
Table of Heat Capacities

	Specific Heat Capacity	
Material	**Btu/(lb_m °F)**	**J/(kg °C)**
Solids		
Aluminum	0.214	895
Brass	0.094	390
Brickwork	0.20	840
Copper	0.094	390
Glass	0.194	812
Gold	0.031	130
Ice	0.504	2,110
Iron, cast	0.130	544
Lead	0.031	130
Sand	0.195	816
Silver	0.056	234
Steel	0.116	485
Stone (typical)	0.20	840
Wood, oak	0.57	2,400
Wood, pine	0.467	1,950

(Continued)

TABLE 10.4 (*Continued*)
Table of Heat Capacities

	Specific Heat Capacity	
Material	**Btu/(lb_m °F)**	**J/(kg °C)**
Liquids		
Aluminum	0.25	1,000
Copper	0.101	423
Alcohol, ethyl	0.58	2,400
Alcohol, methyl	0.60	2,500
Gold	0.0327	137
Lead	0.038	160
Mercury	0.033	140
Nitrogen	0.474	1,980
Oxygen	0.394	1,650
Oil, machine	0.400	1,670
Silver	0.0685	287
Water	1.000	4,184
Gases (confined)		
Air	0.168	703
Ammonia	0.399	1,670
Hydrogen	2.412	10,090
Nitrogen	0.173	724
Oxygen	0.155	649
Steam	0.346	1,450

Note: There may be some variation in these figures depending on the actual temperature, the presence of impurities, etc.

Example 10.9

How much heat is needed to raise the temperature of 198 lb of water in a bathtub from 72.0°F to 110.0°F? How much fuel oil would be needed to produce this temperature change if fuel oil generates 17,500 Btu/lb of heat when burned? (Figure 10.27)

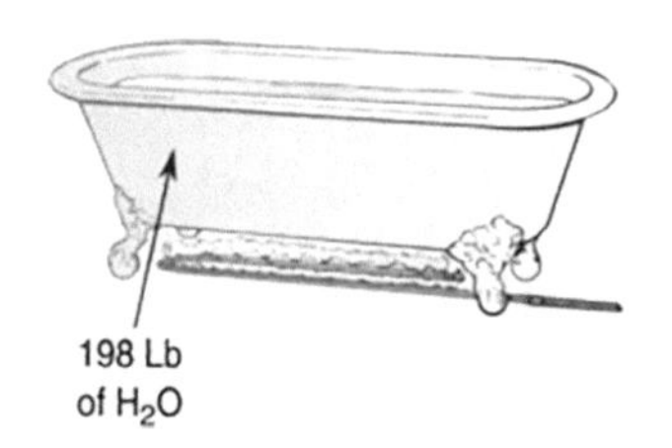

FIGURE 10.27

$$T_i = 72.0°\text{F}$$

$$T_f = 110.0°\text{F}$$

$$w = 198\,\text{lb}$$

$$C_{\text{water}} = 1\frac{\text{Btu}}{\text{Lb°F}}$$

$$Q = ?$$

$$Q = c \cdot w \cdot \Delta T$$

$$Q = \left(1\frac{\text{Btu}}{\cancel{\text{Lb}}\,\cancel{°\text{F}}}\right)(198\,\text{Lb})(110.0°\text{F} - 72.0°\text{F})$$

$$Q = 7{,}524\,\text{Btu}$$

The heat of combustion of fuel oil is 17,500 Btu/lb. To create 7,520 Btu of heat would therefore require:

$$\text{Pounds of fuel oil} = \frac{7{,}520\ \cancel{\text{Btu}}}{17{,}500\frac{\cancel{\text{Btu}}}{\text{lb}}} \leftarrow \text{Note}\frac{1}{\frac{1}{\text{lb}}} = \text{lb}$$

$$\text{Pounds of fuel oil} = 0.430\,\text{lb}$$

Example 10.10

A heater puts out 720 W of heat for 100.0 s, and all this heat goes onto 10.0 kg of steel. How much will the temperature of the steel change? (Figure 10.28)

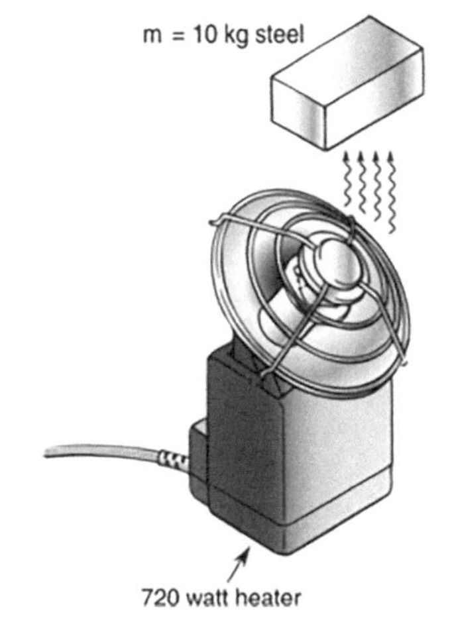

FIGURE 10.28

$$P = 720\,\text{W}$$
$$t = 100.0\,\text{s}$$
$$m = 10.0\,\text{kg}$$
$$C_{steel} = 485\frac{\text{J}}{\text{kg°C}}$$
$$\Delta T = ?$$
$$\Delta T = \frac{Q}{m \cdot C}$$
$$P = \frac{E}{t} \rightarrow E = P \cdot t$$
$$Q = E = P \cdot t$$
$$Q = (720\,\text{W})(100.0\,\text{s})$$
$$Q = 72{,}000\,\text{J}$$
$$\Delta T = \frac{Q}{m \cdot C}$$
$$\Delta T \frac{72{,}000\ \cancel{\text{J}}}{(10.0\ \cancel{\text{kg}})\left(485\frac{\cancel{\text{J}}}{\cancel{\text{kg}}\,\text{°C}}\right)}$$
$$\Delta T = 14.8\text{°C}$$

10.10 CHANGING THE STATE OF A SUBSTANCE

Applying heat to a material doesn't always result in a temperature change. When a material is at its melting or boiling temperature, the addition of heat goes into breaking the intermolecular bonds in the material, not raising its temperature. This breaking of the intermolecular bonds will cause the material to change its *state* from a solid to a liquid (melting) or from a liquid to a gas (vaporizing). The heat needed to cause this change of state is called the *latent heat.*

Mathematically, the equation for the heat needed to change the state of a material is given by:

$$Q = m \cdot L \leftarrow \text{Energy needed to change state}$$

$$\text{Heat}(\text{J}) = \text{mass}(\text{kg}) \times \text{latent heat}(\text{J/kg})$$

10.10.1 Melting and Freezing

A solid changes to liquid when enough heat energy is delivered to it. When the melting temperature is reached, the addition of more heat does not go into raising the temperature of the solid, but into breaking the intermolecular bonds in the solid. When these bonds are broken, the solid turns into liquid. This energy is called the latent heat of fusion.

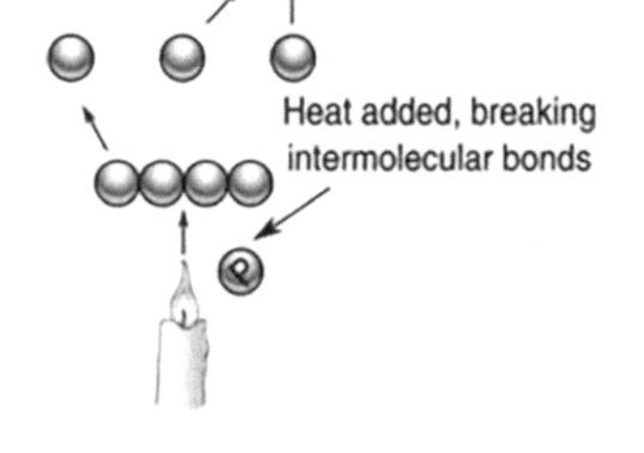

FIGURE 10.29 The latent heat of fusion is the energy needed to break the intermolecular bonds before a solid can change into a liquid.

10.10.2 Latent Heat of Fusion

Latent heat of fusion is the amount of heat needed to melt a unit amount of material when it is at its melting point (Figure 10.29). Conversely, to freeze a liquid, energy has to be removed from it. If the liquid is at its freezing temperature, the amount of heat that must be removed from it to turn it into a solid is equal to its latent heat of fusion. The freezing temperature and melting temperature of a substance are equal (Table 10.5).

TABLE 10.5
Latent Heats of Fusion and Vaporization, Boiling Points, and Freezing Points of Various Materials

		Latent Heat of Fusion			Latent Heat of Vaporization	
Substance	**Melting Point (°C)**	**J/kg**	**cal/g**	**Boiling Point (°C)**	**J/kg**	**cal/g**
Helium	−269.65	5.23×10^3	1.25	−268.93	2.09×10^4	4.99
Nitrogen	−209.97	2.55×10^4	6.09	−195.81	2.01×10^5	48.0
Oxygen	−218.79	1.38×10^4	3.30	−182.97	2.13×10^5	50.9
Ethyl alcohol	−114	1.04×10^5	24.9	78	8.54×10^5	204
Water	0.00	3.33×10^5	79.7	100.00	2.26×10^6	540
Sulfur	119	3.81×10^4	9.10	444.60	3.26×10^5	77.9
Lead	327.3	2.45×10^4	5.85	1,750	8.70×10^5	208
Aluminum	660	3.97×10^5	94.8	2,450	1.14×10^7	2,720
Silver	960.80	8.82×10^4	21.1	2,193	2.33×10^6	558
Gold	1,063.00	6.44×10^4	15.4	2,660	1.58×10^6	377
Copper	1,083	1.34×10^5	32.0	1,187	5.06×10^6	1,210

Example 10.11

How much heat is needed to melt 2.5 kg of ice at its freezing temperature? (Figure 10.30)

$$m = 2.5\text{kg}$$

$$L_{\text{fusion for water}} = 3.33 \times 10^5 \frac{\text{J}}{\text{kg}}$$

$$Q = m \times L$$

$$Q = 2.5\,\cancel{\text{kg}} \times 3.33 \times 10^5 \frac{\text{J}}{\cancel{\text{kg}}}$$

$$Q = 8.33 \times 10^5\,\text{J}$$

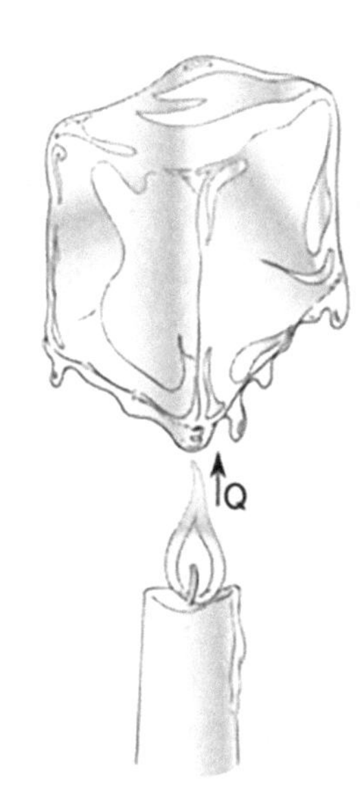

FIGURE 10.30

Example 10.12

How much heat must be removed from 2.5 kg of water at its freezing temperature to turn it into ice? The amount is the same as the amount of heat needed to transform ice into water in the example above:

$$Q = 8.33 \times 10^5\,\text{J}$$

10.10.3 Boiling and Condensing

A liquid changes to a gas when enough heat energy is delivered to it. When the boiling temperature is reached, the addition of more heat goes not into raising the temperature of the liquid, but into breaking the intermolecular bonds in the liquid. When these bonds are broken, the liquid turns into a gas. The energy needed to break these bonds is called the latent heat of vaporization.

10.10.4 Latent Heat of Vaporization

Latent heat of vaporization is the amount of heat needed to vaporize a unit amount of material when it is at its boiling point. Conversely, to condense a gas, in other words, turn it into a liquid, energy has to be removed from it. If the gas is at its condensing temperature, the amount of heat that must be removed to turn it into a liquid is equal to its latent heat of vaporization. The condensing temperature and boiling temperature of a substance are equal.

Example 10.13

How much heat is needed to vaporize 1.5 kg of copper at its boiling point?

$$m = 1.5\text{kg}$$

$$L_{\text{vaporization for copper}} = 5.06 \times 10^6 \frac{\text{J}}{\text{kg}}$$

$$Q = m \times L$$

$$Q = 1.5\,\cancel{\text{kg}} \times 5.06 \times 10^6 \frac{\text{J}}{\cancel{\text{kg}}}$$

$$Q = 7.6 \times 10^6\,\text{J}$$

When calculating the heat needed to change a material's state when its temperature is not at its boiling or freezing temperature, the material's heat capacity must be included in the calculation.

Example 10.14

How many joules of heat must be added to transform 12.0 kg of water from 25°C into steam?

$$m = 12.0\,\text{kg}$$

$$T_{\text{initial}} = 25.0°\text{C}$$

$$T_{\text{boiling}} = 100.0°\text{C}$$

$$Q_{\text{Total}} = Q(\text{heat needed to raise the water to } 100°\text{C}) + Q(\text{latent heat})$$

$$Q(\text{heat needed to raise the water to } 100°\text{C}) = m \cdot c \cdot \Delta T$$

$$Q = 12.0\,\text{kg} \cdot 4{,}184 \frac{\text{J}}{\text{kg°C}} \cdot (100.0°\text{C} - 25.0°\text{C})$$

$$Q = 3.77 \times 10^6\,\text{J}$$

$$Q_{\text{Latent Heat}} = (12.0\,\text{kg})\left(\frac{2.26 \times 10^6\,\text{J}}{\text{kg}}\right)$$

$$Q_{\text{Latent}} = 27.1 \times 10^6\,\text{Jk}$$

$$Q_{\text{Total}} = Q(\text{heat needed to raise the water to } 100°\text{C}) + Q(\text{latent heat})$$

$$Q_{\text{Total}} = 3.77 \times 10^6\,\text{J} + 27.1 \times 10^6\,\text{J}$$

$$Q_{\text{Total}} = 30.87 \times 10^6\,\text{J}$$

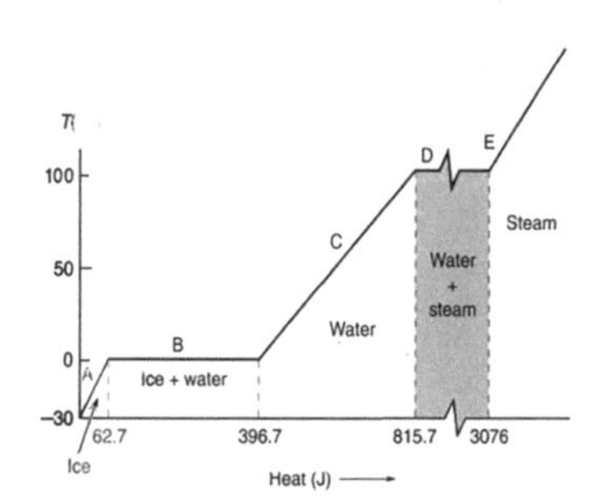

FIGURE 10.31 Graph of the energy needed to change 1 g of ice to liquid and then to steam.

A graph illustrating the energy needed to change 1 g of ice to liquid, then to steam, showing the latent heats is shown in Figure 10.31.

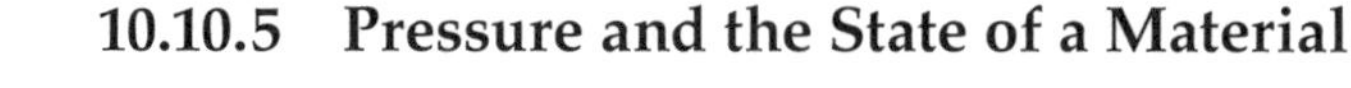

10.10.5 Pressure and the State of a Material

Pressure will also have an effect on what state a material will be in. The pressure and temperature of a material determine whether that material will be a gas, liquid, or solid. By changing these parameters, we can change the state of the material (Figure 10.32). For example, let's look at water:

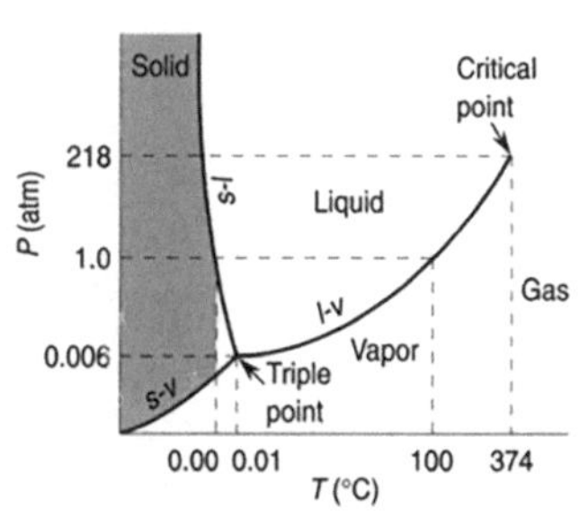

FIGURE 10.32 A phase diagram for water describes the state the water will be in for a given temperature and pressure.

We can cause water to boil below 212°F if the pressure is low enough.

We can cause steam to liquefy above 212°F if the pressure is high enough.

Pressure also has an effect on evaporation and the boiling point of a liquid. Just above the surface of the water where the vapor escapes, water vapor molecules are lurking about. Under high pressure, there is a better chance that an escaping water molecule will be bumped into by these molecules just above the surface. These molecules will therefore push the potential vapor back down into the liquid state, reducing the evaporation or delaying the boiling point (Figure 10.33).

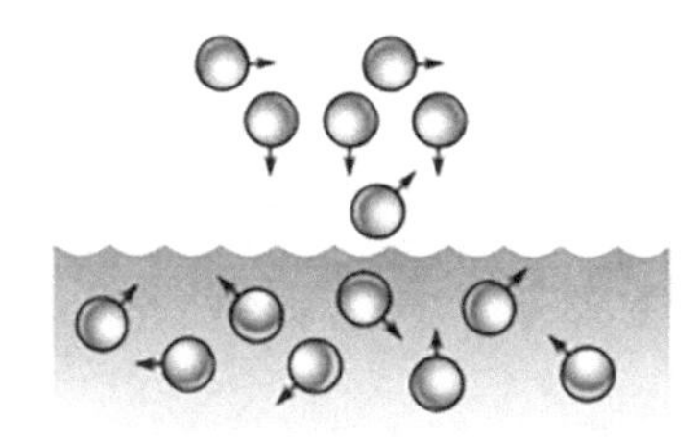

FIGURE 10.33 Under pressure, water molecules cannot escape easily because there is water vapor pushing them back into the liquid state.

10.10.6 Pressure Cooker

An open pot of water boils at approximately 212°F. Water under pressure boils at a higher temperature. This allows food in a pressure cooker to be cooked at a higher temperature and therefore to cook faster (Figure 10.34).

FIGURE 10.34 A pressure cooker delays the boiling temperature using pressure.

10.11 CHAPTER SUMMARY

Units used in this chapter:

Symbol	Unit
t—time	second (s), hour(h)h)
l—thickness or length	meter (m)
h—height	m
A—area	m^2
T—Temperature	Celsius (°C), Fahrenheit (°F), kelvin (K)
E—energy	joules (J), British thermal unit (Btu)
Q—heat	joules (J), Btu, calorie (cal)
P—Power	watts (W)
k—thermal conductivity	$\frac{J}{m \cdot s \cdot °C}$ or $\frac{Btu}{ft \cdot h \cdot °F}$
R-value	$\frac{ft^2 \cdot h \cdot °F}{Btu}$
e—emissivity	No unit
σ—Stephan–Boltzmann constant	$5.67 \times 10^{-8}\left(\frac{W}{m^2K^4}\right)$
m—mass	kg
w—weight	pound (lb)
c—specific heat capacity	$\frac{J}{kg \cdot °C}$ or $\frac{Btu}{lb \cdot °F}$
L—latent heat	J/kg

Temperature scale conversions:

$$T_C = \frac{5}{9}(T_F - 32)$$

$$T_F = \frac{9}{5}T_C + 32$$

$$T_C = T_K - 273$$

$$T_K = T_C + 273$$

Heat is the energy transferred between two systems due to a temperature difference.

Energy conversions:

$$1\,\text{Btu} = 1{,}055\,\text{J}$$

$$1\,\text{Btu} = 778\,\text{ft lb}$$

$$1\,\text{ft lb} = 1.36\,\text{J}$$

$$1\,\text{cal} = 4.186\,\text{J}$$

There are three ways to transfer heat from one place to another: conduction, convection, and radiation.

Conduction: Heat transferred between two materials in contact. Heat conduction is given by the following equation:

$$Q = \frac{k \cdot A \cdot t \cdot (T_2 - T_1)}{l}$$

$$\text{Heat (J)} = \frac{\text{thermal conductivity}\left(\dfrac{\text{J}}{\text{m}\cdot{}^\circ\text{C}\cdot\text{s}}\right)\cdot \text{Area}\left(\text{m}^2\right)\cdot \text{time(s)}\cdot\left(\text{Temp}_2({}^\circ\text{C}) - \text{Temp}_2({}^\circ\text{C})\right)}{\text{length(m)}}$$

Or in the English system,

$$\text{Heat (Btu)} = \frac{\text{thermal conductivity}\left(\dfrac{\text{Btu}}{\text{ft}\cdot{}^\circ\text{F}\cdot\text{h}}\right)\cdot \text{Area}\left(\text{ft}^2\right)\cdot \text{time(h)}\cdot\left(\text{Temp}_2({}^\circ\text{F}) - \text{Temp}_2({}^\circ\text{F})\right)}{\text{length(ft)}}$$

Convection: Heat transferred by the motion of a fluid or gas. A gas or liquid can carry heat to or from an object.

Radiation: Objects can either heat up or cool down by either absorbing or giving off electromagnetic radiation, respectively. An object of surface area A at a temperature T, sitting in an environment of temperature T_0, gives off heat Q at a rate of:

$$\frac{Q}{\Delta t} = e \cdot \sigma \cdot A\left(T^4 - T_0^4\right)$$

$$\frac{\text{Heat(J)}}{\text{time(s)}} = \text{emissivity} \times 5.67 \times 10^{-8}\left(\frac{\text{W}}{\text{m}^2\text{K}^4}\right) \times \text{Area}\left(\text{m}^2\right) \times \left(\text{Temp(K)}^4 - \text{Temp(K)}^4_{\text{Ambient}}\right)$$

Here the constant σ is called the Stephan–Boltzmann constant $= 5.67 \times 10^{-8}\left(\frac{\text{W}}{\text{m}^2\text{K}^4}\right)$

$$P = \frac{Q}{\Delta t} \text{ has units of power or Watts.}$$

Thermal insulation: Thermal insulation is rated by the R-value. A high R-value means the material is a good insulator.

$$R = \frac{l}{k}$$

$$R = \frac{\text{thickness}}{\text{thermal conductivity}}\left(\frac{\text{m}^2 \cdot {}^\circ\text{C} \cdot \text{s}}{\text{J}}\right) \text{or} \left(\frac{\text{ft}^2 \cdot {}^\circ\text{F} \cdot \text{h}}{\text{Btu}}\right)$$

For multiple layers of insulation, the R-value is the sum of the individual R-values.

$$R_{\text{Total}} = \sum_i \frac{L_i}{k_i}$$

Q can be rewritten as:

$$Q = \frac{A \cdot t \cdot \Delta T}{R_{\text{Total}}}$$

10.11.1 Specific Heat Capacity

The amount of heat needed to raise the temperature of a given amount of a substance by 1°C.

$$\Delta T = \frac{Q}{c \cdot m}$$

$$\text{Temperature change}\,(^\circ\text{C}) = \frac{\text{heat (J)}}{\text{specific heat capacity}\left(\frac{\text{J}}{\text{kg}^\circ\text{C}}\right) \cdot \text{mass (kg)}}$$

OR

$$\text{Temperature change}\,(^\circ\text{F}) = \frac{\text{heat (Btu)}}{\text{specific heat capacity}\left(\frac{\text{Btu}}{\text{lb}^\circ\text{F}}\right) \cdot \text{weight (lb)}}$$

The equation for the heat needed to change the state of a material is given by:

$$Q = m \cdot L$$

$$\text{Heat}(\text{J}) = \text{mass}(\text{kg}) \times \text{latent heat}(\text{J/kg})$$

PROBLEM SOLVING TIP

When solving problems dealing with thermal conduction, specific heat, and latent heat, look at the units for k, c, and L. They will tell you what the other units in the equation should be written in.

PROBLEMS

1. Convert 72°F to Celsius and kelvin.
2. What happens to the kinetic energy of a gas as it is heated or cooled?
3. Suppose a heater puts out 1,200 W for 1 min. How many Btu's has it put out?
4. Suppose insulation for your attic has a R-value of 3 and a more expensive insulation has a R-value of 6. What is the difference in heat flow through these two materials?
5. Suppose you build a heated doghouse made from 5 in. thick white pinewood, which has a thermal conductivity of $\frac{0.065\,\text{Btu}}{\text{ft}\,°\text{F}\,\text{h}}$. The doghouse is shaped like a rectangle with dimensions of 36 in. by 24 in. by 28 in. You would like to maintain the temperature inside at 72°F over a period of 8 h when the outside temperature is 32°F. What size heater do you need to install in the doghouse?
6. What if the temperature drops to 0°F in problem 5? How much more heat is needed to keep the house at 72°F?
7. How much heat is needed to heat up a tub containing 25 gal of water from 68°F to 105°F?
8. Suppose a computer puts out 210 W of heat. If this heat is not expelled to the room, the computer will continue to heat up until it is ruined. If a fan removes most of the heat by convection, how does the rest of the heat get out of the computer?
9. What gives off heat faster, a hot object or a warm object?
10. Suppose you have 2 min to drink a hot cup of coffee. You would like the coffee to be as cool as possible before you drink it. Should you pour the cool cream in first and wait for the coffee to cool down, or wait some time and then pour in the cool cream?
11. Suppose you build a heat source using a 100 W light bulb that puts out 20 W of light and 80 W of heat. If all this heat is put into

heating 5.0 kg of water from 22°C to 75°C, how long do you have to wait for the water to reach this temperature?

12. Heat sinks mounted on electronic components dissipate heat by what method? Why are they made with fins?
13. Convert 65°F into Celsius.
14. How much heat is needed to turn 1.0 kg of water at its boiling temperature into steam?
15. Soldering irons are rated in terms of wattage, not temperature. If the iron's wattage is too high, a sensitive electronic component could be damaged by becoming overheated. Why are soldering irons rated in terms of wattage and not temperature?
16. How much heat is needed to convert 22.0 kg of water from 22.0°C into steam?
17. If the atmospheric pressure on a mountaintop is much lower than the pressure at sea level, does water on a mountain top boil at a temperature above or below 100°C? Also, give the reason for your answer.
18. While checking the temperature of an IC, Integrated Circuit, chip the thermometer reads 50°C. What is the temperature in °F?
19. If the aluminum case of a power supply reaches 660°C, it will begin to melt. What is this temperature in °F?
20. A 25 W soldering iron is on for 100 s. How many Btu's of heat did the iron generate?
21. A power transistor is mounted to a heat sink. If the transistor mounting is 0.0005 m thick, has an area of 0.0004 m^2, a thermal conductivity of 300 J/ (s m °C), is on for 500 s, the temperature of the sink is 25°C, and the temperature of the transistor is 52°C, what is the heat flow into the sink?
22. The insulation in an attic is 1 ft thick and has a thermal conductivity of 0.023 Btu/(ft °F h). What is the R-value?
23. An oven is wrapped in an insulating blanket with a R-value of 33 ft^2 °F h/Btu. The total surface area of the oven is 6 ft^2, the oven is on for 3 h with an inside temperature of 450°F sitting in a room of temperature 72°F.

    What is the heat flow out of the oven in Btu?
24. A power transistor bank is mounted on a 1 kg aluminum heat sink. The transistors dissipate 500 W and all this heat flows into the sink. What is the change in temperature of the heat sink after 10 s in Celsius? (Hint $Q = P \times t$)
25. How much heat is needed to melt 1.0 kg of ice at 0°C?
26. The power capability of an IC chip lowers with rising ambient temperature. Suppose the power a chip can dissipate is the derated by

0.57 W/°C above 25°C. Therefore, the formula for the power capability of the chip becomes power = rated power – 0.57 × (temperature – 25°C) for temperatures above 25°C. If a 100 W chip is running at 100°C and is derated by 0.57 W/°C, what is the maximum power the chip can handle?

27. A 10 Ω resistor connected to a 12 V battery is used as a heater in a 10 kg pot of water. If the heater runs for 250 s, what is the change in temperature of the water in Celsius?
28. Water boils at 212°F. What is this temperature in Celsius?
29. How many watts of heat does a 5 Ω resistor generate when 0.5 A is flowing through it?
30. Calculate the heat flow in 1 h through a 0.5 in. diameter aluminum rod 1.5 ft in length with a thermal conductivity of 140 Btu/(ft °F h) and a temperature difference across the ends of 100°F?
31. Heat sinks readily dissipate heat out of an IC chip because they have large surface areas. If a heat sink has five fins measuring 5.0 in. × 1.0 in., what is the total surface area? (Remember each fin has two sides.)
32. A computer is cooled using a fan and a vent hole in the case. Heat also escapes out the sides of the metal case through conduction. If the case is 0.0625 in. thick and one side of the panel measures 18 in. × 24 in., the thermal conductivity is 120 Btu/(ft °F h), and the temperature outside the computer is 72°F and inside is 95°F, and the computer runs for 8 h, how much heat flowed through that one side?
33. Two houses are insulated with insulation having *R*-values of 30 and 20. How many times more does the house having the *R*-20 insulation lose heat versus the house with the *R*-30?
34. A pot of 1 kg of water is being heated on the stove. If the water is to boil, the temperature must rise from 22°C to 100°C. How much heat in joules must be added to the water?
35. If the stove in the previous problem is electric and the burner creates 2,000 W of heat, how long will it take for the water to boil? (Assume all the heat from the burner goes into the water.)
36. How much heat is needed to melt a 0.05 kg ice cube at 0°C?
37. How much heat is needed to turn 0.05 kg water into steam at 100°C?
38. Calculate the radiant power of the sun. The radius of the sun is 7×10^8 m, temperature 5,780 K, and emissivity = 1. The sun radiates into space which has a temperature ~ 0 K.
39. What is the temperature of a star with a radius 10^9 m, radiating 10^{28} W of power? Assume emissivity = 1 and space has a temperature ~ 0 K.

11

Thermodynamics

Heat Engines, Heat Pumps, and Thermal Expansion

Thermodynamics is the physics of heat. Practical applications of thermodynamics include temperature regulation of electronics, heating and air conditioning, automobile engine design, weather prediction, and anywhere else where heat is exchanged.

Why doesn't there exist a car that gets over a 100 miles/gallon? Have the big oil companies been suppressing the patents for this technology in order to suppress the building of such efficient engines? Let's look at the physics behind these ideas (Figure 11.1).

FIGURE 11.1 A cooling tower exchanging heat with the environment.

11.1 HEAT ENGINES

A *heat engine* is any device that borrows heat energy from some reservoir of heat at a high temperature and converts some of it into work plus some wasted heat at a lower temperature (Figure 11.2).

Examples of heat engines are gas, diesel, jet, steam, and rocket engines (Figure 11.3). They all take heat energy from fuel and convert it into some type of work to move a vehicle.

11.1.1 Fuels

Fuels store chemical potential energy. A molecule of gasoline can be thought of as a bunch of atoms separated by compressed springs. The springs are the interatomic forces between atoms. With the addition of heat, the springs will be released from compression, hurling the atoms outward. This is an explosion of the gas molecules, turning the chemical potential into kinetic energy (Figure 11.4). Table 11.1 lists the amount of energies released when these fuels are burned.

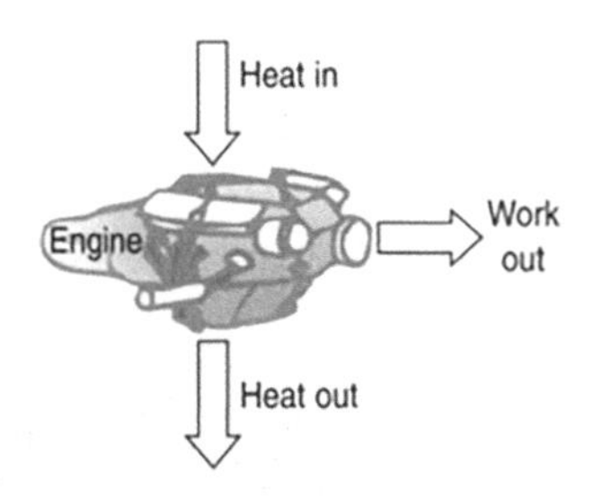

FIGURE 11.2 A heat engine.

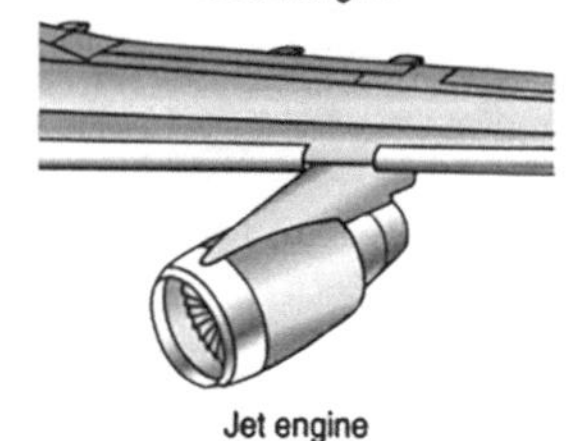

Rocket engine

FIGURE 11.3 Examples of heat engines.

TABLE 11.1
Btu of Heat Produced by the Combustion per Amount of Fuel[a]

Fuel	Heat of Combustion Btu/lb_m	MJ/kg	Air-Fuel Ratio for Complete Combustion (lb_m/lb_m or g/g)
Butane	20,000	46	15.4
Gasoline	19,000	44	14.8
Crude oil	18,000	42	14.2
Fuel oil	17,500	41	13.8
Coke	14,500	34	11.3
Coal, bituminous	13,500	31	10.4

	Heat of Combustion				
	On Mass Basis		On Volume Basis		
Fuel	Btu/lb_m	MJ/kg	Btu/ft^3	kJ/L	Air-Fuel Ratio for Complete Combustion (ft^3/ft^3 or L/L)
Hydrogen	61,400	143	275	10.3	2.4
Methane	23,800	55.2	900	33	9.6
Propane	22,200	51.5	2,400	88	13.7
Natural gas	23,600	54.8	1,000	37	11.9
Coke gas	23,000	53.6	1,300	50	14

Note: 1 MJ/kg = 1,000 kJ/kg = 1,000 J/g
1kJ/L = 1 J/cm^3 = 1 MJ/m^1

[a] Data applies to gases at 1.0 atm pressure and 15.6°C (60°F). Actual values vary slightly depending on geographical origin. Values do not include latent heat in any water vapor formed as a product of combustion.

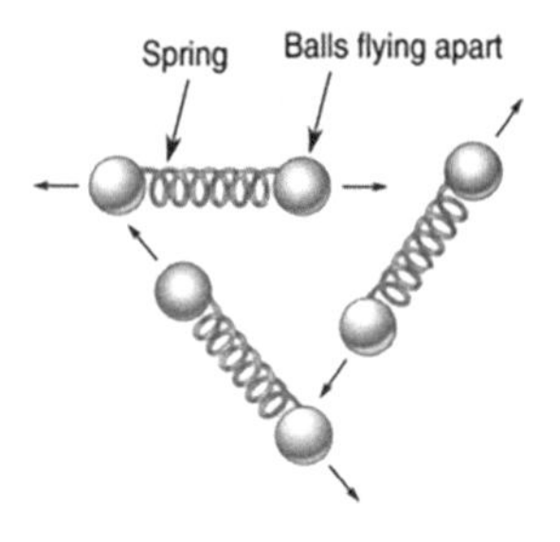

FIGURE 11.4 A chemical reaction releasing heat.

11.1.2 Gas Mileage

The gas mileage of a car is limited by how much energy an engine can extract from a gallon of gas and use it to move the car. We can increase gas mileage by making the car lighter or more aerodynamic or by making a host of other mechanical engine improvements, but there is a limit to how far we can go. A gallon of gas contains about 0.1 million Btu of heat energy. Not all of this energy is used to move the car. The limits are constrained by the laws of thermodynamics.

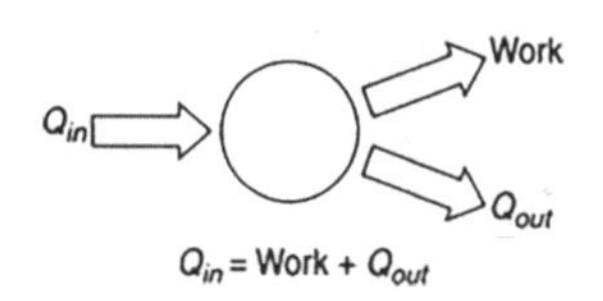

FIGURE 11.5 The first law of thermodynamics.

11.1.3 Laws of Thermodynamics

The *first law of thermodynamics* is about energy conservation (Figure 11.5). You will recall that energy conservation means that the total energy of an isolated system is a constant. Energy is never created or destroyed; it just changes form. Heat is energy. Part of its input into a heat engine goes into performing some useful work, part goes into just heating up the engine, and the rest is just wasted in terms of the exhaust. A more general definition of the first law is:

> First law of thermodynamics: Energy is conserved as it moves through a thermodynamic system.

The first law applied to heat engines implies that the energy gained from a heat source equals the work done plus the heat wasted.

$$Q_{in} = W + Q_{out}$$
$$\text{Heat}_{in}(\text{J}) = \text{Work}(\text{J}) + \text{Heat}_{out}(\text{J})$$

The first law is a statement about the conservation of energy. The energy gained from some fuel source is used to do work by a heat engine plus that wasted in just heating up the environment.

The *second law of thermodynamics* is related to your experience when making chocolate milk (Figure 11.6). You start off with just milk and chocolate syrup. When you mix the two, you get chocolate milk. Now try to separate the chocolate milk into just chocolate syrup and milk. Impossible, you say! This is the second law at work. The second law says it is easier to create a disordered mixture than to try to bring order back into the mixture. A more general way of saying this is:

FIGURE 11.6 The second law of thermodynamics. The universe normally evolves from order, milk and chocolate separated, to disorder, milk and chocolate mixed.

> Second law of thermodynamics: Systems in the universe normally evolve from order to disorder.

When a heat engine burns fuel, it converts the internal energy of the fuel, which, when burned, produces molecules moving about randomly at high speeds into an ordered motion, moving a piston or wheel. All systems in the universe normally evolve from order to disorder. We are asking a heat engine to go against this, to turn randomness into order. The second law implies that it is impossible to build a heat engine that converts all the energy in the fuel into work. Some of the energy will be converted into the orderly motion of a piston, and the rest will go on to just heating the environment, into more disorder. In case of an automobile, most of the energy released from the fuel goes into heating up the cooling water and the exhaust.

11.1.4 Efficiency

The efficiency of an engine is a measure of how much work it can do versus how much heat energy is put in:

$$\text{Efficiency} = \frac{W}{Q_{in}}$$

$$\text{Efficiency} = \frac{\text{Work(J)}}{\text{Heat}_{\text{in}}\text{(J)}}$$

The most efficient engine would be one where all of the fuel's energy is extracted out, performing some useful work. Looking at the first law:

$$Q_{\text{in}} = W + Q_{\text{out}}$$

Solving for W gives:

$$W = Q_{\text{in}} - Q_{\text{out}}$$

Therefore,

$$\text{Efficiency} = \frac{W}{Q_{\text{in}}} = \frac{Q_{\text{in}} - Q_{\text{out}}}{Q_{\text{in}}}$$

To maximize the efficiency, we want to maximize this difference in the numerator. Heat and temperature are related to each other, and it can be shown that efficiency depends mainly on the difference between the input and exhaust temperatures. The larger the temperature difference, the more efficient the heat engine. The *Carnot engine*, named after French scientist N.L. Sadi Carnot, who first proposed it, is ideally the most efficient heat engine possible.

The efficiency of a perfect heat engine is given by:

$$\text{Efficiency} = \left(1 - \frac{T_{\text{out}}(\text{K})}{T_{\text{in}}(\text{K})}\right) \times 100\%$$

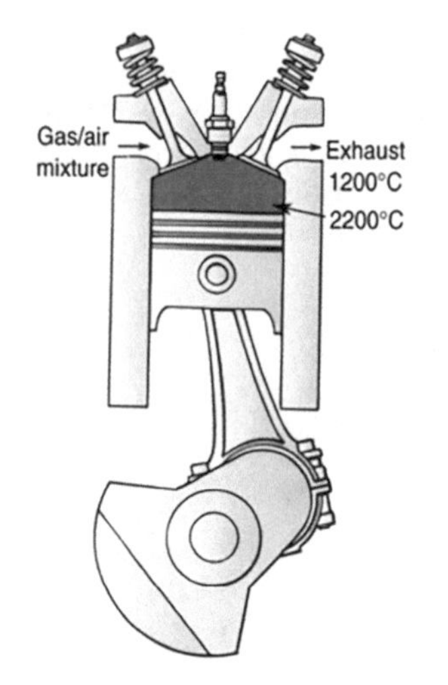

FIGURE 11.7 What is the maximum efficiency of this engine?

This equation says that if the temperature of the heat source of the engine T_{in} is high and the wasted heat that is exhausted into an environment at temperature T_{out} is low, then the efficiency will be high. Fuel burned in an automobile engine raises the temperature inside the combustion chamber to thousands of degrees. The most efficient engine of this type is one where the combustion chamber temperature is very high and the exhaust temperature is very low (Figure 11.7).

Example 11.1

When gas is burned in a car engine, it reaches temperatures of 2,200°C and puts out exhaust at 1,200°C. What is the maximum efficiency of this engine?

$$T_{\text{in}} = 2,200°\text{C}$$

$$T_{\text{out}} = 1,200°\text{C}$$

The temperatures must be converted to kelvin.

$$T_K = T_{°C} + 273$$

$$T_{in} = 2,200°C + 273 = 2,473K$$

$$T_{out} = 1,200°C + 273 = 1,473K$$

$$\text{Efficiency} = \left(1 - \frac{T_{out}(K)}{T_{in}(K)}\right) \times 100\%$$

$$\text{Efficiency} = \left(1 - \frac{1,473K}{2,473K}\right) \times 100\% = 40\%$$

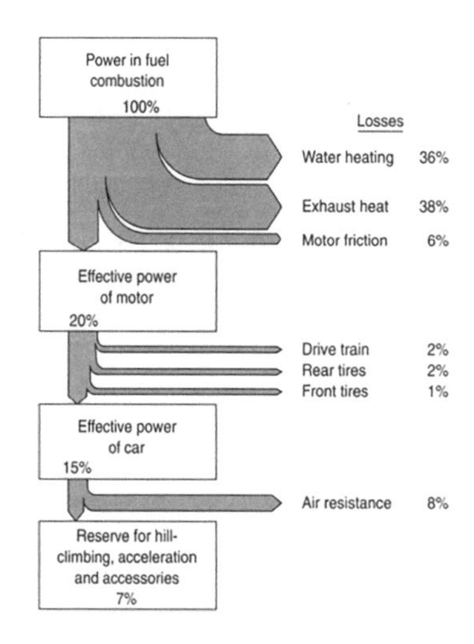

FIGURE 11.8 Most automobile engines have an efficiency of 20%.

From this example, we can see that even a perfect heat engine that has no losses due to practical considerations such as friction will not be 100% efficient. There will still be some heat wasted to the environment (Figure 11.8).

Because the efficiency of any type of heat engine is always less than 100%, it follows that no one will ever build a perpetual motion machine because there will always be a need for some additional energy input without which the machine will eventually slow down and stop.

11.1.5 External Combustion Engine

A steam engine is an external combustion engine. External to the engine, fuel is burned to boil the water to generate steam, which drives the engine. An example of a steam engine with a piston is shown in Figure 11.9. An example of a steam turbine is shown in Figure 11.10.

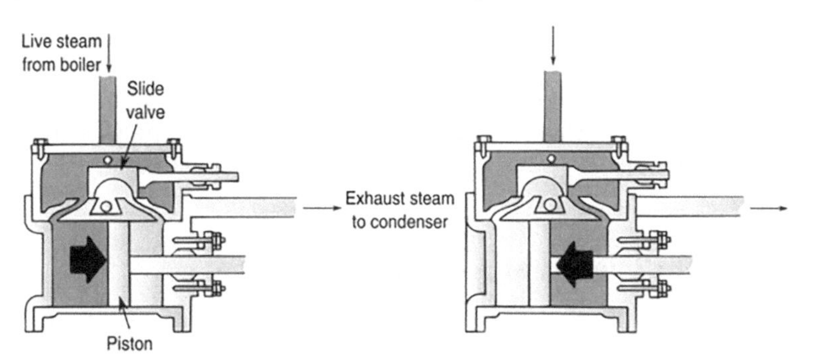

FIGURE 11.9 A steam engine. Valves open and close, alternately pushing the piston back and forth.

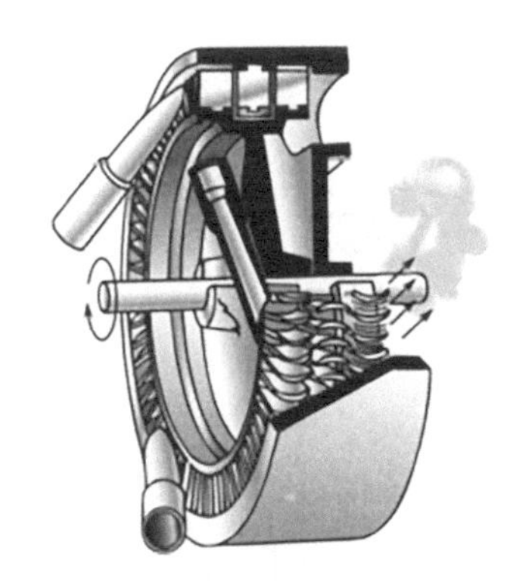

FIGURE 11.10 A steam turbine. The fins on a turbine are designed to extract as much thermal energy as possible from the steam and turn it into rotary motion. After exiting the turbine, the steam is cooled and returns to the boiler to become steam again and repeat the process.

11.1.6 Internal Combustion Engine

Fuel is burned inside the internal combustion engine. Four-stroke engines are illustrated in Figures 11.11 and 11.12.

These engines are called four-stroke because of the four-stroke cycle the engine is taken through.

1. Intake: The piston pulls the gas-air mixture into the cylinder.
2. Compression: The piston rises, compressing the fuel mixture.
3. Power: The fuel ignites, driving the piston down.
4. Exhaust: The piston pushes the burned fuel out.

The cycle then repeats itself, beginning with stroke 1.

The differences between diesel and gasoline engines are as follows:

- Gasoline engines use spark plugs to ignite the fuel. Diesel engines operate under much higher pressures, which ignite the fuel. Compressing any fuel enough will cause it to ignite.
- Diesel engines operate under much higher pressures, 20:1 compression ratio, compared with 8:1 for gasoline. They must be made stronger and therefore heavier than gas engines.
- Diesel engines are more efficient, about 30% efficient, compared with 20% for gas engines.

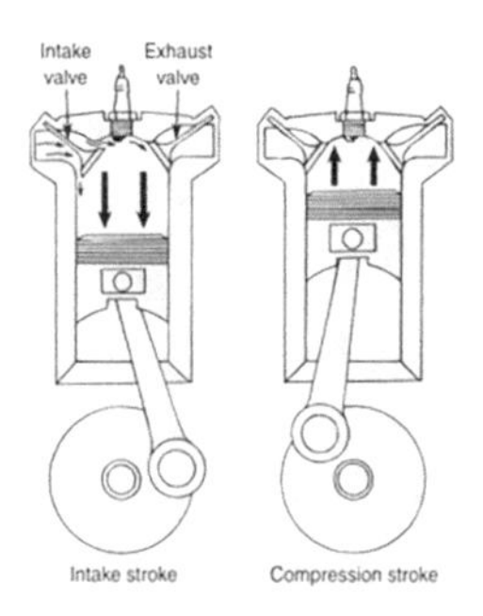

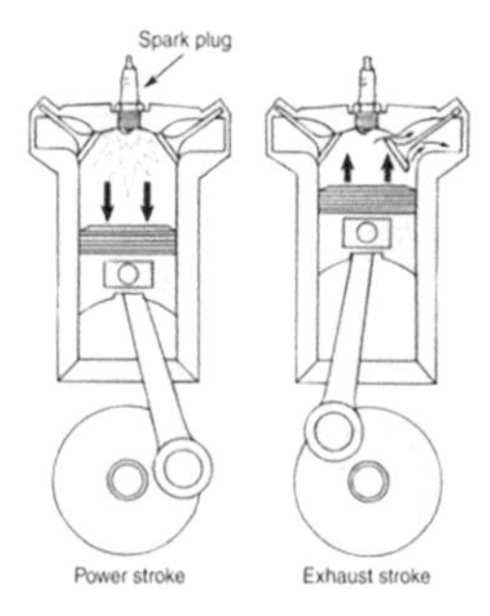

FIGURE 11.11 Four-stroke gasoline engine.

To obtain maximum efficiency/power:

- Increasing the volume of the cylinders can increase the power of an engine; this will allow for more fuel/air mixture to burn. Increasing the amount of fuel/air mixture in the cylinder can also be done by pumping the mixture into the cylinder. This is known as *supercharging.*
- Increasing the fuel/air pressure in a cylinder by reducing the clearance between the piston and the valves can increase the power. There is a limit, however, based on the strength of the material the engine is made of.

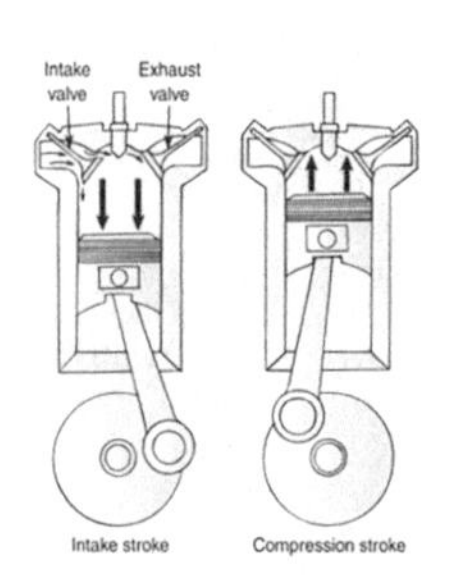

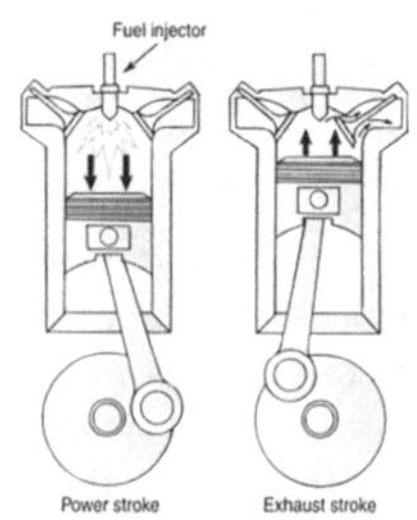

FIGURE 11.12 Four-stroke diesel engine.

11.1.7 The Efficiency of a Human Being

Food is a form of fuel. Its unit of energy is the Calorie. One food Calorie equals 1,000 physics calories (1 C = 1,000 c). One food Calorie equals 4,186 J. *The human body is about 25% efficient,* meaning that only 25% of the energy we get from food is turned into muscular energy. This is about the same efficiency as that of a gasoline automobile engine. Let's analyze the energy needed over the period of one day. A man weighing approximately 155 lb will burn 100 Cal/h just to maintain life (breathing, circulating blood, etc.) in normal waking conditions, and 60 Cal/h while sleeping (Figure 11.13).

FIGURE 11.13 The human body is about 25% efficient meaning, that only 25% of the energy we get from food is turned into muscular energy.

Examples of Calories burned in a 24 h period:

1. 100 Cal/h for 16 waking hours equals 1,600 Cal.
2. Sleeping for 8 h at 60 Cal/h equals 540 Cal.
3. Walking slowly, at 2.5 mph, burns 220 Cal/h. If we approximate the day's activities to walking 5 miles, that would total 440 Cal.

   Total Calories burned = 2,580 Cal

   Therefore, just to maintain this man's weight, he must consume, $Q_{in} = W + Q_{out} = 2{,}580$ Cal every day just to maintain weight.

   Compare 440 Cal burned by walking 5 miles with eating a Big Mac hamburger at 530 Cal. It is clear that if you want to lose weight, reducing your food intake will benefit you a lot more than exercising.

11.1.8 Artificial Leaf

A leaf harnesses energy from the sun, by taking in CO_2 from the atmosphere and sunlight to generate glucose that feed the plant. Plants use only 1% of the energy in the sunlight falling on them to accomplish this. An artificial leaf generates fuel by separating water, H_2O, into hydrogen and oxygen. The hydrogen can be used directly as fuel or the hydrogen can be fed to engineered organisms, which then excretes a wide variety of fuels. Artificial leaves are achieving an efficiency of roughly 10%, that is ten times better than a real leaf! (Figure 11.14)

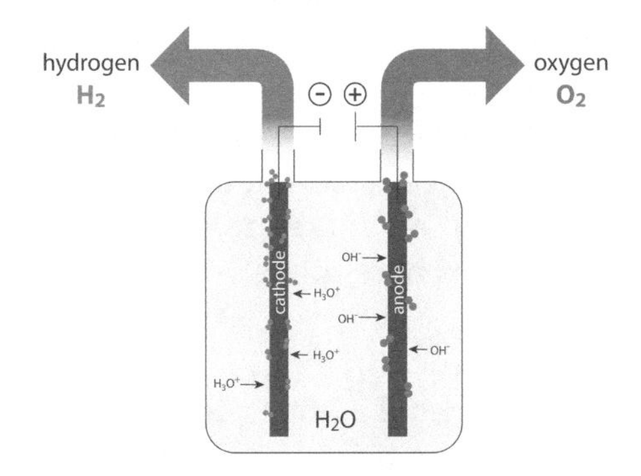

FIGURE 11.14 An artificial leaf consists of a photovoltaic material (the artificial leaf) that creates an electric current when exposed to light that separates the hydrogen from oxygen in water. The collected hydrogen can then be used as fuel.

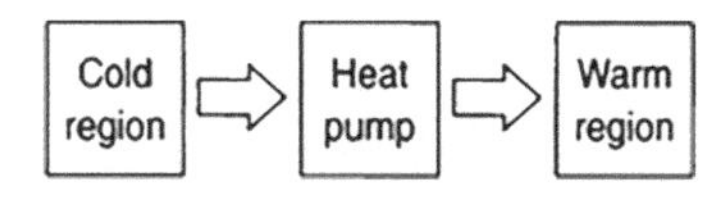

FIGURE 11.15 Heat pumps pump heat from a cold to a hot region.

11.2 HEAT PUMPS

Heat pumps pump heat from a low-temperature region to a high-temperature region (Figure 11.15).

When a gas expands and evaporates, it cools. Anything coming in contact with this gas will cool down as well. This is the principle behind a refrigerator, which is an example of a heat pump (Figure 11.16).

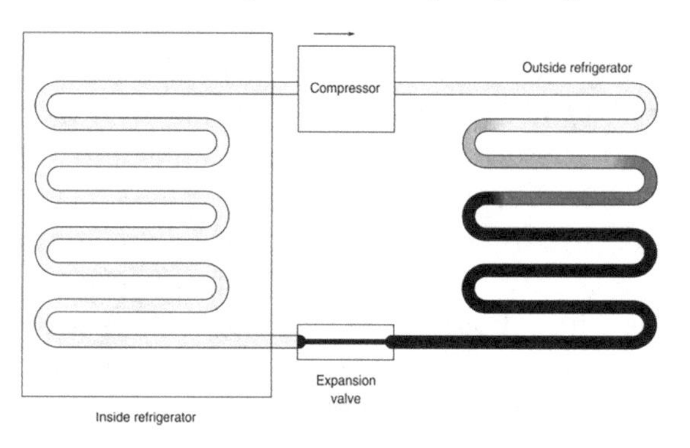

FIGURE 11.16 A refrigerator is a heat pump that draws heat from a low-temperature region to a higher-temperature region.

11.2.1 Refrigeration Cycle

1. A liquid refrigerant, such as ammonia or dichlorodifluoromethane, with a very low boiling point (around—27°F) passes through an expansion valve like the atomizer on a spray gun. As this mist expands and evaporates, it cools (Figure 11.17). This cool gas is run through coils in a refrigerator, cooling the inside.
2. At the other end of this coil is a compressor. The compressor draws in this gas and outputs it at a higher pressure and temperature (Figure 11.18).
3. This high-pressure, high-temperature gas passes through a set of coils on the back of the refrigerator. Passing through the coils, it emits this heat to the room and therefore lowers its temperature. As the gas cools, it turns back into a liquid.
4. The cycle repeats itself.

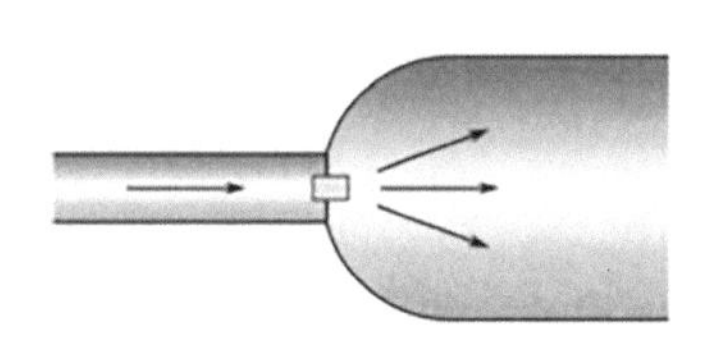

FIGURE 11.17 Expanding gas cools down.

Heat pumps can be used to cool a house, as in the case of an air conditioner (Figure 11.19), or warm a house (see Figure 11.20).

A *Peltier device* is an electronic refrigerator made from several metal-semiconductor junctions in series with an alternating array of n-type and p-type semiconductors (Figure 11.21). Current flowing through this

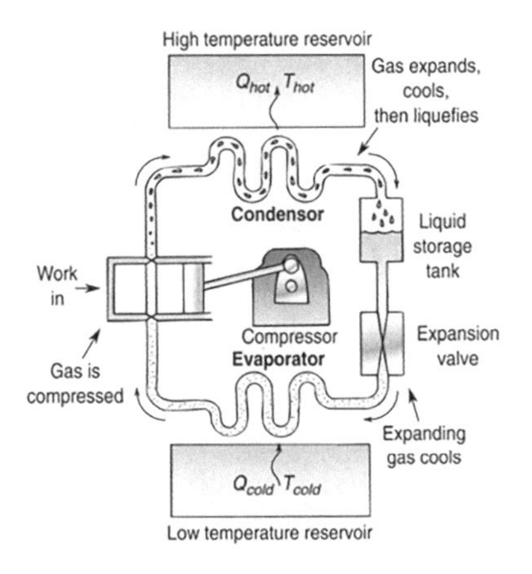

FIGURE 11.18 The gas is compressed, raising its temperature and pressure. The high-pressure, high-temperature gas expands, cools down, and turns back into liquid.

FIGURE 11.19 Air conditioners are basically refrigerators without the insulated box.

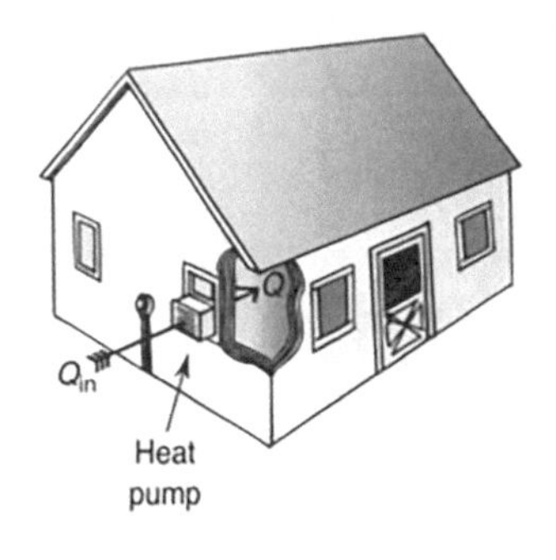

FIGURE 11.20 Heat pumps for heating. A heat pump can absorb heat from the outside during the winter and deliver it to the inside, where it is warmer. It is basically an air conditioner turned around.

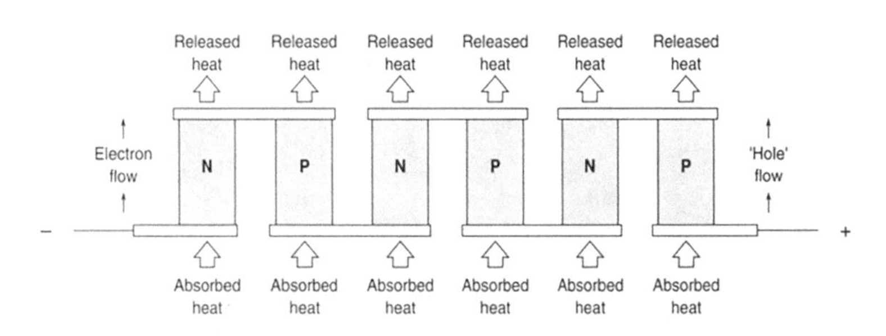

FIGURE 11.21 A Peltier device is an electronic refrigerator made from several metal-semiconductor junctions in series with an alternating array of n-type and p-type semiconductors.

device results in one side getting hot and the other cold. These devices can be found in electronic coolers.

11.3 THERMAL EXPANSION OF GASES, LIQUIDS, AND SOLIDS

Most materials expand when heated. Thermal expansion properties of materials can be exploited to build temperature switches, valves, refrigerators, air conditioners, and a host of other useful devices. Knowledge of how much they expand is important in designing anything that cycles through temperature ranges such as bridges, roads, or even printed circuit boards (Figure 11.22).

FIGURE 11.22 An expansion joint. As the seasons change, the bridge will expand in the heat and shrink in the cold.

11.3.1 Linear Expansion

A change in temperature causes most materials to change length (Figure 11.23). This can be expressed mathematically as:

$$\Delta L = \alpha \cdot L_i \cdot \Delta T$$

$$\text{Length change}(\text{m}) = \text{coefficient of expansion}(1/°\text{C}) \cdot \text{length}_{\text{Initial}}(\text{m}) \cdot \text{temperature change}(°\text{C})$$

In the English system:

$$\text{Length change}(\text{ft}) = \text{coefficient of expansion}(1/°\text{F}) \cdot \text{length}_{\text{Initial}}(\text{ft}) \cdot \text{temperature change}(°\text{F})$$

FIGURE 11.23 Linear thermal expansion.

11.3.2 Area and Volume Expansion

The thermal expansion of an area of a material is twice as big as the linear expansion, and the volume expansion is three times as big (Figure 11.24 and Table 11.2).

$$\Delta A = 2\alpha \cdot A_i \cdot \Delta T \leftarrow \text{Area expansion}$$

$$\Delta V = 3\alpha \cdot V_i \cdot \Delta T \leftarrow \text{Volume expansion}$$

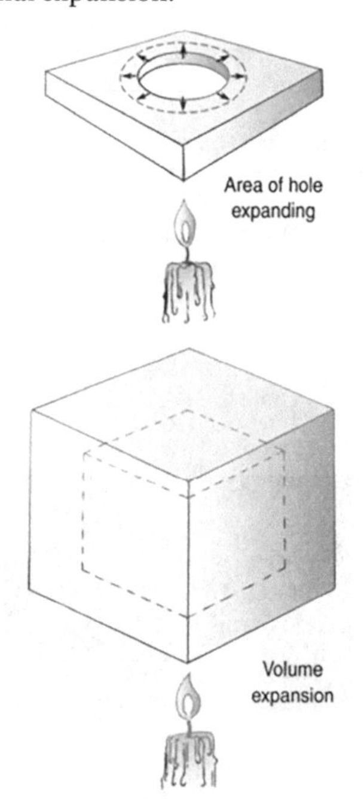

FIGURE 11.24 Area and volume thermal expansion.

TABLE 11.2
Coefficients of Linear Expansion

Material	α (Metric)	α (English)
Aluminum	$2.3 \times 10^{-5}/°C$	$1.3 \times 10^{-5}/°F$
Brass	$1.9 \times 10^{-5}/°C$	$1.0 \times 10^{-5}/°F$
Concrete	$1.1 \times 10^{-5}/°C$	$6.0 \times 10^{-6}/°F$
Copper	$1.7 \times 10^{-5}/°C$	$9.5 \times 10^{-6}/°F$
Glass	$9.0 \times 10^{-6}/°C$	$5.1 \times 10^{-6}/°F$
Pyrex	$3.0 \times 10^{-6}/°C$	$1.7 \times 10^{-6}/°F$
Steel	$1.3 \times 10^{-5}/°C$	$6.5 \times 10^{-6}/°F$
Zinc	$2.6 \times 10^{-5}/°C$	$1.5 \times 10^{-5}/°F$

Example 11.2

A steel tape measure is 100 ft at 72°F. The coefficient of linear expansion for steel is $6.5 \times 10^{-6}/°F$. How much did the tape measure change length at 5°F? (Figure 11.25)

$$L_i = 100\,\text{ft}$$

$$\alpha = 6.5 \times 10^{-6}/°\text{F}$$

$$\Delta T = 72°F - 5°F = 67°F$$

$$\Delta L = \alpha \cdot L_i \cdot \Delta T$$

$$\Delta L = \left(6.5 \times 10^{-6}/°\not{F}\right) \cdot 100\,\text{ft} \cdot 67°\not{F}$$

$$\Delta L = 0.04\,\text{ft}$$

$\Delta L = 0.5\text{in.} \leftarrow$ The length will change by as much as half an inch!

FIGURE 11.25 The length will change by as much as half an inch!

Example 11.3

A brass bushing with an inside diameter of 1.980 in. at 72°F is to be sweat fit onto a shaft of diameter 2.000 in. Sweat fitting is a process of increasing the size of something by heating it. What temperature must the bushing be brought to before it will fit? (Figure 11.26)

Initial diameter = 1.980 in. => the initial radius, $r_i = 1.980\text{in.}/2 = 0.990\text{in.}$

Final diameter = 2.000 in. => the final radius, $r_f = 2.000\text{in.}/2 = 1.000\text{in.}$

$$\alpha_{\text{brass}} = 1.0 \times 10^{-5}/°F$$

$$\Delta A = 2\alpha \cdot A_i \cdot \Delta T$$

$$\Delta T = \frac{\Delta A}{2 \cdot \alpha \cdot A_i}$$

$$\Delta T = \frac{\pi\left(r_f^2 - r_i^2\right)}{2 \cdot \alpha \cdot \pi \cdot r_i^2}$$

$$\Delta T = \frac{r_f^2 - r_i^2}{2 \cdot \alpha \cdot r_i^2}$$

$$\Delta T = \frac{(1.000\text{in.})^2 - (0.990\text{in.})^2}{2(10^{-5}/°F)(0.990\text{in.})^2}$$

$$\Delta T = 1015.2°F$$

$$T_{\text{final}} = T_{\text{initial}} + \Delta T$$

$$T_{\text{final}} = 72°F + 1015.2°F$$

$$T_{\text{final}} = 1087.2°F$$

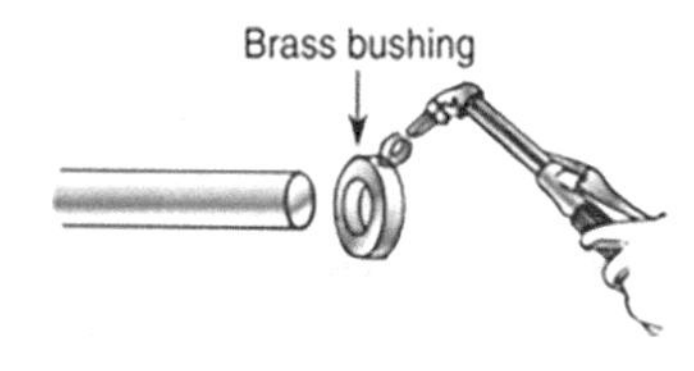

FIGURE 11.26 A thermal switch.

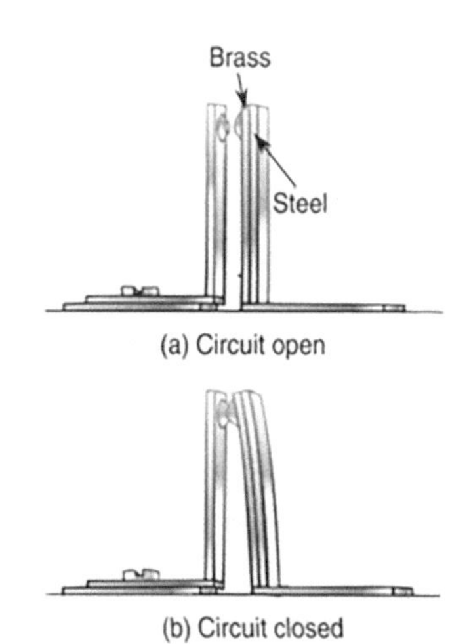

FIGURE 11.27 A thermal switch consists of two dissimilar pieces of metal that have different coefficients of thermal expansion. When heated, the result is a bending of the strip.

11.3.3 Bimetal Strip Thermostat

A *bimetal strip thermostat* is essentially a heat switch consisting of two types of metal bonded together, with electrical connections on either end. As the temperature changes, the two metals expand at different rates, causing the strip to bend and thereby closing the switch (Figure 11.27). Varying the spring tension that is holding the switch closed can set the operating temperature of the switch.

11.3.4 Nitinol Wire

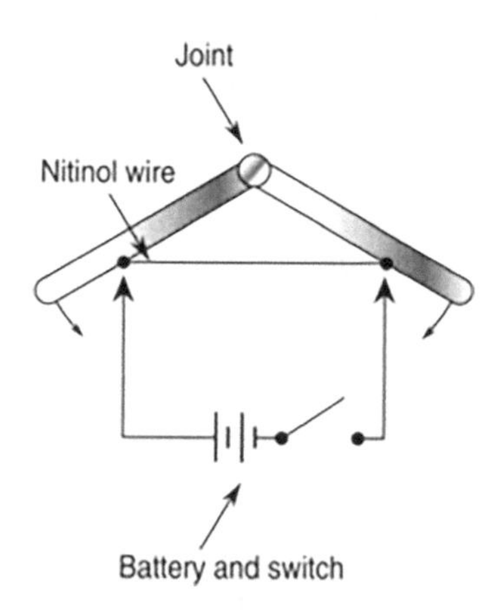

FIGURE 11.28 Nitinol wire contracts when heated, lending itself to actuating arms or levers. Heating takes place when enough current flows through the wire.

Nitinol wire is an alloy of nickel and titanium that is made specifically for actuating mechanical arms. It has a very large negative coefficient of expansion, that is, it contracts when heated. When heated, these wires contract as much as 20%. Running enough current through them will cause them to heat and therefore contract (see Figure 11.28). Turning off the current allows the wire to cool and slowly return to its original length. These wires can be used like motors to move arms or levers.

11.4 CHAPTER SUMMARY

Symbol	Unit
t—time	second (s), hour (h)
L—length	meter (m)
A—area	m^2
V—volume	m^3
T—temperature	Celsius (°C), Fahrenheit (°F), kelvin (K)
W—work	joules(J), British thermal unit (Btu)
Q—heat	joules(J), Btu, calorie (cal)
Efficiency	No unit
α—coefficient of expansion	$\frac{1}{°C}$ or $\frac{1}{°F}$

First law of thermodynamics: Energy is conserved as it moves through a thermodynamic system.

The first law applied to heat engines implies that the energy gained from a heat source equals the work done plus the heat wasted.

$$Q_{in} = W + Q_{out}$$

$$\text{Heat in} = \text{work} + \text{heat out}$$

Second law of thermodynamics: All systems in the universe normally evolve from order to disorder.

The **efficiency** of an engine is a measure of how much work it can do compared with how much heat energy is put in.

$$\text{Efficiency} = \frac{W}{Q_{\text{in}}} = \frac{Q_{\text{in}} - Q_{\text{out}}}{Q_{\text{in}}}$$

Ideal efficiency: The efficiency of a perfect heat engine is given by

$$\text{Efficiency} = \left(1 - \frac{T_{\text{out}}(\text{K})}{T_{\text{in}}(\text{K})}\right) \times 100\%$$

Linear expansion: A change in temperature causes most materials to change length. This can be expressed mathematically as:

$$\Delta L = \alpha \cdot L_i \cdot \Delta T$$

$$\text{Length change}(\text{m}) = \text{coefficient of expansion}\left(1/^\circ\text{C}\right) \cdot \text{length}_{\text{Initial}}(\text{m}) \cdot \text{temperature change}(^\circ\text{C})$$

In the English system:

$$\text{Length change}(\text{ft}) = \text{coefficient of expansion}\left(1/^\circ\text{F}\right) \cdot \text{length}_{\text{Initial}}(\text{ft}) \cdot \text{temperature change}(^\circ\text{F})$$

Area and Volume Expansion

$$\Delta A = 2\alpha \cdot A_i \cdot \Delta T \leftarrow \text{Area expansion}$$

$$\Delta V = 3\alpha \cdot V_i \cdot \Delta T \leftarrow \text{Volume expansion}$$

Heat pumps: Heat pumps pump heat from a low-temperature region to a high-temperature region.

PROBLEM SOLVING TIPS

- When calculating ideal efficiency, remember to calculate in degree kelvin.
- When solving problems dealing with thermal expansion, look at the units of the coefficient of thermal expansion when determining what units the other variables in the equation should be in.

PROBLEMS

1. If an engine performs 125,000 J of work using a fuel containing 620,000 J of energy, how much energy is wasted? What is the efficiency?
2. Why can't a perpetual motion machine ever be invented?
3. Compare two engines, one with an exhaust temperature of 1,600°C and the other with an exhaust temperature of 1,100°C. Which engine is more efficient?
4. Determine your calorie intake for one day and approximate the amount of work you perform. What is your efficiency?
5. A piece of brass 12.00 in. long is heated and changes temperature by 123°C. What is its change in length?
6. A piece of steel has a hole in it with a diameter of 1.00 in. A pipe is to fit into it with a diameter of 1.05 in. How high must the steel be heated so that the pipe will fit into the hole?
7. What is the purpose of an expansion joint on a bridge?
8. How does a thermostat work?
9. A thermal circuit breaker uses a bimetal strip to open a circuit in the event of an overload. How does this work?
10. How does an air conditioner work? Provide a diagram.
11. Give an example of the second law of thermodynamics from everyday life.
12. Suppose you connect an electric motor to a generator and use the output from the generator to run the motor. Will this work? What do the laws of thermodynamics say about this?
13. If a heat engine takes in 2,900 J of heat and expels 1,800 J, how much work did it perform? What is its efficiency?
14. What is the difference between an internal and an external combustion engine?
15. What does the first law of thermodynamics say?
16. How could you use an air conditioner to heat your house in the winter?
17. Gasoline can produce 19,000 Btu of heat/pound of fuel. If an engine burns 2 lb of gasoline and performs 8,500 Btu of work, how much energy was wasted?
18. What is the efficiency of the engine in the previous problem?
19. An electric motor consumes 5,000 J of energy producing 4,000 J of work. What is its efficiency?

20. If the temperature in the combustion chamber of a lawn mower engine reaches 2,000°C and the exhaust temperature is 1,100°C, what is its maximum possible efficiency?
21. If a human being is 25% efficient and eats a Big Mac hamburger of 540 Cal, how much work can he perform using this burger for fuel?
22. If we consider a perfect heat engine, a Carnot engine, running at 25% efficiency with an internal temperature at 1,900°C, what would the exhaust temperature be? Remember to convert the temperature into kelvin!
23. If an aluminum computer case 14 in. long warms up 25°F, what is its change in length?
24. A thermal switch made from two brass rods is separated by 0.05 cm before heating and close when heated up by 100°C. Assuming both the rods are of the same length and both are heated uniformly, what was their original length before heating? Remember both are expanding!
25. A square steel plate measuring 25 cm on a side is heated up by 500°C. What is its change in area?
26. A piece of Nitinol wire contracts when heated; it has a negative coefficient of expansion. If the contraction is 0.012 in./°F, how many inches will a 12 in. piece of wire contract if heated up by 10°F?
27. From Table 11.1, find out which fuel has the highest heat of combustion.
28. Where is most of the energy wasted in an automobile?
29. Cogeneration of electrical energy is the generation of electrical energy by burning fuel in a generator and using the wasted heat to warm a living space. If the generator is 25% efficient and burns 2,500 Btu of energy, how much energy is available for heat?
30. Your mom yells at you for having a messy room. How would you use the second law of thermodynamics in your defense to explain the condition of your room?
31. If an automobile loses a lot of its energy in heating up the water in the radiator, why don't they eliminate the radiator?
32. Name two ways to increase the power in an internal combustion engine.
33. A combination of diet and exercise is the best way to lose weight. If you could lower the efficiency of a human being, would this help in their weight loss?

34. A Peltier device is an electronic refrigerator. If it generates 40 W of cooling power and is 30% efficient, how much electrical power does it consume?
35. A 100 ft long house has aluminum siding put on it one winter. The summer comes and the house warms up 70°F from the winter. Do you have to worry about the siding falling off the house? Do a calculation.
36. From Table 11.2, find out which material has the lowest coefficient of expansion.

12 Electric Force

There are four forces in nature: gravity, two types of nuclear forces, the strong and weak nuclear forces, and electricity and magnetism. Electricity and magnetism may seem like two distinct forces; however, they are intimately related. The electric force is what holds electrons in their orbits around the nucleus. In this chapter, we will look at where the electrical force comes from and some applications of it in technology.

12.1 CHARGE

Consider an atom. At its center there are particles called neutrons and protons, and surrounding it are particles called electrons (Figure 12.1).

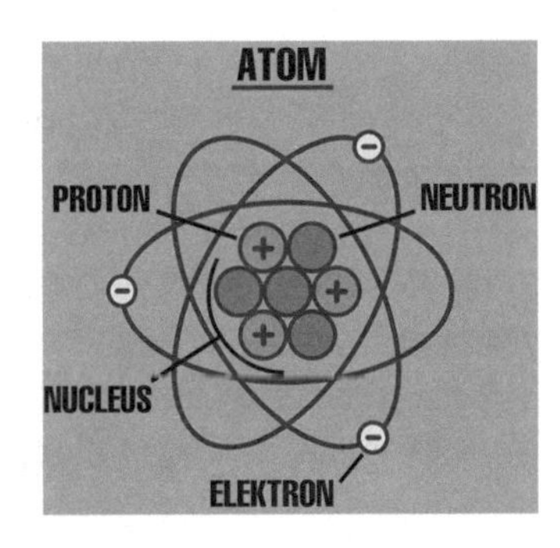

FIGURE 12.1 The atom consists of protons, electrons, and neutrons.

The protons and electrons have a property known as charge. There are two types of charges, positive and negative. The charge on an electron is negative and that of the proton is positive.

The unit of charge is the coulomb.

One coulomb of charge is equivalent to 6.25×10^{18} electrons bunched together.

One electron, e^-, therefore has a charge of $1/(6.25 \times 10^{18}) = -1.6 \times 10^{-19}$ coulombs.

One proton has a charge of $+1.6 \times 10^{-19}$ coulombs.

Particles that have different charges are attracted to each other, and particles with like charges are repelled by each other. This is similar to the way people behave, "opposites attract, likes repel." The charge of a particle is therefore the origin of the electrical force. If we place two charges near each other at rest and let go, they will begin to move. This change in motion is called an acceleration. According to Newton, if something is accelerating there must be a force on it. It's been determined by experiments that force between two charges depends on the amount of charge on each object and is inversely related to the distance squared between the objects (Figure 12.2).

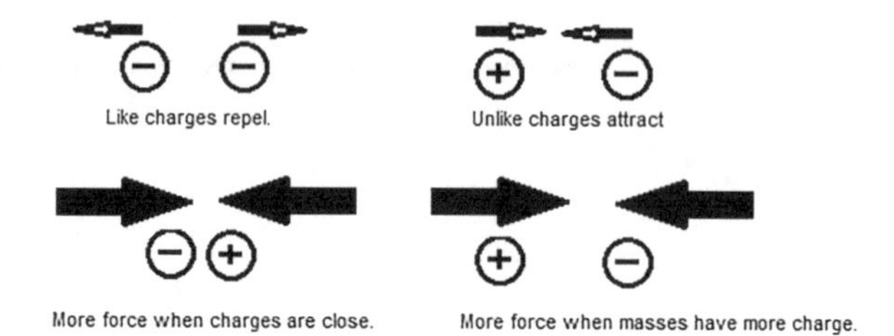

FIGURE 12.2 Like charges repel each other and unlike charges attract each other. The closer the charges are to each other, the stronger the force between them. The larger the charge on a mass, the larger the force.

Coulomb's law expresses the force between charges:

$$F = k \cdot \frac{q_1 \cdot q_2}{r^2}$$

$$\text{Force(N)} = \left(9 \times 10^9 \text{N} \frac{\text{m}^2}{\text{C}^2}\right) \frac{(\text{charge on 1 in Coulombs(C)})(\text{charge on 2 in Coulombs(C)})}{(\text{distance between each body in meters(m)})^2}$$

The constant $k = 9 \times 10^9 \text{N} \frac{\text{m}^2}{\text{C}^2}$ and is also equal to $k = \frac{1}{4\pi\varepsilon_0}$ where ε_0 is called the permittivity, and is a constant equal to $8.85 \times 10^{-12} \frac{\text{C}^2}{\text{Nm}^2}$. After plugging in for ε_0 it follows, $k = 9 \times 10^9 \text{N} \frac{\text{m}^2}{\text{C}^2}$.

Example 12.1

Two masses are charged and placed 0.1 m apart. For mass 1 assume q_1 = m1 C and m_1 = 1 kg, and mass 2 has q_2 = −m2 C and m_2 = 2 kg. Find the force on each mass and their accelerations (Figure 12.3).

Finding the force on each mass:

$$F = k \cdot \frac{q_1 \cdot q_2}{r^2} = 9 \times 10^9 \text{N} \frac{\text{m}^2}{\text{C}^2} \frac{(0.001\text{C})(-0.002\text{C})}{(0.1\text{m})^2}.$$

$F = -1.8 \times 10^6$ N ← Force on each mass directed toward each other.

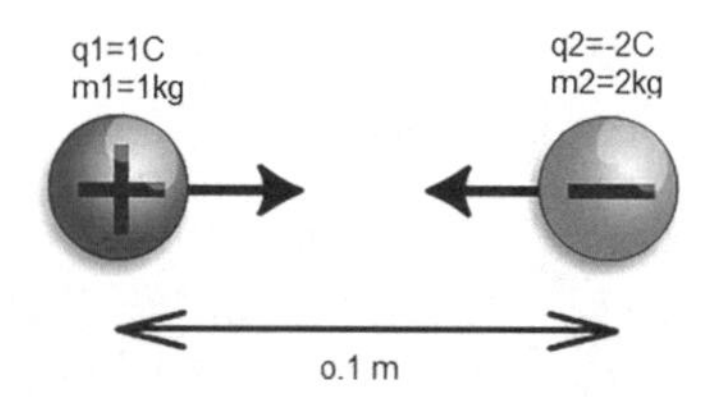

FIGURE 12.3 Find the force on each mass and their acceleration.

Finding the magnitude of acceleration of each mass we use $F = m \cdot a$.

$$a_1 = \frac{F}{m_1} = \frac{1.8 \times 10^6 \text{N}}{1\text{kg}}$$

$$a_1 = 1.8 \times 10^6 \frac{\text{m}}{\text{s}^2} \leftarrow \text{Directed toward mass 2.}$$

$$a_2 = \frac{F}{m_2} = \frac{1.8 \times 10^6 \text{N}}{2\text{kg}}$$

$$a_2 = 9 \times 10^5 \text{ m/s}^2 \leftarrow \text{Directed toward mass 1.}$$

12.2 THE ELECTRIC FIELD

To explain how charges can influence each other at a distance, the concept of the electric field was invented. An electric field is something that surrounds the space around the charge in which another charge reacts to. This is similar to the gravitational field, where one body can influence the other at a distance (Figure 12.4).

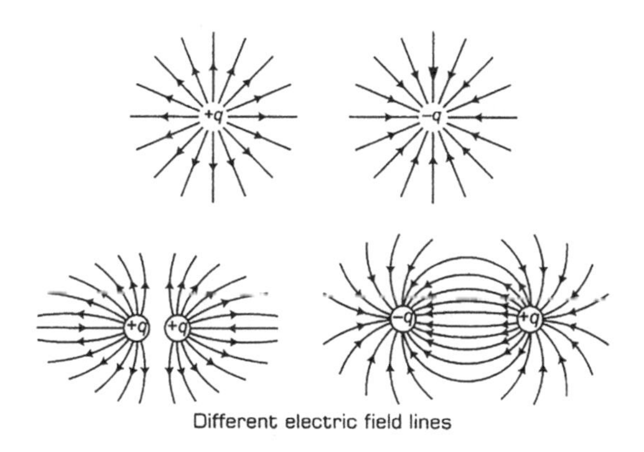

FIGURE 12.4 Electric field lines of positive and negative charges and between charges.

The electric field E has a magnitude and direction and is therefore vector. The force on a charge will be in the same direction as the electric field if the charge is positive and in the opposite direction if the charge in negative (Figure 12.5).

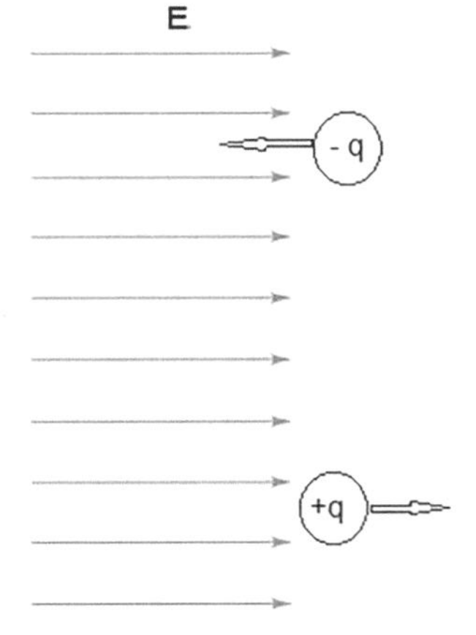

FIGURE 12.5 The force on a charge in an electric field E.

The strength of an electric field is measured in Newton/Coulomb. The force on a charge q due to the electric field E is given by:

$$F = q \cdot E$$

$$\text{Force(N)} = (\text{Charge(C)}) \cdot (\text{Electric field (N/C)})$$

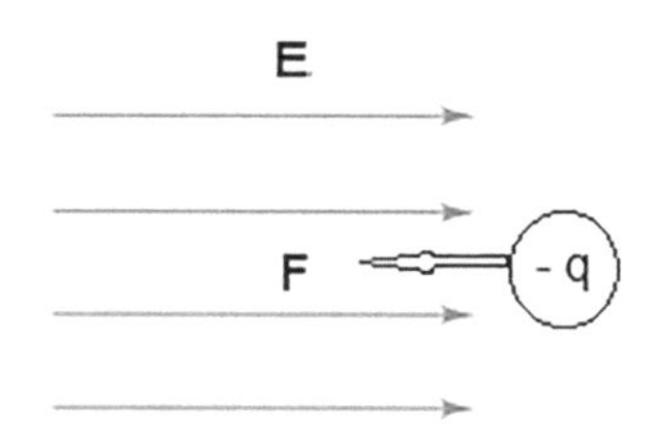

FIGURE 12.6 What is the force on the electron?

Example 12.2

An electron is placed in an electric field of strength 2 N/C pointing in the positive x direction. What is the force on it? (Figure 12.6)

$$E = 2\text{N/C}$$

$$q = -1.6 \times 10^{-19}\,\text{C}$$

$$F = q \cdot E$$

$$F = \left(-1.6 \times 10^{-19}\,\text{C}\right)\left(2\text{N/C}\right)$$

$$F = -3.2 \times 10^{-19}\,\text{N}$$

12.3 THE ELECTRIC FIELD OF A POINT CHARGE

We can figure out the electric field of a point charge, such as an electron, by way of comparison.

Let's compare the electric force law to Coulomb's law.

$$F = k \cdot \frac{q_1 \cdot q_2}{r^2}$$

$$F = q \cdot E$$

Equating the two, we see that the electric field due to a point charge is:

$$F = q_1 \cdot E = k \cdot \frac{q_1 \cdot q}{r^2}$$

$$E_{\text{point charge}} = k \cdot \frac{q}{r^2}$$

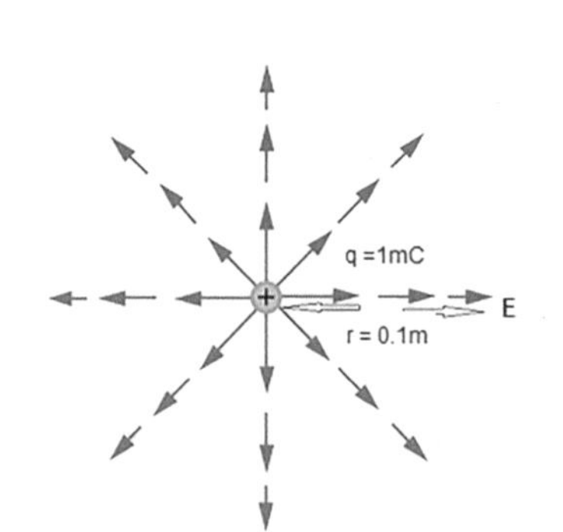

FIGURE 12.7 What is the strength of the electric field 0.1 m away from a 1 mC charge?

Example 12.3

What is the strength of the electric field 0.1 m away from a 1 mC charge? (Figure 12.7)

$$E_{\text{point charge}} = k \cdot \frac{q}{r^2}$$

$$E_{\text{point charge}} = 9 \times 10^9 \text{N}\frac{\text{m}^2}{\text{C}^2} \cdot \frac{10^{-3}\text{C}}{(0.1\text{m})^2}$$

$$E_{\text{point charge}} = 9 \times 10^8 \text{N}$$

12.4 WORK AND THE ELECTRIC FIELD

We learned in a previous chapter that the change in energy in a body is equal to the work done on it. $W = F \cdot d$ (when the force is in the same direction as motion).

If we substitute the electrical force $F = q \cdot E$ into W we get:

$$W = q \cdot E \cdot d$$

$$\text{Work(J)} = (\text{charge(C)}) \cdot (\text{Electric field(N/C)}) \cdot (\text{distance(m)})$$

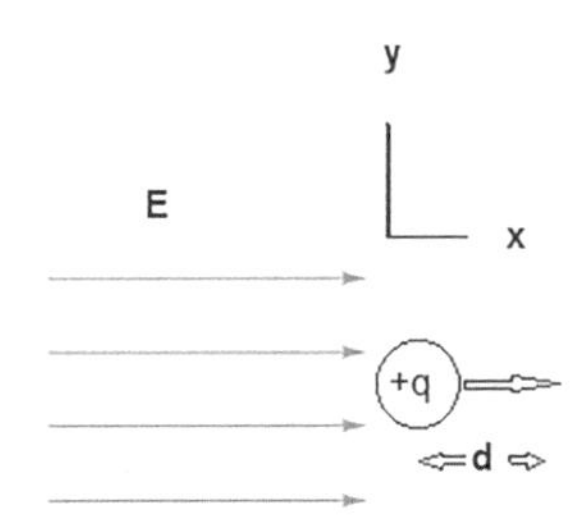

FIGURE 12.8 What is the work done on this charge?

Example 12.4

A proton moves a distance of 0.01 m in the positive x direction through an electric field of strength 0.1 N/C pointing in the positive x direction. What is the work done on this charge? (Figure 12.8)

$$q = 1.6 \times 10^{-19}\,\text{C}$$
$$d = 0.01\text{m}$$
$$E = 0.1\text{N/C}$$
$$W = q \cdot E \cdot d$$
$$W = (1.6 \times 10^{-19}) \cdot (0.1\text{N/C}) \cdot (0.01\text{m})$$
$$W = 1 \cdot 6 \times 10^{-22}\,\text{J}$$

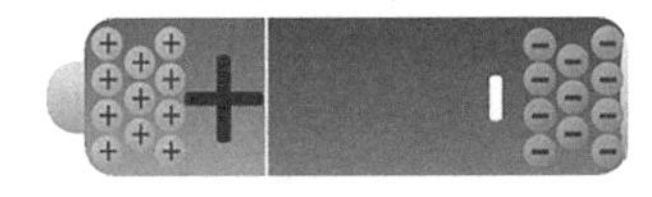

FIGURE 12.9 A simple battery consists of two halves, one half is full of negative charge and the other has a deficit of negative charge, courtesy of sparkfun.com.

12.5 GENERATING AN ELECTRIC FIELD

A uniform electric field can be generated between two metal parallel plates by placing an equal amount of charge on each plate but of opposite signs. So, from where do we get the charge to place on the plates? We can get this charge from a battery. A simple battery consists of two halves, one half is full of negative charges and the other has a deficit of negative charges (Figure 12.9).

Consider two parallel plates connected to a battery. The electrons stored in the negative side of the battery are repelled by each other, so if a wire is connected to the negative terminal of a battery, the electrons will "crawl" up the wire to get away from the other electrons in the battery. These electrons will begin to fill the negative plate with charge. Meanwhile, on the other plate the "free" electrons on it will start to leave this plate to get away from the negative charge they "feel" on the negative plate. This exodus of negative charge has only one place to go, into the positive side of the battery. This will result in this plate to be effectively charged positive, that is, if we take away negative charge from a charge neutral region, that region is effectively charged positive (Figure 12.10).

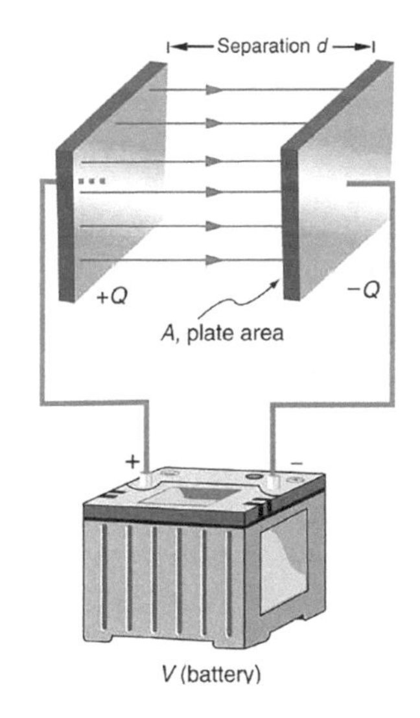

FIGURE 12.10 A uniform electric field can be generated between two metal parallel plates by placing an equal amount of charge on each plate but of opposite signs by using a battery.

Example 12.5

Suppose we place a negative charge next to the positive plate, and move it toward the negative plate. Calculate the work done on an electron by pushing it a distance

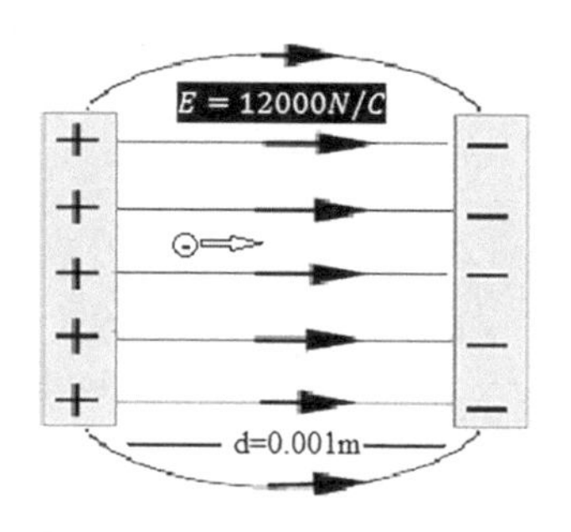

FIGURE 12.11 Calculate the work done and the increase in potential energy of the charge.

d through an electric field to the negative plate, and find its increase in potential energy (Figure 12.11).

Assume:

$E = 12{,}000$ N/C pointing in the positive x direction and $x = 0$ at the +Q plate.

$d = 0.001$ m

$q = -1.6 \times 10^{-19}$C

$$W = q \cdot E \cdot d$$

$$W = \left(-1.6 \times 10^{-19}\text{C}\right) \cdot \left(12{,}000 \frac{\text{N}}{\text{C}}\right) \cdot (0.001\text{m})$$

$$W = -1.92 \times 10^{-18}\,\text{J}$$

Recall from the chapter on energy that the change in the change in Potential Energy = – Work performed, therefore:

$$\Delta U = -W = 1.92 \times 10^{-18}\,\text{J}$$

12.5.1 Electrical Potential Energy and Voltage

The electric field is related to something we are more familiar with, that is, voltage. *Voltage* is effectively the "electrical pressure" that drives the charges to move in a circuit. A battery at a certain voltage, connected to a circuit, results in charges moving through the circuit, or current. The charges, electrons, stored in the battery are repelled by each other through the electric field generated by each charge. A battery connected up to a circuit, therefore, sets up an electric field in the wires of the circuit, which in turn delivers a force on the electrons to push them through the wires. The battery therefore acts as a source of charge and a type of "electrical pressure" to drive the charge through the circuit. A battery is a store of electrical energy. The electrical potential energy stored in a battery depends on how much charge is stored in it as well as what the voltage of the battery is.

$$U = q \cdot V$$

$$\text{Potential Energy (J)} = (\text{charge(C)}) \cdot (\text{Voltage(V)})$$

Example 12.6

A fully charged AA battery has a voltage of 1.5 V and 10^4 coulombs of charge in it. How much electrical potential energy is stored in it? (Figure 12.12)

$$U = q \cdot V$$

$$U = \left(10^4\,\text{C}\right)(1.5\text{V})$$

$$U = 1.5 \times 10^4\,\text{J}$$

FIGURE 12.12 AA battery contains ~ 10^4 J of stored energy.

Compare this to the energy stored in one cup of gasoline at about 8×10^6 J.

12.6 ELECTRIC FIELD AND VOLTAGE

Let's see how the electric field and the voltage are related.

Recall the work energy theorem $W = -\Delta U$.

Plugging in for W and U gives:

$$W = -\Delta U$$
$$q \cdot E \cdot d = -\Delta(q \cdot V)$$

Solving for E gives:

$$\text{Electric field}\left(\frac{\text{V}}{\text{m}}\right) = -\frac{\text{Change in voltage(V)}}{\text{distance(m)}}$$

The electric field can have units of N/C or V/m.

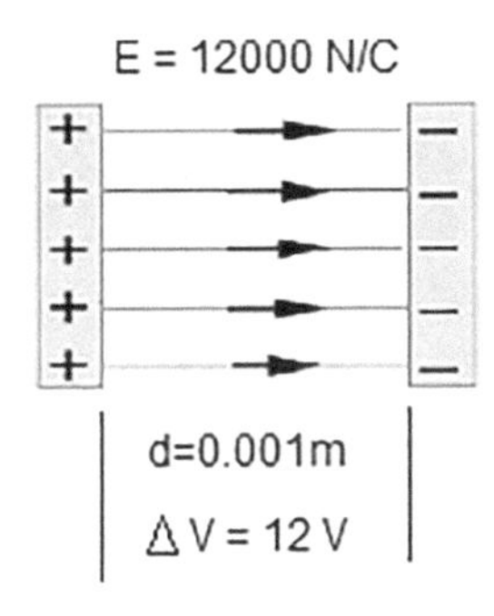

FIGURE 12.13 Find the voltage difference between the plates.

Example 12.7

Consider the previous example where two parallel plates had an electric field E = 12,000 N/C pointing in the positive x direction and x = 0 at the +Q plate. Find the voltage difference between the plates (Figure 12.13).

Assume:
d = 0.001m
E = 12,000 N/C or we can write E = 12,000 V/m

$$\Delta V = -E \cdot d = -\left(12{,}000\frac{\text{V}}{\text{m}}\right) \cdot (0.001\text{m})$$

$\Delta V = -12\text{V}$ (The voltage dropped 12V going from the positive plate to the negative plate.)

12.6.1 Armed and Dangerous, Building an Electron Gun

An electron gun can be built by accelerating electrons from one plate of a capacitor to another plate with a hole in it, Figure 12.14. The plates have a battery connected to them to create an electric field between them that in turn accelerates the electrons. The source of the electrons to be accelerated will come from a heated wire, the filament. The filament will heat up by running electricity through it from a battery. Once the filament becomes hot enough it will begin to "boil" off electrons. These electrons will pass through the negative plate, which is made from a metal screen. The electrons that make it through the screen will enter the region where the electric field exits. These electrons are accelerated toward the positive plate and some will pass through the hole in the plate at a high velocity. Now, let's do the numbers and calculate this velocity.

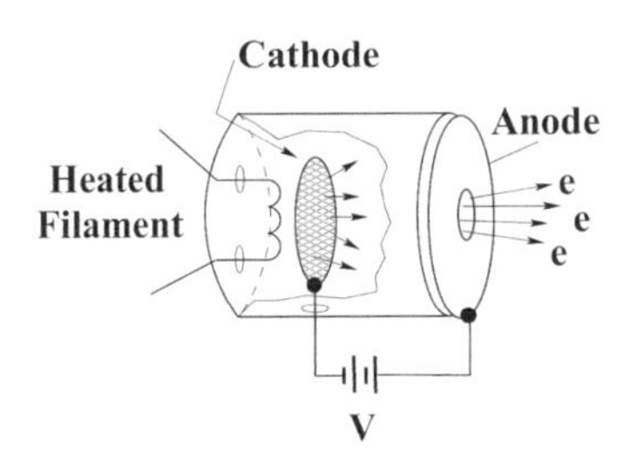

FIGURE 12.14 An electron gun.

Example 12.8

Calculate the speed of the electrons exiting an electron gun, given the following:

$V = 12\,\text{V}$ (voltage between the pates)
$d = 0.02\,\text{m}$ (distance between the plates)
$q = -1.6 \times 10^{-19}\text{C}$ (charge of an electron)
$m = 9.11 \times 10^{-31}\,\text{kg}$ (mass of an electron)

We will assume that the electrons start with an initial velocity of zero to simplify the calculation. We will also assume that the system sits in a vacuum so that the electrons do not "bump" into air molecules. We can use energy conservation to determine the electrons' exit velocity,

$$\Delta K = -\Delta U$$

$$\frac{1}{2}mv^2 = -q \cdot \Delta V$$

$$v = \sqrt{-\frac{2q \cdot \Delta V}{m}} = \sqrt{-\frac{(2)\left(-1.6\times 10^{-19}\,\text{C}\right)(12\text{V})}{9.11\times 10^{-31}\,\text{kg}}}$$

$$v = 2.05 \times 10^{6}\,\frac{\text{m}}{\text{s}} \text{ (This is 0.68\% the speed of light!)}$$

FIGURE 12.15 The largest particle accelerator in the world called the LHC, Large Hadron Collider, located in Geneva, Switzerland.

The Large Hadron Collider (LHC) is currently the world's largest and most powerful particle accelerator. The LHC consists of a 27 km ring of magnets wrapped around two pipes under vacuum that the accelerated particles travel through in opposite directions. Situated throughout the ring are a number of accelerating devices to boost the energy of the particles as they move through this beamline. The acceleration of these particles is done using electric fields that are oscillating at just the right frequency and phase to boost the particles to move at near the velocity of light. It is analogous to pushing someone on a swing; if you push at just the right time they will move higher and higher. When the particles reach a particular energy, the magnets steer the two beams of particles that are moving in opposite directions into each other. The result of these colliding particles is recorded with large detectors surrounding the collision zone. The 2015 Nobel Prize in Physics was awarded based on a particle called the Higgs Boson discovered at the LHC (Figure 12.15).

12.7 CAPACITANCE

The ability to store charge is very useful and used throughout modern technology. The capacity to store charge is known as **capacitance**, and a device that is designed to store charge is called a **capacitor**. Capacitors are used for regenerative breaking in electric cars, memories in computers, microphones, computer touch screens, accelerometer sensors in smartphones, and the list goes on. The schematic symbol for a capacitor in electronics is drawn as a set of parallel plates. Figure 12.16 shows some types of capacitors used in electronics.

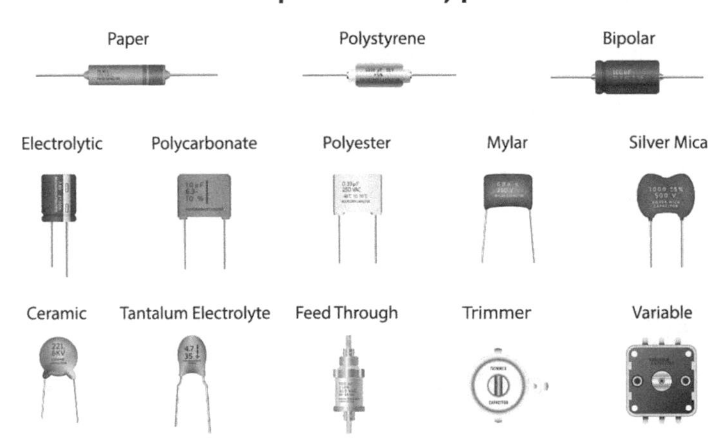

FIGURE 12.16 Capacitors used in electronics.

The charge a body can store depends on its capacitance, which depends on its geometry and its voltage.

$$q = C \cdot V$$

$$\text{Charge}(\text{C}) = (\text{Capacitance}(\text{F})) \cdot (\text{Voltage}(\text{V}))$$

The unit of capacitance is the farad. One farad is a very large unit and normally the capacitance of something is declared in units such as:

1 microfarad (1 μF) = $10\ ^{6}$ F
1 picofarad (1 pF) = 10^{-12} F

Example 12.9

How much charge can a 10 μF capacitor store if it has a voltage of 5 V across it?

$$q = C \cdot V$$
$$q = (10\quad \text{F}) \cdot (5\text{V})$$
$$q = 50\quad \text{C}$$

Parallel plate capacitors consist of two parallel conductive plates separated by a small distance with air or a nonconductive material, called a dielectric, separating the plates. (Figure 12.17)

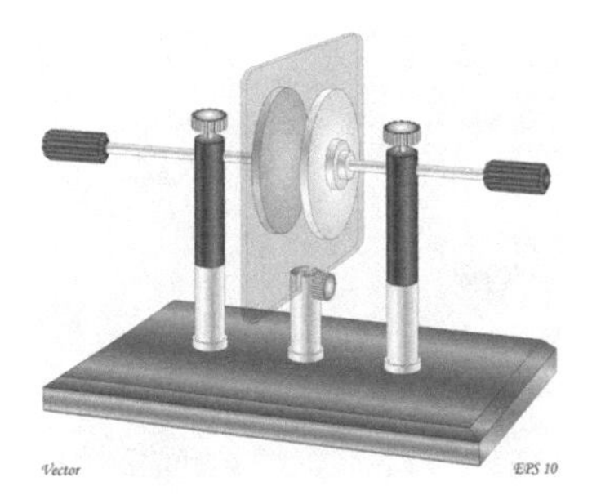

FIGURE 12.17 A parallel plate capacitor.

The capacitance of these parallel plates, or the ability to hold charge depends on the area A of the plates, the separation distance of the plates d, and the permittivity ε of the dielectric, i.e., the material between the plates. The permittivity is the material's ability to "resist" an electric field. This is all summarized by the following equation:

$$C = \frac{\varepsilon \cdot A}{d}$$

$$\text{Capacitance}(\text{F}) = \frac{\left(\text{permittivity}\left(\frac{\text{C}^2}{\text{Nm}^2}\right)\right) \cdot \left(\text{Area}\left(\text{m}^2\right)\right)}{\text{distance}(\text{m})}$$

The permittivity of a material is given by $\varepsilon = \varepsilon_r \cdot \varepsilon_0$ where ε_r is called the relative permittivity and $\varepsilon_0 = 8.85 \times 10^{-12} \frac{\text{C}^2}{\text{Nm}^2}$.

Example 12.10

A capacitor consists of two parallel plates of area $A = 0.01\ \text{m}^2$ separated by a distance $d = 10^{-4}$ m with Mylar, which has a relative permittivity $\varepsilon = 3.2$. Find its capacitance.

$$C = \frac{\varepsilon \cdot A}{d} = \frac{(3.2)\left(8.85 \times 10^{-12} \frac{\text{C}^2}{\text{Nm}^2}\right) \cdot \left(0.01\text{m}^2\right)}{10^{-4}\,\text{m}}$$

$$C = 2.8 \times 10^{-9}\,\text{F}$$

12.7.1 Energy Storage

The energy storage capability of capacitors continues to rise with advances in technology. Electric cars, trains, backup power for modern electronics are some of the uses of modern capacitors. The energy storage capability of a capacitor is given by:

$$U = \frac{1}{2} \cdot C \cdot V^2$$

$$\text{Energy}(\text{J}) = \frac{1}{2}(\text{Capacitance}(\text{F})) \cdot (\text{Voltage}(\text{V}))^2$$

Example 12.11

A 10 F capacitor is charged to a voltage of 5 V. How much energy is stored in it? How does this energy stored compare to a AA battery?

$$U = \frac{1}{2} \cdot C \cdot V^2 = \frac{1}{2}(10\text{F}) \cdot (5\text{V})^2$$

$U = 125$ J of energy stored in the capacitor.

Compare this stored energy to 10^4 J of energy stored in a AA battery. The AA battery has about 100 times more stored energy than this capacitor!

Capacitors are not going to replace batteries yet for large, long-term energy storage, such as in an electric car. Their energy density, that is, how much energy thay can store per weight is much smaller than that for batteries (Figure 12.18).

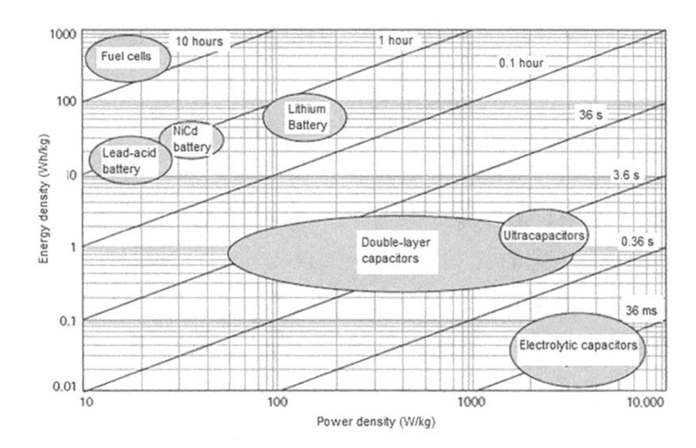

FIGURE 12.18 Graph of the energy densities of batteries and capacitors along with their charging times, courtesy of the US Defense Logistics Agency.

12.7.2 Don't Be Shocked by the Following

In the winter, when the air is dry, walking across a rug sometimes results in you receiving an electrical shock; this is called static electricity. What is happening here is your feet are rubbing electrons off of the carpet. The electrons lost by the carpet have now accumulated on your body, which now has a negative charge. It is possible to reach a voltage of tens of thousands of volts relative to your surroundings. Now if you touch, say a door knob, the excess electrons on your body will discharge to the knob.

Example 12.12

Suppose you walk across the rug accumulating charge as you walk. When reaching for the doorknob you observe a spark 0.5 cm long leaving your hand. Being the curious person you are you look up the **breakdown voltage** of air, that is, the maximum voltage difference between two objects separated by some distance in air before an arc occurs. A quick Google search comes up with a breakdown voltage of air 3 kV/mm and that the human body roughly has a capacitance of 100–200 pF; let's call it 150 pF. What was the accumulated charge on your body?

$$d = 0.5\,\text{cm} = 5\,\text{mm}$$

$$C = 150\,\text{pf}$$

$$V = \left(3\frac{\text{kV}}{\text{mm}}\right)5\,\text{mm}$$

$$V = 15\,\text{kV} = 15{,}000\,\text{V}$$

$$q = C \cdot V = \left(15\times10^{-12}\,\text{F}\right)\cdot\left(15{,}000\,\text{V}\right)$$

$$q = 2.25\times10^{-7}\,\text{C}$$

Is this a lot of charge? Let's see how many electrons this would be. The charge of one electron is -1.6×10^{-19} C so that the number of these extra electrons on your body is:

$$\text{Number} = \left(\frac{1 \text{ electron}}{1.6 \times 10^{-19} \text{C}}\right)\left(2.25 \times 10^{-7} \text{C}\right)$$

$$\text{Number} = 1.4 \times 10^{12} \text{ electrons} (1.4 \text{ trillion electrons!})$$

12.7.3 Van de Graaff Generator

Imagine if you could automate the collecting of charge as in the case of walking across a rug in the winter. A Van de Graaff generator is a device that does just this. It consists of a metal sphere acting as a capacitor and a moving belt between two rollers, see Figure 12.19. The lower roller has a motor connected to it to drive the belt. There is a metal "comb" that rubs the belt, as in the case of you walking across the rug in the winter, and becomes charged negative, and the belt becomes positive. This positive charge is transported upward to a metal sphere. The sphere is connected to another comb rubbing the belt. The electrons on the sphere are attracted to the positive belt. The belt removes electrons from a metal sphere by transferring this charge off of the sphere through the comb to the belt. The removal of negative charge results in the sphere becoming positively charged. Tabletop Van de Graaff generators can typically have voltages of hundreds of thousands of volts.

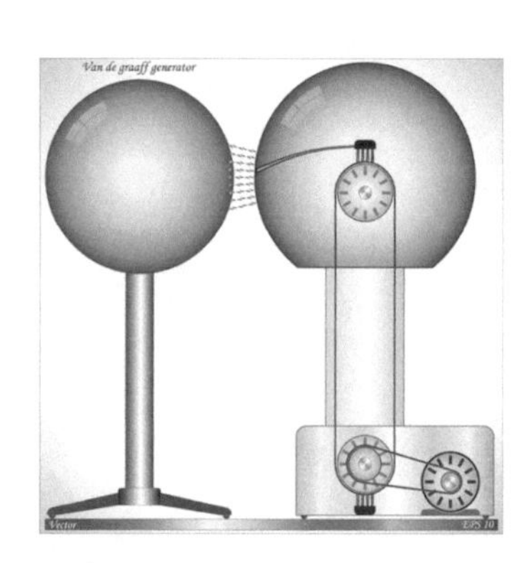

FIGURE 12.19 A Van de Graaff generator.

12.7.4 Catching a Lightning Bolt (Don't Try This at Home!)

Capacitors are able to get charged and discharged much more quickly as compared to batteries. Could a capacitor be used to collect energy from lightning? Lightning strikes, over the entire earth, roughly 44 times/s. One bolt of lightning contains roughly a billion joules of energy delivered in several microseconds. Recall that power $P = E/t$, therefore we can approximate the power in a single lightning bolt to be:

$$P = \frac{E}{t} = \frac{10^{9} \text{ J}}{10^{-6} \text{ s}}$$

$P = 10^{15}$ watts (This is only an estimate since bolts vary according to voltage, current, and duration.)

Wow! Looks like free energy if we could just capture and store it. However, harvesting energy from lighting is not easy. It is sporadic, and varies in location and intensity. Methods to capture this energy range from indirect energy conversion, such as capturing the hydrogen that disassociates from rapidly heated water by the lightning bolt, to direct conversion using lightning arrestors to capture part of the strike and store the energy in a capacitor. So far, no company has developed a practical commercial system (Figure 12.20).

FIGURE 12.20 Harvesting energy from lighting is not easy.

12.8 CHAPTER SUMMARY

Symbols used in this chapter:

Symbol	Unit
t—time	second (s)
r—distance	meter (m)
d—distance	meter (m)
A—area	m^2
v—velocity	m/s
a—acceleration	m/s^2
m—mass	kg
F—force	newton (N)
U—potential Energy	joule (J)
U—energy	joule (J)
q—charge	coulomb (C)
k—Coulomb's law constant	$9\times10^9\,N\frac{m^2}{C^2}$
E—electric field	N/C or V/m
ε—permittivity of dielectric	$\frac{C^2}{Nm^2}$
ε_0—permittivity of empty space	$\frac{C^2}{Nm^2}$
V—voltage	volt (V)
W—work	joule (J)
C—capacitance	farad (F)

Coulomb's law describes the force between two charges.

$$F = k\cdot\frac{q_1\cdot q_2}{r^2}$$

$$\text{Force(N)} = \left(9\times10^9\,N\frac{m^2}{C^2}\right)\frac{(\text{charge on 1 in Coulombs(C)})(\text{charge on 2 in Coulombs(C)})}{(\text{distance between each body in meters (m)})^2}$$

The electric field is the medium through which charges interact with each other. The force on a charge q due to the electric field E is given by:

$$F = q \cdot E$$

$$\text{Force(N)} = (\text{Charge(C)}) \times (\text{Electric field (N/C)})$$

Work and the Electric Field

$$W = q \cdot E \cdot d$$

$$\text{Work(J)} = (\text{charge(C)}) \times (\text{Electric field(N/C)}) \times (\text{distance(m)})$$

The electrical potential energy stored in a battery depends on how much charge is stored in it as well as what the voltage of the battery is.

$$U = q \cdot V$$

$$\text{Energy}_{\text{Potential}}\ (\text{J}) = (\text{charge(C)})(\text{voltage(V)})$$

Electric Field and Voltage

$$E = -\frac{V}{d}$$

$$\text{Electric field}\left(\frac{\text{V}}{\text{m}}\right) = -\frac{\text{voltage(V)}}{\text{distance(m)}}\text{(The electric field can have units of N/C or V/m)}$$

Capacitance is a measure of how much charge a body can store.

$$q = C \cdot V$$

$$\text{Charge(C)} = (\text{Capacitance(F)}) \cdot (\text{Voltage(V)})$$

Parallel plate capacitors consist of two parallel conductive plates separated by a small distance with air or a nonconductive material, called a dielectric, separating the plates.

The capacitance of these plates or the ability to hold charge is given by the following:

$$C = \frac{\varepsilon \cdot A}{d}$$

$$\text{Capacitance(F)} = \frac{\left(\text{permittivity}\left(\frac{\text{C}^2}{\text{Nm}^2}\right)\right) \cdot \left(\text{Area}\left(\text{m}^2\right)\right)}{\text{distance(m)}}$$

The permittivity of a material is given by $\varepsilon = \varepsilon_r \cdot \varepsilon_0$ where ε_r is called the relative permittivity and $\varepsilon_0 = 8.85 \times 10^{-12} \frac{\text{C}^2}{\text{Nm}^2}$.

The energy storage capability of a capacitor is given by:

$$U = \frac{1}{2} \cdot C \cdot V^2$$

$$\text{Energy}(\text{J}) = \frac{1}{2}(\text{Capacitance}(\text{F})) \cdot (\text{Voltage}(\text{V}))^2$$

PROBLEMS

1. What is the force between an electron and a proton separated by 10^{-10} m in a hydrogen atom?
2. A hydrogen atom at first seems like an unstable system in the sense that the electron should "crash" into the proton because they are electrically attracted to each other. Why does this not happen?
3. Suppose you build a hover board by charging a board you stand on and a platform below it with an equal charge. What would the charge have to be to support the weight of a 712 N person, 0.001 m above the ground?
4. Two electrons separated by 1 m will accelerate away from each other at what rate?
5. What is the electric field generated by a proton at a distance of 0.01 m?
6. What is the force on an electron placed 0.01 m from a proton?
7. A charge of +2 C is placed in an electric field of strength 2 N/C, pointing in the positive x direction. What is the force on it?
8. An evil scientist charges the earth and the moon equally with a charge in hopes of separating them even more, just to see what would happen. There would now be both an electrical force and a gravitational force on them. Find the charge so that the net force between them, $F_{\text{Net}} = F_{\text{Electric}} - F_{\text{Gravitation}} = 0$. Use $m_{\text{earth}} = 5.972 \times 10^{24}$ kg, $m_{\text{moon}} = 7.35 \times 10^{22}$ kg.
9. What is the electric field between the parallel plates of a capacitor if the plates are separated by 1 mm and have a voltage of 100 V across them?
10. A hydrogen atom is placed between the plates separated by 0.001 m of a capacitor. What voltage would have to be placed across the plates to separate the electron from the proton in the hydrogen atom?
11. How fast will a charge of 0.1 C and mass 0.01 kg be moving if released from rest in an electric field of 0.1 N/C after 1 s?
12. Let's build a particle collider by accelerating charges at each other between the parallel plates of a capacitor. How many times faster will the electron be moving relative to the proton when they collide?
13. How much work is done on a charge of +2 C moved 1cm in the positive x direction through an electric field of 10 N/C pointing

in the negative x direction? What is the change in the charge's potential energy?

14. What is the permeability of a dielectric with a relative permeability = 5?
15. What is the energy stored in a AA battery holding a charge of 3,600 C?
16. Compare the energy content stored in the following batteries, D-cell, AA, 9 V. Assume a AA battery has 3,600 C of charge, a D-cell has 20 times the of charge of a AA battery, and a 9 V battery has half the of charge of a AA battery.
17. How much charge can a capacitor of 10 μF store if it has 5 V across it?
18. How does the capacitance change if the plate separation increases?
19. How does the capacitance change if the dielectric permeability increases?
20. What is the voltage on a capacitor of 10 μF if it has 0.1 μC of charge on it?
21. What is the capacitance of a parallel plate capacitor with plates of area 0.001 m^2, plate separation of 1 mm, and a dielectric of relative permittivity $\varepsilon = 2.6$?
22. If the plates of a capacitor are doubled in area, how does this change the capacitance?
23. How much energy can a capacitor of 10 F store if charged to 3.3 V?
24. A 2 F capacitor stores 25 J of energy. What is the voltage across it?
25. How many capacitors with a capacitance of 2 F and a voltage of 5 V would it take to equal the energy contained in a stick of dynamite at 10^6 J/stick?
26. A defibrillator delivers a 250 J of energy to a patient at 500 V. How much charge is delivered?
27. What would happen if a capacitor has too much voltage applied across it?
28. Suppose an electric car is powered by a 12 V battery containing 4.1 MJ of energy. If it takes 150 lb or 668 N of force to move the car at a constant rate, how far can the car travel? Assume an efficiency of 70%.
29. Suppose a cloud in a thunderstorm has a voltage relative to the earth of 4×10^5 V and is 4×10^3 m above the earth. What is the electric field in the atmosphere?
30. How many AA batteries holding 3,600 C of charge would equal the energy in a bolt of lightning at 10^9 J?

13 Electricity

Electricity is the movement of charge through a conductor. How we manipulate this movement is the basis of modern technology. This chapter will cover how to control the flow of charge and some of its applications (Figure 13.1).

FIGURE 13.1 An oscilloscope showing a changing electrical signal.

13.1 BATTERIES

Electricity is the movement of charge or electrons flowing through a conductor, Figure 13.2. Recall that like charges repel and unlike charges attract, Figure 13.3. To make an electron move, place another electron by it. Two electrons placed near each other will try to move apart.

Charges never actually touch each other, but they experience each other through an electric field emanating from them. To get charge to flow, place a charge at the end of a conductor (Figure 13.4). So, where do we get these charges from? One source of charge is a battery.

A *battery* is a storehouse of electrons with one side full of electrons (the negative side) and the other side (the positive side) with fewer electrons (Figure 13.5). Connecting a wire to the negative side of the battery allows an escape route for the electrons to get away from each other. As the electrons escape, they begin to build up in the wire, and the flow from the battery will stop (Figure 13.6).

To continue the flow, a path to the positive side (where there are fewer electrons) needs to be made. Connecting a light bulb or anything else

FIGURE 13.2 Electricity is the movement of charge through a conductor.

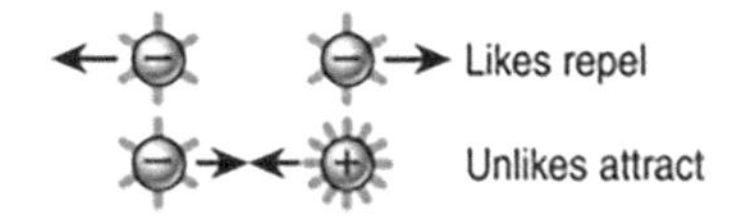

FIGURE 13.3 Like charges repel and unlike charges attract.

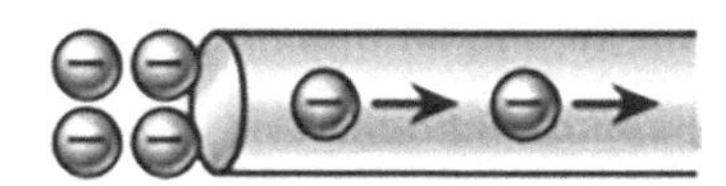

FIGURE 13.4 To get electricity to flow, place a charge at the end of the conductor.

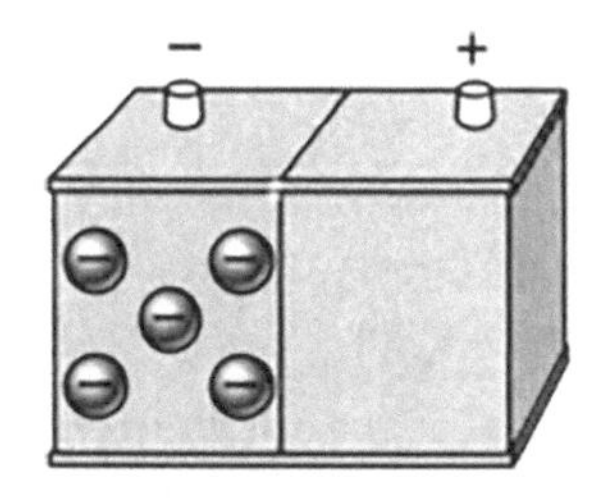

FIGURE 13.5 A battery is a storehouse of electrons.

between the two battery terminals will complete the path back to the battery, Figure 13.7.

Power Supply: A power supply is like a battery that never runs out of power, because it is plugged into the wall.

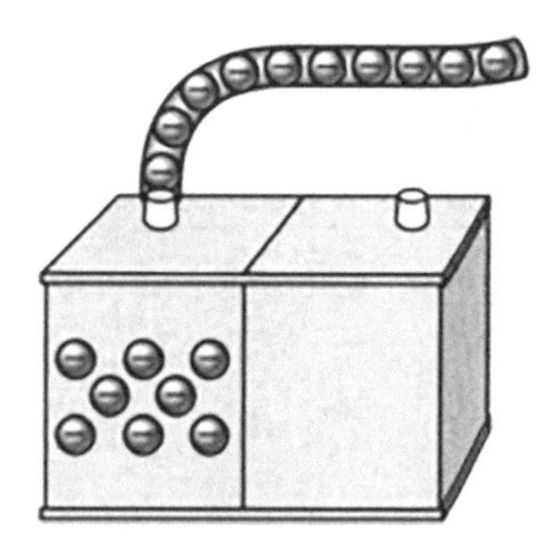

FIGURE 13.6 Wire connected to a battery will fill up with electrons, then the flow will stop.

13.2 VOLTAGE

Voltage can be thought of as electrical pressure. To get water to flow through a hose we must apply pressure on it. To get charge to flow through a wire, we must apply electrical pressure on it. Volt (V) is the unit of voltage. A battery is a source of electrical pressure and charge. The larger the voltage, the more it can push charge through a circuit.

13.3 CURRENT

Current is the flow rate of charge through a conductor, Figure 13.8. Water flowing through a pipe flows at some rate like gal/min. Likewise, electron flow through a wire at a rate of (number of electrons)/s is measured in the unit amperes or amps, abbreviated as A.

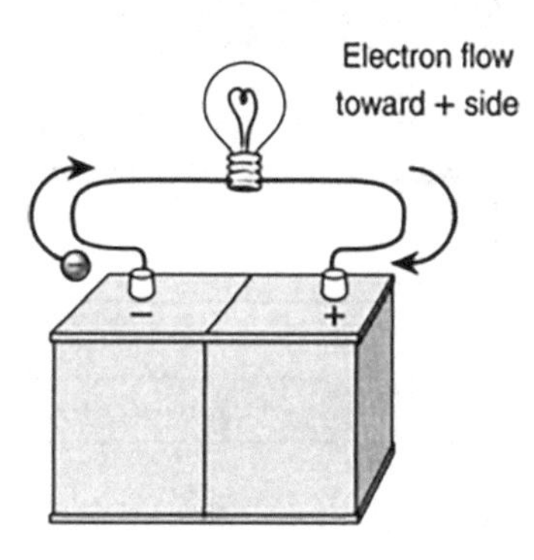

FIGURE 13.7 A complete path will allow the electrons to flow back to the battery.

$$I = \frac{q}{t}$$

$$\text{Current(A)} = \frac{\text{charge(C)}}{\text{time(s)}}$$

$$1\,\text{ampere} = 1\,\text{coulomb/second} \rightarrow 1\,\text{A} = 1\frac{\text{C}}{\text{s}}$$

One coulomb of charge is equivalent to 6.25×10^{18} electrons bunched together.

One electron, e^-, therefore has a charge of $1/(6.25 \times 10^{18}) = -1.6 \times 10^{-19}$ coulombs.

Example 13.1

A charge of 10 C flows through a circuit every 2 s. What is the current flowing through the circuit?

$$I = \frac{q}{t} = \frac{10\text{C}}{2\text{s}}$$

$$I = 5\frac{\text{C}}{\text{s}} = 5\,\text{A}$$

Typical currents of devices around the home are:

Hair dryer ~10 A
Lamp ~1 A
LED, light-emitting diode ~ 0.01 A

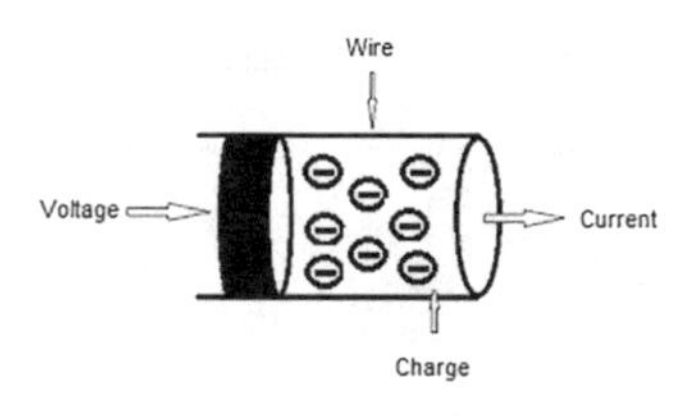

FIGURE 13.8 Voltage can be thought of as electrical pressure pushing charge through a wire.

13.3.1 AC versus DC

DC: Direct current only flows in one direction, Figure 13.9. An example of a DC voltage is a battery. An example of a DC circuit is a battery connected to a light bulb. Current flows out of one terminal, passes through the light, and flows back into the other terminal always in the same direction.

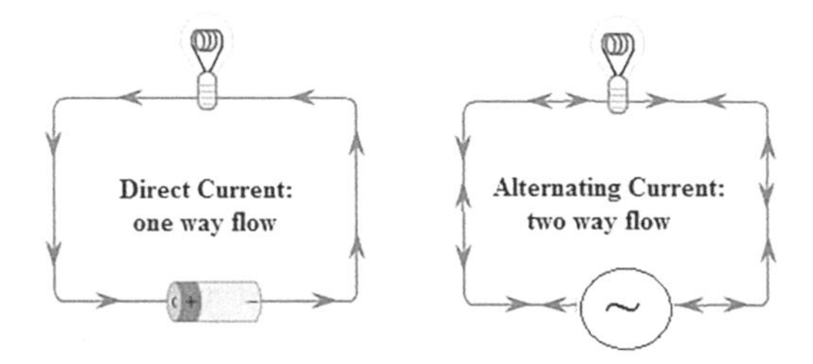

FIGURE 13.9 AC/DC circuits.

AC: Alternating current flows in two directions, Figure 13.9. Your house outlets have AC voltage. An example of an AC circuit is a circuit with a source of voltage that alternates the polarity of the voltage at its two terminals. The two terminals "take turns" at being positive while the other is negative. The current will alternately flow out of, say, terminal 1 through the circuit into terminal 2, then the current will reverse and flow out of terminal 2 and into terminal 1.

AC current is used to transmit electrical power over long distances. The reason is that unlike DC power, an AC voltage can be easily raised or lowered using a transformer. The big, gray cylindrical objects on the phone poles are transformers. They lower the voltage from a large voltage to 220 and 110 V for your home, Figure 13.10.

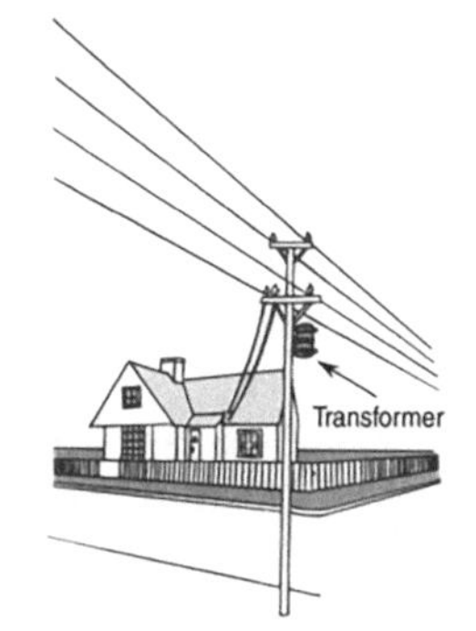

FIGURE 13.10 AC power is transmitted to your home. The transformer lowers the voltage and delivers it to your house.

13.4 RESISTANCE

Resistance means opposition to current flow and is measured in a unit called ohms (Ω) (Figure 13.11).

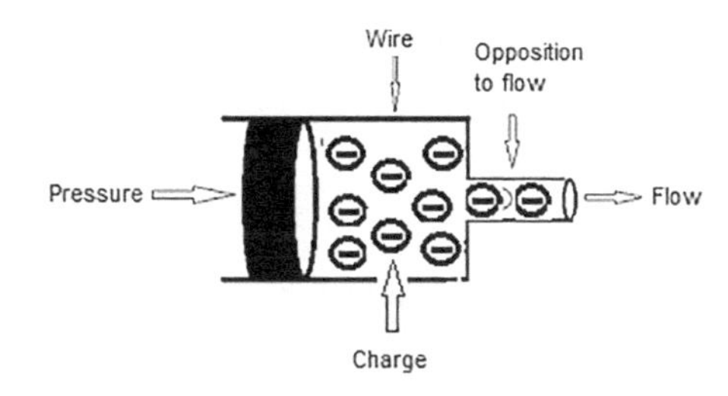

FIGURE 13.11 Resistance means opposition to current flow.

TABLE 13.1
Resistivities

Material	Resistivity (ρ) (Ω m)
Insulators	
Teflon	1.0×10^{23}
Quartz	7.5×10^{17}
Rubber	7.5×10^{17}
Glass	7.5×10^{17}
Conductors	
Nichrome alloy	1.6×10^{-6}
Lead	2.2×10^{-7}
Iron	9.7×10^{-8}
Tungsten	9.7×10^{-8}
Aluminum	2.7×10^{-8}
Gold	2.2×10^{-8}
Copper	1.7×10^{-8}
Silver	1.6×10^{-8}
Graphene	1.0×10^{-8}

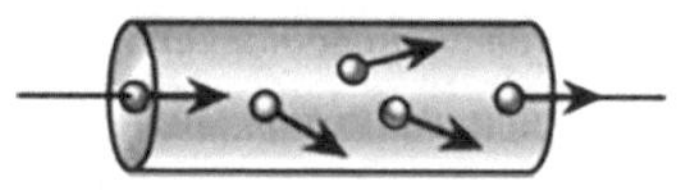

FIGURE 13.12 A resistor scatters electrons flowing through it, causing a reduction in flow of the circuit.

TABLE 13.2
Table of Metric Prefixes

1 Giga = 1 G = 1×10^{9}
1 Mega = 1 M = 1×10^{6}
1 kilo = 1 k = 1×10^{3}
Unity = 1 = 1×10^{0}
1 milli = 1 m = 1×10^{-3}
1 micro = 1 μ = 1×10^{-6}
1 nano = 1 n = 1×10^{-9}
1 pico = 1 p = 1×10^{-12}

Resistivity of a material (ρ) is a measure of its ability to pass electrical current. Conductors have a small opposition to flow or a low resistivity (Table 13.1).

Conductors: Conductors are materials that allow electricity to flow easily. Materials such as copper and aluminum are conductors because they have low resistivities.

Insulators: Insulators are materials that prevent electricity from flowing. Materials such as glass and rubber are insulators because they have high resistivities.

Semiconductors: Semiconductors are materials that fall somewhere between the conductivities of conductors and insulators. Different types of semiconductors can be put together in various combinations to either enhance or limit the flow of current, such as in the case of transistors or diodes.

13.4.1 Resistors

Resistors reduce the flow of current in a circuit. Materials vary in their abilities to pass current. They are analogous to the faucet in plumbing as they reduce the flow of water. Inside a resistor there is a material that causes the electrons to scatter when they move through it. The value of the resistance is determined by controlling the concentration of this scattering material in the resistor. Higher scattering means a higher resistance, Figure 13.12. Resistors come in a variety of shapes and sizes. Some resistors can be used as sensors of temperature, force, humidity, magnetic fields, etc. because these quantities can affect the resistor's resistance.

Units: Many units in electronics have prefixes in front of them, see Table 13.2.

13.5 SCHEMATIC SYMBOLS

Schematic symbols are shorthand symbols used to represent components in electronics. For example, rather than drawing a picture of a battery, a symbol is used instead. Schematic symbols save time and space when drawing a circuit. Some of the symbols are given below (Figure 13.13).

13.6 CIRCUIT

The word *circuit* comes from the word *circle*. The electrons complete a path that returns them to their origin, similar to a circle.

Open circuit, or open circle, means the path is open and therefore no flow of electrons.

Closed circuit, or closed circle, means the path is complete and a flow is possible.

Short circuit, means the current does not flow along the intended path because a shortcut is taken (Figure 13.14).

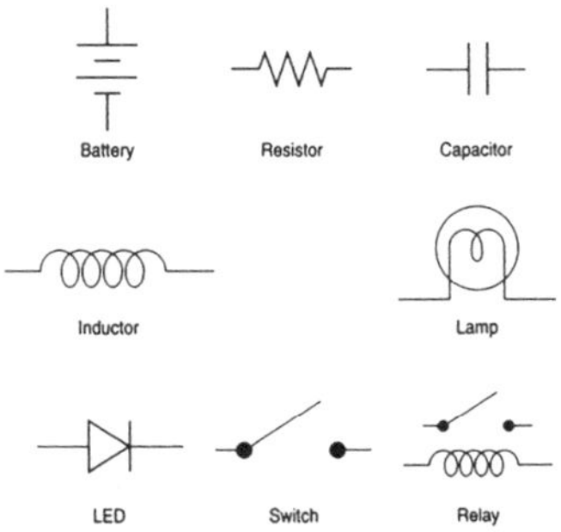

FIGURE 13.13 Schematic symbols.

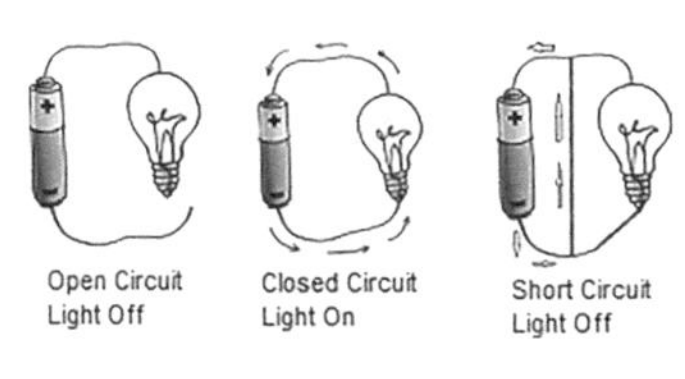

FIGURE 13.14 Open, closed, and short circuits.

13.7 OHM'S LAW

Voltage, current, and resistance are related to each other in a circuit by Ohm's law,

$$\mathrm{I} = \frac{V}{R}$$

$$\text{Current}(\mathrm{A}) = \frac{\text{Voltage}(\mathrm{V})}{\text{Resistance}(\Omega)}$$

This equation says that if the electrical pressure (voltage) is big and the opposition to the flow (resistance) is small, the flow (current) will be large (Figure 13.15).

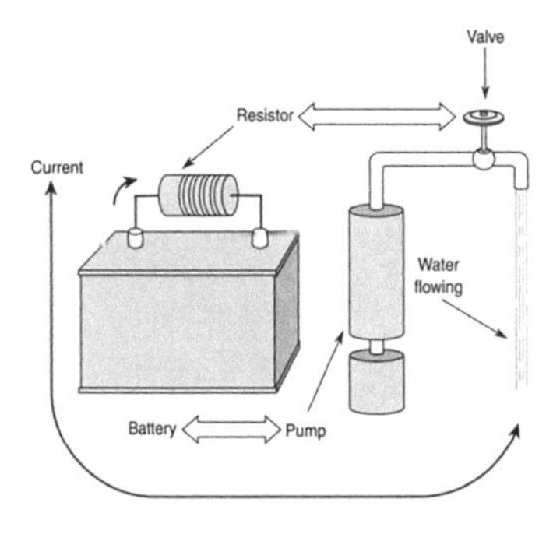

FIGURE 13.15 A plumbing analogy to electricity.

Three versions of Ohm's law are:

$$I = \frac{V}{R}$$

$$V = I \cdot R$$

$$R = \frac{V}{I}$$

Example 13.2

A light bulb has a resistance of 2 Ω and is connected to a 12 V battery. What is the current flowing through it? (Figure 13.16)

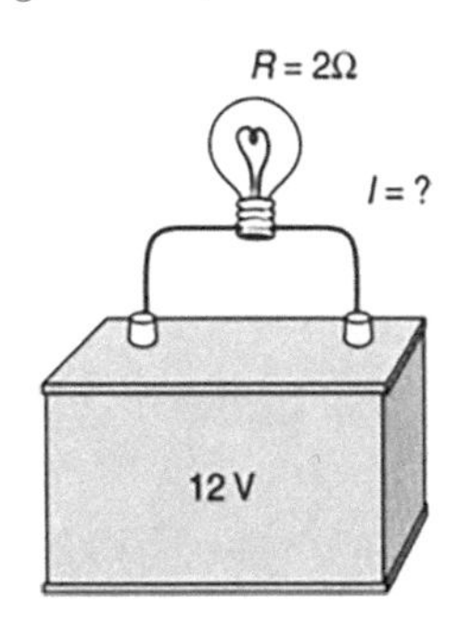

FIGURE 13.16 What is the current flowing through the bulb?

$$I = \frac{V}{R} = \frac{12\text{V}}{2\Omega}$$

$$I = 6\text{A}$$

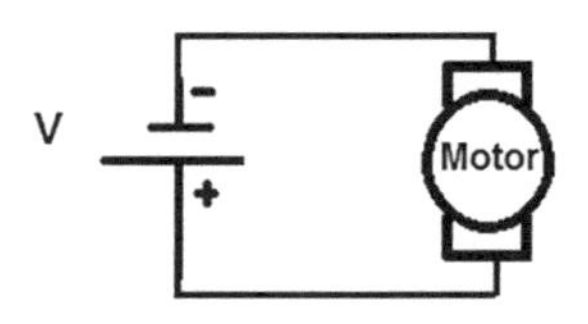

FIGURE 13.17 What is the supply voltage to the motor?

Example 13.3

A motor has a resistance of 8 Ω and has 4 A running through it. What is the supply voltage? (Figure 13.17)

$$V = I \cdot R$$

$$V = (4\text{A}) \cdot (8\Omega)$$

$$V = 32\text{V}$$

13.8 SERIES CIRCUITS

A series circuit is a circuit where each component is connected together in a way such that there is only one path for the current to take, Figure 13.18.

FIGURE 13.18 A series circuit.

A SERIES CIRCUIT HAS THE FOLLOWING PROPERTIES:

- The current is the same through every component in a series circuit.

$$I_{\text{Source}} = I_{R_1} = I_{R_1} = \ldots$$

- The sum of the voltages across each component equals the source voltage.

$$V_{\text{Source}} = \sum_i V_{R_i} = V_{R_1} + V_{R_2} + V_{R_3} + \ldots$$

- The total resistance in series circuit equals the sum of the individual resistances.

$$R_{\text{Series}} = \sum_i R_i = R_1 + R_2 + R_3 + \ldots$$

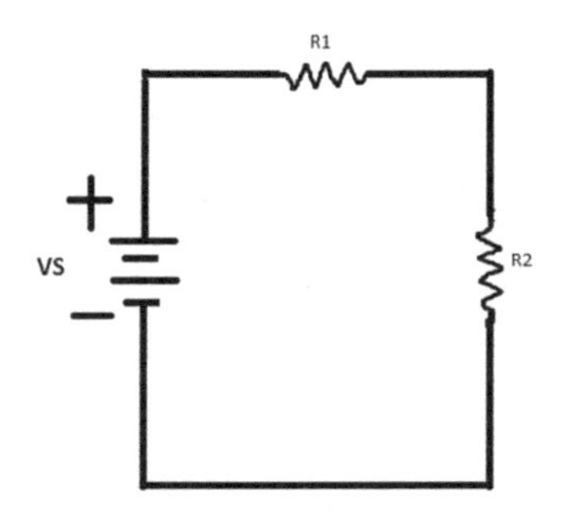

FIGURE 13.19 A series circuit consists of two resistors, 1 and 2 kΩ, in series with a 9V source.

Example 13.4

A series circuit consists of two resistors, 1 and 2 kΩ, in series with a 9V source (Figure 13.19).

a. Find the total resistance of the circuit.

$$R_{\text{Series}} = \sum_i R_i = 1\text{k}\Omega + 2\text{k}\Omega$$

$$R_{\text{Series}} = 3\text{k}\Omega$$

b. Find the current in the circuit.

Using Ohm's law $I = \frac{V}{R}$, we use the source voltage for V and the total resistance for R.

$$I = \frac{9\,\text{V}}{3 \times 10^3\,\Omega}$$

$I = 3 \times 10^{-3}\,\text{A} = 3\,\text{mA}$ ←This is the current through each component, resistors and source.

c. Find the voltage across each component.

Using Ohm's law and solving for V gives $V = IR$.

$$V_{R_1} = I \cdot R = \left(3 \times 10^{-3}\,\text{A}\right) \cdot \left(3 \times 10^3\,\Omega\right)$$

$$V_{R_1} = 9\,\text{V}$$

$$V_{R_2} = I \cdot R = (3 \times 10^{-3}\,\text{A}) \cdot (2 \times 10^3\,\Omega)$$

$$V_{R_2} = 6\,\text{V}$$

d. Verify that the sum of the voltage drops equals the source voltage.

$$\sum_i V_{R_i} = V_{R_1} + V_{R_2} = 3\,\text{V} + 6\,\text{V} = 9\,\text{V} \leftarrow \text{This is the source voltage.}$$

Example 13.5

A series circuit consists of a resistor in series with a LED and a battery with a voltage of 5 V. This LED needs 2.1 V across it and 15 mA through it to run properly. What resistance should you choose the resistor to achieve this? (Figure 13.20)

Applying the rules for a series circuit:

$$V_{\text{Source}} = \sum_i V_{R_i} = V_R + V_{\text{LED}}$$

$$V_R = V_{\text{Source}} - V_{\text{LED}}$$

$$V_R = 5\text{V} - 2.1\text{V}$$

$$V_R = 2.9\text{V}$$

The current is the same through every component in a series circuit, so the current through the resistor equals the current through the LED, $I = 15\,\text{mA}$.

Applying Ohm's law:

$$R = \frac{V}{I} = \frac{2.9\text{V}}{15 \times 10^{-3}\,\text{A}}$$

$$R = 193.3\ \Omega$$

This is not a standard resistor size, but anything close to it will do the job.

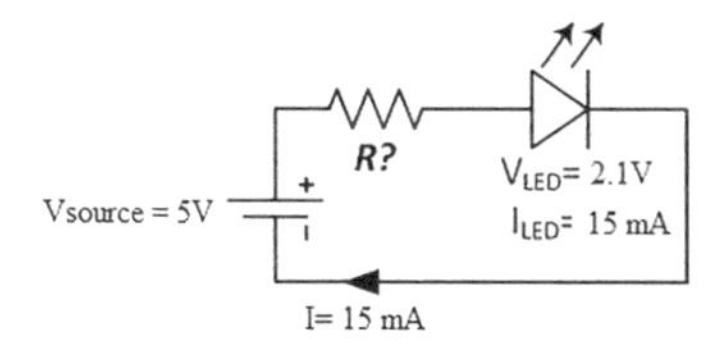

FIGURE 13.20 A series circuit consists of a resistor in series with a LED and a battery with a voltage of 5V.

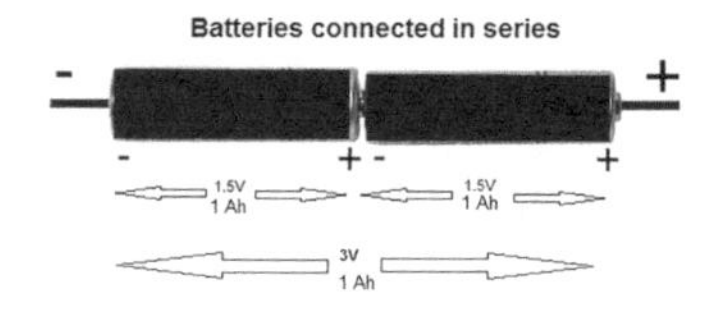

FIGURE 13.21 Batteries connected in series increases the total voltage, but the amp-hour rating stays the same.

13.8.1 Batteries in Series

Connecting a set of batteries together in series means the batteries are connected together with the negative of one terminal connected to the positive of the other terminal, Figure 13.21. The voltage across the batteries will now equal the sum of the individual battery voltages.

$$V_{\text{Batteries in series}} = \sum_i V_i = V_{\text{Battery1}} + V_{\text{Battery2}} + \ldots$$

Example 13.6

How would you connect together a set of AA batteries to run a 3 V motor?

AA batteries have a voltage of 1.5 V batteries, so connecting the batteries in series will produce a voltage of 3 V.

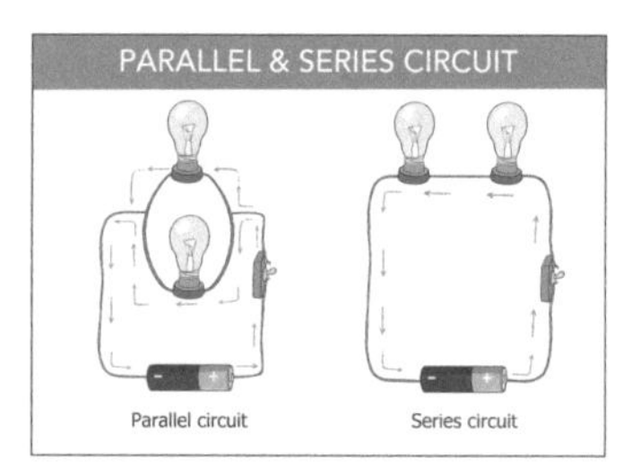

FIGURE 13.22 A parallel circuit compared to a series circuit.

13.9 PARALLEL CIRCUITS

A parallel circuit is a circuit where each component is connected together in a way such that there are multiple paths for the current to take, see Figure 13.22.

A PARALLEL CIRCUIT HAS THE FOLLOWING PROPERTIES:

- The voltage across every component in a parallel circuit is the same.

$$V_{\text{Source}} = V_{R_1} = V_{R_2} = \ldots$$

- The total resistance of connecting resistors in parallel obeys the following formula.

$$R_{\text{Parallel}} = \left(\sum_i \frac{1}{R_i}\right)^{-1} = \left(\frac{1}{R_1} + \frac{1}{R_2} + \frac{1}{R_3} + \ldots\right)^{-1}$$

- The sum of the currents through each component equals the current out of the source.

$$I_{\text{Source}} = \sum_i I_{R_i} = I_{R_1} + I_{R_2} + I_{R_3} + \ldots$$

The current out of the source can also be found by dividing the source voltage by the total resistance in the circuit.

$$I_{\text{Source}} = \frac{V_{\text{Source}}}{R_{\text{Total}}}$$

Example 13.7

A parallel circuit consists of three resistors, 90, 45, and 180 Ω, in parallel with a 9V source (Figure 13.23).

a. Find the total resistance of the circuit.

$$R_{\text{Parallel}} = \left(\sum_i \frac{1}{R_i}\right)^{-1} = \left(\frac{1}{90\,\Omega} + \frac{1}{45\,\Omega} + \frac{1}{180\,\Omega}\right)^{-1}$$

$$R_{\text{Parallel}} = 25.7\,\Omega$$

b. Find the voltage across each component.

The voltage across each component is the same in a parallel circuit. The source voltage equals 9V, so the voltage across each resistor equals 9V.

$$V_{R_1} = V_{R_2} = V_{R_3} = 9\,\text{V}$$

c. Find the current in each resistor in the circuit.

The voltage across each component is the same in a parallel circuit and using Ohm's law,

$$I_{R_1} = \frac{V}{R_1} = \frac{9\,\text{V}}{90\,\Omega}$$

$$I_{R_1} = 0.1\,\text{A}$$

$$I_{R_2} = \frac{V}{R_2} = \frac{9\,\text{V}}{45\,\Omega}$$

$$I_{R_2} = 0.2\,\text{A}$$

$$I_{R_3} = \frac{V}{R_3} = \frac{9\,\text{V}}{180\,\Omega}$$

$$I_{R_3} = 0.05\,\text{A}$$

d. Find the current out of the source.

The current out of the source equals the sum of the currents through each component.

$$I_{\text{Source}} = \sum_i I_{R_i} = 0.1\,\text{A} + 0.2\,\text{A} + 0.05\,\text{A}$$

$$I_{\text{Source}} = 0.35\,\text{A}$$

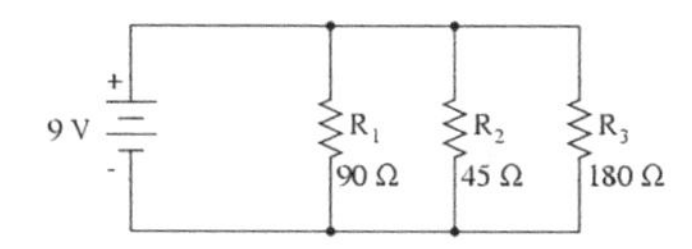

FIGURE 13.23 A parallel circuit consists of three resistors, 90, 45, and 180 Ω, in parallel with a 9V source.

Example 13.8

A parallel circuit consists of two resistors, 90 and 45 Ω, in parallel with an 18V source.

a. Find the total resistance of the circuit.

$$R_{\text{Parallel}} = \left(\sum_i \frac{1}{R_i}\right)^{-1} = \left(\frac{1}{90\,\Omega} + \frac{1}{45\,\Omega}\right)^{-1}$$

$$R_{\text{Parallel}} = 30\,\Omega$$

b. Find the current out of the battery, in other words, find I_{Source}.

$$I = \frac{V_{Source}}{R_{Total}}$$

$$I = \frac{18\,V}{30\,\Omega}$$

$$I = 0.6\,A$$

This result for the current out of the battery can also be obtained by determining the current through each resistor and then adding these currents.

Parallel vs. Serial, A Christmas Gone Bad: While shopping for Christmas lights you've narrowed it down to a set of lights that are wired in series and another set wired in parallel. You buy the series wired lights because they are cheaper. Christmas arrives, you plug in the tree, and nothing! Checking each bulb, you find one of the bulbs has burned out. You conclude this is why none of the lights are turning on, you have an open circuit! Thinking back on your decision to buy the series lights, you realize had you bought the light wired in parallel, a bad bulb would not have shut off the whole string.

13.10 POWER

Electrical power is the rate at which electrical energy is being used. The electrical power consumed by a component in a circuit equals the current through the component times the voltage across the component. Power is measured in the unit watts, abbreviated as W.

$$P = I \cdot V$$

$$\text{Power}(\text{W}) = \text{Current}(\text{A}) \cdot \text{Voltage}(\text{V})$$

We can obtain different versions of the power law by substituting Ohm's law into it.

$$P = I \cdot V = \left(\frac{V}{R}\right) \cdot V = \frac{V^2}{R}$$

$$P = I \cdot V = I \cdot (I \cdot R) = I^2 \cdot R$$

We have three versions of the power formula:

$$P = I \cdot V$$

$$P = \frac{V^2}{R}$$

$$P = I^2 \cdot R$$

Which one you use depends on the information given to you.

Example 13.9

A 9V battery is connected to a radio that draws 0.1 amps. What is the power consumed by this radio? (Figure 13.24)

$$V = 9\text{V}$$
$$I = 0.1\text{A}$$
$$P = ?$$
$$P = I \times V$$
$$P = 0.1\text{A} \times 9\text{V}$$
$$P = 0.9\text{W}$$

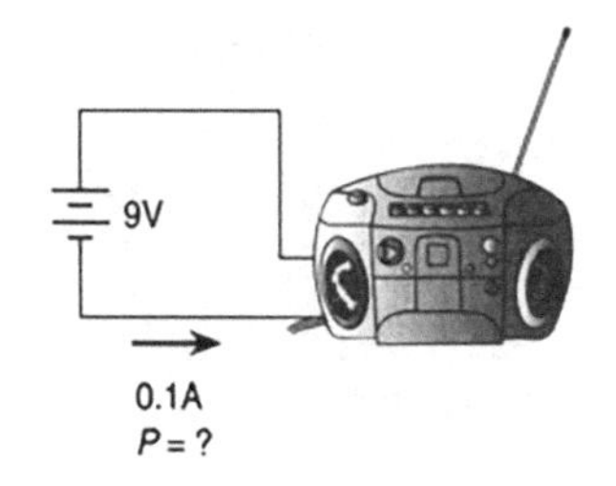

FIGURE 13.24 What is the power consumed by this radio?

Example 13.10

A 10 kΩ resistor has 10V across it with 1 mA of current running through it. Find the power dissipated by the resistor in three different ways.

$$P = I \cdot V = \left(1 \times 10^{-3}\text{A}\right) \cdot \left(10\text{V}\right) = 0.01\text{W}$$

$$P = \frac{V^2}{R} = \frac{(10\text{V})^2}{10 \times 10^3\Omega} = 0.01\text{W}$$

$$P = I^2 \cdot R = \left(1 \times 10^{-3}\text{A}\right)^2 \cdot \left(10 \times 10^3\Omega\right) = 0.01\text{W}$$

No matter which formula you choose you will get the same result!

Load: The electrical load of a circuit is the component connected up to a circuit for which the circuit is supposed to power. A heavy load means the load draws a lot of current and therefore a lot of power.

13.10.1 Electrical Energy

The power company charges you by how much electrical energy you use. The power meter on your house actually records the amount of energy you use, not power. Recall that energy = power × time. The meter measures the power consumed and records for how long that amount of power is drawn from the utility company. The power is recorded in kilowatts and time in hours, therefore the energy is measured by the meter in kWh.

$$E = P \cdot t$$

$$\text{Energy}(\text{kWh}) = \text{Power}(\text{kW}) \cdot \text{time}(\text{h})$$

To calculate the number of kilowatt-hours, use:

$$\text{kWh} = \text{kW} \cdot \text{h}$$

FIGURE 13.25 A household power meter measures energy used in kWh.

Example 13.11

Suppose you blow-dry your hair with a hair dryer that is rated at 1.2 kW. If you operate the dryer for 1/2 h, the number of kWh used.

$$\text{kWh} = (1.2\text{kW})(1/2\text{h})$$
$$\text{kWh} = 0.6\text{kWh}$$

13.10.2 Energy Cost

The power company charges a certain rate, cents per kWh, for energy consumed (Figure 13.25). To determine the cost for the use of so many kWh consumed, apply the following equation:

$$\text{Cost}(\text{cents}) = \text{rate}(\text{cents/kWh}) \times \text{kWh}$$

Example 13.12

On average, the power companies charge 12 cents/kWh. Suppose you leave a 100 W light bulb on overnight for 8 h. How much will this cost you?

$$\text{Rate} = 12\frac{\text{cents}}{\text{kWh}}$$
$$100\,\text{W} = 0.1\text{kW}$$
$$\text{kWh} = (0.1\text{kW})\cdot(8\text{h})$$
$$\text{kWh} = 0.8\text{kWh}$$
$$\text{Cost} = \text{rate} \times \text{kWh}$$
$$\text{Cost} = 12\frac{\text{cents}}{\text{kWh}} \times 0.8\text{kWh}$$
$$\text{Cost} = 9.6\,\text{cents}$$

So, if your dad yells at you about leaving the light on all night, just toss him a dime!

13.11 BATTERIES AND AMP-HOUR RATING

The amp-hour rating of a battery is how much charge a battery can store. How much charge a battery can store is determined by its amp-hour rating. The more charge it can store, the longer a battery can last when supplying a certain current. This number is normally printed on rechargeable batteries only.

$$\text{Ah} = I(\text{A}) \times t(\text{h})$$

Example 13.13

A battery has a 10 Ah rating. How long will it last if 10, 5, and 2.5 A are drawn out of it?

At I = 10 A, it will last for 1 hour.

At I = 5 A, it will last for 2 hours.

At I = 2.5 A, it will last for 4 hours.

The more current the battery is delivering, the shorter the life of the battery.

13.11.1 Amp-Hour Rating and Batteries in Serial and Parallel

13.11.1.1 Batteries in Series

You connect batteries together in series to get a larger voltage, but amp-hour rating of the batteries stays the same.

13.11.1.2 Batteries in Parallel

A larger amp-hour rating can be produced by connecting a set of batteries together in parallel, that is, the positive terminals are connected together and the negative terminals are connected together, Figure 13.26.

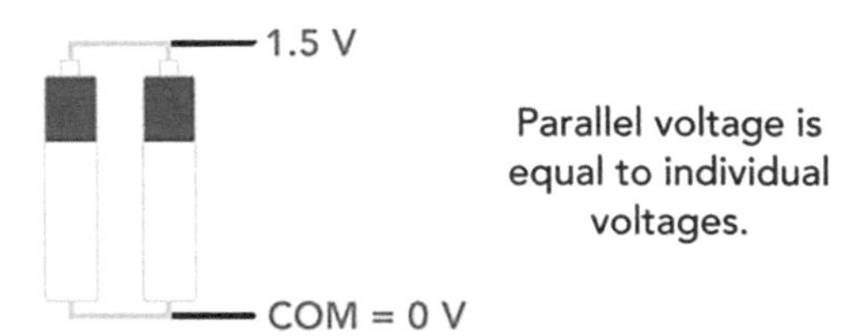

FIGURE 13.26 A larger amp-hour rating can be produced by connecting a set of batteries together in parallel.

The amp-hour rating of this set of batteries in parallel will now equal the sum of the individual battery's amp hour ratings. Connect batteries together in parallel to get a larger amp-hour rating, longer lifetime, but the voltage of the batteries stays the same.

$$\mathrm{Ah}_{\text{Batteries in parallel}} = \sum_i \mathrm{Ah}_i = \mathrm{Ah}_{\text{Battery1}} + \mathrm{Ah}_{\text{Battery2}} + \ldots$$

Example 13.14

The amp-hour rating can be doubled by connecting two batteries together in parallel. The amp-hour voltage is doubled; however, the voltage stays the same. Doubling the amp-hour of the batteries means they will last twice as long.

Example 13.15

How would you connect together a set of AA batteries so that a motor that requires one AA battery will stay running twice as long? (Table 13.3)

TABLE 13.3
Battery Types

Parameter	Lead Acid	NiCd	NiMH	Alkaline	Li-Ion	Li-Polymer
Cell voltage (V)	2.0	1.2	1.2	1.5	3.6	3.7
Relative cost	Low	Moderate	High	Very low	Very high	Very high
Internal resistance	Low	Very low	Moderate	Varies	High	Low
Self discharge (%/month)	2%–4%	15%–30%	18%–20%	0.3%	6%–10%	5%
Cycle life (charge cycles to reach 80% of rated capacity)	500–2,000	500–1,000	500–800	Low	1,000–1,200	>1,000
Overcharge tolerance	High	Medium	Low	Medium	Very low	Very low
Energy density by volume (Wh/L)	70–110	100–120	135–180	220	280–320	~400
Energy density by weight (Wh/kg)	30–45	45–50	55–65	80	90–110	130–200

Connect the batteries in parallel, see Figure 13.27.

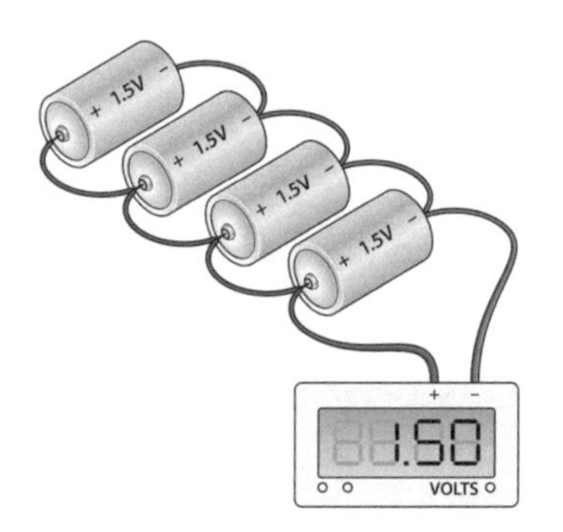

FIGURE 13.27 Batteries connected in parallel.

Power Supplies: A power supply is like a battery that never runs out of power, because it is plugged into the wall. Some of them are adjustable in voltage, and they all have a limit as to how much current they can put out at a given voltage.
Compliance: The maximum current a power supply can put out before the output voltage begins to drop.

Example 13.16

A 5 V power supply has a compliance of 1.5 amps. If a circuit draws more than 1.5 A from the supply, the voltage will drop below 5 V, possibly damaging the supply.

13.12 CHAPTER SUMMARY

Symbols used in this chapter:

Symbol	Unit
t—time	second (s)
q—charge	coulomb (C)
I—current	ampere (A)
Ah—amp-hour	amp-hour (Ah)
V—voltage	volt (V)
R—resistance	ohm (Ω)
P—power	watt (W)
kWh—kilowatt-hours	kilowatt-hours (kWh)

Current is the flow of charge through a conductor.

$$I = \frac{q}{t}$$

$$\text{Current}(\text{A}) = \frac{\text{charge}(\text{C})}{\text{time}(\text{s})}$$

$$1\,\text{ampere} = 1\,\text{coulomb/second} \rightarrow 1\,\text{A} = 1\frac{\text{C}}{\text{s}}$$

Amp-Hour: The amp-hour rating of a battery is how much charge a battery can store.

$$\text{Ah} = I(\text{A}) \times t(\text{h})$$

Batteries in Series: You connect batteries together in series to get a larger voltage, but amp-hour rating of the batteries stays the same.

$$V_{\text{Batteries in series}} = \sum_i V_i = V_{\text{Battery1}} + V_{\text{Battery2}} + \ldots$$

Batteries in Parallel: You connect batteries together in parallel to get a larger amp-hour rating, longer lifetime, but the voltage of the batteries stays the same.

$$Ah_{\text{Batteries in parallel}} = \sum_i Ah_i = Ah_{\text{Battery 1}} + Ah_{\text{Battery 2}} + \ldots$$

Resistance means opposition to flow and it has the unit ohms (Ω).

Ohm's Law: Voltage, current, and resistance are related to each other by Ohm's law,

$$I = \frac{V}{R}$$

$$\text{Current}(\text{A}) = \frac{\text{Voltage}(\text{V})}{\text{Resistance}(\Omega)}$$

This equation says that if the electrical pressure (voltage) is big and the opposition to the flow (resistance) is small, the flow (current) will be large.

Three versions of Ohm's law are:

$$I = \frac{V}{R}$$

$$V = I \cdot R$$

$$R = \frac{V}{I}$$

Power: Electrical power is the rate at which electrical energy is being used. Power is measured in the unit watts, abbreviated as W.

$$P = I \cdot V$$

$$\text{Power} = \text{Current}(\text{A}) \cdot \text{Voltage}(\text{V})$$

We can obtain different versions of the power law by substituting Ohm's law into it.

$$P = I \cdot V = \left(\frac{V}{R}\right) \cdot V = \frac{V^2}{R}$$

$$P = I \cdot V = I \cdot (I \cdot R) = I^2 \cdot R$$

We have three versions of the power formula:

$$P = I \cdot V$$

$$P = \frac{V^2}{R}$$

$$P = I^2 \cdot R$$

Which one you use depends on the information given to you.

Electrical Energy: Electrical energy is measured in kilowatt-hours.

$$\text{Energy} = \text{Power} \times \text{Time} = (\text{kilowatts}) \times (\text{hours}) \text{ or kWh.}$$

$$\text{kWh} = \text{kW} \times \text{hours}$$

Series Circuits: A series circuit is a circuit where each component is connected together in a way such that there is only one path for the current to take.

Key points about a series circuit:

- The current is the same through every component in a series circuit.

$$I_{\text{Source}} = I_{R_1} = I_{R_1} = \cdots$$

- The sum of the voltage drops across each component equals the source voltage.

$$V_{\text{Source}} = \sum_i V_{R_i} = V_{R_1} + V_{R_2} + V_{R_3} + \ldots$$

- The total resistance in series circuit equals the sum of the individual resistances.

$$R_{\text{Series}} = \sum_i R_i = R_1 + R_2 + R_3 + \ldots$$

Parallel Circuits: A parallel circuit is a circuit where each component is connected together in a way that there are multiple paths for the current to take.

Key points about a parallel circuit:

- The voltage across every component in a parallel circuit is the same.

$$V_{\text{Source}} = V_{R_1} = V_{R_2} = \cdots$$

- The sum of the currents through each component equals the current out of the source.

$$I_{\text{Source}} = \sum_i I_{R_i} = I_{R_1} + I_{R_2} + I_{R_3} + \ldots$$

- The total resistance of connecting resistors in parallel obeys the following formula.

$$R_{\text{Parallel}} = \left(\sum_i \frac{1}{R_i}\right)^{-1} = \left(\frac{1}{R_1} + \frac{1}{R_2} + \frac{1}{R_3} + \ldots\right)^{-1}$$

PROBLEM SOLVING TIPS

- When solving problems dealing with Ohm's law, do not calculate in kΩ or mV or mA, convert all values to their base units, volts, ohms, and amps before plugging into Ohm's law.

PROBLEMS

1. What makes current flow in a closed circuit?
2. If a flashlight bulb has a resistance of 30 Ω and is connected to two D-cells, creating a voltage of 3 V, what is the current flowing through the bulb?
3. What is the power consumed in problem 2?
4. If a car battery has an amp-hour rating of 96 Ah and each time it is started, it draws 250 A for a period of 5 s, how many times can it start the car without the battery being recharged? Assume that the charging system on the car is not working.
5. Go around your house and determine the total kWh of the energy consumed over the period of a few hours. Compare this with the meter reading on the outside of your house. Does it agree?
6. Suppose you wish to change the brightness of a light. How could you do this using a resistor?
7. Design a test system to measure the power consumed by a motor driving an elevator.
8. Suppose your car battery is dead and the battery has an amp-hour rating of 100 Ah. How long will it take to fully charge this dead battery if the charger is set to a charge rate of 5 amps?
9. What is the resistance of a 1 and a 2 kΩ in parallel?
10. What is the resistance of a 1 and a 2 kΩ in series?
11. If the power company charges 10 cents/kWh, how much does it cost you over a year's time to blow-dry your hair for 5 min in a day with a 1 kW hair dryer? Be careful of the units!
12. A series circuit consists of a resistor in series with a LED and a battery with a voltage of 9 V. This LED needs 1.5 V across it and 10 mA through it to run properly. What resistance should you choose the resistor to achieve this?
13. Ten coulombs of charge flow through a circuit every 5 s. What is the current through the circuit?
14. How long can an electric car travel if the 12 V motor draws 30 A, and the batteries contain 500 Ah of charge?

15. How far does the car travel in problem 14 if it is traveling at 55 mph? Hint: Look at the units.
16. A 12 V, 6 Ah battery is connected to a 10 Ω light bulb. What is the current?
17. A 12 V, 6 Ah battery is connected to a 10 Ω light bulb. What is the power dissipated?
18. A 12 V, 6 Ah battery is connected to a 10 Ω light bulb. How long will the bulb stay lit?
19. A 110 V motor draws 5 A. What is the power?
20. A 110 V motor draws 5 A and runs for 5 h. How many kWh were used?
21. A 110 V motor draws 5 A and runs for 5 h. If energy costs 8 cents/kWh, what is the cost?
22. Two 1.5 V, 3 Ah D-cells are connected in parallel with a 1.5 Ω light bulb. How long will it stay lit?
23. Two 1.5 V, 3 Ah D-cells are connected in series with a 1.5 Ω light bulb. How long will it stay lit?
24. A motor is running on 12 V and has 2 A flowing through it. What is the electrical power of the motor?
25. A motor is running on 12 V and has a resistance of 3 Ω. What is the electrical power of the motor?
26. At a rate of 9 cents/kWh, how much does it cost in cents to leave a 60 W light bulb on for 8 h?
27. If the light in problem 1 ran for 8 h every day for 1 year, 365 days, how much would the bill be in cents?
28. The bulb in a flashlight with 2 D-cells (each D-cell is 1.5 V) in series has a resistance of 30 Ω. What is the current drawn?
29. If the D-cells in problem 3 have an amp-hour rating of 2 Ah, how long will the light stay lit?
30. Three D-cells with an amp-hour rating of 3.5 Ah are placed in parallel, what is the amp-hour value of this set?
31. If you have two identical resistors and place them in parallel, how does total resistance compare to the resistance of the individual resistors?
32. If you have two identical resistors and place them in series, how does total resistance compare to the resistance of the individual resistors?
33. A 10 Ω resistor connected to a 12 V battery is used as a heater. How many watts of heat are generated?
34. An electric heater costs 9 cents/kWh to use versus 4 cents/kWh for a heater running on natural gas. How many more times expensive is it to run the electric heater?

35. A motor is drawing 0.5 A from a 12 V battery. What is the resistance of the motor?
36. A series circuit consists of two resistors, 1 and 2 kΩ, in series with a voltage source. What is the current flowing out of the voltage source if the 1 kΩ resistor has 3 mA flowing through it.
37. A parallel circuit consists of two resistors, 1 and 2 kΩ, in parallel with a 12 V source. Find the current through each resistor, and the current out of the source.
38. A switch is opened and current stops flowing through a circuit. Is this an open or closed circuit?
39. A switch is closed and current begins flowing through a circuit. Is this an open or closed circuit?
40. What does it mean for a circuit to have a short circuit?

14

Magnetism

Magnetism and electricity are closely related. One can be generated from the other and both have many applications. This chapter will discuss magnetism and its applications, as well as how to calculate and measure the strength of magnetic fields (Figure 14.1).

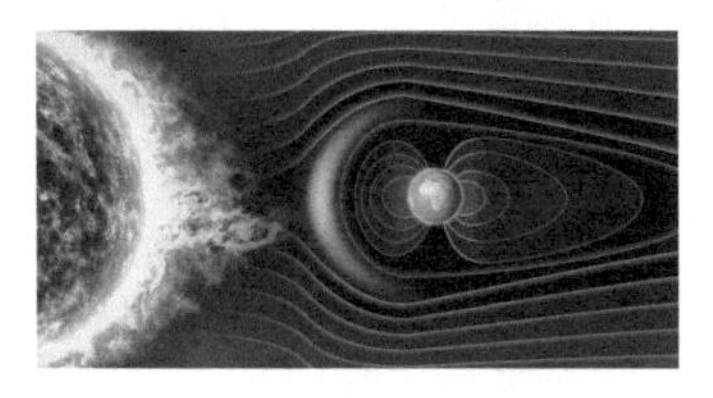

FIGURE 14.1 The earth's magnetic field protecting the earth from the solar wind.

Materials are magnetic because of the way the electrons in materials interact with one another. Atoms have nuclei made of protons and neutrons and "orbiting" the nuclei are electrons. Electrons themselves act like little magnets because of a property called spin. Two electrons with spins oriented the same way will be magnetic and spins oriented in the opposite direction will cancel each other and therefore will be nonmagnetic. Most atoms are nonmagnetic because they have an equal number of electrons with spin up and spin down in an orbit, Figure 14.2. A magnetic atom, however, does not have an equal number of electrons with spin up and spin down. Iron is an example of a magnetic material. Each magnetic atom in a bar magnet contributes to the overall magnetism of the bar.

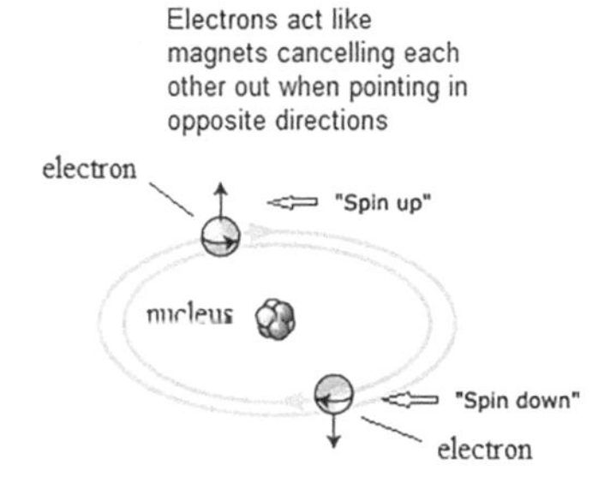

FIGURE 14.2 Most atoms are nonmagnetic because they have an equal number of electrons with spin up and spin down in an orbit.

Materials that can be magnetized and retain their magnetism are called *permanent magnets*. Not all materials are magnetic (Figure 14.3). For example, copper and stainless steel are nonmagnetic. Permanent magnets are usually made from iron, steel, or alloys such as Alnicol (aluminum, nickel, and cobalt).

All magnets have a north pole and a south pole. The like poles repel and the unlike poles attract one another (Figure 14.4). This repulsion of

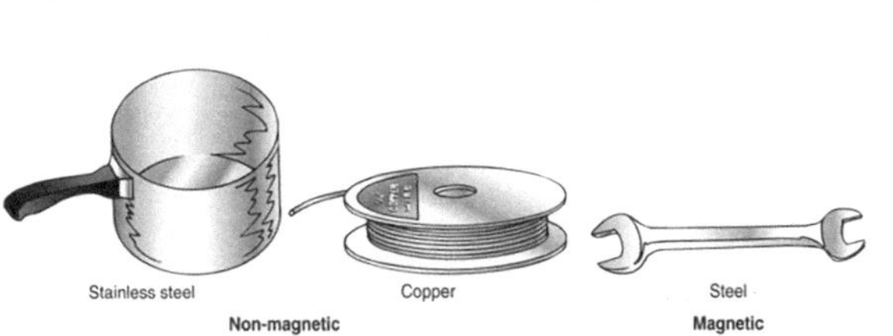

FIGURE 14.3 Some magnetic and nonmagnetic materials.

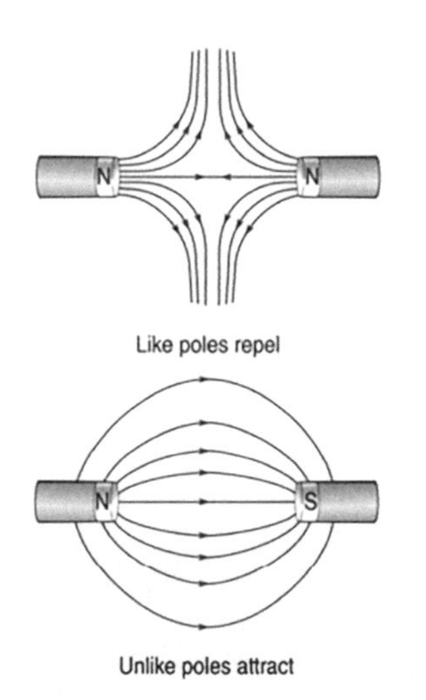

FIGURE 14.4 Like poles repel. Unlike poles attract.

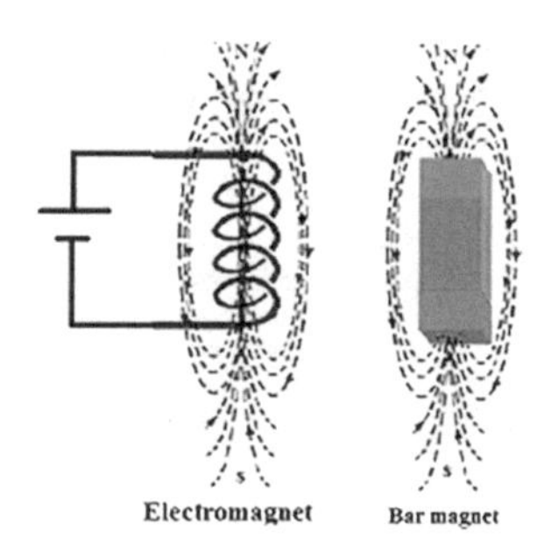

FIGURE 14.5 Passing a current through a coil of wire produces a magnetic field similar to a field from a bar magnet.

FIGURE 14.6 An electromagnet used at a junkyard.

like poles has many applications, such as magnetic bearings and magnetically levitated trains.

14.1 ELECTROMAGNETS

Current moving through a wire will produce a magnetic field (Figure 14.5). Passing a current through a coil of wire produces a magnetic field similar to a field from a bar magnet.

14.1.1 Applications of Electromagnets

14.1.1.1 Electromagnet Crane

By turning the current on or off, the field can be turned on or off. A crane with an electromagnet can easily pick up thousands of pounds (Figure 14.6).

14.1.1.2 Relay

A relay is an electromagnetic switch that consists of a coil and a switch (Figure 14.7). When a current passes through the coil, it generates a magnetic field, which pulls the switch closed.

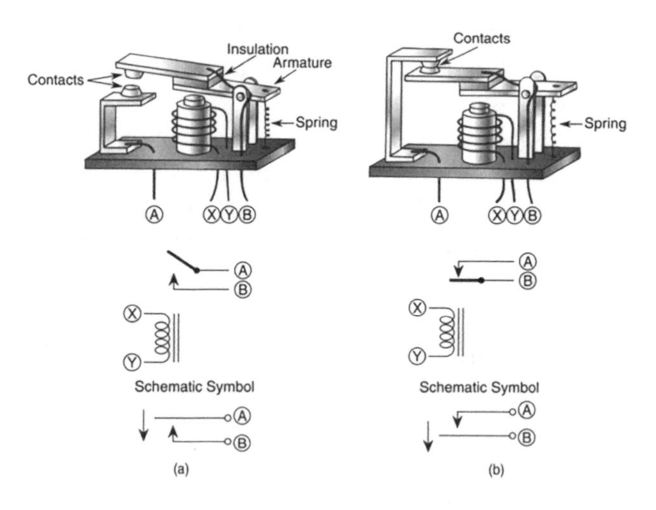

FIGURE 14.7 A relay is an electromagnetic switch.

14.1.1.3 Solenoid Actuator

A solenoid actuator is a solenoid with a movable cylinder that can be extended or retracted with a magnetic field when the coil is energized. The solenoid actuator can be used to open/close valves, door locks, and many other devices that require a linear actuation (Figures 14.8 and 14.9).

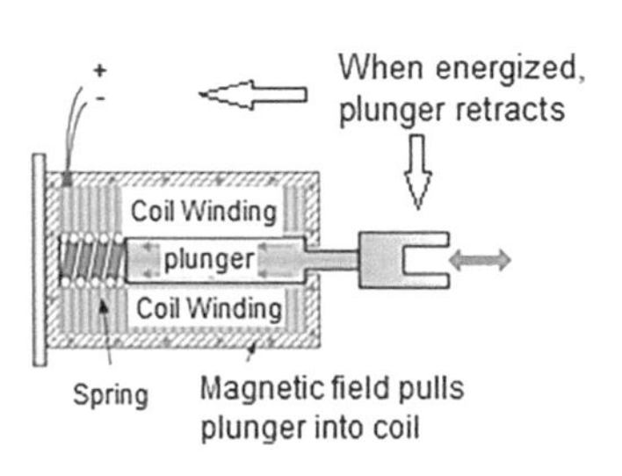

FIGURE 14.8 A solenoid actuator.

14.1.1.4 Speaker

A speaker consists of a cone attached to a coil loosely wrapped around a permanent magnet. When an alternating current (AC) flows through the coil, a magnetic field is generated and interacts with the permanent magnet, forcing the cone to move back and forth creating sound (Figure 14.10).

14.1.1.5 Superconducting Magnets

Generating magnetic fields using standard electromagnets is limited because the field strength is proportional to the current. The higher the current, the greater the heating of the wire, $P = I^2 R$. Big electromagnets need to be cooled with water. To obtain higher magnetic fields, a superconducting wire is used instead of copper wire. Superconducting wire generates virtually no heat. The only problem with it is that it has to be cooled to very low temperatures using liquid nitrogen or helium and it requires a costly cryogenic system (Figure 14.11).

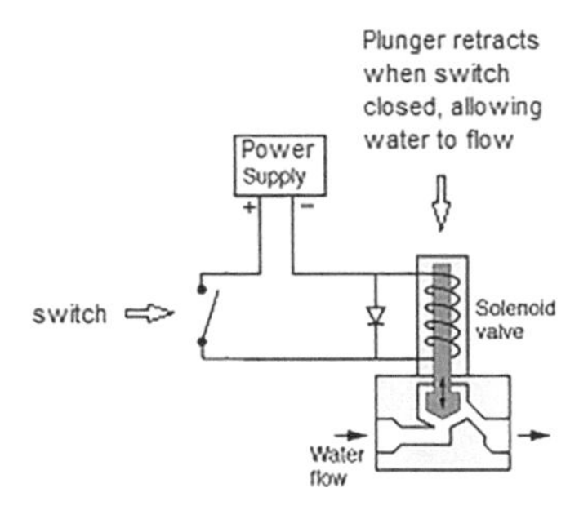

FIGURE 14.9 An electronic solenoid valve.

14.1.1.6 Magnetically Levitated Trains

A magnetically levitated train consists of superconducting magnets placed on the bottom of the train and normal coils placed in the track

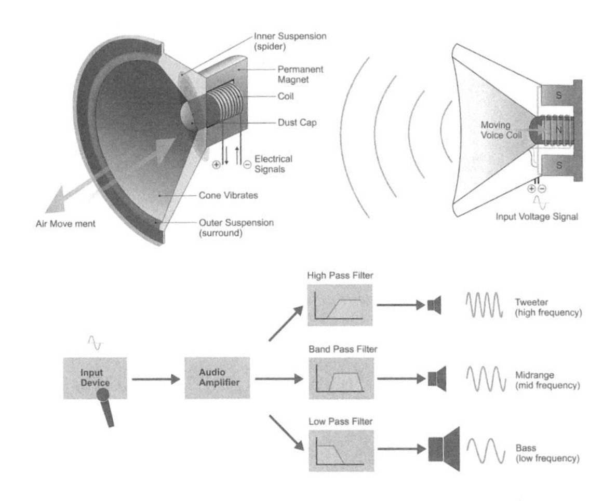

FIGURE 14.10 A speaker.

FIGURE 14.11 A superconducting magnet.

(Figure 14.12). When the train passes over the coils in the track, a current is induced in these coils. This current generates a magnetic field that opposes the superconducting magnet's field. The like poles of the two magnets repel, and the train floats.

Hover Boards use electromagnets that produce a changing electromagnetic field. This field interacts with a metallic surface under the board producing electrical current in the metal that in turn produce its own magnetic field that repels the board (Figure 14.13).

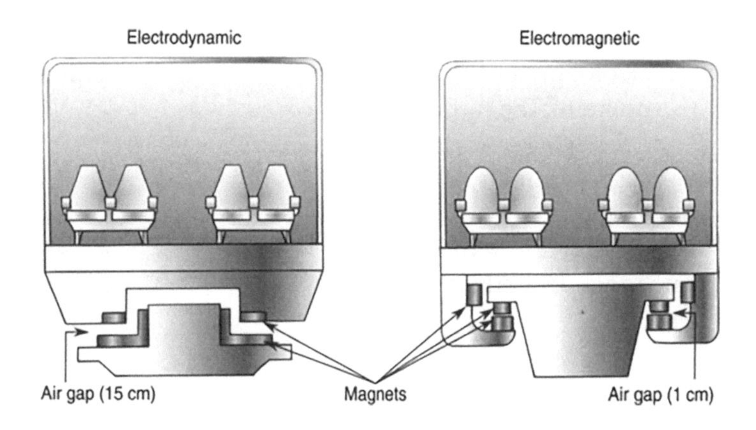

FIGURE 14.12 A magnetically levitated train.

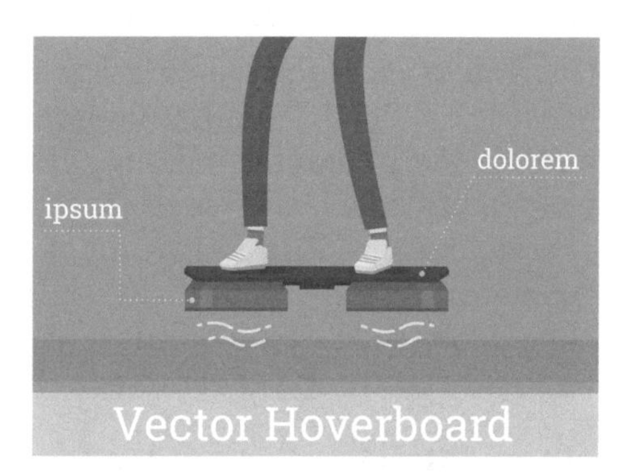

FIGURE 14.13 A hover board floats on a magnetic field.

14.2 EARTH'S MAGNETIC FIELD

Inside the earth there is molten iron. In this state, the iron is like a soup, with atoms missing some of their electrons together with those electrons moving freely about. Because the earth is spinning, so is the molten iron. A magnetic field is generated from this rotating mixture similar to that of an electromagnet (Figure 14.14).

14.2.1 Compass

A compass is a bar magnet with its magnetic north pole painted red, or the tip of an arrow painted on it. It is balanced on a needle, so it is free to rotate. The compass' magnetic north pole points to a magnetic south pole, because like poles repel and unlike poles attract. The fact that compasses point to geographic North brings us to the conclusion that the geographic North Pole of the earth is actually a magnetic south pole! (Figures 14.15 and 14.16).

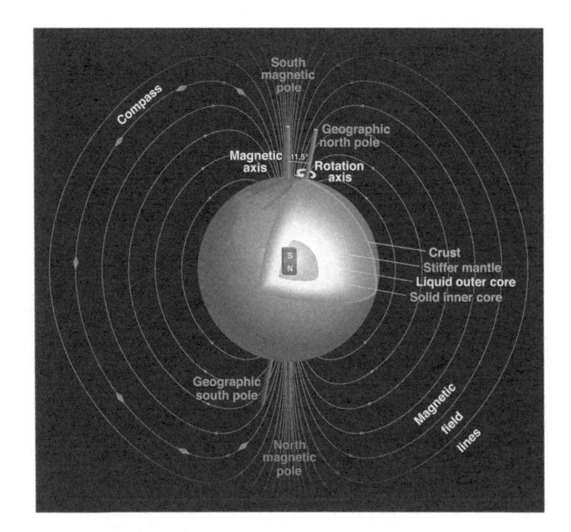

FIGURE 14.14 A magnetic field is generated inside the earth because of electric currents in the molten iron in its core, similar to an electromagnet.

FIGURE 14.15 The earth's geographical North Pole is actually a magnetic south pole.

FIGURE 14.16 Adafruit's HMC5883L Breakout—Triple-Axis Magnetometer Compass Sensor measures magnetic fields in the x, y, and z directions using a magnetoresistive sensor. These sensors are made from a nickel-iron thin film resistive strip that changes resistance in the presence of a magnetic field, courtesy of Adafruit.com.

14.2.2 Magnetic Field of a Current Carrying Wire

Passing a current through a wire will generate a magnetic field. The magnetic field will surround the wire as shown in Figure 14.17.

The unit for magnetic field is the tesla. A tesla is the measure of the concentration of magnetic field line in a given area. The earth's magnetic field has a strength of approximately 5.2×10^{-5} tesla. Another unit of magnetic field is the gauss.

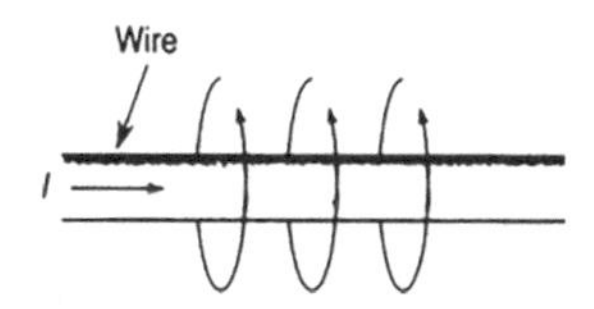

FIGURE 14.17 A magnetic field is generated by passing a current through a wire.

$$1\text{ tesla}(\text{T}) = 10^4\text{ gauss}$$

For a straight current carrying wire, the field strength at a distance r in air from the wire is given by:

$$B = \frac{\mu_0 \cdot I}{2\pi r}$$

$$\text{Magnetic field (T)} = \frac{4\pi \times 10^{-7} \times \text{current (A)}}{2\pi \times \text{distance (m)}}$$

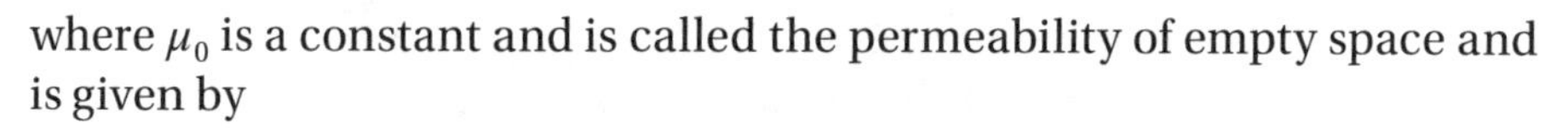

where μ_0 is a constant and is called the permeability of empty space and is given by

$$\mu_0 = 4\pi \times 10^{-7}\,\frac{\text{T}\cdot\text{m}}{\text{A}}.$$

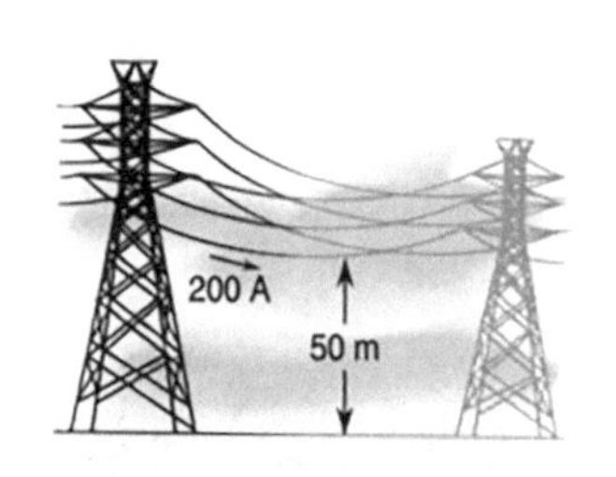

FIGURE 14.18 What is the strength of a magnetic field 50 m from a power line carrying 200 A of current?

Example 14.1

What is the strength of a magnetic field 50 m from a power line carrying 200 A of current? (Figure 14.18)

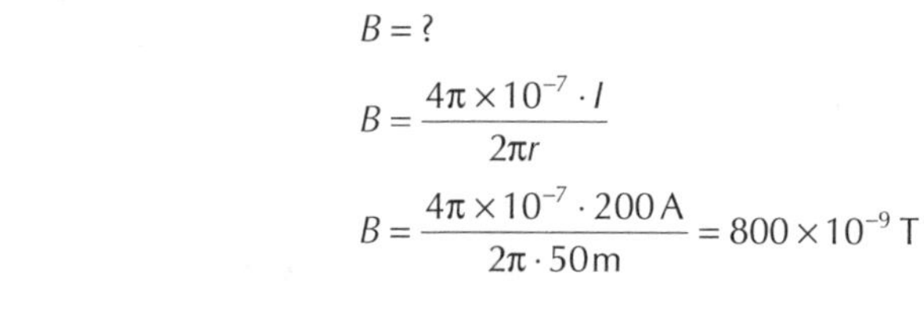

$$I = 200\text{ A}$$
$$r = 50\text{m}$$
$$B = ?$$
$$B = \frac{4\pi \times 10^{-7} \cdot I}{2\pi r}$$
$$B = \frac{4\pi \times 10^{-7} \cdot 200\text{ A}}{2\pi \cdot 50\text{m}} = 800 \times 10^{-9}\text{ T}$$

14.3 MAGNETIC FIELD INSIDE A COIL

Passing a current through a coil of wire produces a magnetic field similar to the field from a bar magnet. The strength of the magnet can be enhanced if a magnetic material is placed inside the coil such as a nail. Permeability is a measure of the ability of a material to enhance the strength of a magnetic field (Table 14.1). The permeability of a material μ is given by:

$$\mu = \mu_r \cdot \mu_0$$

$$\text{permeability} = (\text{relative permeability}) \cdot 4\pi \times 10^{-7}\,\frac{\text{T}\cdot\text{m}}{\text{A}}$$

TABLE 14.1
Table of Relative Permeability

Material	μ_r
Air	1
Copper	1
Aluminum	1
Nickel	100
Iron	5,000
Mumetal (at 1 kHz)	20,000
Super permalloy (at 1 kHz)	100,000

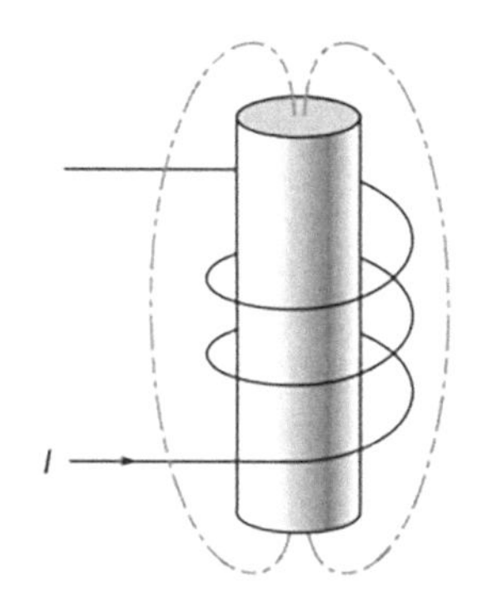

FIGURE 14.19 A magnetic field generated by an electromagnet.

For a long, tightly turned coil of length L with N turns, the magnetic field is given by (Figure 14.19):

$$B = \mu \cdot I \cdot \frac{N}{l}$$

$$\text{Magnetic field (T)} = \text{permeability}\left(\frac{\text{T}\cdot\text{m}}{\text{A}}\right)\cdot \text{current (A)} \cdot \frac{\text{number of turns}}{\text{coil length(m)}}$$

Example 14.2

A current of 5.0 A passes through a coil of length 0.10 m with 4,500 turns. What is the strength of the field inside the coil with an air core, $\mu_r = 1$? (Figure 14.20)

$$\mu = \mu_r \cdot \mu_0$$

$$\mu = 1 \cdot 4\pi \times 10^{-7}\ \frac{\text{T}\cdot\text{m}}{\text{A}}$$

$$\mu = 4\pi \times 10^{-7}\ \frac{\text{T}\cdot\text{m}}{\text{A}}$$

$$B = \mu \cdot I \cdot \frac{N}{l}$$

$$B = \left(4\pi \times 10^{-7}\ \frac{\text{T}\cdot\text{m}}{\text{A}}\right)\cdot(5\,\text{A})\cdot\left(\frac{4{,}500}{0.1\text{m}}\right)$$

$$B = 0.28\,\text{T}$$

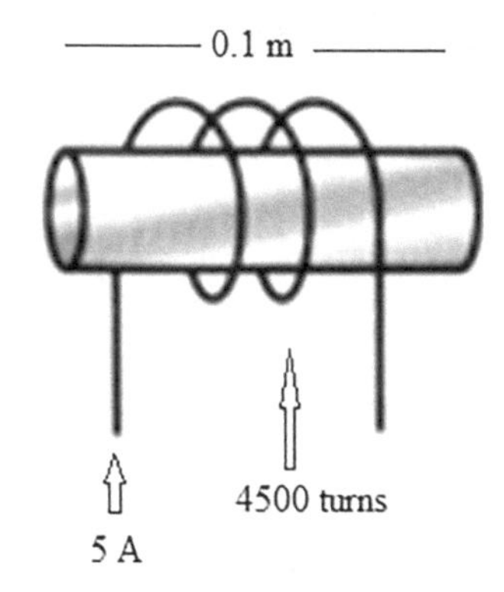

FIGURE 14.20 What is the strength of the field inside the coil?

14.4 ELECTROMAGNETIC INDUCTION

Electromagnetic induction is the generation of electricity from magnetism. Magnetic fields can push electrons through a conductor just like a battery can. In order for the electricity to be generated, however, the magnetic field must be changing in some way near the wire. Rotating a coil or wire near the magnet or moving a magnet back and forth will accomplish this (Figure 14.21).

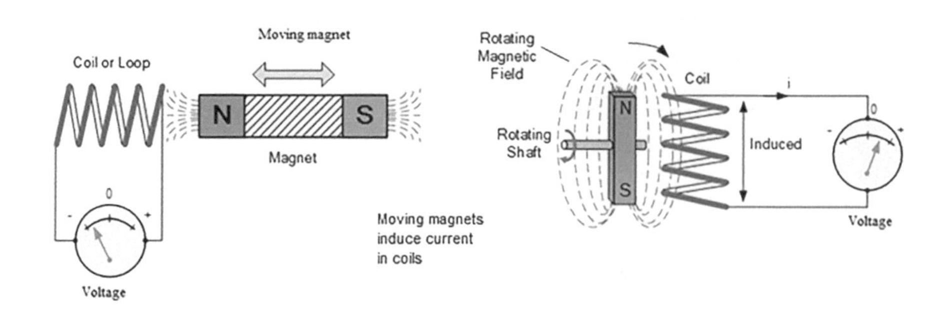

FIGURE 14.21 Rotating a coil or wire near the magnet or moving a magnet back and forth will generate electricity.

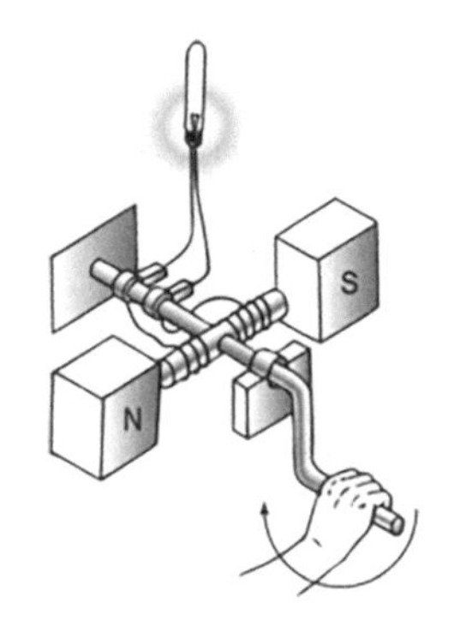

FIGURE 14.22 An electrical generator.

14.5 AN ELECTRICAL GENERATOR

A simple electrical generator consists of a rotating coil in a magnetic field (Figure 14.22). As the coil rotates, the penetration of the magnetic field through the coil changes. By induction, this changing magnetic field through the coil induces an AC current in it.

14.5.1 Alternators

An automobile alternator consists of a rotating coil inside a set of three fixed coils (Figure 14.23). The rotating coil is connected to a source of current, the car battery, and therefore generates a magnetic field. This rotating magnetic field induces an AC current into the fixed coils. The AC output is then rectified and fed to the battery.

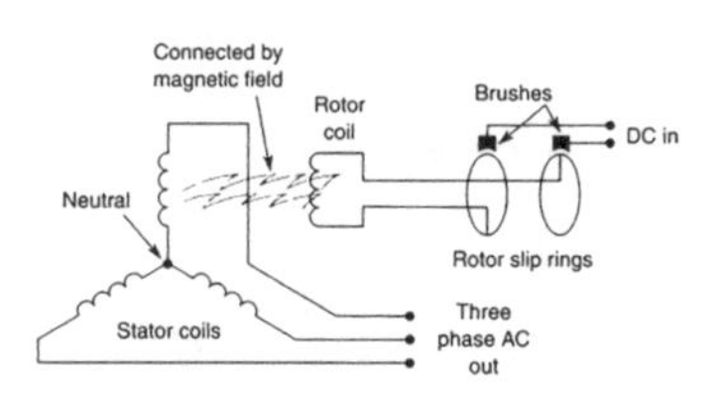

FIGURE 14.23 An automobile alternator.

14.6 MOTORS

14.6.1 DC Motors

A permanent-magnet DC (direct current) electric motor consists of a rotating coil with current running through it, centered around a permanent magnet. Current running through the coil produces its own magnetic field, which creates an attraction to the fixed magnet, causing it to spin.

The current through the coil is made to change direction halfway through a rotation to keep it spinning. Otherwise, it would get stuck in one place. The rotating coil is called the *rotor,* the permanent magnet is called the *stator,* and the device that changes the current's direction is called the *commutator.* Current is delivered by a set of brushes that rub the commutator (Figure 14.24). The stator magnetic field strength can be adjustable; this is called the field coil as shown in Figure 14.25.

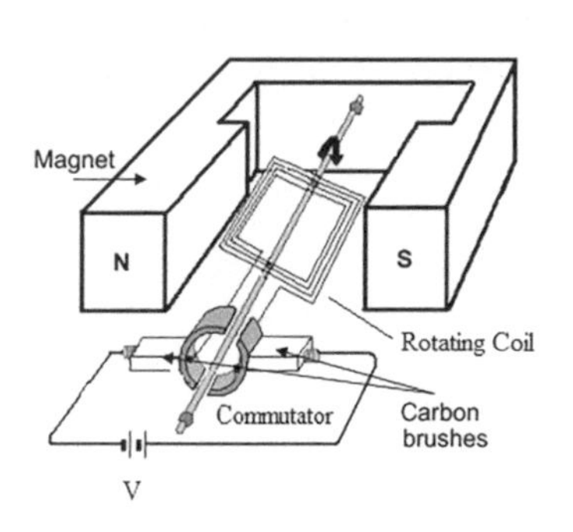

FIGURE 14.24 A DC motor.

A DC motor can act as a generator (Figure 14.26). Spinning the motor's shaft will generate a voltage.

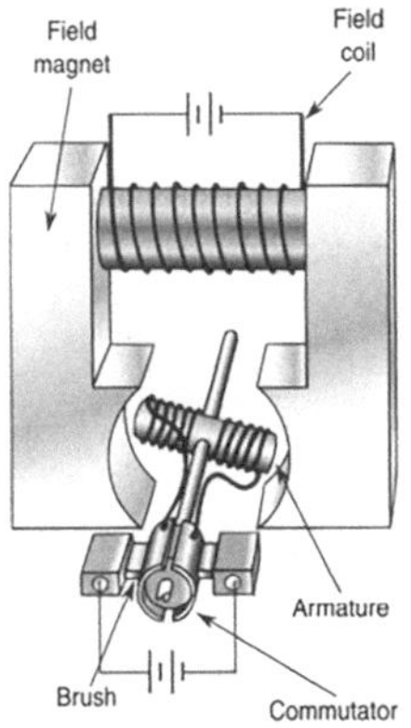

FIGURE 14.25 A DC motor with a field coil.

14.6.2 AC Motors

The most commonly used AC motor is the AC induction motor, which has no brushes or commutators. The stator consists of coils with AC current passing through them, producing a rotating magnetic field (Figure 14.27). The rotor has a squirrel cage-type construction (Figure 14.28). As the magnetic field

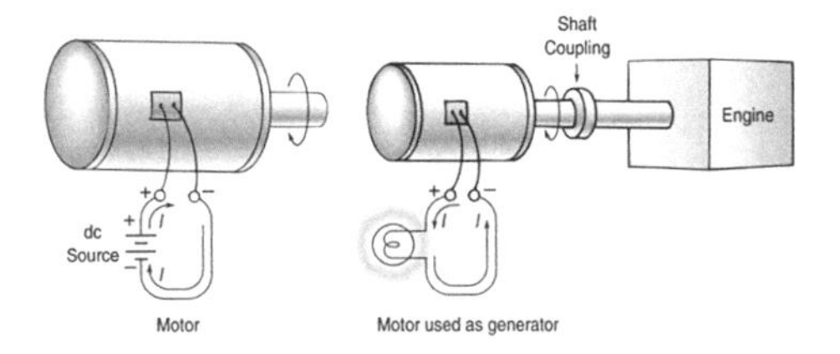

FIGURE 14.26 A DC motor used to generate electrical power.

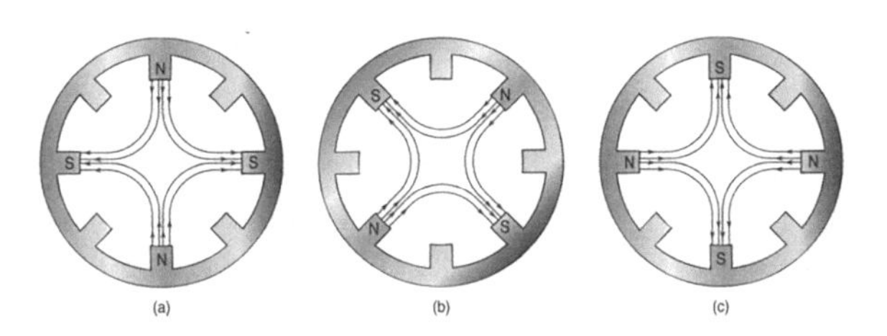

FIGURE 14.27 The stator produces a rotating magnetic field.

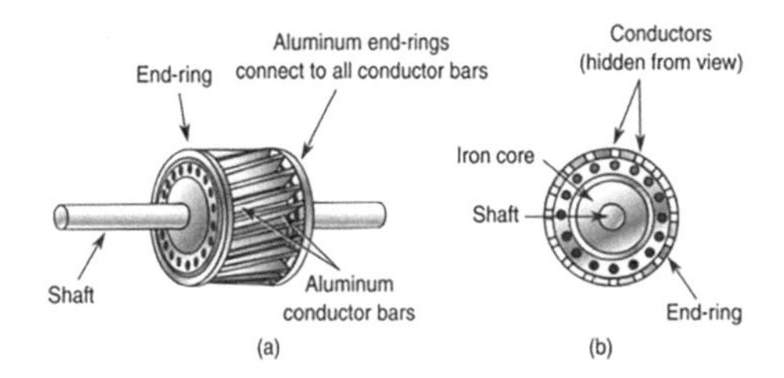

FIGURE 14.28 The rotor is made like a squirrel cage.

rotates around the squirrel cage, a current is induced into the coil below the field. This, in turn, produces its own magnetic field and generates a torque with the stator's magnetic field, causing the rotor to rotate.

14.7 ELECTRIC METERS

The electric meter on your house is an energy meter that measures the number of kilowatt-hours used (Figure 14.29). It works on the same principle as an AC induction motor. There are two electromagnets, one connected across the source voltage to the house and the other connected in series with source. Between the two electromagnets is an aluminum disk that is free to rotate. Because of the way the two electromagnets are connected, they are out of phase with one another by 90° and produce a rotating magnetic field. These electromagnets induce currents into the aluminum disk, which produces a magnetic field. This magnetic field

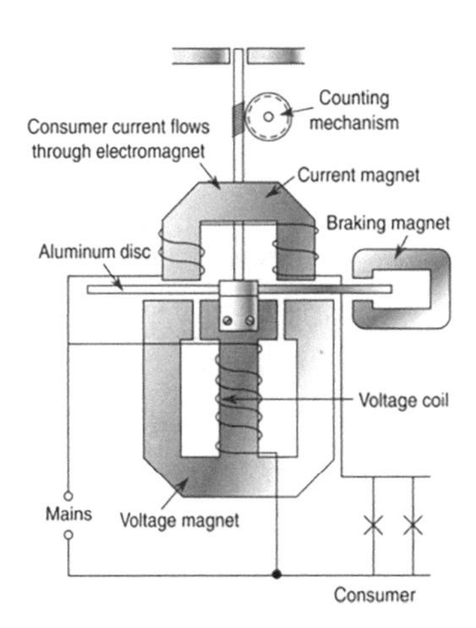

FIGURE 14.29 An electric meter or an energy meter.

interacts with the electromagnet's field, causing the disk to rotate. The number of rotations the motor makes is proportional to the current, voltage, and the time the motor is running for (recall $E = I \times V \times t$). A counting mechanism records this number in terms of kilowatt-hours (kWh).

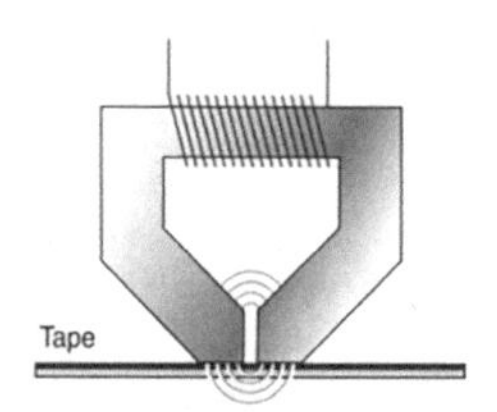

FIGURE 14.30 A tape recorder's head storing its magnetic field on the magnetic tape.

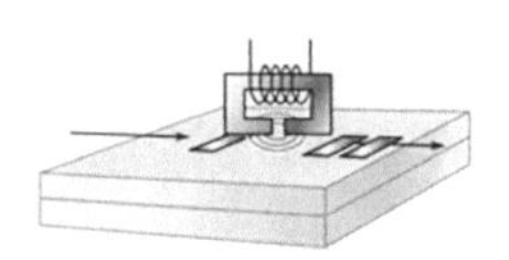

FIGURE 14.31 Reading a magnetic tape.

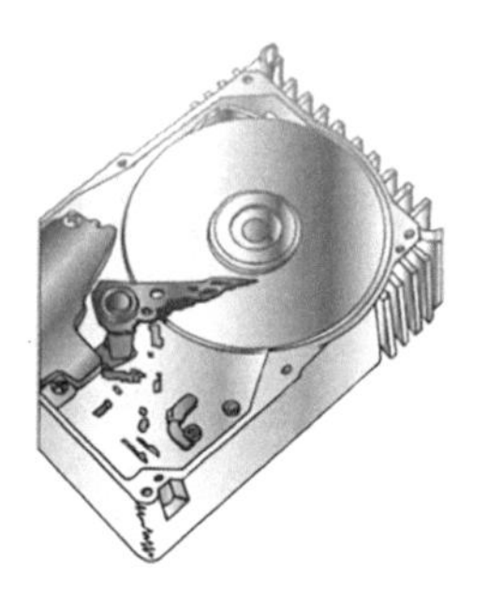

FIGURE 14.32 A hard disk drive.

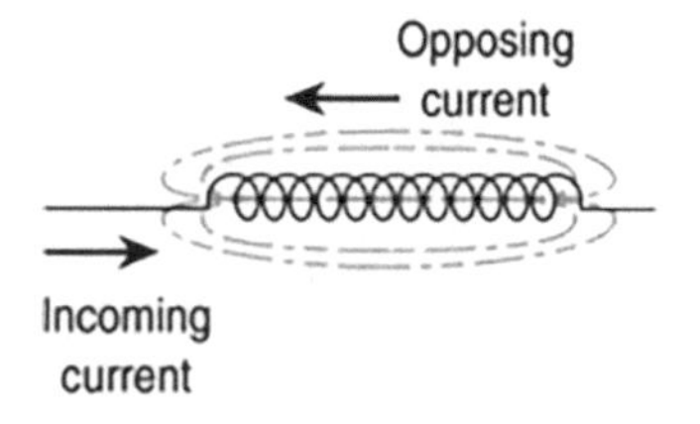

FIGURE 14.33 The inductor opposes a rapidly-changing current by using magnetic fields generated in the coil to produce a countercurrent flow, which then opposes the incoming current.

14.8 MAGNETIC MEMORY

Magnetic tape data storage may seem like a technology of the past, but recent advances have made it so that a tape cartridge holding over a kilometer of tape can store hundreds of terra bytes of data in a smaller space than a hard drive takes up. This has been accomplished by using magnetic particles that are just a few nanometers in size rather than hundreds of nanometers used in the older tape.

The magnetic tape is made by applying a mixture of ferric oxide powder and glue to a piece of plastic tape. Iron oxide (FeO) is common rust. Ferric oxide (Fe_2O_3) is another oxide of iron. Both are *ferromagnetic* materials that, when exposed to a magnetic field, will remain magnetized. The magnetic tape is passed over an electromagnet called a *head*. An electrical signal from a microphone is sent into the head, generating a magnetic field. The ferric oxide in the tape lines up in size and direction with the magnetic field of the head when it passes under it. The tape therefore stores the electrical signal applied to the head in the form of magnetism (Figure 14.30).

To play back what was recorded, the head is used as a magnetic field sensor. As the tape is passed over the head, the magnetic field in the tape induces a current into the head. The original electrical signal that magnetized the tape is now recreated (Figure 14.31).

14.8.1 Hard Disks

A hard disk on a computer stores information in a similar way as a magnetic tape. It accomplishes this by storing the information in magnetic domains and on a high-speed rotating disk. The disk rotates at thousands of inches/second. The head moves radially across the disk, enabling it to reach and retrieve any stored information at a very high speed (Figure 14.32).

14.8.2 Inductors

Inductors are devices that oppose a changing current. An inductor is simply a coil of wire. When a current is changing through an inductor, the magnetic field generated by this current will change. By electromagnetic induction, a countercurrent will be induced into the coil, which will oppose the incoming current (Figure 14.33). The inductor will therefore act like a resistor and reduce the current flow. Inductance is a measure of the inductor's ability to fight this changing current. Inductors can be used as surge protectors (Figure 14.34).

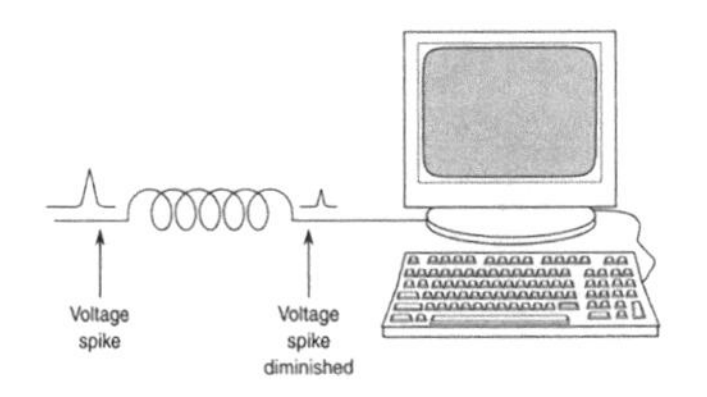

FIGURE 14.34 Inductors can be used as surge protectors, preventing voltage spikes from entering your computer.

The inductance of an inductor depends on a number of factors, the area of the coil A, the length of the coil l, the number of turns in the coil N, and the permeability μ. The unit of inductance is the henry, H.

The inductance of an inductor is given by the following formula:

$$L = \frac{\mu \cdot A \cdot N^2}{l}$$

$$\text{Inductance(H)} = \frac{\text{permeability} \cdot \text{Area}\left(\text{m}^2\right) \cdot \text{Number of turns}^2}{\text{length (m)}}$$

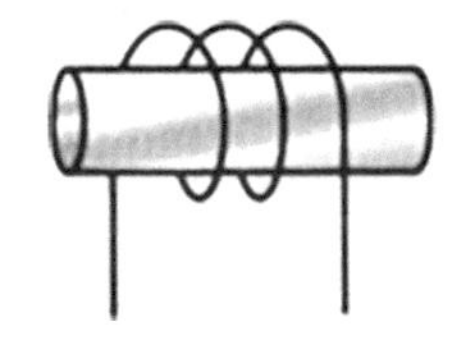

FIGURE 14.35 An inductor.

Example 14.3

An inductor is made from three loops or turns of wire around a cylinder of a material that has a relative permeability $\mu = 1{,}000$ and a radius $r = 0.01\text{m}$. The coil is spaced out so that it has a length of 0.02 m. Find its inductance (Figure 14.35).

Finding the area of the coil A,

$$A = \pi r^2 = \pi(0.01\text{m})^2$$

$$A = 3.14 \times 10^{-4}\,\text{m}^2$$

Finding the permeability.

$$\mu = \mu_r \cdot \mu_0$$

$$\mu = 1{,}000 \cdot 4\pi \times 10^{-7} \frac{\text{T} \cdot \text{m}}{\text{A}}$$

$$\mu = 12.57 \times 10^{-4} \frac{\text{T} \cdot \text{m}}{\text{A}}$$

$$L = \frac{\left(12.57 \times 10^{-4} \frac{\text{T} \cdot \text{m}}{\text{A}}\right) \cdot \left(3.14 \times 10^{-4}\,\text{m}^2\right) \cdot \left(3^2\right)}{0.02\text{m}}$$

$$L = 1.78 \times 10^{-4}\,\text{H}$$

14.9 TRANSFORMERS

A transformer is a device that transforms a small AC voltage into a large AC voltage or large AC voltage into a small AC voltage. It consists of two coils next to one another. The primary of a transformer is the input and the secondary of a transformer is the output (Figure 14.36). When current passes through the primary, it generates a magnetic field, which passes

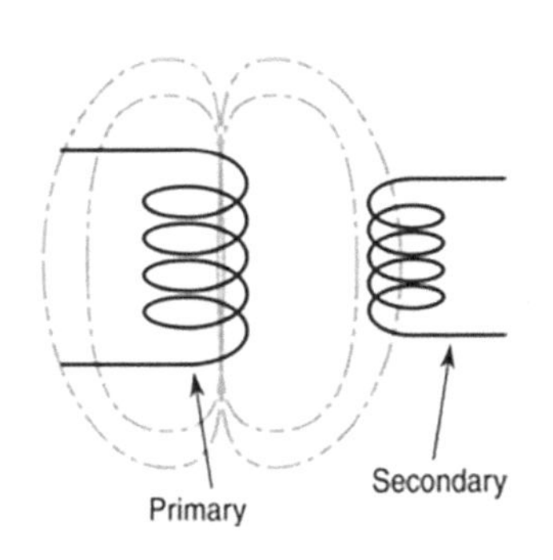

FIGURE 14.36 A transformer.

through the secondary coil. The magnetic field pushes on the electrons in the second secondary, by electromagnetic induction, creating the current in the secondary coil.

The relationship between the input voltage and the output voltage depends on the ratio of the number of turns of the secondary, N_s, to the number of turns of the primary, N_p.

$$V_s = \frac{N_s}{N_p} V_p$$

$$\text{Voltage}_{\text{Secondary}} = \frac{\text{Turns}_{\text{Secondary}}}{\text{Turns}_{\text{Primary}}} \text{Voltage}_{\text{Primary}}$$

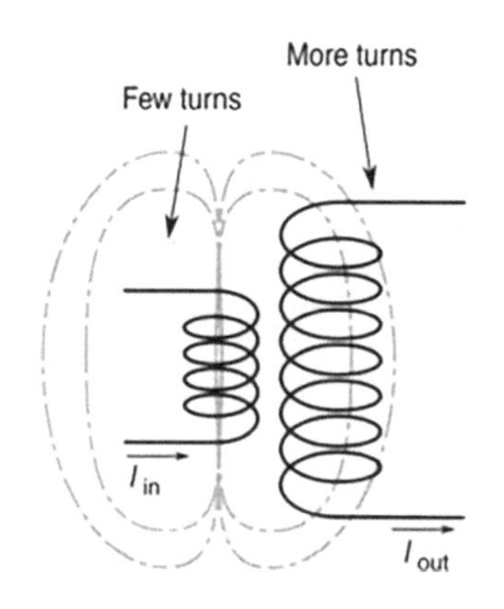

FIGURE 14.37 A step-up transformer.

A transformer with more turns in the secondary than in the primary will step up the input voltage, and is called a step-up transformer. A transformer with more turns in the primary than in the secondary will step down the input voltage, and is called a step-down transformer (Figures 14.37 and 14.38).

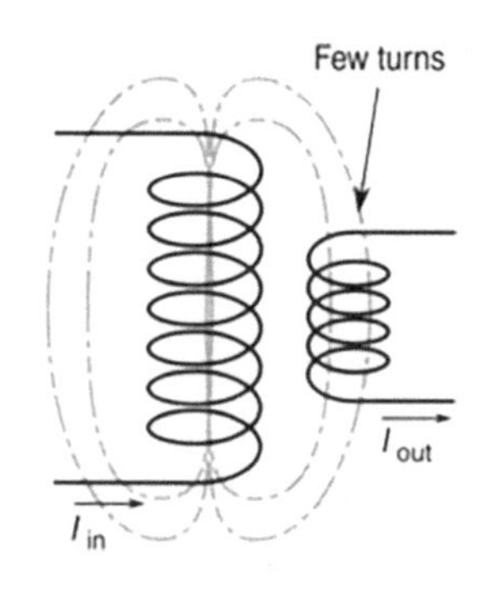

FIGURE 14.38 A step-down transformer.

Example 14.4

A transformer consists of a primary with 100 turns with an applied AC voltage of 12 V and a secondary of 200 turns. What is the voltage out of the secondary?

$$V_s = \frac{N_s}{N_p} V_p = \frac{200}{100} 12\,\text{V}$$

$$V_s = 24\,\text{V}$$

14.10 INDUCTIVE HEATING

Current running through metal produces heat. If a varying magnetic field is placed below a cooking pot, electromagnetic induction will generate a current in the pot, causing it to heat up (Figure 14.39).

FIGURE 14.39 Inductive heating.

14.11 MAGNETIC RESONANCE IMAGING

Magnetic resonance imaging (MRI) is a noninvasive technique for "looking" inside a body without going inside, or sending something through the body like X-ray (Figure 14.40).

MRI works on the principle that different body tissues respond differently to varying magnetic fields. By measuring these different responses over an area, a two- or three-dimensional image can be formed.

Let's look in detail at how MRI works. When an atom is placed in a strong, constant magnetic field, its nucleons act like small magnets themselves and will line up either parallel or antiparallel to the field (Figure 14.41).

In addition to the constant field, a small varying magnetic field is placed perpendicular to the constant field. This will cause the nucleons to slightly precess around their original direction. At just the right frequency, the precession will be large and will absorb a lot of energy from the varying magnetic field. This is the *resonant frequency*. The resonant frequency is different for different materials being studied, and this is how we can distinguish among materials (Figure 14.42).

Inside the MRI machine is a large superconducting magnet of 0.5–2T in strength, smaller field-shaping magnets, and RF (radio frequency) coils that generate a time-varying magnetic field. The large magnet does the overall aligning of the nucleons in the patient. The shaping magnets select the area of the body to scan, and the RF coils sweep through frequencies, measuring the resonant frequencies of the tissue under inspection. All this information is collected by a computer and put together to form an image (Figure 14.43).

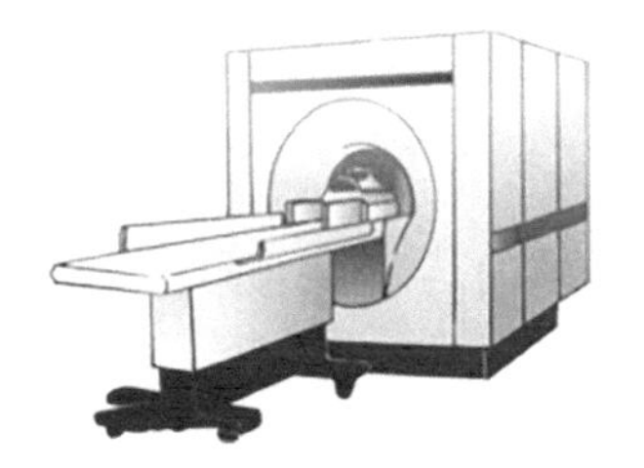

FIGURE 14.40 A MRI machine.

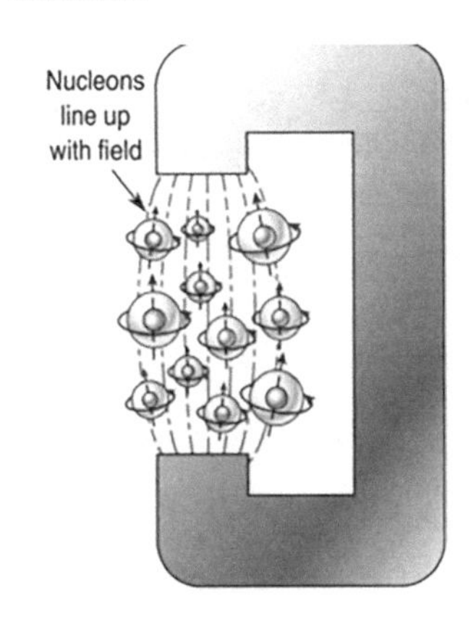

FIGURE 14.41 Nucleons in an atom line up parallel or antiparallel to the applied field.

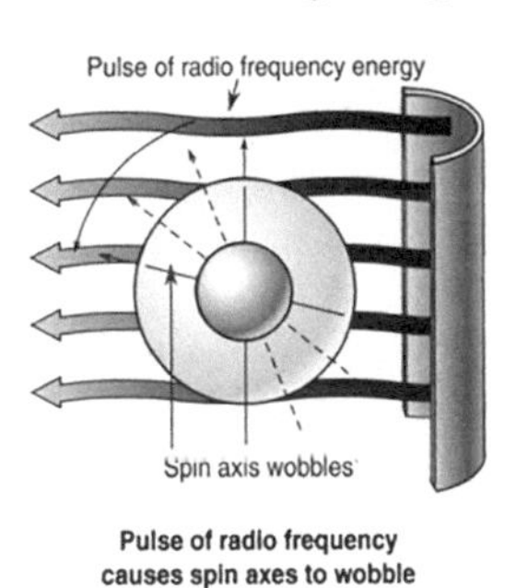

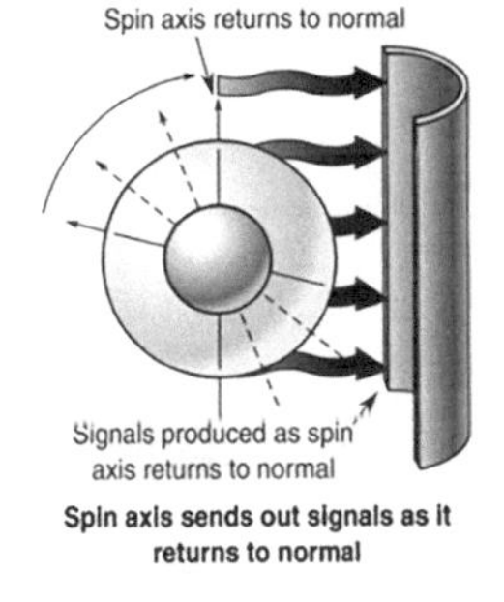

FIGURE 14.42 Determining the resonant frequency of the nucleons.

FIGURE 14.43 A MRI image.

14.12 MEASURING MAGNETIC FIELDS

14.12.1 Hall Probe

A Hall probe is the most common method of measuring magnetic fields. A Hall probe consists of a semiconductor material with a current running through it. If a magnetic field penetrates the Hall probe, the current flowing inside the probe will be pushed to one side, as shown (Figure 14.44). This will produce a voltage across the probe because the charge will be concentrated on one side. This voltage is proportional to the magnetic field strength.

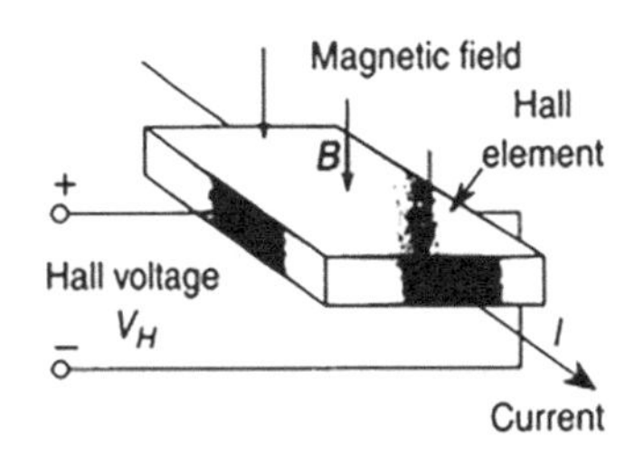

FIGURE 14.44 A Hall probe responds to a magnetic field by generating a voltage proportional to the field strength.

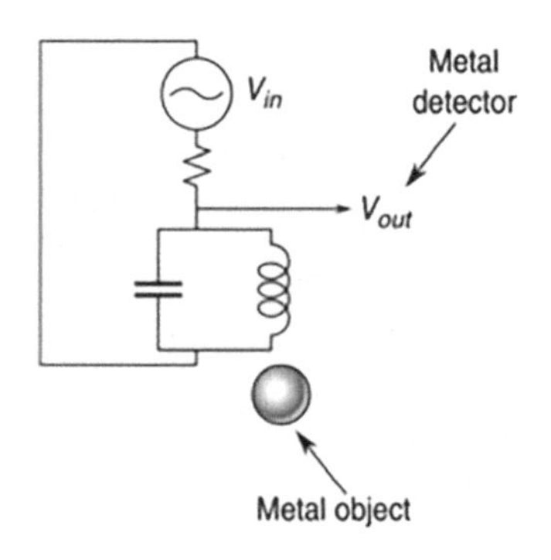

FIGURE 14.45 A simple metal detector consists of a resonant frequency RLC circuit.

14.12.2 Metal Detectors

A simple metal detector consists of a resonant frequency RLC (Resistor, Inductor, Capacitor) circuit (Figure 14.45). As the circuit is brought near some metal, the inductance of the inductor will change and therefore so will the resonant frequency. This frequency can be monitored in some way with a meter or headphone.

14.13 CHAPTER SUMMARY

Symbols used in this chapter:

Symbol	Unit
t—time	second (s)
r—distance	meter (m)
l—length	meter (m)
B—magnetic field	tesla (T)
I—current	ampere(A)
μ and μ_0—permeability	$4\pi \times 10^{-7}\ \frac{\mathrm{T \cdot m}}{\mathrm{A}}$
μ_r—relative permeability	No unit
N—turns	No unit
V—voltage	volt (V)
L—inductance	henry (H)

Permanent magnets are materials that can be magnetized and retain their magnetism.

Ferromagnetic materials, when exposed to a magnetic field, will remain magnetized.

Electromagnets: Current moving through a wire will produce a magnetic field. Passing a current through a coil of wire produces a magnetic field similar to a field from a bar magnet.

Tesla and gauss are the units of magnetic field.

$$1\ \text{tesla}(\mathrm{T}) = 10^4\ \text{gauss}$$

For a straight wire with current running through it, the field strength at a distance is given by:

$$B = \frac{\mu_0 \cdot I}{2\pi r}$$

$$\text{Magnetic field}\ (\mathrm{T}) = \frac{4\pi \times 10^{-7} \times \text{current}(\mathrm{A})}{2\pi \times \text{distance}\ (\mathrm{m})}$$

where μ_0 is called the permeability and is given by

$$\mu_0 = 4\pi \times 10^{-7} \frac{\text{T} \cdot \text{m}}{\text{A}}$$

Magnetic field inside a coil: Passing a current through a coil of wire produces a magnetic field similar to a field from a bar magnet.

For a long, tightly turned coil of length L with N turns, the magnetic field is given by:

$$B = \mu \cdot I \cdot \frac{N}{l}$$

$$\text{Magnetic field (T)} = \text{permeability}\left(\frac{\text{T} \cdot \text{m}}{\text{A}}\right) \cdot \text{current (A)} \cdot \frac{\text{number of turns}}{\text{coil length(m)}}$$

The **permeability** of a material μ is given by:

$$\mu = \mu_r \cdot \mu_0$$

$$\text{permeability} = (\text{relative permeability}) \cdot 4\pi \times 10^{-7} \frac{\text{T} \cdot \text{m}}{\text{A}}$$

Electromagnetic induction means generating electricity from magnetism.

Inductors are devices that oppose a changing current.

The unit of inductance is the henry, H. The inductance of an inductor is given by the following formula:

$$L = \frac{\mu \cdot A \cdot N^2}{l}$$

$$\text{Inductance(H)} = \frac{\text{permeability} \cdot \text{Area}\left(\text{m}^2\right) \cdot \text{Number of turns}^2}{\text{length (m)}}$$

A **transformer** is a device that transforms a small AC voltage into a large AC voltage or large AC voltage into a small AC voltage.

$$V_s = \frac{N_s}{N_p} V_p$$

$$\text{Voltage}_{\text{Secondary}} = \frac{\text{Turns}_{\text{Secondary}}}{\text{Turns}_{\text{Primary}}} \text{Voltage}_{\text{Primary}}$$

PROBLEM SOLVING TIP

- Pay attention to units to ensure that the final answer is dimensionally correct.

PROBLEMS

1. What is the origin of the earth's magnetic field, and how would you measure it?
2. Explain how electricity can be generated from magnetism and vice versa?
3. Two magnets with like poles facing one another will repel. How can this fact be used to build a magnetic bearing?
4. What is a relay and how does it work?
5. What is the strength of a magnetic field 1.0 m from a wire carrying a current of 20.0 A?
6. A reed switch consists of a glass tube containing a switch made from two pieces of spring metal separated from each other. When a magnet is brought near the field it pulls the switch closed. How can this be used as a proximity switch?
7. What is the magnetic field inside a long, thin coil of length 0.08 m, 2,300 turns, and with a current of 0.50 amps flowing through it?
8. How does the motor work in a toy slot car?
9. Suppose you build an electrical generator from a car alternator powered by a 3 hp lawn mower engine. If the efficiency of the alternator is 90% and the alternator is putting out 12 V, how much current can it deliver? Note 1 hp = 746 W.
10. If a Hall probe has a sensitivity of 22 gauss/mV, and is putting out 5,000 mV, what is the field strength?
11. How does the electric meter on your house measure the electrical energy you use?
12. How does a magnetic tape, floppy disk, or credit card work, and why is it a bad idea to bring magnets near them?
13. Design a windmill that generates electricity from a DC motor and stores it in a battery.
14. What is the magnetic field at the ground, 6.0 m below a power line carrying 350 A?
15. Suppose you want to test the effect of magnetic fields on a gerbil. To generate the field, you make a coil from a toilet paper roll measuring 0.025 m in radius and wrap 500 turns of wire around it. If

the wire has a resistance of 1.5 Ω and is connected to a 12 V battery, what is the magnetic field on the gerbil?

16. What current is needed to generate a 1.0×10^{-7} T magnetic field, 1.0 m away from a straight wire?
17. A Hall probe sits in a magnetic field of strength 0.2 T. If the probe has a sensitivity of 100 mV/T, what voltage (in mV) does the probe put out?
18. What is the strength of a magnetic field inside a solenoid of length 0.03 m with 2,000 turns and with a current of 0.1 A flowing through it?
19. The police have a new weapon in their arsenal against crime, a type of magnetic cannon. The magnetic cannon consists of a solenoid mounted on the front of the police car. During a car chase, a large current is dumped into the solenoid generating a large magnetic pulse, which passes into the car being chased and "zaps" the car's computer, disabling the vehicle. Suppose 500 A is dumped into the cannon, which has 500 turns and is 0.1 m long. What is the magnetic field generated?
20. The earth has a magnetic field strength of 5.2×10^{-5} T. How many times stronger is the strength of the magnetic field in the previous problem compared to the earth's field?
21. What is the magnetic field 0.2 m away from the power cord of a 1,100 W, 110 V hair dryer?
22. A Hall probe measures the strength of a magnetic field. If the probe has a sensitivity of 10.0 V/T, and a field strength of 0.02 T is being measured, what is the voltage measured on the probe?
23. An electromagnet is used in a buzzer. The buzzer consists of a coil of length 0.01 m and 80 turns of wire. The coil has an iron core that increases its permeability by a factor of 1,000. What is the strength of the magnetic field if 0.1 A is flowing through the coil?
24. The automatic flusher on urinals consists of a sensor to detect your presence and a spool valve, which is a type of solenoid valve that controls the flow of water. The solenoid is a coil with a rod that moves in and out of it. When current is applied to the solenoid, the rod retracts into the solenoid allowing the water to flow. If the coil is 0.05 m long with 1,000 turns, and 0.1 A flows through it, what is the magnetic field in the center of the coil?
25. Another unit for magnetic field is the gauss. 1 T = 10,000 gauss. The earth's magnetic field is 5.2×10^{-5} T. What is this field in gauss?
26. If the current in a straight wire increases two times, how many times does the magnetic field surrounding it increase?

27. If the distance a magnetic field being measured from a wire carrying current increases two times, how many times does the field decrease?
28. A motor turns electrical energy into rotational energy. If a motor has an efficiency of 90% and consumes 5,000 J of energy, what rotational energy is generated?
29. A generator turns rotational energy into electrical energy. If the generator has an efficiency of 90% and creates 40,000 J of electrical energy, what was the input of rotational energy?
30. An electromagnet creates a lifting force of 200 lb when a current of 100 A is running through it. How many amps are needed to lift 300 lb?
31. How close does one have to be near a wire carrying 15 A in order to measure a magnetic field of 0.25 T?
32. An induction stove creates heat in a metal pot by generating a current in it through electromagnetic induction. If the resistance across the pot is $2 \times 10^{-3}\ \Omega$, and a current of 300 A is flowing through the pot, how many watts of heat is being created in the pot?
33. A solenoid is an electromagnetic coil with a core that can move in and out. If the field inside a coil is 0.35 T and the coil has a length of 0.01 m and 5 A is running through it, how many turns does it have?
34. The electrical energy meter on every house measures the current, voltage, and time the current is flowing for, to determine how much energy is being used, and therefore how much to charge you. If 50 A is being drawn at 110 V for 500 s, how much energy was used?
35. What is the resistance of the coil in a superconducting magnet at its operating temperature?
36. An inductor is made from 1,000 loops or turns of wire around a cylinder of a material that has a permeability $\mu = 50$ and a radius $r = 0.015$ m. The coil is spaced out so that it has a length of 0.025 m. Find its inductance.
37. How many turns is required to make a 2 mH inductor with a core of radius 0.001 m, permeability $\mu = 100$, and length equal to 0.02 m?
38. A transformer consists of a primary with 500 turns with an applied AC voltage of 14 V and a secondary of 1,500 turns. What is the voltage out of the secondary?
39. What would the turns' ratio $\frac{\text{Turns}_{\text{Secondary}}}{\text{Turns}_{\text{Primary}}}$ have to be to double the input voltage?
40. A transformer works on the principle of electromagnetic induction, which says a changing magnetic field passing through a loop of wire will induce a current in it. What is the output from a transformer that is supplied with constant voltage into the primary?

15

Waves

This chapter will describe waves in general, but will focus on sound waves in particular. Common types of waves are water, light, sound, and radio waves (Figure 15.1). Although they are very different from one another, they share many characteristics. With the whole world going wireless, knowledge of waves and their properties is very important.

FIGURE 15.1 Water waves.

To create a wave, you must disturb the medium. In case of water, when you tap it with a stick, the disturbance will propagate outward (Figure 15.2). The actual water molecules touched by the stick do not travel outward. Disturbing the water merely pushes on the water adjacent to it, and then that water pushes on the water next to it, and so on. It is like dominoes falling over and causing the others to fall over. There is motion, but there is no transport of matter (Figure 15.3). It is just the disturbance that moves. The wave carries energy but no matter.

FIGURE 15.2 The creation of a water wave.

FIGURE 15.3 Dominoes behave in a wavelike manner, propagating a disturbance. No one domino gets transported; only the energy is transported.

15.1 VELOCITY, FREQUENCY, AND WAVELENGTH

Creating waves on water requires disturbing the water. For example, tapping the water with a stick at some rate or frequency will produce a wave with the same frequency. This wave will travel outward with some velocity producing a series of crests and troughs. The period of the wave is the time it takes for the wave to travel a distance from crest to crest (Figure 15.4).

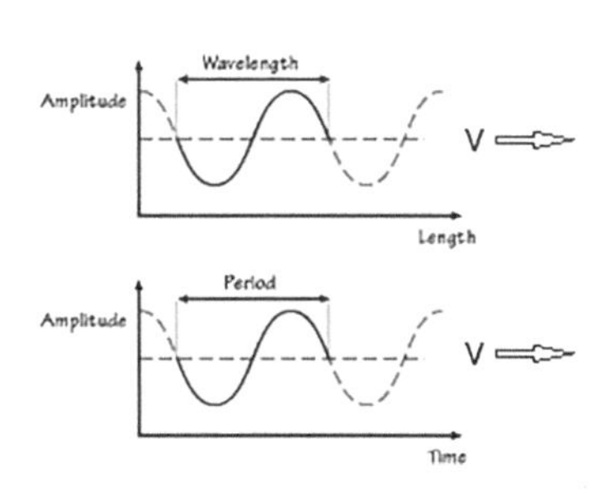

FIGURE 15.4 A wave of wavelength λ, a period T.

The frequency of the wave is the rate at which the wave "wiggles". The frequency of the waves is related to the period of the wave by:

$$f = \frac{1}{T}$$

$$\text{Frequency}(\text{Hz}) = 1/\text{period}(\text{s})$$

Frequency has units of hertz, Hz or cycles/second.

$$1\,\text{Hz} = 1\frac{\text{cycle}}{\text{s}} = \frac{1}{\text{s}}$$

The wavelength of a wave is the distance between crests (Figure 15.4). A wave's wavelength (λ), frequency (f), and speed (v) are related. All waves, including light, sound, and radio can be described by the following equation:

$$v = \lambda \cdot f$$

$$\text{Velocity}(\text{m/s}) = \text{wavelength}(\text{m}) \times \text{frequency}(\text{Hz})$$

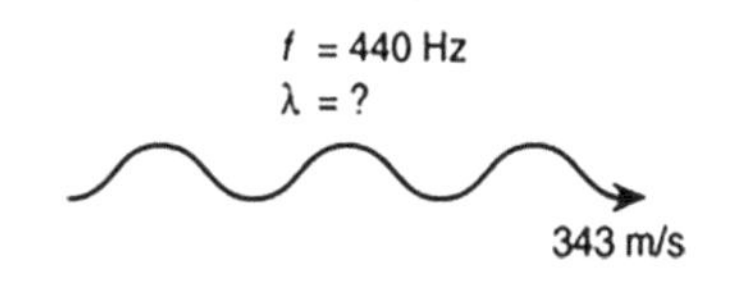

FIGURE 15.5 What is the wavelength of this wave?

Example 15.1

The musical note A has a frequency of 440 Hz (Figure 15.5). If the sound wave is traveling at 343 m/s, what is its wavelength?

$$f = 440\,\text{Hz}$$

$$v = 343\,\text{m/s}$$

$$\lambda = ?$$

$$v = \lambda \cdot f$$

$$\lambda = \frac{v}{f} \leftarrow \text{solving for } \lambda \text{ gives:}$$

$$\lambda = \frac{343\frac{\text{m}}{\text{s}}}{440\,\text{Hz}}$$

$$\lambda = \frac{343\frac{\text{m}}{\not{\text{s}}}}{440\frac{1}{\not{\text{s}}}}$$

$$\lambda = 0.78\,\text{m}$$

15.2 WAVE TYPES

There are two types of waves, transverse and longitudinal.

15.2.1 Transverse Waves

Transverse waves oscillate perpendicular to the direction of travel of the wave (Figure 15.6). Light is a transverse wave (Figure 15.7).

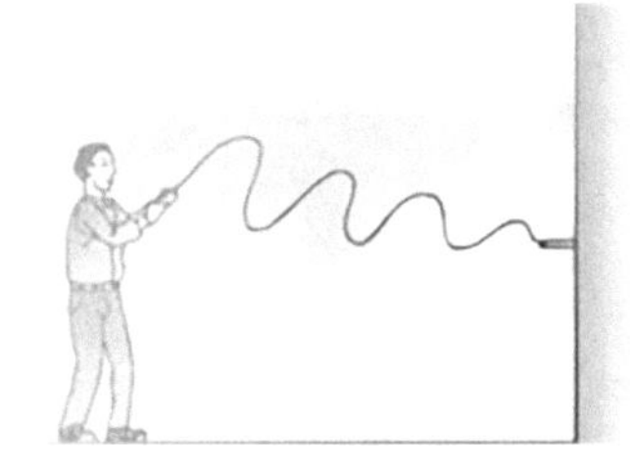

FIGURE 15.6 A jump rope demonstrating transverse waves.

15.2.2 Longitudinal Waves

Longitudinal waves oscillate back and forth in the direction the wave is traveling (Figure 15.8). Sound is a longitudinal wave.

Some waves can be both transverse and longitudinal, as in the case of water (see Figure 15.9).

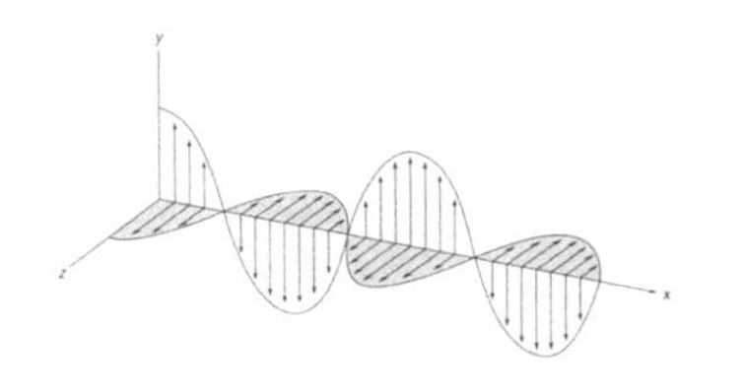

FIGURE 15.7 Light is a transverse wave. It is made up of electric and magnetic fields that oscillate perpendicular to the direction of travel.

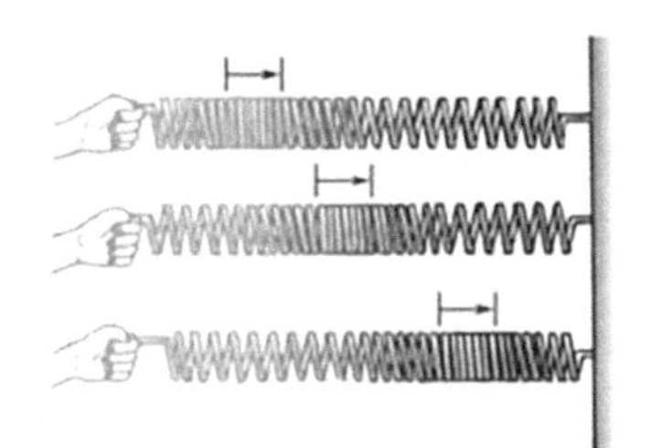

FIGURE 15.8 Longitudinal waves oscillate back and forth in the direction the wave is traveling.

FIGURE 15.9 Water is both a longitudinal and a transverse wave. A cork will bob up and down (the transverse component) and will also move back and forth (the longitudinal component).

15.3 GENERAL WAVE PROPERTIES

15.3.1 Refraction

Refraction is the bending of a wave when it moves from one medium into another. Waves refract because the speed of the wave changes when the wave moves from one medium into another. For example, light travels slower through water than through air. A wave front, striking the surface of water from air, will bend because the speed of the wave changes when the wave enters the water. The wave front in the water will lag behind the incoming wave. As a result, the wave will bend (Figure 15.10).

Sound waves exhibit refraction when moving from warm air of relatively low density to a cooler, denser air (Figure 15.11). The speed of sound in warm air is faster than in cool air. As a result, the wave that enters the cool air falls behind the wave in the warm air. The sound wave is effectively bent because of this slowing down.

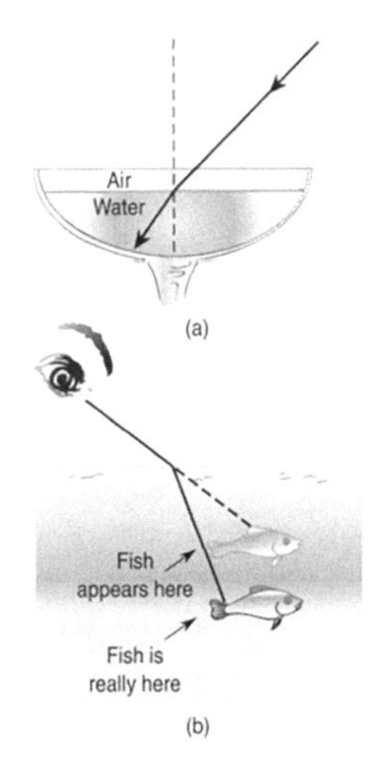

FIGURE 15.10 Light moving from air to water. (a) Light bends because it slows down in the water and cannot keep up with the incoming wave. (b) Objects appear shallower in the water because of refraction.

15.3.2 Superposition

Multiple waves can travel through the same medium without affecting each other. The overall wave will be the sum of all the waves. This is called

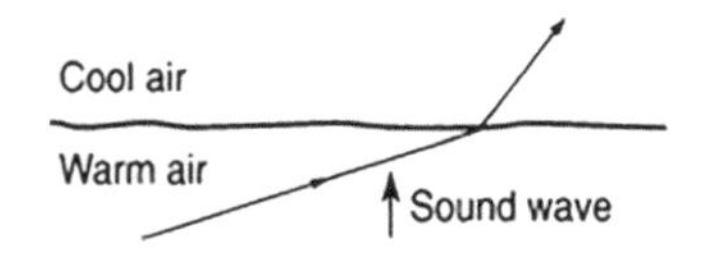

FIGURE 15.11 Sound waves bend when entering a different temperature region.

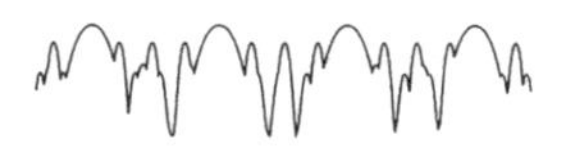

FIGURE 15.12 A superposition of many signals on one wire.

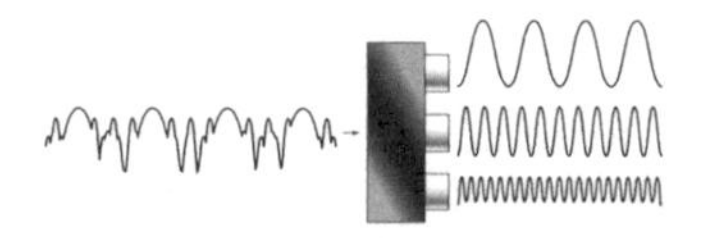

FIGURE 15.13 A filter separating out the individual waves.

superposition. Light, sound, and water waves all obey superposition. For example, multiple electrical signals of different frequencies can be sent down a single wire, and at the end of the wire the signal made up of all the individual signals can be filtered to isolate each individual signal. If you looked at the waveform coming out of this wire, it would look like one big messy wave (Figure 15.12). The individual waves can be separated out with a filter because they operate at different frequencies (Figure 15.13).

15.3.3 Constructive Interference

Waves that are in sync with one another combine to make a bigger wave. This is known as *constructive interference* (Figure 15.14).

15.3.4 Destructive Interference

Waves that are out of sync with one another subtract to make a smaller wave. This is known as *destructive interference* (Figure 15.15).

Noise canceling earbuds listen to incoming sound and create another sound 180° out of phase to combine with the incoming sound. The two waves destructively interfere with one another leaving the listener in silence (Figure 15.16).

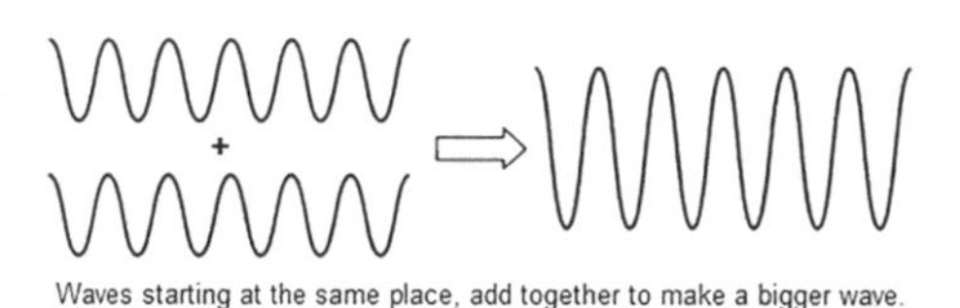

FIGURE 15.14 Constructive interference.

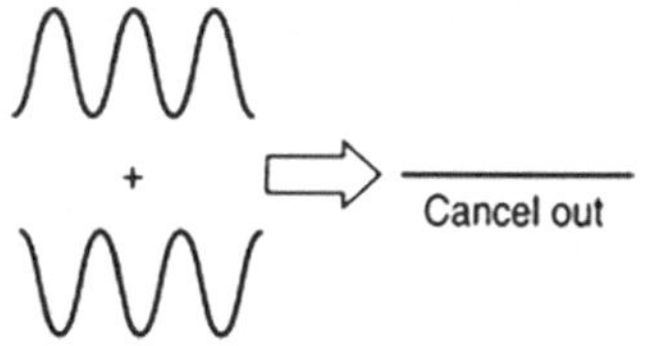

FIGURE 15.15 Destructive interference. Pictured are two waves 180° out of phase.

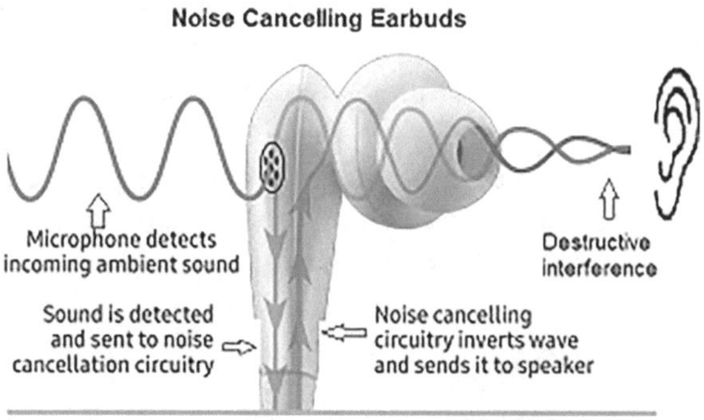

FIGURE 15.16 Noise canceling earbuds use destructive interference to cancel out noise.

15.3.5 Tone

Multiple waveforms, acting together, make up the *tone* of a sound (Figure 15.17). The waves making up one sound may have different frequencies, phases, and intensities.

FIGURE 15.17 Most sound waves are a composite of many different waves.

15.3.6 Resonance

The *resonance frequency* is the frequency at which if a system is stimulated with will produce a maximum response. For example, a singer can shatter a glass if the volume and frequency of his or her voice are just right (Figure 15.18). Antennas emit the maximum power at their designed resonance frequencies.

FIGURE 15.18 Example of resonance: maximum energy transfer.

15.3.7 Diffraction

Diffraction is the bending of waves around obstacles. Diffraction occurs most strongly when the size of the obstacle is the size of the wavelength of the wave (Figure 15.19).

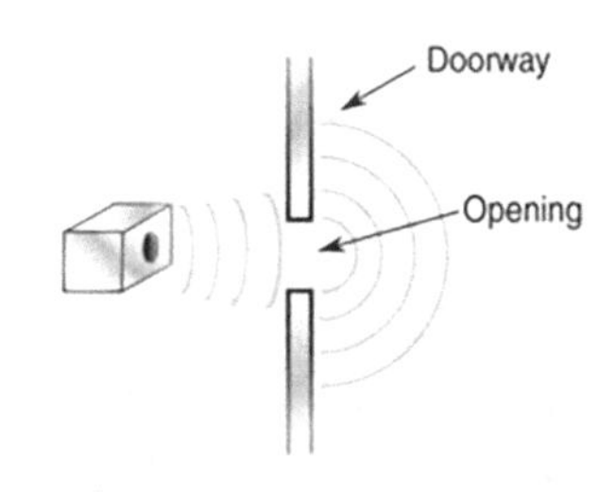

FIGURE 15.19 An example of diffraction is sound traveling through openings such as doorways.

15.4 DISTANCE AND VELOCITY MEASUREMENTS WITH WAVES

15.4.1 Radar, Lidar, and Sonar

Radar and sonar use radio and sound waves, respectively, to determine the distance to an object (Figures 15.20 and 15.21). When a wave is emitted, a clock is started and it times how long it takes for the wave to travel, reflect off an object, and return. Radar is typically used for long-distance

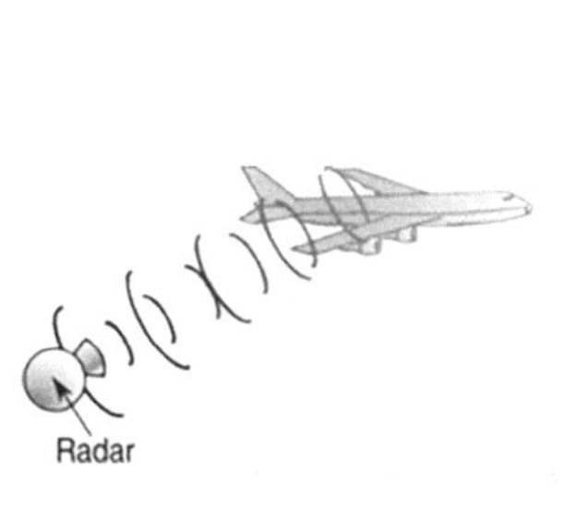

FIGURE 15.20 Radar.

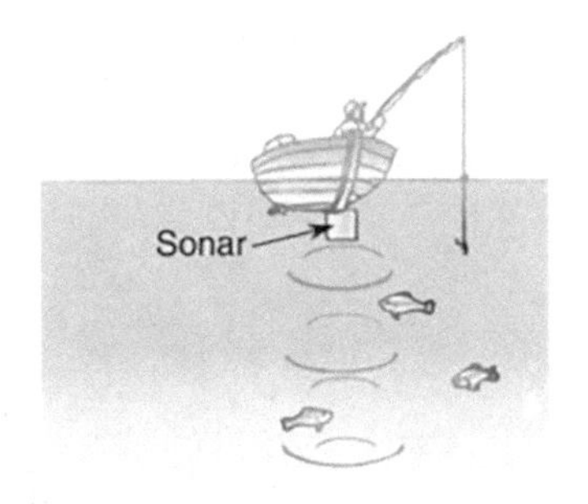

FIGURE 15.21 Sonar.

determinations, such as by air traffic controllers to locate airplanes. Lidar works on the same principle as a radar; it only uses light instead of radio waves. Sonar moves much more slowly than radar and attenuates over a shorter distance, which is why it is better used for short-distance applications. Underwater, radar fails because radio waves are heavily absorbed in water. Sonar is therefore used by submarines and fishermen.

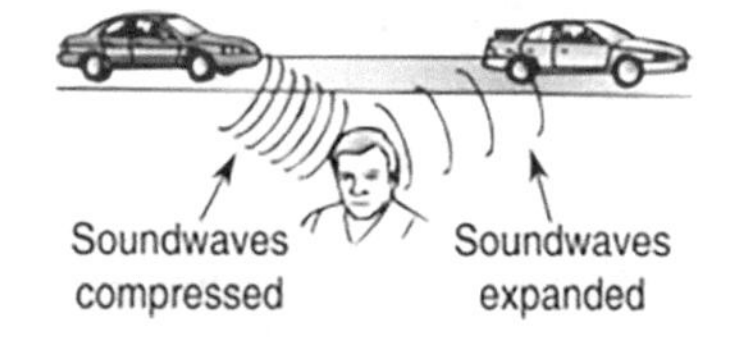

FIGURE 15.22 The Doppler shift of approaching and receding sources.

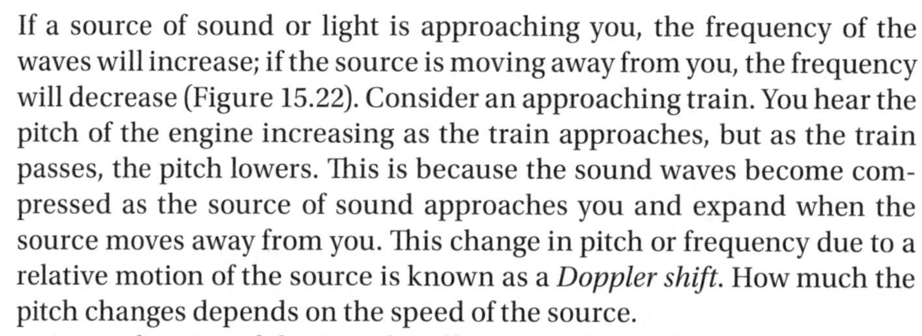

15.4.2 Doppler Shift

If a source of sound or light is approaching you, the frequency of the waves will increase; if the source is moving away from you, the frequency will decrease (Figure 15.22). Consider an approaching train. You hear the pitch of the engine increasing as the train approaches, but as the train passes, the pitch lowers. This is because the sound waves become compressed as the source of sound approaches you and expand when the source moves away from you. This change in pitch or frequency due to a relative motion of the source is known as a *Doppler shift*. How much the pitch changes depends on the speed of the source.

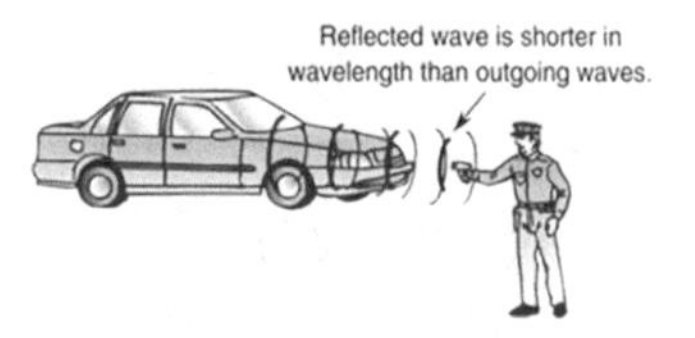

FIGURE 15.23 Police radar using the Doppler effect to determine speed.

An application of the Doppler effect is a police *radar gun*, which consists of a microwave radio source emitting a beam of radio waves at a given frequency (Figure 15.23). The frequency of radio waves increases after the waves bounce off oncoming objects. This is because the wave becomes compressed. The reflected wave returns to the gun, and its frequency is compared with the frequency of the original wave. The difference in the frequencies of the two waves is a measure of how fast the car is moving. Doppler radar is used to determine the speed of clouds (Figure 15.24).

FIGURE 15.24 Weather forecasters use Doppler radar to determine the speed of clouds.

Sound waves exhibit the Doppler effect. Ultrasonic transducers are common ways to measure the distance to an object and can also be used to determine its speed (Figure 15.25). Speed can be determined by bouncing an ultrasonic sound wave off of a moving object, then measuring the change in frequency between the outgoing and incoming reflected waves.

Mathematically, the frequency shift from the Doppler effect for light and sound is given by:

For light,

$$f_o = f_s \cdot \frac{\sqrt{1 - v_s^2/c^2}}{1 \pm v_s/c} \Leftarrow + \text{ sign for receding sources, } - \text{ sign for approaching source.}$$

For sound,

$$f_o = f_s \cdot \left(\frac{v \pm v_o}{v \mp v_s} \right) \Leftarrow \frac{-}{+} \text{ sign for a source or observer moving away from each other.}$$

$$\frac{+}{-} \text{ sign for a source or observer moving toward each other.}$$

where

f_o = frequency of the observer
f_s = frequency of the source
c = speed of light
v = speed of sound
v_s = speed of the source
v_o = speed of the observer

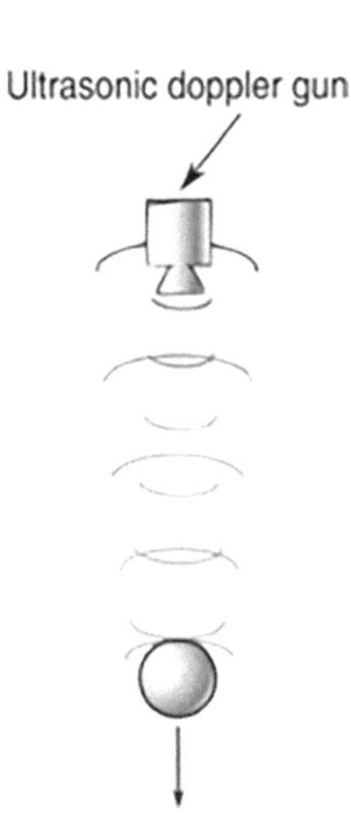

FIGURE 15.25 The Doppler effect can be applied to ultrasonic waves to determine speed.

Example 15.2

A train is approaching you at 22.0 m/s and is blowing a whistle at a frequency of 44$\bar{0}$ Hz. Suppose the speed of sound is 345 m/s. What does the frequency of the whistle sound like to you? Assume you are standing still, $v_o = 0$ (Figure 15.26).

$f_s = 440\text{Hz}$

$v = 345\text{m/s}$

$v_s = 22.0\text{m/s}$

$v_o = 0$

$f_o = ?$

$$f_o = f_s \cdot \left(\frac{v}{v - v_s} \right) \leftarrow \text{Source is approaching, so we choose the – sign.}$$

$$f_o = 44\bar{0}\text{ Hz} \left(\frac{345\text{m/s}}{345\text{m/s} - 22.0\text{m/s}} \right) = 47\bar{0}\text{ Hz}$$

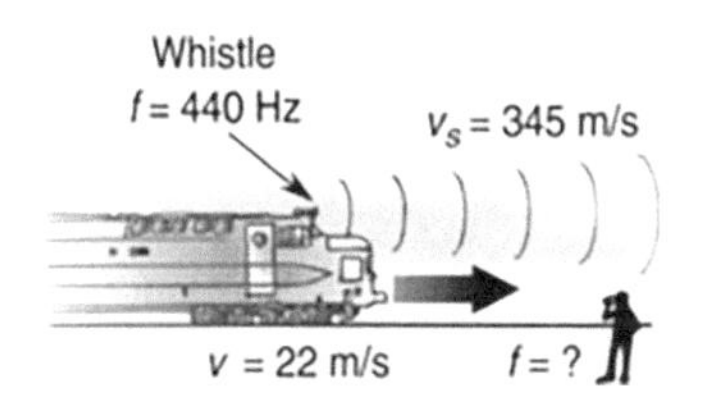

FIGURE 15.26 What does the frequency of the whistle sound like to you?

Example 15.3

Some police radar guns use a microwave transmitter that sends out frequencies in the X-band of 10.525 GHz or the K-band centered at 24.150 GHz. Suppose a police officer is shooting radar with a radar gun that sends out a frequency of 24.150 GHz (Figure 15.27). If a car is approaching at 100 mph, what does the return frequency measure? Use c = 186,000 miles/s for the speed of light.

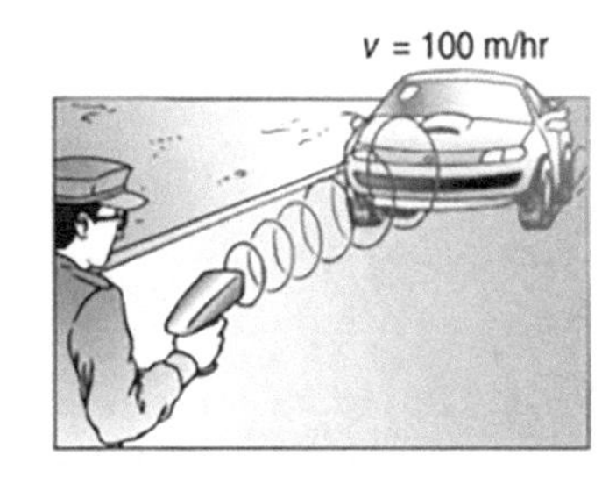

FIGURE 15.27 What does the return frequency measure?

$$f_s = 24.150\,\text{GHz}$$

$$c = 186{,}000\,\text{miles/s}$$

$$v_s = 100.0\,\text{miles/h}$$

$$f_o = ?$$

$$1\,\text{h} = 3{,}600\,\text{s}$$

$$186{,}000\frac{\text{miles}}{\cancel{\text{s}}}\left(\frac{3{,}600\,\cancel{\text{s}}}{1\,\text{h}}\right) = 6.70 \times 10^8\,\text{miles/h}$$

$$f_o = f_s \cdot \frac{\sqrt{1 - v_s{}^2/c^2}}{1 - v_s/c} \Leftarrow - \text{ sign for approaching.}$$

$$f_o = 24.150\,\text{GHz} \cdot \left(\frac{\sqrt{1 - \dfrac{(100.0\,\text{miles/h})^2}{(6.70 \times 10^8\,\text{miles/h})^2}}}{1 - \dfrac{100.0\,\text{miles/h}}{6.70 \times 10^8\,\text{miles/h}}} \right)$$

$$f_o = 24.149996\,\text{GHz}$$

Note: We have neglected the rules of significant digits in this example.

15.5 SOUND WAVES

Sound is a longitudinal wave. Sound waves need a medium through which to travel, such as air, solids, or liquids. To create a sound wave, we need to somehow disturb the medium. For example, a stereo speaker creates sound by moving its cone back and forth (Figure 15.28). As the speaker cone moves forward, it compresses the air; when it moves backward, it creates a slight vacuum. The speaker is alternately compressing and decompressing the air. The volume of air disturbed by the speaker does not travel outward. It merely pushes on the air next to it, and then that air pushes on the air next to it. It is just the disturbance that moves.

FIGURE 15.28 A speaker cone moves back and forth, creating sound. Sound is the compression and decompression of air traveling out from the source.

15.5.1 Speed of Sound

Sound typically travels fastest through solids, then liquids, and slowest through gases (Table 15.1).

The speed of sound changes with temperature, pressure, and humidity of the air. At 72°F in dry air, at 1 atmosphere (atm) of pressure, the velocity is:

$$v_s = 345\frac{\text{m}}{\text{s}} \text{ or } v_s = 1{,}132\frac{\text{ft}}{\text{s}} \text{ or } v_s = 772\frac{\text{miles}}{\text{hour}}$$

TABLE 15.1
The Speed of Sound Through Media

	Speed	
Medium	**m/s**	**ft/s**
Aluminum	6,420	21,100
Brass	4,700	15,400
Steel	5,960	19,500
Granite	6,000	19,700
Alcohol	1,210	3,970
Water (25°C)	1,500	4,920
Air, dry (0°C)	331	1,090
Vacuum	0	0

In general, the speed of sound changes with temperature in air according to:

$$v = 331\frac{\text{m}}{\text{s}} + \left(0.6\frac{\text{m/s}}{°\text{C}}\right) \cdot T_c$$

where v is in m/s and T_c is in Celsius, the speed of sound increases with temperature.

Example 15.4

What is the velocity of sound in air at 31°C?

$$T_c = 31°\text{C}$$

$$v = 331\frac{\text{m}}{\text{s}} + \left(0.6\frac{\text{m/s}}{°\text{C}}\right) \cdot T_c$$

$$v = 331\frac{\text{m}}{\text{s}} + \left(0.6\frac{\text{m/s}}{\cancel{°\text{C}}}\right) \cdot 31\cancel{°\text{C}}$$

$$v = 350\,\text{m/s}$$

15.5.2 Pitch

The *pitch* of sound is the same as the frequency. In terms of generating sound with a speaker, the pitch is determined by how fast the speaker

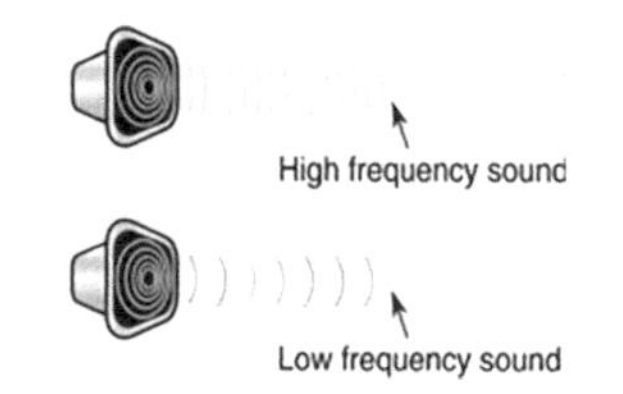

FIGURE 15.29 A speaker generating high- and low-frequency sounds.

cone is moving back and forth. If the cone is oscillating back and forth at a high frequency, the pitch will be high (Figure 15.29).

The human ear can hear sounds that range from a frequency of 20 Hz to 20 kHz. Sounds below this range are called *infrasonic*. Sounds above this range are called *ultrasonic*.

Low-frequency sounds spread out from a source, that is, they are not very directional (Figure 15.30). They are not readily absorbed, and as a result, they travel through walls easily. This is familiar to anyone living next door to someone with a stereo.

High-frequency sounds are more directional than low-frequency sounds (Figure 15.31). Because of this, in the ultrasonic region, there are many applications where the waves can be focused.

FIGURE 15.30 Low-frequency sounds spread out from a source.

FIGURE 15.31 High-frequency sounds are directional.

15.5.3 Ultrasound Imaging

Obstetricians use *ultrasound imaging* to see a fetus while it is still inside the mother (Figure 15.32). Sound waves, with frequencies of 3.5–7 MHz, are sent into the mother's abdomen. Sound travels at about 1,540 m/s in soft tissues. Quartz crystals generate the sound and also act as the receiver of the returning signals. These waves are reflected and absorbed by the soft tissues. The reflected waves returning to the ultrasound transducer are then put together to form the image of the fetus. The higher-frequency waves have higher resolution than the low-frequency waves, but they do not penetrate as deeply.

15.5.4 Ultrasonic Tape Measure

An *ultrasonic tape measure* measures the distance to objects by bouncing an ultrasonic wave off the object and timing how long it takes for the wave to come back. The longer it takes, the farther away the object (Figures 15.33 and 15.34).

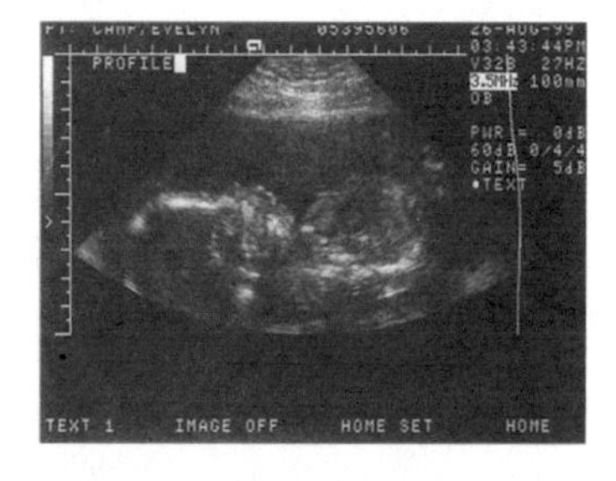

FIGURE 15.32 Ultrasound imaging.

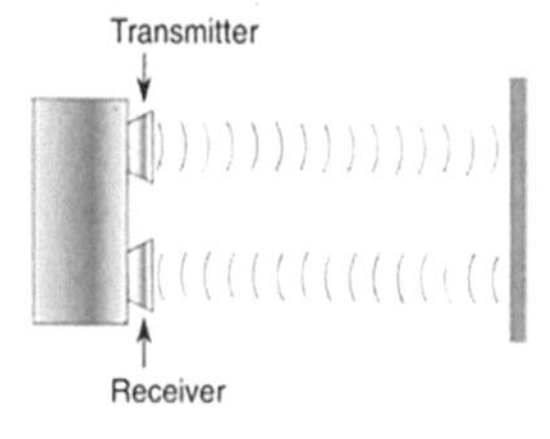

FIGURE 15.33 The ultrasonic tape measure works by sending a sound wave and timing how long it takes to return after bouncing off an object whose distance is being measured.

FIGURE 15.34 Ultrasonic Rangefinder courtesy of Adafruit.

15.5.5 Ultrasonic Cleaning

Ultrasonic cleaning uses sound waves above 20 kHz to vibrate dirt loose from objects being cleaned (Figure 15.35). The sound causes the object to vibrate at the same frequency, shaking the dirt and grime loose.

FIGURE 15.35 Ultrasonic cleaning.

15.5.6 Sound Pressure

A sound wave is essentially a pressure wave, that is, a compression and decompression of air that travels. When this pressure wave impacts the eardrum, the eardrum sends a signal to the brain, and this signal is interpreted as sound. There is energy in this pressure wave. The rate at which this energy passes through a given area is given by the intensity.

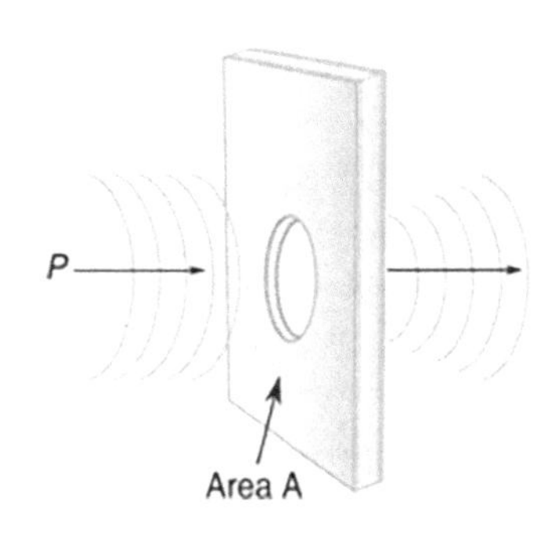

FIGURE 15.36 Sound of power *P* passing through an area *A*.

15.5.7 Sound Intensity

The intensity or the amount of power delivered by waves per unit area is given by (Figure 15.36):

$$I = \frac{P}{A}$$

$$\text{Intensity}\left(\text{W/m}^2\right) = \frac{\text{Power (W)}}{\text{Area}\left(\text{m}^2\right)}$$

If we consider a point source of sound radiating isotropically (equally in all directions) outward, then the sound radiates outward spherically and the intensity at a distance *r* from the source is given by (Figure 15.37),

$$I = \frac{P}{A}$$

$$I = \frac{P}{4\pi r^2}$$

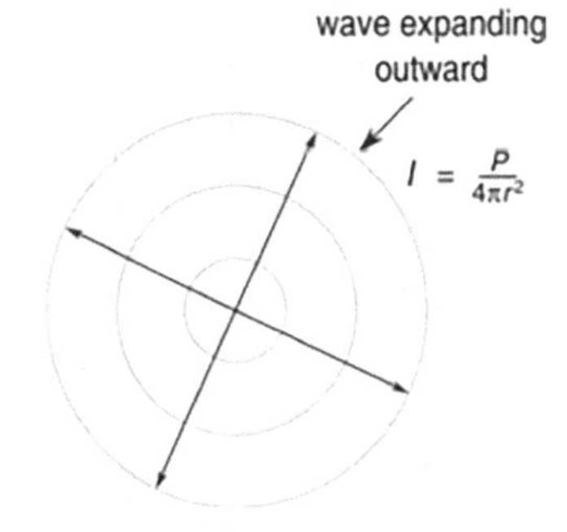

FIGURE 15.37 Sound radiating isotropically (equally in all directions) outward.

Example 15.5

A 1.0 W point source of sound radiates outward isotropically. What is the intensity 3.0 m away from the source?

$$P = 1.0\text{W}$$

$$r = 3.0\text{m}$$

$$I = \frac{P}{4\pi r^2}$$

$$I = \frac{1.0\,\text{W}}{4\pi(3.0\,\text{m})^2}$$

$$I = 8.8 \times 10^{-3}\,\frac{\text{W}}{\text{m}^2}$$

15.5.8 Sound Intensity Level

The threshold intensity for hearing is $I_o = 1 \times 10^{-12}\,\text{W/m}^2$. A sound that has twice the intensity as another does not sound twice as loud to the human ear. The ear's response to sound is said to be logarithmic. A sound that is twice as intense as another does not sound twice as loud but only slightly louder. The volume control on a stereo is logarithmic, which takes into account the logarithmic nature of our hearing.

Sound intensity level is measured in decibels (dB) and is given by:

$$SL = 10 \cdot \log\left(\frac{I}{1\times10^{-12}\,\text{W/m}^2}\right)$$

$$\text{Sound intensity level } (d\beta) = 10 \cdot \log \left(\frac{\text{Intensity } (\text{W/m}^2)}{\text{hearing intensity threshold}(1\times10^{-12}\,\text{W/m}^2)}\right)$$

FIGURE 15.38 A decibel meter.

Sound levels can be measured with a decibel meter (Figure 15.38). From the sound intensity level equation, we can say that a sound that is 10^N (where N is an integer) times more intense than the threshold of hearing means that $SL = N \times 10$ dB.

- A sound that is ten times more intense than threshold means $SL = 10$ dB.
- A sound that is 100 times more intense than threshold means $SL = 20$ dB.
- A sound that is 1,000 times more intense than threshold means $SL = 30$ dB.

Another way of thinking in decibels is the *rule of 3 dB*, which says that for every 3 dB increase in sound intensity level, the intensity (I) doubles. See Table 15.2 for typical intensity levels.

TABLE 15.2
Typical Sound Levels

Jet takeoff (60 m)	120 dB	
Construction site	110 dB	Intolerable
Shout (1.5 m)	100 dB	
Heavy truck (15 m)	90 dB	Very noisy
Urban street	80 dB	
Automobile interior	70 dB	Noisy
Normal conversation (1 m)	60 dB	
Office, classroom	50 dB	Moderate
Living room	40 dB	
Bedroom at night	30 dB	Quiet
Broadcast studio	20 dB	
Rustling leaves	10 dB	Barely audible
	0 dB	

Example 15.6

Let's compare the intensity and the sound intensity levels of two sounds. According to the sound level table, the sound level of normal conversation is 60 dB. How does the intensity of this sound compare with the intensity of a sound at sound intensity level 66 dB?

Increasing a sound from 60 to 63 dB means the intensity has doubled according to the rule of 3 dB. Increasing a sound from 63 to 66 dB means the intensity has doubled again. Therefore, the intensity must have increased 2 times 2, or 4 times, rising from 60 to 66 dB.

Example 15.7

What is the sound intensity level of a sound with intensity 10^5 times as intense as the threshold for hearing?

$$I = 10^5 \times I_o$$

$$SL = N \times 10\,\text{dB}$$

$$SL = 5 \times 10\,\text{dB}$$

$$SL = 50\,\text{dB}$$

A sound that is 100,000 times more intense than threshold is only 50 dB larger!

15.5.9 Microphones

Most microphones in use are dynamic air pressure sensors. A diaphragm responds to a sound wave by moving back and forth. A sensor behind the diaphragm measures this motion and converts it into an electrical signal. Any type of sensor that is capable of measuring a displacement in a diaphragm can be used in a microphone. Listed below are four types of microphones (Figures 15.39–15.42).

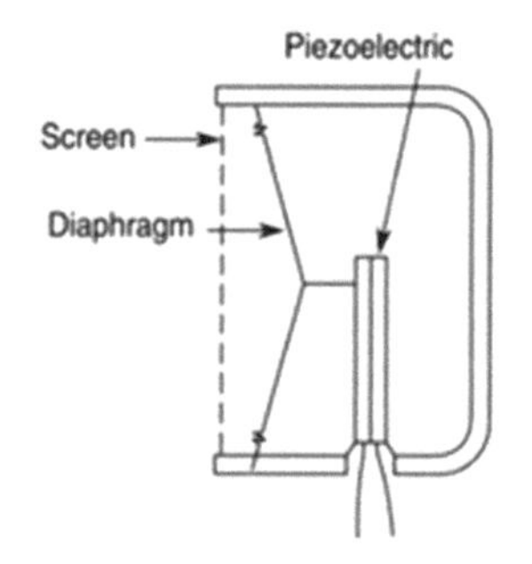

FIGURE 15.39 A crystal microphone. As the diaphragm moves back and forth from a sound wave, it causes the crystal to be flexed. Deforming this piezoelectric crystal produces a voltage across it proportional to the level of the sound wave.

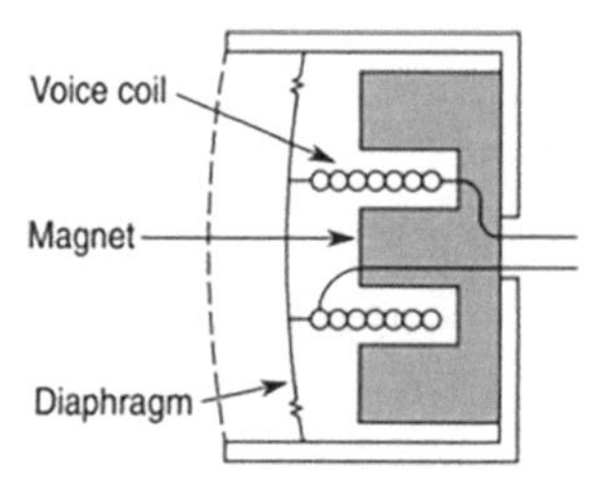

FIGURE 15.40 A dynamic or moving coil microphone. As the diaphragm moves back and forth from a sound wave, the attached coil moves through a magnetic field, inducing a voltage into it.

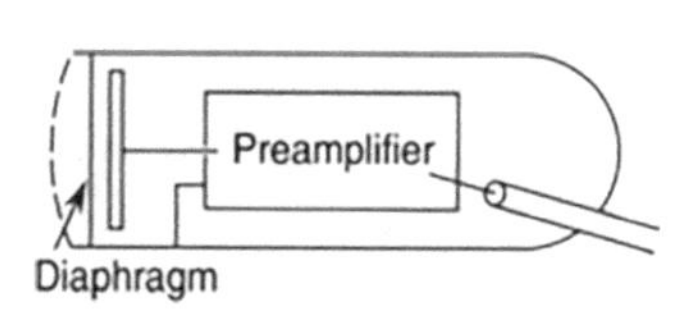

FIGURE 15.41 A condenser microphone measures sound levels using a capacitor. The diaphragm acts as a movable plate on a capacitor that responds to sound pressure.

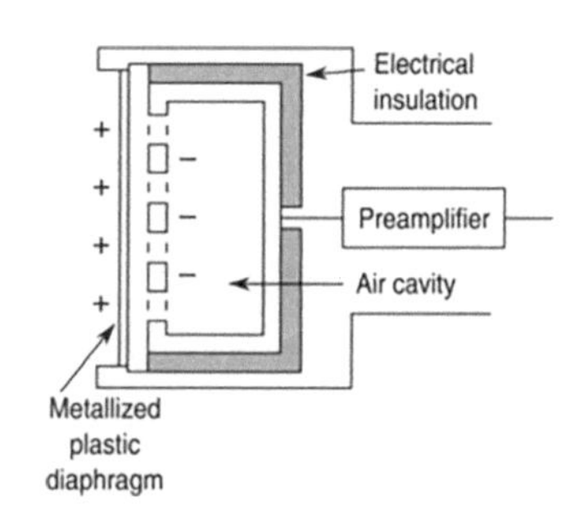

FIGURE 15.42 An electret-condenser microphone is essentially a condenser microphone with a permanent charge placed on the diaphragm at manufacturing. A built-in preamplifier, such as a field effect transistor (FET), amplifies the small signals.

15.5.10 Gravitational Waves

Gravitational waves are ripples in space-time that travel at the speed of light caused by some violent and energetic event such as the merger of two black holes. Albert Einstein predicted such waves in 1916, in his theory of gravity, General Relativity. In 2015, approximately 100 years after Einstein's prediction, these waves were first detected. The source of this detection turned out to be two black holes spiraling into one another. General Relativity predicts that when a mass is accelerated it will produce waves on the space surrounding it, think of ripples on a pond when a rock is tossed in. These waves are extremely weak and require extremely large accelerated masses, and a very sensitive instrument to measure them. This first detection of gravitational waves took place on twin Laser Interferometer Gravitational-wave Observatory (LIGO) detectors, located in Livingston, Louisiana and Hanford, Washington, USA (Figure 15.43). The signals measured in these detectors agree with the prediction of the merger of two black holes about 29 and 36 times the mass of the sun 1.3 billion years ago; it took 1.3 billion years traveling at the speed of light to reach us.

At each observatory, a 4 km long L-shaped set of tubes under vacuum, called an interferometer, uses laser light that travels back and forth bouncing off mirrors at each end. The laser light from the two arms are then combined together. Destructive and constructive interferences will take place between the two beams if the distance between the arms change. Einstein's theory predicts that the distance between the mirrors

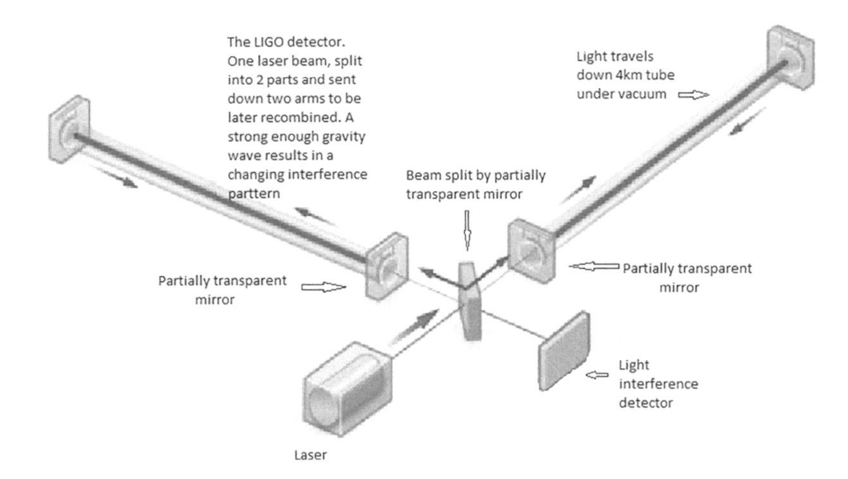

FIGURE 15.43 Laser Interferometer Gravitational-wave Observatory (LIGO).

will change when a gravitational wave passes through the detector. A change as small as 10^{-19} m can be measured with these detectors. This is smaller than the diameter of a proton!

15.6 CHAPTER SUMMARY

Symbols used in this chapter.

Symbol	Unit
t—time	second (s)
T—period	second (s)
d—distance	meter (m)
r—radius	meter (m)
A—area	m^2
v—velocity	m/s
λ—wavelength	meter (m)
f—frequency	hertz (Hz)
c—speed of light	3×10^8 m/s
T—temperature	Celsius (°C)
I—intensity	watts/meters^2 (W/m^2)
P—power	watts (W)
SL—sound level	decibels (dB)

$$f = \frac{1}{T}$$

$$\text{Frequency(Hz)} = 1/\text{period(s)}$$

All waves, including light, sound, and radio can be described by the following equation:

$$v = \lambda \cdot f$$

$$\text{Velocity}(\text{m/s}) = \text{wavelength}(\text{m}) \times \text{frequency}(\text{Hz})$$

Transverse waves: Transverse waves oscillate perpendicular to the direction of travel of the wave. Light is a transverse wave.

Longitudinal waves: Longitudinal waves oscillate in the direction of travel of the wave. Sound is a longitudinal wave.

Refraction: Refraction means the bending of a wave when it moves from one medium into another.

Superposition: Multiple waves can travel through the same medium without affecting each other.

Constructive interference: Waves that are in step or in phase with one another add to make a bigger wave.

Destructive interference: Waves that are out of step or out of phase with one another subtract to make a smaller wave.

Tone: The tone of a sound describes the way multiple waveforms making up the sound wave act together.

Resonance: The resonance frequency is the frequency at which if a system is stimulated with will produce a maximum response.

Diffraction: Diffraction is the bending of waves around obstacles.

Doppler shift: Doppler shift is the shifting of the frequency of a wave because of a relative motion between the source and the observer. Mathematically, the frequency shift from the Doppler effect for light and sound is given by:

For light,

$$f_o = f_s \cdot \frac{\sqrt{1 - v_s^{\,2}/c^2}}{1 \pm v_s/c} \Leftarrow + \text{ sign for receding sources,}$$

$$- \text{ sign for approaching source.}$$

For sound,

$$f_o = f_s \cdot \left(\frac{v \pm v_o}{v \mp v_s} \right) \Leftarrow \frac{-}{+} \text{ sign for a source or observer moving away from each other.}$$

$$\frac{+}{-} \text{ sign for a source or observer moving toward each other.}$$

where:

f_o = frequency measured by the observer
f_s = frequency of the source
c = speed of light
v = speed of sound
v_s = speed of the source
v_o = speed of the observer

In general, the speed of sound changes with temperature in air according to:

$$v = 331\frac{\text{m}}{\text{s}} + \left(0.6\frac{\text{m/s}}{{}^\circ\text{C}}\right)\cdot T_c$$

Pitch: The pitch of sound is the same as the frequency.

Sound Intensity: The intensity or the amount of power delivered by waves per unit area is given by:

$$I = \frac{P}{A}$$

$$\text{Intensity}\left(\text{W/m}^2\right) = \frac{\text{Power(W)}}{\text{Area}\left(\text{m}^2\right)}$$

If we consider a point source of sound radiating isotropically (equally in all directions) outward, then the sound radiates outward spherically and the intensity at a distance r from the source is given by,

$$I = \frac{P}{A}$$

$$I = \frac{P}{4\pi r^2}$$

15.6.1 Sound Intensity Level

The threshold intensity for hearing is $I_o = 1 \times 10^{-12}$ W/m^2. A sound that has twice the intensity as another does not sound twice as loud to the human ear. The ear's response to sound is said to be logarithmic. A sound that is twice as intense as another does not sound twice as loud but only slightly louder. The volume control on a stereo is logarithmic, which takes into account the logarithmic nature of our hearing.

Sound intensity level is measured in decibels (dB) and is given by:

$$SL = 10 \cdot \log\left(\frac{I}{1\times10^{-12}\,\text{W/m}^2}\right)$$

$$\text{Sound intensity level } (d\beta) = 10 \cdot \log \left(\frac{\text{Intensity } (\text{W/m}^2)}{\text{hearing intensity threshold}\left(1\times10^{-12}\,\text{W/m}^2\right)}\right)$$

PROBLEM SOLVING TIPS

When solving problems involving the formula $v = \lambda \cdot f$ be sure to convert f into hertz.

PROBLEMS

1. If a water wave comes onto the shore every 5 s, what is the frequency of the wave?
2. If you are listening to a radio station that broadcasts at 96 MHz, what is the wavelength of this wave?
3. If a signal from a satellite takes 0.01 s to travel to earth, how far away is the satellite?
4. How can you create an electromagnetic wave?
5. How does a stereo speaker produce sound?
6. How are fiber optic cables able to carry much more information than a copper wire of the same size?
7. How can many conversations take place at the same time on a phone line and not get mixed together?
8. What is the difference between a transverse wave and a longitudinal wave?
9. A police officer is shooting his radar gun in the X-band at an oncoming car traveling at 12 mph. What is the change in frequency of the returned signal?
10. An ultrasonic and a laser ranger are used to determine the distance to an object 25 m away. How long does it take for each signal to go out and return after reflecting off this object?
11. What is the speed of sound in air at 28°C?
12. When wood wind instruments warm up they become "sharp". This is because the air inside the instrument warms up. Why would this make the instrument sound sharp?

13. Is the speed of sound faster in warm air or cold air?
14. What is the sound intensity level of a sound that is 10^5 times more intense than the threshold of hearing?
15. A 0.010 W point source of sound radiates outward equally in all directions. What is the intensity 3.0 m away from the source?
16. What is the sound intensity level of the sound in problem 15?
17. One sound is twice as intense as another sound. What is their relative intensity in dB?
18. An increase in one octave means a sound has doubled in frequency. If a 1 kHz sound rises 3 octaves, what is the new frequency?
19. An infrared LED (light-emitting diode) emits light with a wavelength of 940 nm. What is its frequency?
20. A tachometer measures the speed of rotation of a shaft. A simple tachometer can be made by attaching a piece of tape to a shaft, and listening with a microphone to the tape slap a fixed object as the shaft rotates. By observing the microphone's output on an oscilloscope, the period of the signal can be measured. The rpm (revolutions per minute) of the shaft is given by rpm = 60/period(s). If the frequency of the pattern on the scope is 50 Hz, what is the shaft's rpm?
21. Electrical power in the U.S. operates at 60 Hz. What is the period of this signal?
22. A car playing its radio is approaching you at 80 ft/s. If the frequency of the sound as heard by the driver is 5 kHz, what frequency does a pedestrian hear as the car is approaching her? Assume the speed of sound is 1,132 ft/s.
23. What is the speed of a train receding from you if frequency of the whistle the engineer hears is twice as high as what you hear standing on the track? Assume the speed of sound is 343 m/s.
24. What is the velocity of sound inside your lungs at 37°C?
25. A long-range microphone uses a parabolic microphone of area $0.8\,m^2$ to collect the sound. If it is listening to a sound with an intensity of $10^{-5}\,W/m^2$, how much power is being gathered?
26. A source of sound that radiates outward spherically has an intensity of $10^{-3}\,W/m^2$ at a distance of 4 m. What is its power?
27. What is the sound intensity level of a sound with an intensity of $10^{-8}\,W/m^2$?
28. Using Table 15.1, find out for which material is the speed of sound the fastest.
29. Why can't sound travel through a vacuum?

30. Does the speed of sound increase with air temperature or decrease?
31. What is the frequency of sound waves used in ultrasound imaging?
32. How does a dynamic microphone work?
33. How does a police radar gun work?
34. A sound that increases by 3 dB in sound intensity level increases by what intensity?
35. What is the sound level of a normal conversation?
36. What sounds are more directional, low or high frequency sounds?
37. What happens at the resonance frequency of a system?

16

Light

This chapter will discuss electromagnetic waves, with a focus on light and its applications. The ideas contained in this chapter should help you understand the basics behind optical storage of information, that is, compact discs (CDs), fiber optics used in telecommunications, as well as a host of other devices that use light.

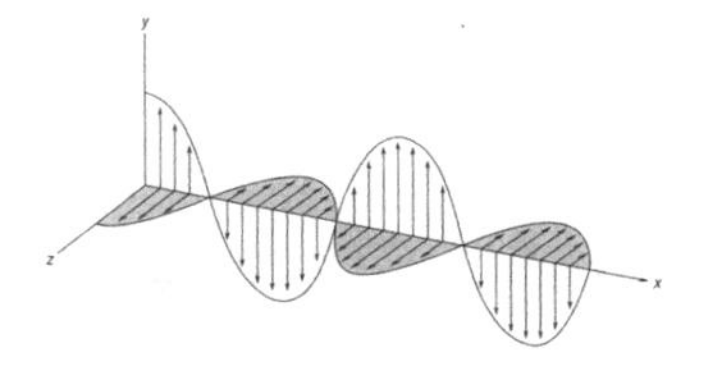

FIGURE 16.1 Light is a wave made of electric and magnetic fields that oscillate perpendicular to the direction of travel.

16.1 ELECTROMAGNETIC WAVES

Electromagnetic waves are different than water and sound waves because they can travel through a vacuum, that is, they need no medium in which to travel through. Electromagnetic waves are made of electric and magnetic fields that change with time and oscillate perpendicular to the direction of travel (Figure 16.1).

The electromagnetic spectrum consists of all the electromagnetic waves of various frequencies (Figure 16.2). If the electromagnetic fields oscillate in the right frequency range, we can see them. Light is just part of a whole spectrum of electromagnetic waves. At other frequencies they may be radio waves, X-rays, microwaves, and so on.

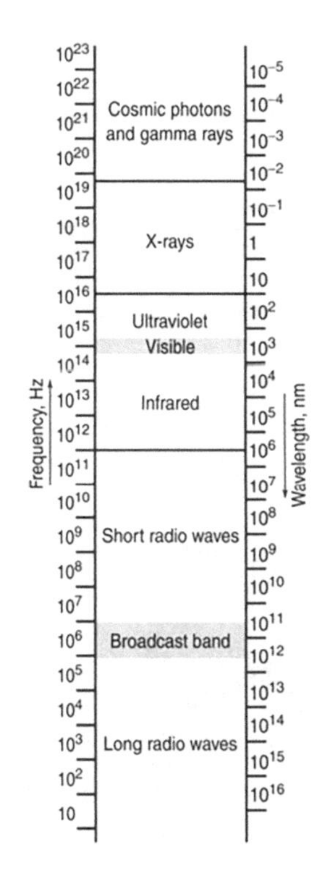

FIGURE 16.2 The electromagnetic spectrum.

16.2 THE VISIBLE SPECTRUM

The *visible spectrum* is the part of the spectrum that we can see. Different colored lights correspond to different wavelengths of light.

16.2.1 White Light

White light is light composed of many different colors (Figure 16.3). The colors can be separated out using a prism.

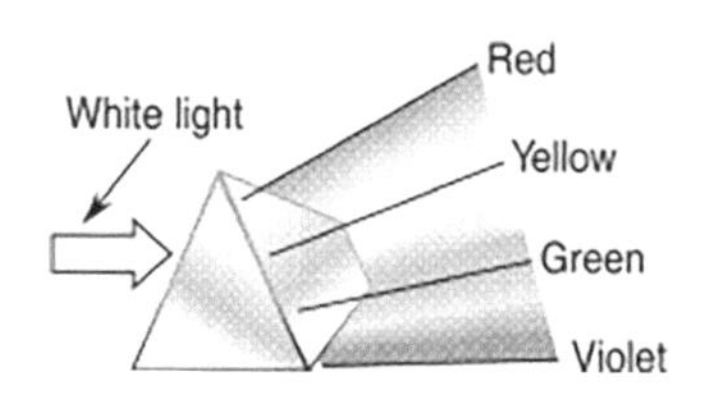

FIGURE 16.3 White light is composed of many different colors.

16.2.2 Colors

In order to see an object, light must be reflected off it. If you shine white light on an object and all the colors that compose the white light are absorbed except, for example, green, the object will appear green. Paints are essentially color absorbers. The color of the paint is the color from white light that does not get absorbed (Figure 16.4).

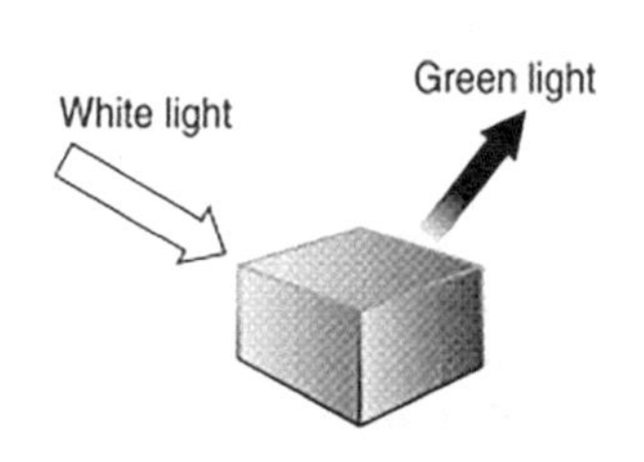

FIGURE 16.4 The color of an object is the result of the color of the white light shining on the object that does not get absorbed.

16.2.3 Infrared

At wavelengths slightly longer than visible light lies the infrared part of the spectrum. *Infrared radiation* is otherwise known as *thermal radiation* or *radiant heat* (Figure 16.5).

16.2.4 Ultraviolet

At wavelengths slightly shorter than visible light lies the ultraviolet part of the spectrum. Ultraviolet rays, or UV rays, cause sunburn. Welders wear goggles to protect their eyes from UV rays (Figure 16.6).

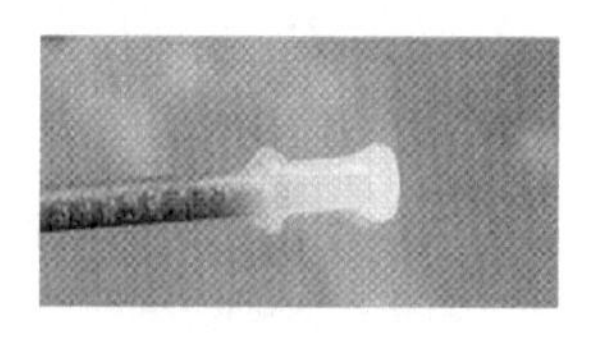

FIGURE 16.5 The infrared part of the spectrum is associated with the transfer of heat.

16.3 GENERATION OF ELECTROMAGNETIC WAVES

Electromagnetic waves can be generated in different ways to create specific regions of the electromagnetic spectrum.

16.3.1 Radio Waves

Wiggling an electric charge will generate radio waves. To understand this better, consider an electron with its electric radial field lines emanating from it. If a charge is wiggled up and down, these lines will be moved up and down, similar to a jump rope being snapped, generating a wave traveling down a rope (Figure 16.7). Like the jump rope, the electromagnetic wave will be transverse, but unlike the jump rope, no medium is needed for the wave to travel on. Accompanying this wave will also be a magnetic field (this will be explained in the next chapter). The radio waves generated will have the same frequency as the frequency the electrons are wiggled at. A radio transmitter, along with its antenna, is effectively an electron wiggler.

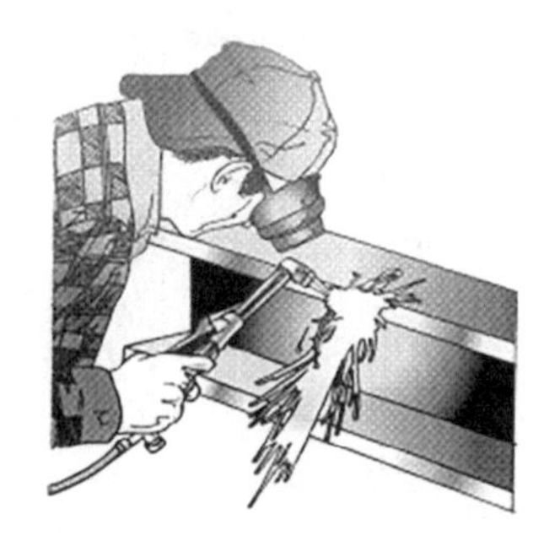

FIGURE 16.6 Large amounts of UV light are produced in welding or cutting metals.

16.3.2 Microwaves

Accelerating electrons in a circular path at a particular rate will generate microwaves. A *magnetron* is a device that generates microwaves using a magnet to accelerate a current of electrons into a circle, using a magnetic

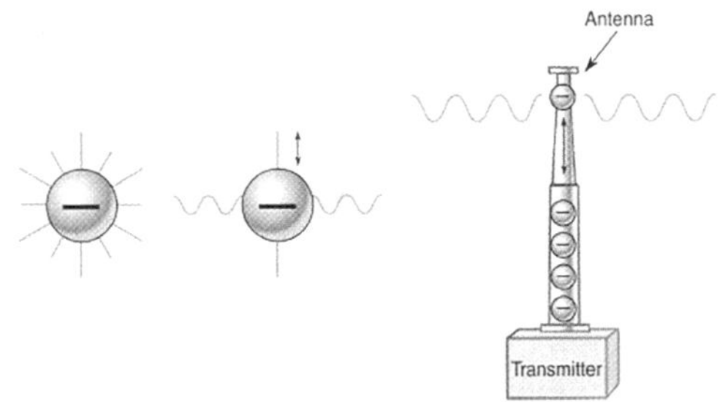

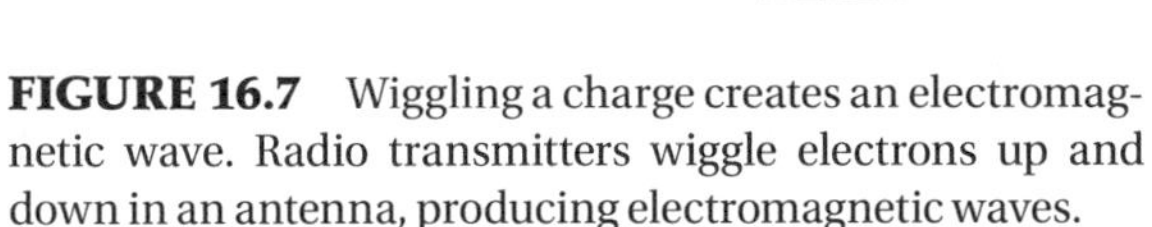
FIGURE 16.7 Wiggling a charge creates an electromagnetic wave. Radio transmitters wiggle electrons up and down in an antenna, producing electromagnetic waves.

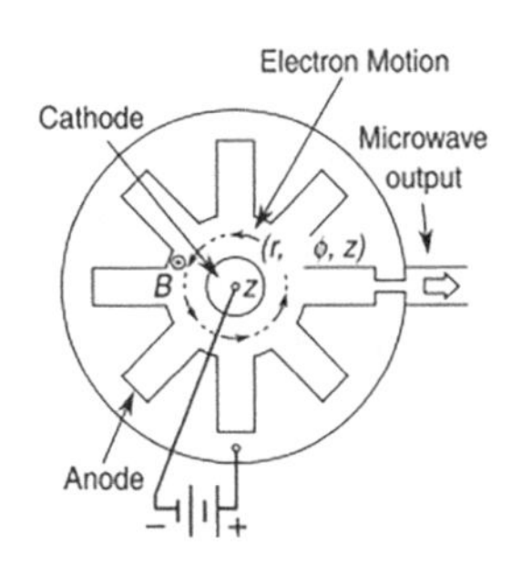

FIGURE 16.8 Generating microwaves using a magnetron.

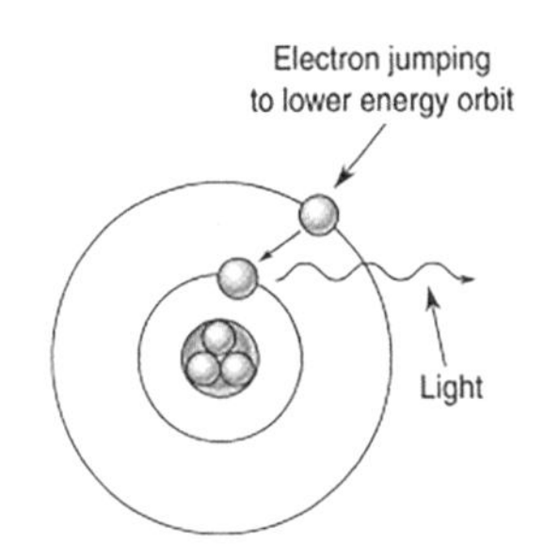

FIGURE 16.9 Light emitted from atoms results from electrons jumping from one energy orbit to a lower one.

field (Figure 16.8). Accelerating electrons like this causes them to radiate electromagnetic waves in the microwave region of the spectrum. Every microwave oven contains a magnetron.

16.4 LIGHT

Electromagnetic waves also can come from atoms. These waves are much shorter than radio waves. Orbiting every atom are electrons, some of which have more energy than others. If an atom is given energy, or *excited*, the electrons orbiting the atom begin jumping orbits. When an electron jumps from a high-energy orbit to a lower-energy orbit, the atom gives off an electromagnetic wave called a *photon*, or a packet of energy (Figure 16.9). The photon has a frequency, or color, related to the energy difference between the orbits. If the energy range between the two orbits is just right, visible light will come out. *Sodium vapor lights* illustrate this clearly. The photons coming out of sodium vapor as a result of the electrons jumping orbits correspond to a wavelength of 590 nm, or yellow light.

FIGURE 16.10 Light is produced in incandescent lamps by heating of the filament using an electric current. This heat is energizing so many electrons in so many different ways that the output is composed of many frequencies, or white light.

Incandescent light bulbs generate light by heating the filament inside with an electric current. This heat energizes many different electrons in the filament in many different ways. Photons of various wavelengths will be produced, resulting in white light (Figure 16.10). Most incandescent light bulbs emit only 10% of their energy in the visible part of the spectrum.

Fluorescent lights contain a low-pressure mercury vapor (Figure 16.11). At one end of the tube is a filament that is heated and at the other end is an electrode at a larger voltage, compared with the first. The hot filament "boils" off electrons and these are accelerated toward the high-voltage end, striking mercury atoms on route, which causes the mercury atom to excite, then de-excite, giving off UV light. The UV light then strikes a fluorescent coating on the inside of the tube, exciting these atoms, which,

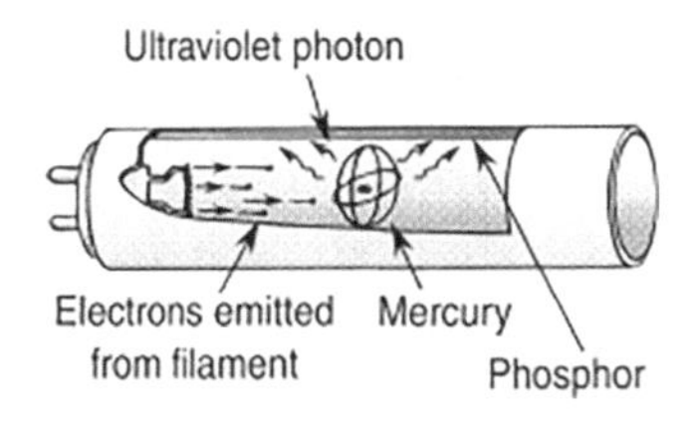

FIGURE 16.11 A fluorescent light.

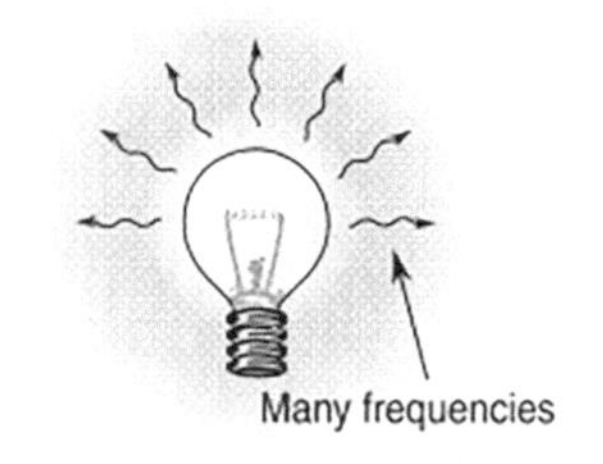

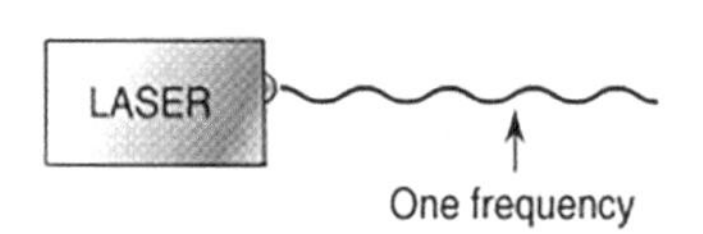

FIGURE 16.12 White light contains many frequencies. Laser light contains one frequency and so is monochromatic.

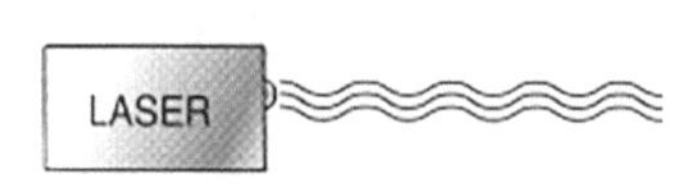

FIGURE 16.13 White light is incoherent. Laser light is coherent.

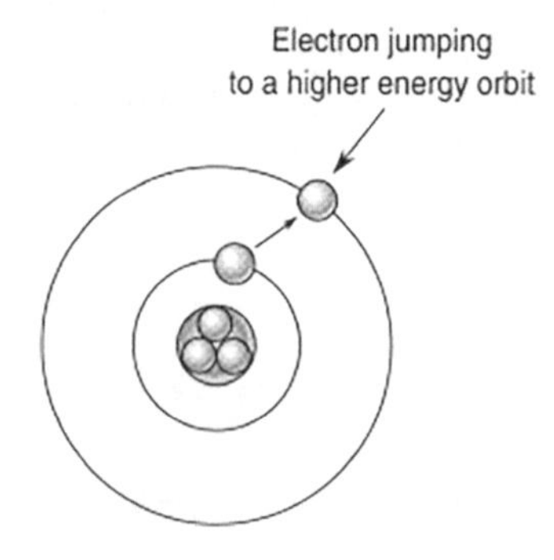

FIGURE 16.14 Electron excitation.

in turn, give off visible light. Fluorescent lights are much more energy efficient than incandescent light, putting out about 40% of their light in the visible part of the spectrum.

16.4.1 Lasers

Laser light is different than light from a light bulb. Light from a light bulb is white light, that is, of many frequencies. A laser light contains one frequency; in other words, it is monochromatic (Figure 16.12).

Light waves coming out of a light bulb are all out of sync with one another. The waves are said to be incoherent. Laser light is coherent, so that all the waves of the light are in step with one another (Figure 16.13).

The mechanism by which gas lasers and solid-state lasers work is the same.

The process consists of the following steps:

Step 1: Electrons in their orbits around their atoms are excited to higher-energy orbits (Figure 16.14). Excitation can be done in various ways using light, electrical discharge, chemical reactions, and so on.

Step 2: Excited electrons begin to jump down to lower-energy orbits. The electrons spend different amounts of time in the various orbits before jumping to the next lower orbit. If the electrons spend a lot of time in one particular orbit, soon there will be a buildup of electrons there. This is called *population inversion* because a large number of the electrons of that atom now reside at that level (Figure 16.15).

Step 3: When electrons start to jump from a highly populated level to a lower level, light is emitted. This affects the other electrons sitting on the other atoms, causing them to jump down also. This is called *stimulated emission* (Figure 16.16). The light emitted is all in step with the light that stimulated the emission and is of the same wavelength because the energy differences are all the same.

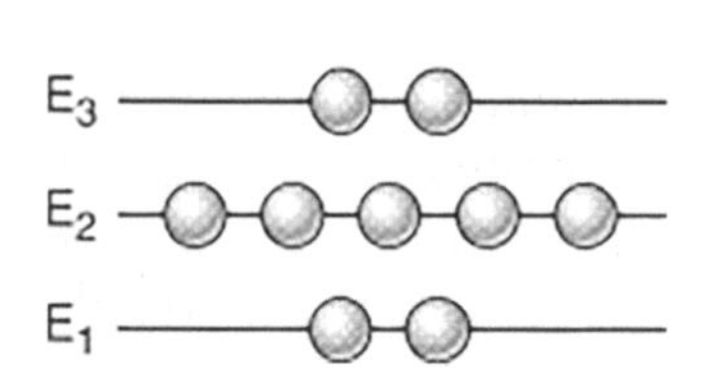

FIGURE 16.15 Population inversion.

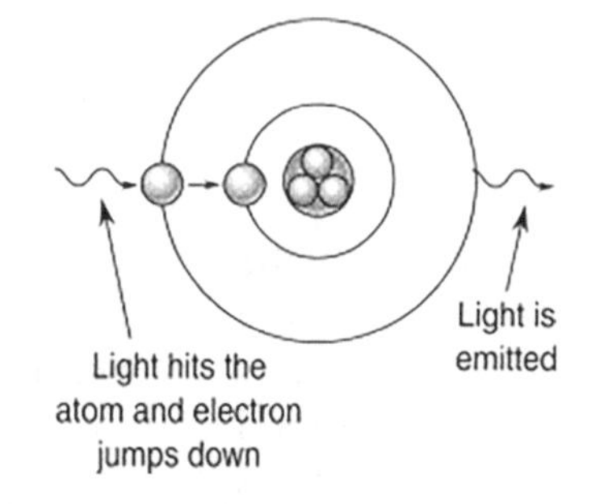

FIGURE 16.16 Stimulated emission.

This whole process is contained in the acronym *laser*, which stands for Light Amplification by Stimulated Emission of Radiation.

16.5 THE SPEED OF ELECTROMAGNETIC WAVES

Light travels at 186,000 miles/s or 3×10^8 m/s in a vacuum. A convenient conversion is that light travels 1 ft in 1 nanosecond (ns).

Example 16.1

How long does it take for a radio signal to reach a satellite orbiting 22,000 miles above the earth (Figure 16.17)?

$$d = 22{,}000\,\text{miles}$$

$$v = 18{,}600\,\frac{\text{miles}}{\text{s}}$$

$$v = \frac{d}{t}$$

$$t = \frac{d}{v} \Leftarrow \text{After some algebra.}$$

$$t = \frac{22{,}000\,\cancel{\text{miles}}}{186{,}000\,\frac{\cancel{\text{miles}}}{\text{s}}} = 0.12\,\text{s}$$

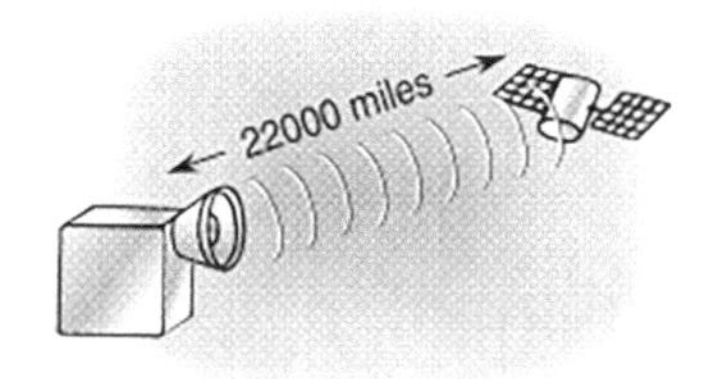

FIGURE 16.17 How long does it take for a radio signal to reach a satellite orbiting 22,000 miles above the earth?

16.5.1 Speed, Frequency, and Wavelength

The velocity, wavelength, and frequency of any wave are given by the formula:

$$v = \lambda \cdot f$$

In the case of light,

$$v = c = 186{,}000\,\text{miles/s or } 3 \times 10^8\,\text{m/s}$$

Therefore,

$$c = \lambda \cdot f$$

$$\text{Speed of light (m/s)} = \text{wavelength (m)} \times \text{frequency (Hz)}$$

Example 16.2

WBEZ radio station in Chicago broadcasts at 91.5 MHz. What is the wavelength in miles and feet of this radio wave?

$$f = 91.5\text{MHz} = 91.5 \times 10^6\,\text{Hz}$$
$$c = 186{,}000\,\text{miles/s}$$
$$c = \lambda \cdot f$$
$$\frac{c}{f} = \frac{\lambda \cdot f}{f} \leftarrow \text{solving for } \lambda$$
$$\lambda = \frac{c}{f}$$
$$\lambda = \frac{186{,}000\,\text{miles/s}}{91.5 \times 10^6\,\text{Hz}}$$
$$\lambda = 0.00203\,\text{miles}$$
$$0.00203\,\cancel{\text{miles}}\left(\frac{5{,}280\,\text{ft}}{1\,\cancel{\text{miles}}}\right) = 10.7\,\text{ft}$$

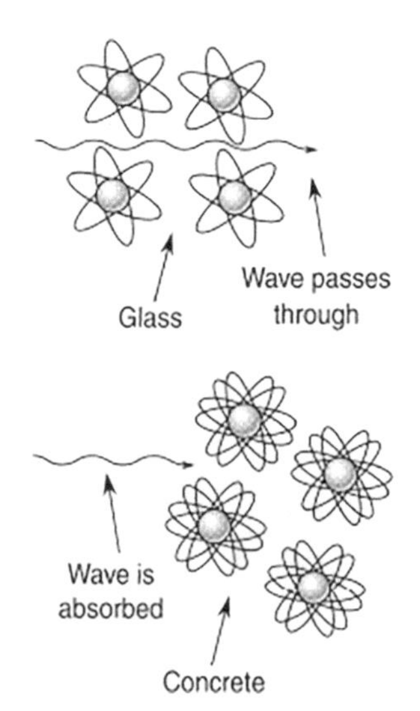

FIGURE 16.18 Light passes through glass and not through an opaque material because the molecules composing the glass do not absorb the light.

16.6 OPTICS

Optics is concerned with generating, gathering, bending, and focusing light for some useful purpose.

16.6.1 Opacity

Why can you see through glass and not through a concrete wall or a non-transparent material? If a material is not transparent, it is *opaque*. Light passes through glass and not through an opaque material because the molecules composing the glass do not absorb the light (Figure 16.18). They allow it to pass, unlike an opaque material, which absorbs the light.

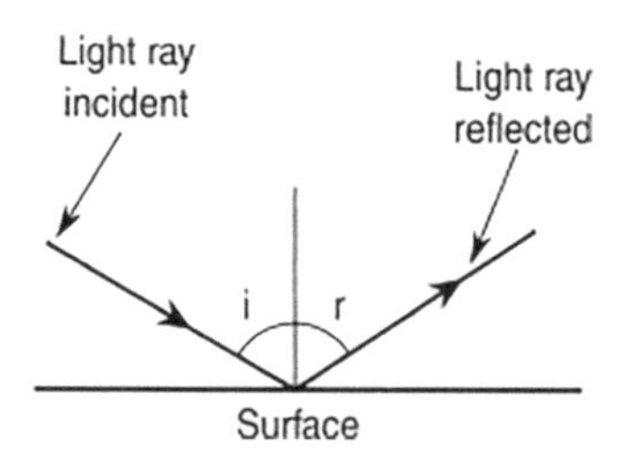

FIGURE 16.19 The angle of incidence equals to the angle of reflection.

16.6.2 Reflection of Light

The angle of the incident ray equals the angle of the reflected ray.

Light must be reflected off an object to be able to see it. A light ray reflecting off a surface will reflect at the same angle at which it struck the surface (Figure 16.19). If the incident and reflected angles are measured with respect to a normal or perpendicular to the surface, then,

$$\theta_i = \theta_r$$

$$\text{angle incident} = \text{angle reflected}$$

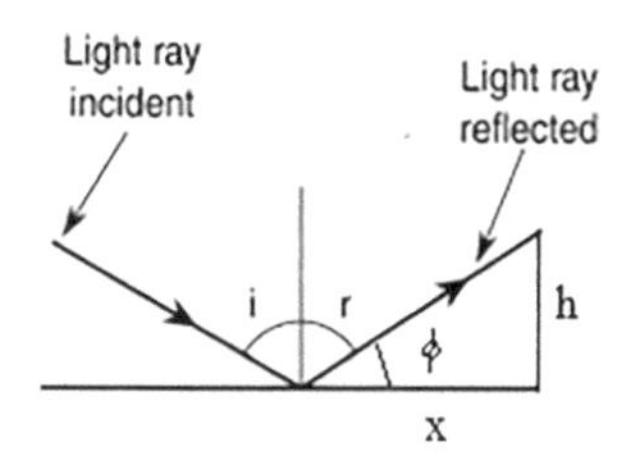

FIGURE 16.20 Where will the reflected beam hit the wall?

Example 16.3

A laser beam strikes a mirror on a floor 2 ft from a wall, at an incident angle of 30° with respect to the mirror's normal (Figure 16.20). Where will the reflected beam hit the wall?

$$\theta_i = \theta_r$$

$$\theta_r = 30°$$

$$\phi = 90° - 30°$$

$$\phi = 60°$$

$$h = x \cdot \tan(\phi)$$

$$h = (2\text{ft}) \cdot \tan(60°)$$

$$h = 3.46\,\text{ft above the floor}$$

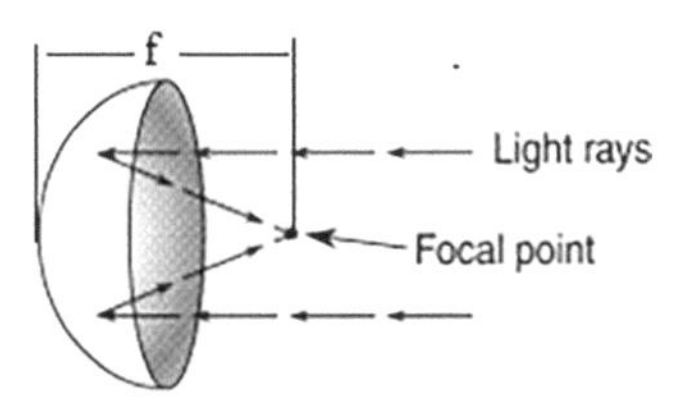

FIGURE 16.21 Cut out a section of a ball, coat the inside with something reflective, and you have a spherical mirror.

16.6.3 Spherical Mirrors

Cut out a section of a ball, coat the inside with something reflective, and you have a spherical mirror (Figure 16.21).

16.6.4 Focal Point

The focal point of a spherical mirror is the point where all the reflected rays pass through (Figure 16.21). A makeup mirror is an example of a spherical mirror (Figure 16.22).

For a spherical mirror, the focal point is a distance from the mirror equal to half the mirror's radius.

$$f_{\text{Spherical}} = \frac{R}{2}$$

$$\text{Focal point} = \frac{\text{radius of curvature of mirror}}{2}$$

FIGURE 16.22 Spherical mirrors, such as a makeup mirror, can be used to magnify images. They can also be used to focus light into a small area. Parallel incoming rays will focus on a point a distance from the mirror equal to half the mirror's radius.

Example 16.4

A spherical mirror of radius 20 cm will focus sunlight at what distance from the mirror?

$$R = 20\,\text{cm}$$

$$f_{\text{Spherical}} = \frac{R}{2}$$

$$f_{\text{Spherical}} = \frac{20\,\text{cm}}{2}$$

$$f_{\text{Spherical}} = 10\ \text{cm}$$

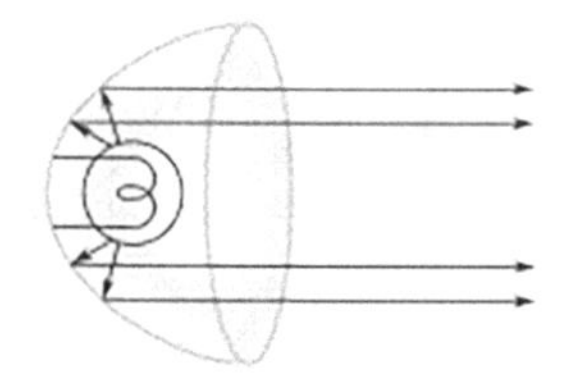

FIGURE 16.23 Flashlights and automobile headlights are made with parabolic reflectors.

16.6.5 Parabolic Mirror

Flashlights and automobile headlights are made with parabolic reflectors (Figure 16.23). A mirror with a parabolic shape can focus light much more sharply than a spherical mirror. Any light source at the focal point will be reflected off the mirror into a parallel of beam light. Conversely, any incoming parallel beam of light will be focused on a point. Most big telescopes are made with parabolic mirrors (Figure 16.24).

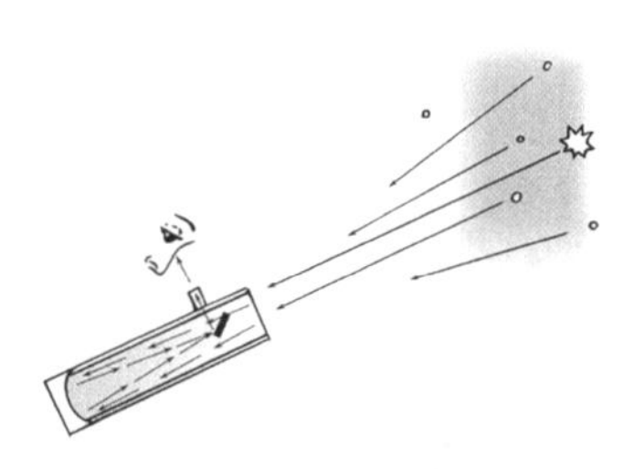

FIGURE 16.24 Many telescopes are made with parabolic mirrors.

16.6.6 Corner Reflector

Coat the inside corner of a box where the sides meet the top with something reflective and you have a corner reflector (Figure 16.25).

The astronauts left a corner reflector on the moon. It is designed so that any incoming light ray will be reflected back parallel to it (Figure 16.26).

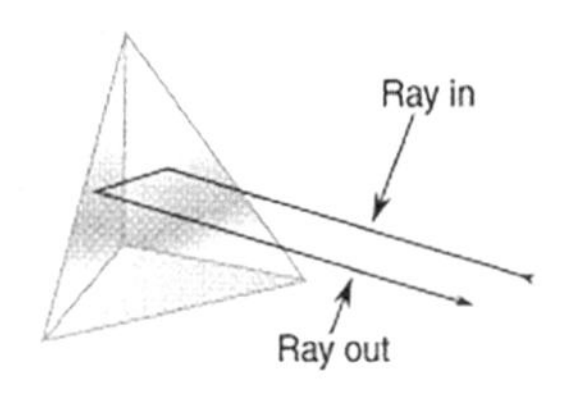

FIGURE 16.25 A corner reflector will take an incoming light ray and reflect it back, parallel to itself.

16.7 DIFFRACTION AND REFRACTION OF LIGHT

Diffraction is the bending of waves around an obstacle (Figure 16.27). An example of diffraction is the outline formed around shadows. The light waves bend around an obstacle and, through interference, form the outline.

Refraction is the bending of waves when they pass from one medium into another. An example of this is light passing from air into water. Light travels slower through water than it does through air. As a result, the wave front that enters the water will fall behind the wave in the air. The light is effectively bent because of this slowing down of the light (Figure 16.28).

The *index of refraction* (n) is a measure of how many times slower light travels through a material than through a vacuum (Table 16.1).

FIGURE 16.26 The distance to the moon is routinely measured by shooting a laser from earth to a corner reflector on the moon and timing how long it takes to come back.

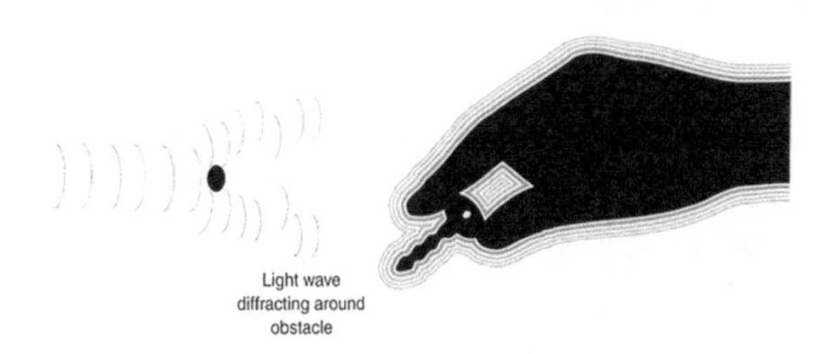

FIGURE 16.27 Light waves diffracting or bending around an obstacle.

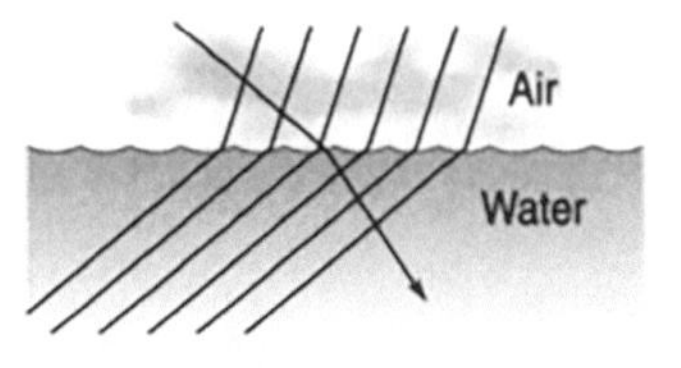

FIGURE 16.28 Light is bent or refracted when moving from air to water because the wave front slows down in the water.

TABLE 16.1
Indices of Refraction Measured at a Wavelength of 589 nm

Material	$n = \frac{c}{v}$
Vacuum	1.0000
Air (at STP)	1.0003
Water	1.33
Ethyl alcohol	1.36
***n* for Various Types of Glass**	
Fused quartz	1.46
Crown glass	1.52
Light flint	1.58
Lucite or Plexiglas	1.51
Sodium chloride	1.53
Diamond	2.42

$$n = \frac{c}{v}$$

$$\text{Index of refraction} = \frac{\text{speed in a vacuum}}{\text{speed in a medium}}$$

The index of refraction is the ratio of the speed of light in a vacuum to the speed of light through some material. It is a measure of how much slower light travels in a given medium than in a vacuum. The index of refraction has no units.

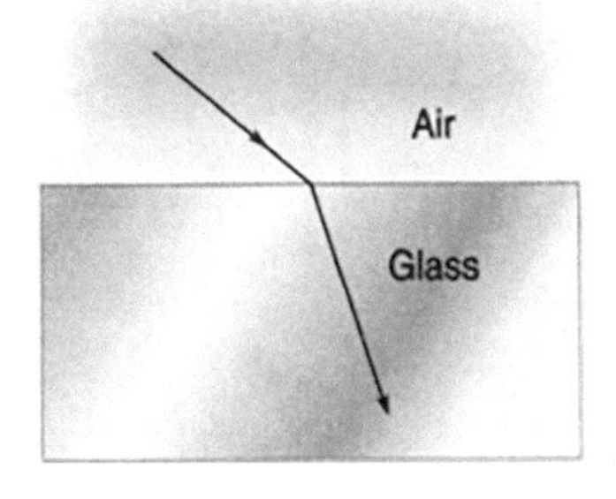

FIGURE 16.29 What is the speed of light in that lens?

Example 16.5

A light wave enters a glass lens with an index of refraction 1.2 (Figure 16.29). What is the speed of light in that lens?

$$n = 1.2$$

$$v_{\text{vacuum}} = 3 \times 10^8 \frac{\text{m}}{\text{s}}$$

$$v_{\text{glass}} = ?$$

$$n = \frac{\text{speed in a vacuum}}{\text{speed in a medium}}$$

$$\text{Speed in medium} = \frac{\text{speed in vacuum}}{n}$$

$$\text{Speed in glass} = \frac{3 \times 10^8 \frac{\text{m}}{\text{s}}}{1.2} = 2.5 \times 10^8 \frac{\text{m}}{\text{s}}$$

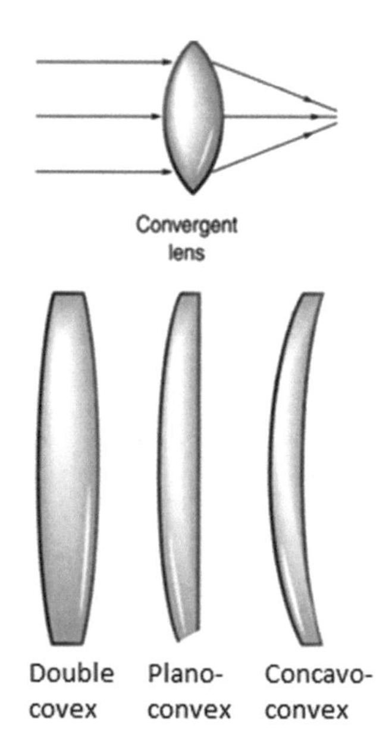

FIGURE 16.30 Convergent lens.

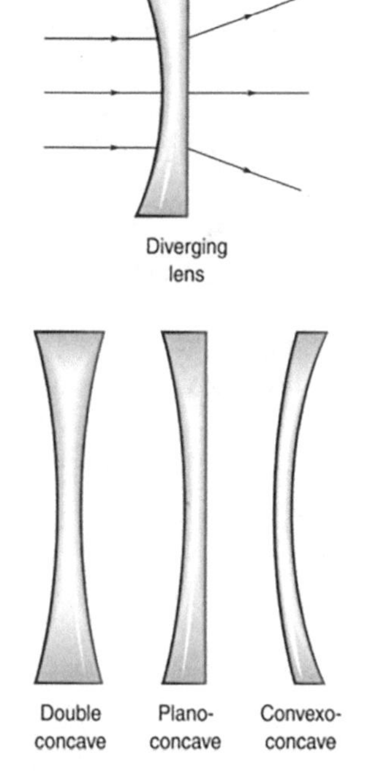

FIGURE 16.31 Divergent lens.

16.8 LENSES

We can use the refractive properties of glass to make lenses bend and focus light. *Convergent lenses* are lenses that focus light (Figure 16.30). An example is a magnifying glass.

Divergent lenses are lenses that spread out light (Figure 16.31). An example of a divergent lens is the eyeglass prescription for someone who is nearsighted (Figure 16.32). The eye focuses an image short of the retina, where the image is supposed to form. The divergent lens corrects this by extending the focal length onto the retina.

16.8.1 Thin Lens Equation

A convergent lens will focus a parallel set of light rays on a point called the focal point on an axis, called the principle axis, that is centered on and runs through the lens (Figure 16.33).

A parallel set of light rays can be obtained from a faraway source such as the sun. To obtain the focal length f, focus the sunlight on a point and measure the distance from lens to where the light is focused (Figure 16.33).

A lens can be used to enlarge or reduce the size of an object being viewed through the lens. The distance from the lens where the image forms d_i is related to the focal length f and the distance from the object to the lens d_o (Figure 16.34).

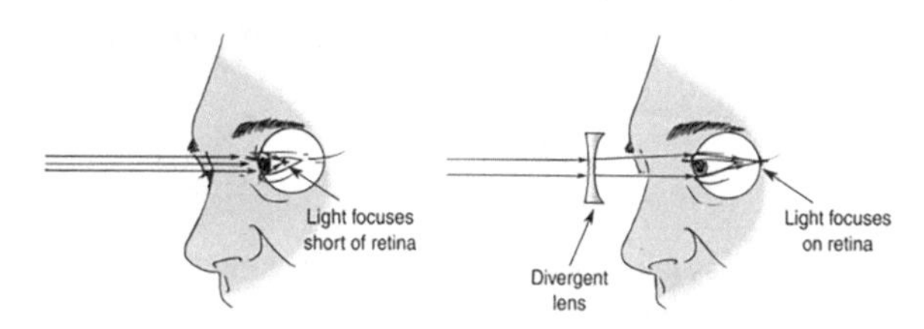

FIGURE 16.32 A corrective divergent lens for nearsightedness.

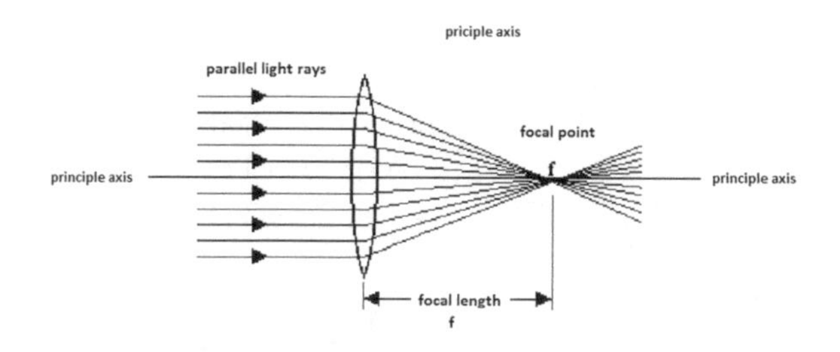

FIGURE 16.33 A convergent lens.

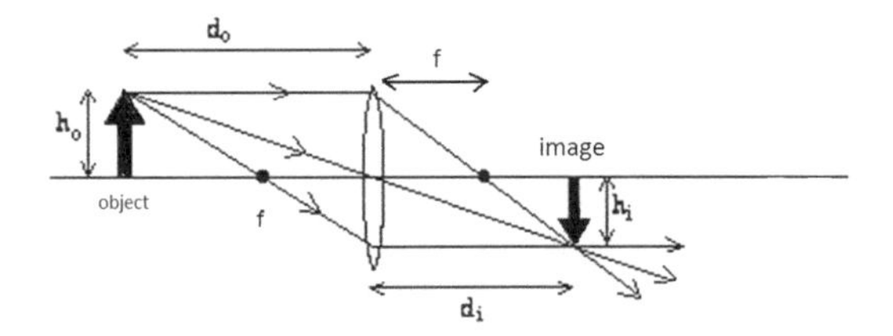

FIGURE 16.34 An object viewed though a lens will appear inverted.

$$\frac{1}{f}=\frac{1}{d_i}+\frac{1}{d_o}$$

$$\frac{1}{\text{focal length(m)}}=\frac{1}{\text{distance}_{\text{image}}\text{(m)}}+\frac{1}{\text{distance}_{\text{object}}\text{(m)}}$$

Other variations of this formula are:

$$f=\frac{1}{\frac{1}{d_i}+\frac{1}{d_o}} \quad d_i=\frac{1}{\frac{1}{f}-\frac{1}{d_o}} \quad d_o=\frac{1}{\frac{1}{f}-\frac{1}{d_i}}$$

To generate a light ray diagram as in the figure, follow the steps:

1. Draw a ray from the top of the object parallel to the axis to the lens. Where the ray strikes the lens draw a ray through the focal point.
2. Draw a ray from the top of the object through the center of the lens.
3. Draw a ray from the top of the object though the focal point on the object side to the bottom of the lens, then continue this line parallel to the axis.
4. An image will be formed where all the lines meet from steps 1–3.

Example 16.6

Given the focal length of a converging lens equals 20 cm, and the object is 30 cm from the lens, where does the image form?

$$d_i=\frac{1}{\frac{1}{f}-\frac{1}{d_o}}=\frac{1}{\frac{1}{20\text{ cm}}-\frac{1}{30\text{ cm}}}$$

$$d_i = 60\text{ cm}$$

16.8.2 Magnification

The magnification is a measure of how much larger or smaller the image is compared to the object. The magnification can be found by taking the ratio of the object distance d_o to the image distance d_i.

$$m = -\frac{d_i}{d_o}$$

$$\text{Magnification} = -\frac{\text{distance}_{\text{image}}(\text{m})}{\text{distance}_{\text{object}}(\text{m})}$$

- The magnification has no unit.
- A negative m means the image is inverted.

The magnification relates the size of the object and image. Magnification is the ratio of the height of the image h_i to that of the object h_o. The height of something is always greater than or equal to zero, so the absolute value of m, $|m|$, is used in the following:

$$|m| = \frac{h_i}{h_o}$$

$$|\text{Magnification}| = \frac{\text{height}_{\text{image}}}{\text{height}_{\text{object}}}$$

Example 16.7

Given the focal length of a converging lens equals 20 cm, and a 2 cm tall object is 30 cm from the lens, where does the image form, and what is the magnification and the height of the image?

$$d_i = \frac{1}{\frac{1}{f} - \frac{1}{d_o}} = \frac{1}{\frac{1}{20\,\text{cm}} - \frac{1}{30\,\text{cm}}}$$

$$d_i = 60\,\text{cm}$$

$$m = -\frac{d_i}{d_o} = -\frac{60\,\text{cm}}{30\,\text{cm}}$$

$$m = -2 \leftarrow \text{Negative means the image is inverted.}$$

$$h_i = |m| \cdot h_o = |-2| \cdot (2\text{cm})$$

$$h_i = 4\,\text{cm}$$

Example 16.8

A 5 cm wide object 20 cm from a lens is to be imaged onto a 0.5 cm wide charge-coupled device (CCD) chip. Find a lens with the appropriate focal length and determine the distance from the lens to the CCD chip (Figure 16.35).

The required magnification

$$m = \frac{h_i}{h_o} = \frac{0.5\text{cm}}{5\text{cm}} = 0.1$$

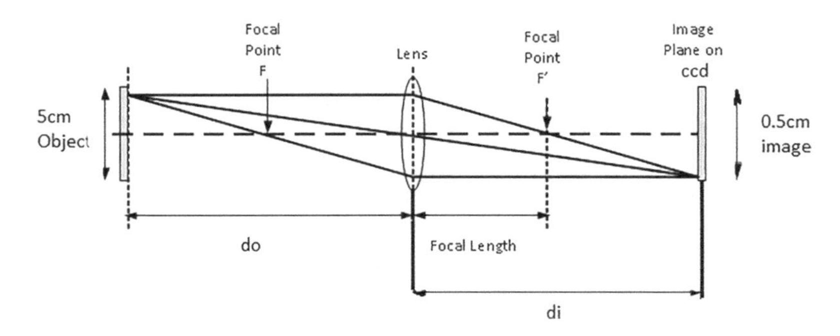

FIGURE 16.35 Find a lens with the appropriate focal length and determine the distance from the lens to the CCD chip.

Using the absolute value of m, $|m| = \frac{d_i}{d_o}$, we get:

$$d_i = m \cdot d_o = (0.1)(20\,\text{cm})$$
$$d_i = 2\,\text{cm}$$

We can now find the focal length of the lens:

$$f = \frac{1}{\frac{1}{d_i} + \frac{1}{d_o}} = \frac{1}{\frac{1}{2\,\text{cm}} + \frac{1}{20\,\text{cm}}}$$
$$f = 1.82\,\text{cm}$$

16.8.3 *f*-number

The *f*-number relates the focal length to the diameter of a lens.

$$f\text{-number} = \frac{f}{D}$$

$$f\text{-number} = \frac{\text{focal length(m)}}{\text{diameter(m)}}$$

Example 16.9

A lens has a focal length of 2 cm and a diameter of 1 cm. What is its *f*-number?

$$f\text{-number} = \frac{f}{D} = \frac{2\,\text{cm}}{1\,\text{cm}}$$
$$f\text{-number} = 2$$

16.8.4 Compound Lenses

Using two or more lenses together can result in much greater magnification (Figures 16.36 and 16.37).

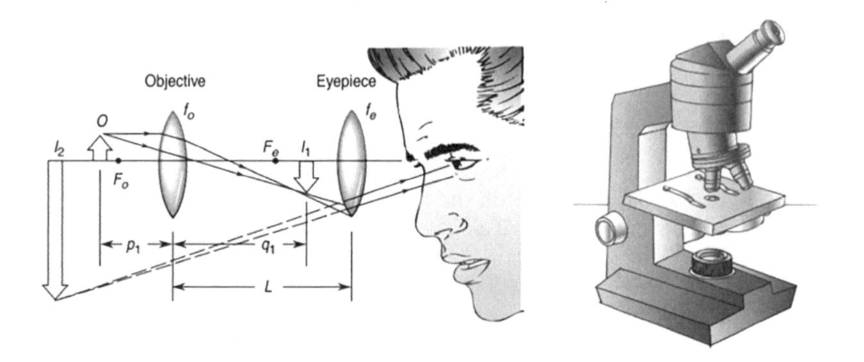

FIGURE 16.36 The microscope.

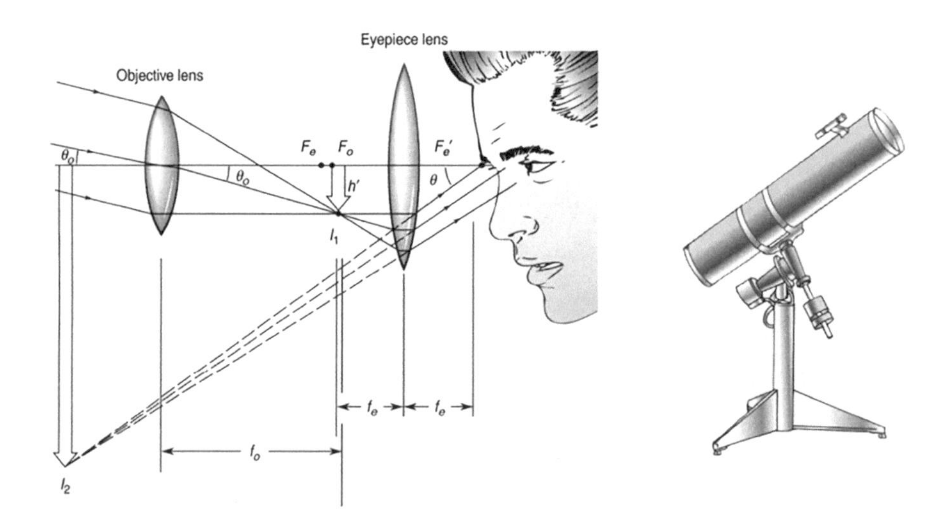

FIGURE 16.37 The telescope.

16.8.5 Aberration

Aberration is a deviation from ideal focusing of light by a lens. As a result, the image is not sharply focused. Some of the problems may lie with the geometry of the lens. Other aberrations may occur because different wavelengths of light are not refracted equally. This last problem is called *dispersion*.

16.8.6 Dispersion

Light refracts differently depending on its wavelength. A prism separates white light into colors because the speeds of the different colors of white light are different (Figure 16.38). As a result, they are bent at different angles, with violet bending the most and red the least.

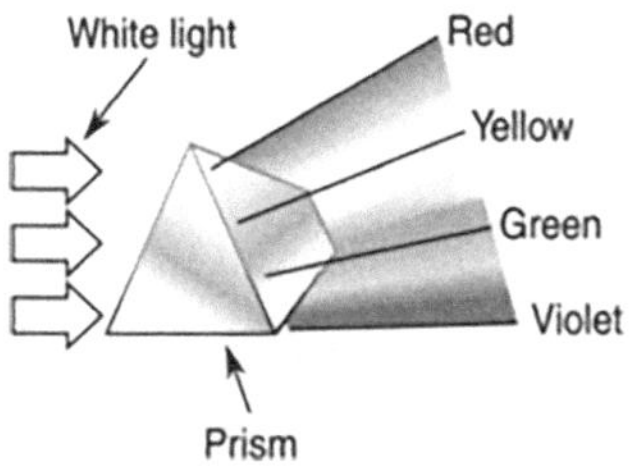

FIGURE 16.38 A prism separating the colors of white light.

16.9 INTENSITY AND ILLUMINATION

16.9.1 Luminous Intensity

The luminous intensity of a light is a measure of how much light a source emits in a given area. One unit of luminous intensity is candela, a unit that is related to candles. The brightness of light sources in the past was compared to the brightness of candlelight. Light emanating from a light emitting diode (LED) is concentrated in a light cone of about 30°. A typical LED puts out 40 millicandelas (mcd). Ultra-bright LEDs can put out 23,000 mcd. Another unit of luminous intensity is the lumen (lm).

For a light source that puts out light uniformly in all directions, the conversion from candelas to lumens is:

$$1\,\text{cd} = 4\pi\,\text{lm} \leftarrow \text{For light sources that emit light in all directions uniformly}$$

A common 40 W incandescent light puts out about 500 lumens. A 40 W fluorescent light puts out about 2,300 lumens, making it much more efficient than an incandescent bulb.

16.9.2 Illumination

Illumination is the measure of a light's apparent brightness at some distance away from the source. The unit of illumination is lm/m^2 or lux.

$$1\,\frac{\text{lm}}{\text{m}^2} = 1\,\text{lux}$$

If the light radiates equally in all directions, the illumination is given by:

$$E = \frac{I}{4\pi r^2}$$

$$\text{Illumination}(\text{lux}) = \frac{\text{intensity}(\text{lm})}{4\pi \cdot \text{distance}(\text{m})^2}$$

FIGURE 16.39 A light source that emits light uniformly in all directions creates a sphere of light.

The light moves outward from a source that emits uniformly in all directions, creating a sphere of light (Figure 16.39). Think of the light as painting a sphere from the inside. The larger the sphere, the thinner the layer of light paint for a given amount of light. The light will appear to be dimmer to you because you are looking at only a small section of the sphere, painted thinly.

Example 16.10

A light source radiates equally in all directions with an intensity of 500.0 lumens (Figure 16.40). What is the illumination 3.0 m from the source?

$$I = 500.0\,\text{lm}$$

$$r = 3.0\,\text{m}$$

$$E = ?$$

$$E = \frac{I}{4\pi r^2}$$

$$E = \frac{500.0\,\text{lm}}{4\pi(3.0\,\text{m})^2} = 4.4\frac{\text{lm}}{\text{m}^2} = 4.4\,\text{lux}$$

FIGURE 16.40 What is the illumination 3.0 m from the source?

16.10 OPTICAL DEVICES

16.10.1 Optical Storage

Compact discs store binary information on them in the form of bumps embedded in the plastic (Figure 16.41). The bumps are created on a thin coating of aluminum, sandwiched between two pieces of plastic. On a CD that can store 650 MB of data, the bumps are 0.5 microns wide, 0.97 microns long, and 125 nm high. They are arranged in a spiral that, if stretched out, would be 5 miles long!

The bumps reflect laser light that shines on them differently than on the flat areas. Electronics, along with a light sensor, measure the reflected light and interpret the information.

Digital video discs (DVDs) store 7.5 times more data than a CD, that is, 4.7 GB. They accomplish this by making the bumps smaller and closer together, allowing them to carry more data in the same space.

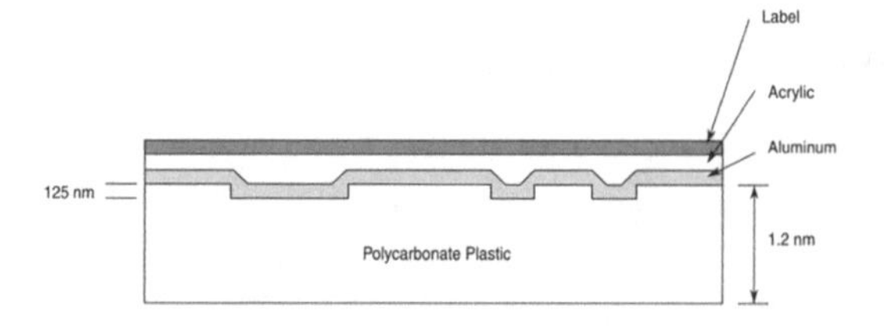

FIGURE 16.41 The bumps on a compact disc store binary ones and zeros.

16.10.2 Photodiodes, Resistors, and Transistors

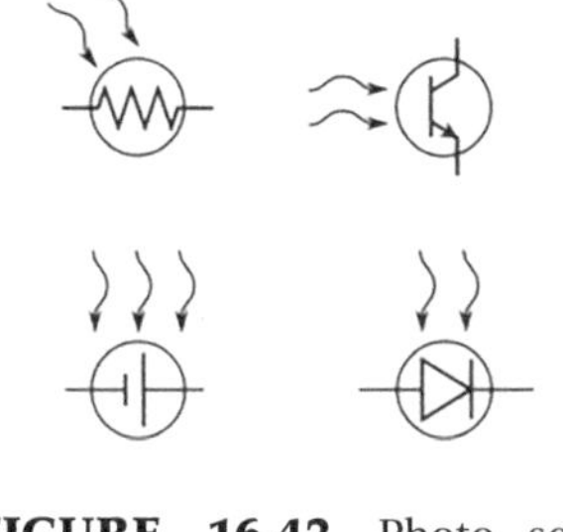

FIGURE 16.42 Photo sensors.

Resistors, diodes, and transistors can be made photosensitive and therefore can be activated by light (Figure 16.42).

Photoresistors are made from a light-sensitive material such as cadmium sulfide (Figure 16.43). Their resistance lowers with light level (Figure 16.44). They are used to detect light levels in cameras, streetlights, and so forth.

Photodiodes are essentially small solar cells in a small package (Figure 16.45). They are used to detect light by putting out a voltage proportional to the light level.

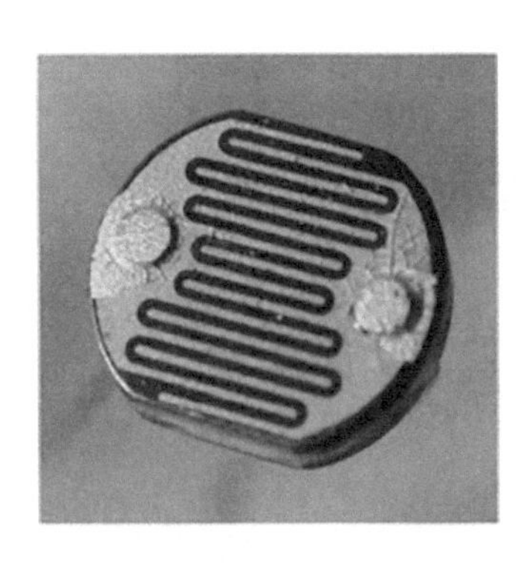

FIGURE 16.43 A photoresistive cell.

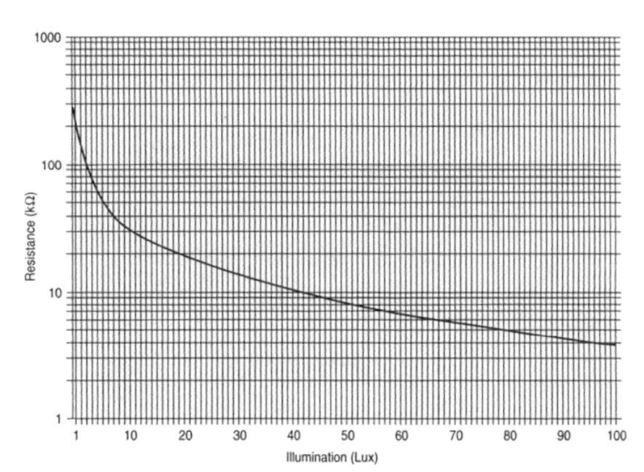

FIGURE 16.44 Typical response curve of a photo cell.

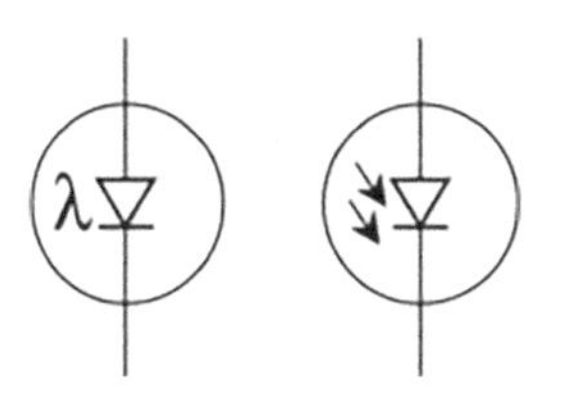

FIGURE 16.45 Schematic symbol of a photodiode.

Phototransistors have a light-sensitive base (Figure 16.46). Instead of a wire connected to the base, a window is present, allowing light through to turn on the transistor. Phototransistors have a larger output than a photodiode due to the amplifying nature of a transistor.

FIGURE 16.46 A phototransistor.

16.10.3 Charge-Coupled Device

Digital cameras use a device called a CCD to record an image (Figure 16.47).

Figure 16.48 shows how a CCD works. Light is focused on to an array of capacitors which are light sensitive. A charge builds up on each capacitor proportional to the light intensity. Each capacitor element is called a pixel. The pixels are then read by a device and sent to a processor to construct the image.

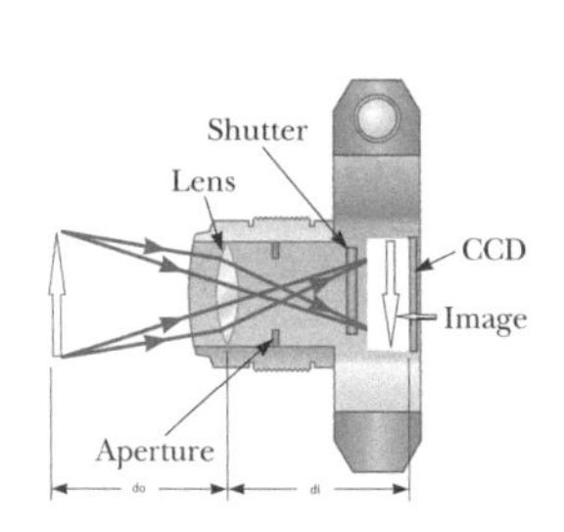

FIGURE 16.47 A digital camera using a CCD sensor.

Example 16.11

A 1 megapixel CCD contains a square array of 1,000 × 1,000 light sensitive elements contained in a package typically smaller than 5 × 5 micron². A 2 cm tall object is to be imaged with a lens onto 1 megapixel CCD, see Figure 16.48. What is the resolution of this image?

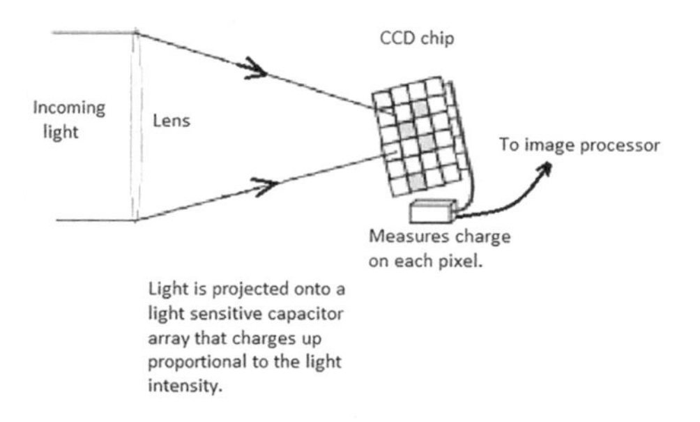

FIGURE 16.48 How a CCD works.

$$\text{Resolution per pixel} = \frac{\text{Object height}}{\text{Number of vertical pixels}} = \frac{2\,\text{cm}}{1{,}000\,\text{pixels}}$$

$$\text{Resolution per pixel} = 0.002\frac{\text{cm}}{\text{pixel}}$$

At least 2 pixels should be imaged, thus the resolution is:

$$\text{Resolution} = \left(0.002\frac{\text{cm}}{\text{pixel}}\right)(2\,\text{pixels})$$

$$\text{Resolution} = 0.004\,\text{cm}$$

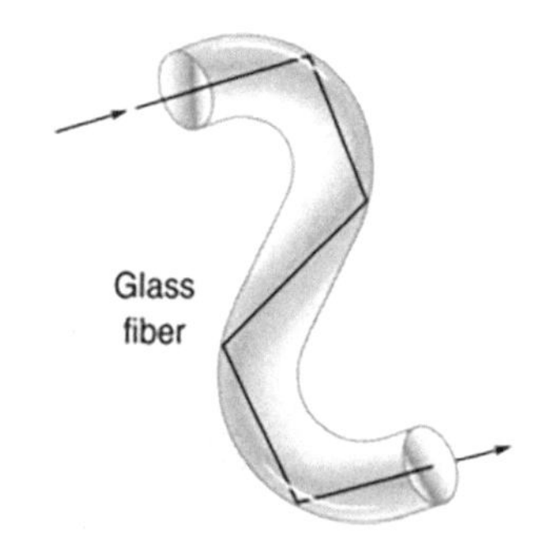

FIGURE 16.49 A light pipe.

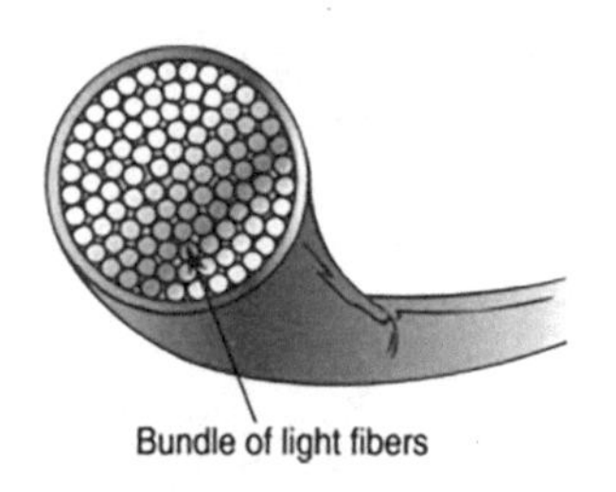

FIGURE 16.50 A fiber optic cable consists of many fibers.

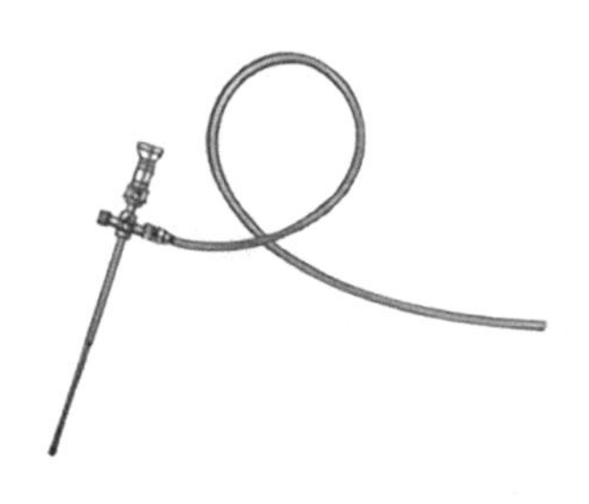

FIGURE 16.51 Fiber optics is used in medical procedures, such as laparoscopic surgery.

16.10.4 Fiber Optics

A *fiber optic* cable is a light pipe (Figure 16.49).

Light can be routed around in a light pipe similar to the way water can be routed through plumbing. Light traveling through a fiber optic cable is internally reflected, bouncing its way through the cable until it exits at the end. A few fiber optic cables in communications can replace big bundles of copper wire (Figure 16.50). Signals running through copper wire dissipate heat, resulting in degradation of the signals. In fiber optics, this problem is much smaller, allowing for a much higher transmission density without degradation (Figure 16.51).

16.10.5 Solar Energy

Solar energy comes from the sun, warms us, makes plants grow, and drives the weather. The average illumination from sunlight is about 1,000 W/m^2 at the earth's surface.

This energy can be captured and used to heat our homes or create electricity (Figure 16.52).

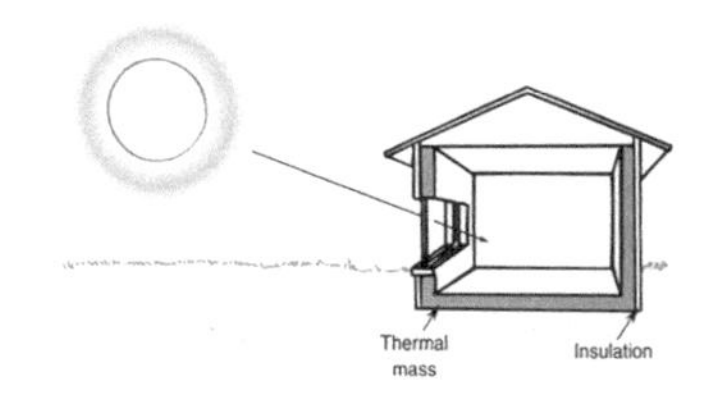

FIGURE 16.52 The average illumination from sunlight is about 1,000 $\frac{W}{m^2}$ at the earth's surface.

Solar cells convert sunlight directly into electricity (Figure 16.53). They are basically a very thin diode, a PN junction, with one side transparent to light. As light makes its way to the junction, it knocks free electrons on the P side, and they are transported to the N side because of the electric field present in every PN junction. A grid is placed over the solar cell to collect these electrons on the front, and a metal film is placed on the back to act as the positive terminal. The cell now acts like a battery, deriving its energy from light instead of some chemical reaction.

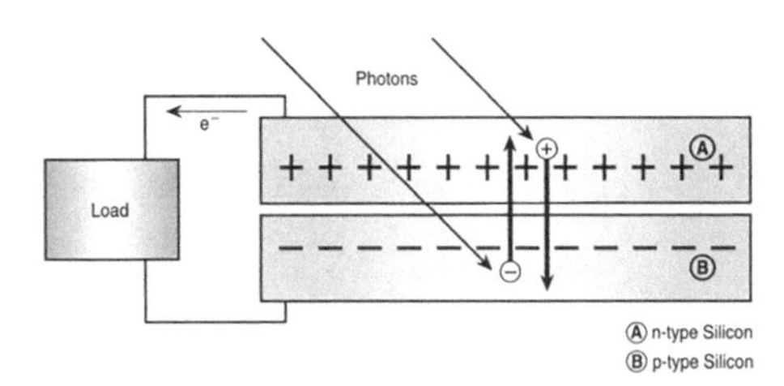

FIGURE 16.53 A solar cell.

The solar cell cannot convert all of the light energy into electricity because it is only about 20% efficient. Solar cells are sensitive to only a small band of wavelengths in sunlight. Seventy percent of the sunlight falling on the cell has wavelengths that are either too short or too long. Other losses are due to the high resistance of the semiconductors the cell is made of. To enhance efficiency, a cell is coated with a nonreflective agent to prevent the sunlight from just reflecting off the cell. A lens may be used to concentrate a large area of light onto a solar cell, increasing its light-gathering power (Figure 16.54). Currently priced at about 12 cents/kWh on average, the cost of solar power is on par with electricity from the utility companies.

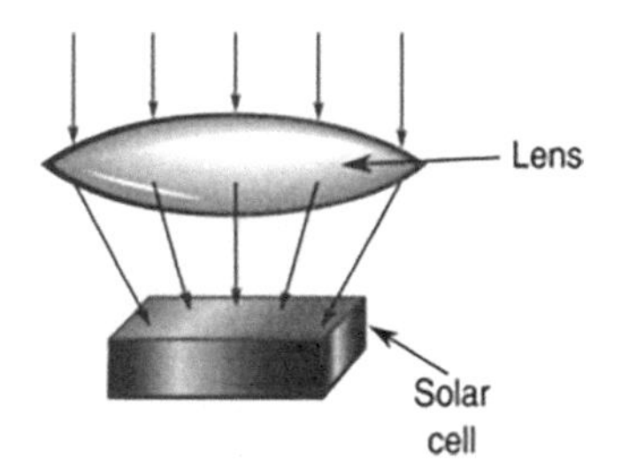

FIGURE 16.54 Lenses help collect more light to focus on the solar cell.

16.11 LIGHT SCATTERING

Our blue sky, during the day, is a result of the way sunlight scatters off of air molecules; this is called *Rayleigh scattering.* The air molecules scatter blue light the most and red light the least. Therefore, when you look up into the sky, you see blue everywhere. The rest of the colors contained in the white light pass right on through the atmosphere. This results in the sky appearing blue (Figure 16.55).

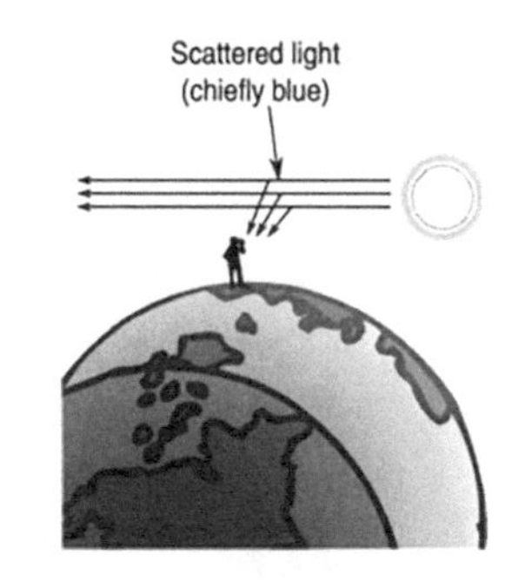

FIGURE 16.55 Blue light scatters more than the other colors in white light, giving the sky its blue color.

The reddening of the sky at night is also a result of this scattering. When the sun sets near the horizon, light must travel a greater distance through the atmosphere to reach earth than at noon. By the time the light

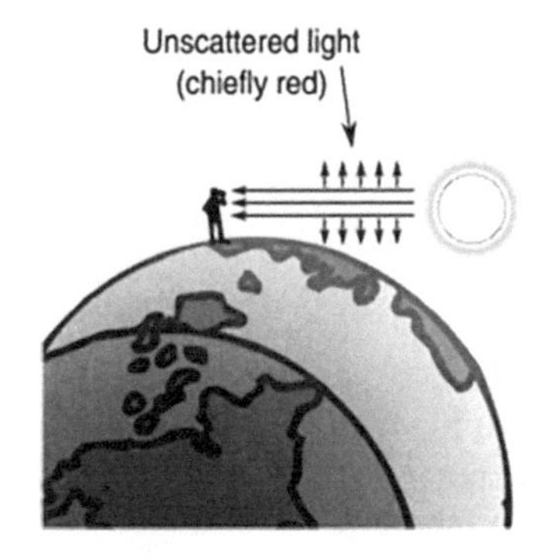

FIGURE 16.56 The sky appears red in the evening because light must travel a longer distance through the atmosphere, resulting in all colors except for red being scattered away.

has reached earth, all the colors have been scattered away except for red. As a result, the sky appears red (Figure 16.56).

16.12 CHAPTER SUMMARY

Symbols used in this chapter

Symbol	Unit
t—time	second (s)
d—distance	meter (m)
D—diameter	meter (m)
h—height	meter (m)
v—velocity	m/s
c—speed of light	3×10^8m/s
λ—wavelength	meter (m)
f—frequency	hertz (Hz)
f—focal length	meter (m)
m—magnification	no unit
n—index of refraction	no unit
f—number	no unit
I—intensity	lumens (lm), candela (cd)
E—illumination	lux

Light: Light is a wave made of electric and magnetic fields that change with time.

Light speed: Light travels at a speed of 186,000 miles/s or 3×10^8m/s in a vacuum.

$$c=186{,}000\,\text{miles/s or } 3\times10^8\,\text{m/s}.$$

$$c=\lambda\cdot f$$

$$\text{Speed of light}=\text{wavelength}\times\text{frequency}$$

Laser: Laser stands for Light Amplification by Stimulated Emission of Radiation.

Luminous intensity: The luminous intensity (I) of a light is a measure of how much light a source emits in a given area. It is measured in candelas. For a light source that emits light uniformly in all directions, the conversion from candelas to lumens is:

$$1\,\text{cd}=4\pi\,\text{lm}\leftarrow\text{For light sources that emit light in all directions uniformly}$$

Illumination is a measure of a light's apparent brightness at some distance away from the source.

The unit of illumination is lm/m^2 or lux.

$$1\ \frac{\text{lm}}{\text{m}^2} = 1\ \text{lux}$$

If the light radiates equally in all directions, the illumination is given by:

$$E = \frac{I}{4\pi r^2}$$

$$\text{Illumination(lux)} = \frac{\text{intensity(lm)}}{4\pi \cdot \text{distance(m)}^2}$$

Reflection of light: The angle at which a light ray reflects off a surface is equal to the angle at which it originally struck the surface. This is called the incident angle and is equal to the reflected angle:

$$\theta_i = \theta_r$$

$$\text{angle incident} = \text{angle reflected}$$

Spherical mirror: The focal point is a distance from the mirror equal to half the mirror's radius.

$$f_{\text{Spherical}} = \frac{R}{2}$$

$$\text{Focal point} = \frac{\text{radius of curvature of mirror}}{2}$$

Diffraction is the bending of waves around an obstacle. The **index of refraction** (n) is a measure of how many times slower light travels through a material than through a vacuum.

$$n = \frac{c}{v}$$

$$\text{Index of refraction} = \frac{\text{speed in a vacuum}}{\text{speed in a medium}}$$

Convergent lenses are lenses that focus light. An example is a magnifying glass.

Divergent lenses are lenses that spread out light.

Dispersion: Dispersion is the wavelength dependence of refraction.

16.12.1 Thin Lens Equation

A convergent lens will focus a parallel set of light rays on a point called the focal point on an axis, called the principle axis, that is centered on and runs through the lens. The distance from the lens where the image forms d_i is related to the focal length f and the distance from the object to the lens d_o.

$$\frac{1}{f} = \frac{1}{d_i} + \frac{1}{d_o}$$

$$\frac{1}{\text{focal length(m)}} = \frac{1}{\text{distance}_{\text{image}}\text{(m)}} + \frac{1}{\text{distance}_{\text{object}}\text{(m)}}$$

Other variations of this formula are:

$$f = \frac{1}{\frac{1}{d_i} + \frac{1}{d_o}} \quad d_i = \frac{1}{\frac{1}{f} - \frac{1}{d_o}} \quad d_o = \frac{1}{\frac{1}{f} - \frac{1}{d_i}}$$

16.12.2 Magnification

The magnification is a measure of how much larger or smaller the image is compared to the object. The magnification can be found by taking the ratio of the object distance d_o to the image distance d_i.

$$m = -\frac{d_i}{d_o}$$

$$\text{Magnification} = -\frac{\text{distance}_{\text{image}}\text{(m)}}{\text{distance}_{\text{object}}\text{(m)}}$$

The absolute value of the magnification is ratio of the height of the image h_i to that of the object h_o.

$$|m| = \frac{h_i}{h_o}$$

$$|\text{Magnification}| = \frac{\text{height}_{\text{image}}}{\text{height}_{\text{object}}}$$

***f*-number** relates the focal length to the diameter of a lens.

$$f\text{-number} = \frac{f}{D}$$

$$f\text{-number} = \frac{\text{focal length(m)}}{\text{diameter(m)}}$$

PROBLEM SOLVING TIPS

- When solving problems dealing with frequency, wavelength, and the speed of light, remember to convert megahertz into hertz.

PROBLEMS

1. List the different colors of the visible and near visible parts of the electromagnetic spectrum and their applications.
2. Ham radio operators sometimes bounce radio signals off the moon. If the moon is about 250,000 miles away, how long does it take the signal to reach the moon?
3. How long is the wavelength of an AM radio wave of frequency 550 kHz?
4. How does an atom emit light?
5. List the various steps involved in the workings of a laser.
6. Look at Figure 16.44. Does a photoresistor's resistance increase or decrease with light level?
7. Suppose a 100 W light bulb puts out 1,250 lm of light. What is its illumination 2.0 m away?
8. How can you use the illumination equation to find the distance to a source of light with a known intensity?
9. Suppose you take a parabolic reflector with a collection area of $3\,m^2$ and place a solar cell at its focus. How many watts of sunlight will be focused onto the solar cell if the average illumination is $1{,}000\,W/m^2$? How many watts of electricity will be generated if the cell's efficiency is 20%?
10. How do lenses bend light?
11. What is the difference between ultraviolet and infrared lights in terms of wavelength?
12. When white light shines on an object and the object appears to be of a certain color, what has happened to the other colors in the white light?
13. Why can we see through glass?

14. If a ¼ wave antenna is designed to work with a radio station operating at 93 MHz, how long should the antenna be?
15. If light strikes a flat mirror at an angle of 30° with respect to a perpendicular to the mirror, what is the angle of the reflected light?
16. How long does it take for a radio signal to reach a satellite 20,000 miles above the earth's surface?
17. A radar determines the distance to an object by sending out a radio wave and timing how long it takes for the wave to propagate outward, bounce off an object, and return. If the round-trip travel time is 5.0×10^{-5} s, how far away is the object in meters?
18. What is the approximate wavelength of a FM radio station operating at 96 MHz?
19. The inside of a microwave oven is 12 in. wide. If one wavelength of a microwave spans this distance, what is the frequency of the magnetron (the source of the microwaves)?
20. The efficiency of an incandescent light bulb is only about 10% (only 10% of the energy it consumes is turned into visible light). The remaining 90% is dissipated as heat. How many watts of visible light does a 60 W light produce?
21. Light strikes a flat mirror at an angle of 30° with respect to the perpendicular to the surface. What is the angle of reflection?
22. The index of refraction of a material is a measure of how many times light slows down when traveling through it as compared to a vacuum. If the index of refraction is 1.4, what is the speed of light in that medium?
23. If the intensity of an isotropic light source (a source that emits uniformly in all directions) is 300 lm and is viewed 4.0 m away, what is the illumination?
24. What is the distance to a light source that has a luminous intensity of 450 lm and an illumination of 0.001 lux?
25. The electrical power out of a solar cell depends on its efficiency, the illumination of the sun upon it, and its effective area (the area of sunlight it can collect). If the area of the cell is $0.09\,m^2$ and the illumination of the sun is $1{,}000\,W/m^2$, and the efficiency of the cell is 20%, what is the output power of the cell?
26. Light travels 25 m through water which has an index of refraction 1.33. How long did it take to travel this distance?
27. If the speed of light is 186,000 miles/s, how many ft/s is this?
28. Electronic signals travel near the speed of light. If a signal travels at 95% the speed of light through a cable, how many feet does it travel in 1 ns?

29. Diamond has an index of refraction 2.42. What is the speed of light in a diamond in m/s?
30. A 40 W fluorescent light emitting 500 lm of light at 40% efficiency wastes 60% of its energy in generating heat. How much of the 40 W is turned into heat?
31. An isotropic light source (emits light equally in all directions) emits 500 lm. How many candelas is this?
32. Two identical light sources are 2 and 3 m away from an observer. If the observer measures 8.0 lux, what is the intensity of these light sources?
33. A one-quarter wave antenna has a length that is ¼ the length of the radio wave it is transmitting. If the frequency of the transmission is 10.0 MHz, how long is the antenna?
34. A source of light emits 100 W/m^2 of light. A solar cell can collect this light and turn it into electricity. By using a lens, more light can be gathered and focused onto the cell. If a lens is capable of gathering 0.3 m^2 of light and focusing it onto the cell, how many watts of light are focused onto the cell?
35. What is the focal point of a spherical mirror with a radius of 12 in.?
36. The focal length of a converging lens equals 25 cm and the object is 35 cm from the lens. Where does the image form?
37. The focal length of a converging lens equals 15 cm, and the image is 20 cm from the lens. What is the distance from the object to the lens?
38. An object is 60 cm from a lens, and the image forms 20 cm from the lens. What is the focal length?
39. A 2 cm tall object is 25 cm from a lens, and the image forms 35 cm from the lens. What is the magnification and how tall is the image?
40. A lens has a focal length of 4 cm and a diameter of 2 cm. What is its *f*-number?
41. What is the radius of curvature for a spherical mirror with a focal length of 30 cm?
42. A laser beam strikes a flat mirror at an angle of 20° with respect to the normal. The mirror is on the floor reflecting the beam onto the wall 2 m above the floor. How far from the wall did the laser beam strike the mirror?

17

Data Acquisition-Sensors and Microcontrollers

This chapter will cover data acquisition in experimental physics. Microcontrollers offer an inexpensive way to acquire data from analog and digital sensors. Interfacing microcontrollers to sensors along with data analysis using Excel will be the focus of this chapter.

FIGURE 17.1 Photosensitive sensors called photocells or LDR change their resistance with light level, courtesy of adafruit.com.

17.1 RESISTIVE SENSORS

Physical quantities such as light, temperature, force, and humidity can be measured by sensors, whose resistances are sensitive to these quantities. Examples include resistive temperature devices (RTDs), light sensitive resistors (photoresistors or photocells), force sensors, and strain gauges (Figures 17.1–17.4). All of these sensors exhibit a dramatic drop in resistance as the light, temperature, force, and humidity increase (Figure 17.5).

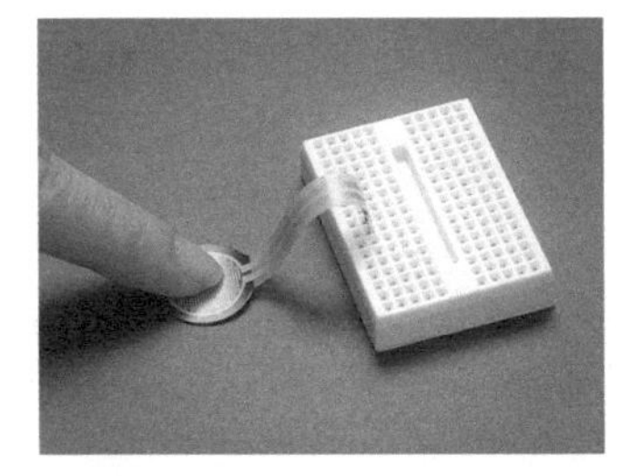

FIGURE 17.2 Force sensitive resistors change their resistance when they are deformed, i.e., bent, stretched, or compressed because of a force, courtesy of adafruit.

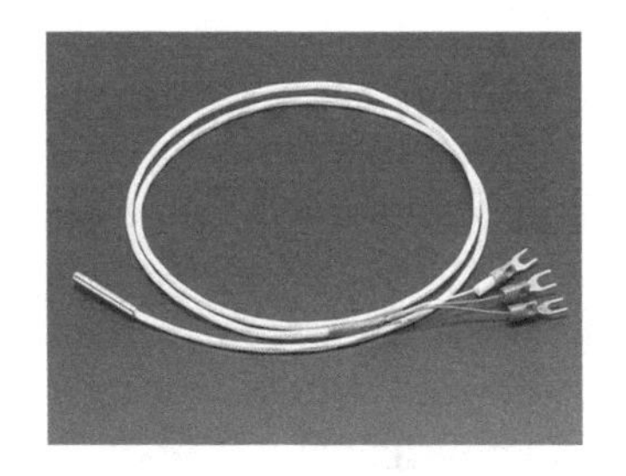

FIGURE 17.3 RTD changes resistance with a temperature change, courtesy of adafruit.com.

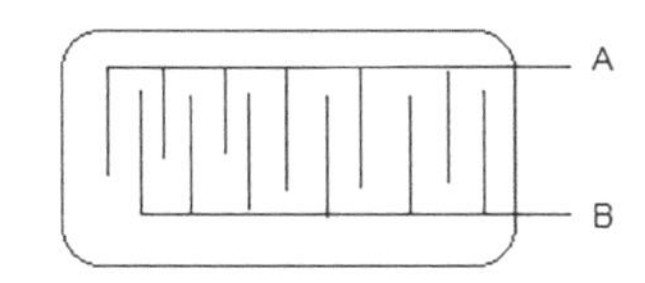

FIGURE 17.4 A resistive humidity sensor. Resistance between leads *A* and *B* changes with humidity.

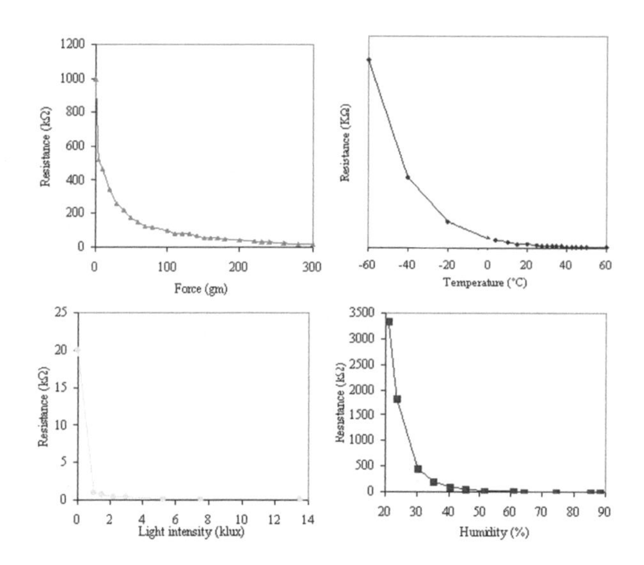

FIGURE 17.5 Plot of resistance vs. temperature, force, light, and humidity for resistive sensors.

Example 17.1

A strain gauge is used to measure a force. Its resistance changes by 100 Ω/lb. What is the change in resistance when measuring 0.09 lb?

This problem is effectively asking for a conversion from pounds to ohms. We know, of course, that it does not make sense in terms of dimensional analysis to convert pounds into ohms, but we can think of them as effectively equal in the context of this problem.

$$100\ \Omega = 1\,\text{lb}$$

$$0.09\,\cancel{\text{lb}}\left(\frac{100\,\Omega}{1\,\cancel{\text{lb}}}\right) = 9\,\Omega$$

17.1.1 Measuring Resistive Sensors

Many sensors respond to measuring a quantity with a change in resistance. A data acquisition device typically reads voltage on their input, not resistance. Running a current through the resistive sensor will produce a voltage according to Ohm's law, $V = I \cdot R$. We can now read this voltage and determine what physical quantity the sensor is measuring. To convert a resistance to a voltage, we can place the resistive sensor into a voltage divider.

Voltage Divider: A voltage divider is a very handy series circuit that can be used to change a sensor with a resistive output to a sensor with voltage output (Figure 17.6).

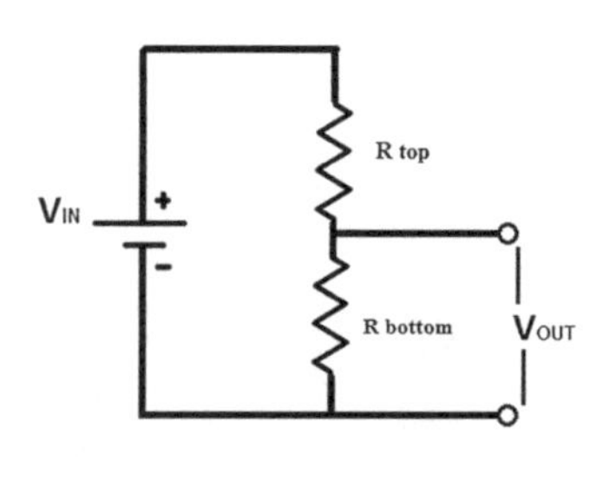

FIGURE 17.6 A voltage divider.

The voltage output from the divider is:

$$V_{out} = \frac{R_{bottom}}{R_{bottom} + R_{top}} V_{in}$$

From the voltage divider formula, we can see that if R_{bottom} increases then V_{out} increases and if R_{top} increases then V_{out} decreases:

$$R_{bottom} \uparrow \Rightarrow V_{out} \uparrow$$

$$R_{top} \uparrow \Rightarrow V_{out} \downarrow$$

Example 17.2

Given a 9 V source, a 10 and a 5 kΩ resistor connected as shown in Figure 17.6, find the output voltage.

$$V_{out} = \frac{R_{bottom}}{R_{bottom} + R_{top}} V_{in}$$

$$V_{out} = \frac{5\ \text{k}\Omega}{5\ \text{k}\Omega + 10\ \text{k}\Omega} 9\ \text{V}$$

$$V_{out} = 3\ \text{V}$$

17.1.2 Humidity Sensor

A humidity sensor is called a hygrometer. The resistive hygrometer consists of two sets of electrical traces, not touching, laid out like combs interlaced with one another. In presence of moisture, the water will bridge the gap between the traces and the resistance between the two will drop (Figure 17.7).

FIGURE 17.7 A humidity sensor.

Example 17.3

A resistive hygrometer is placed in a voltage divider circuit to act as a water alarm. The sensor is placed in the top half of the divider. When exposed to water, the sensor's resistance will drop, and hence the voltage out will increase (Figure 17.8).

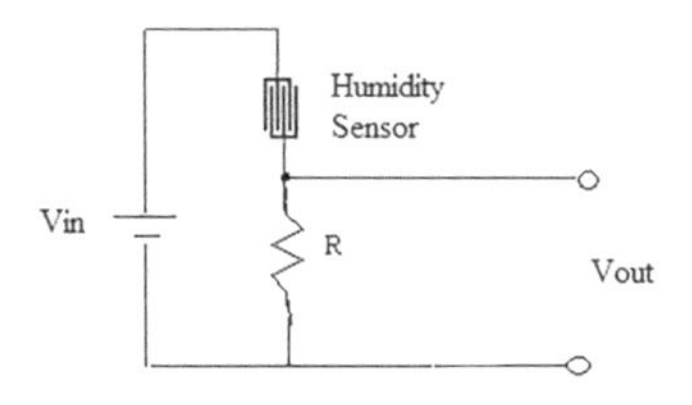

FIGURE 17.8 A resistive hygrometer is placed in a voltage divider

17.1.3 Photoresistors

A photoresistor changes resistance with light level. The brighter the light, the lower the sensor's resistance; the dimmer the light, the higher the sensor's resistance (Figure 17.9). Placing the sensor in the bottom half of the circuit results in the output voltage to be high when the sensor is dark, and low when the sensor is illuminated. Placing the sensor in the top half of the circuit results in the output voltage to be low when the sensor is dark, and high when the sensor is illuminated.

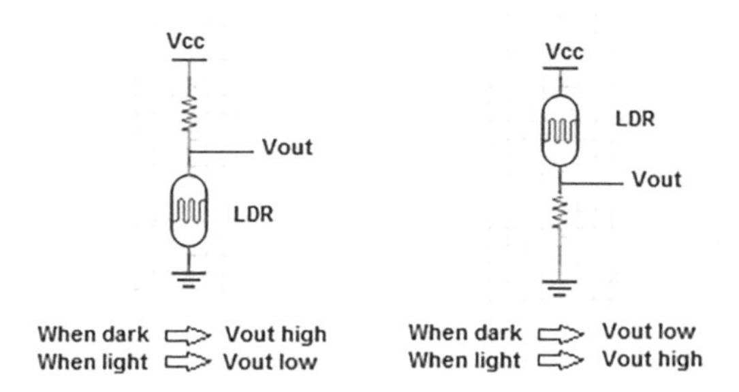

FIGURE 17.9 The placement of a photoresistor in a voltage divider determines when the output will be high or low with light level.

Example 17.4

A laser and a photoresistor used to construct a burglar alarm (Figure 17.10).

A laser is placed on one side of the door and a light sensor on the other. When an object interrupts the beam, the light level at the sensor drops and its resistance changes. This resistance change can be turned into a voltage change. To turn the change of the sensor's resistance into a change in voltage, the sensor is placed in a voltage divider circuit.

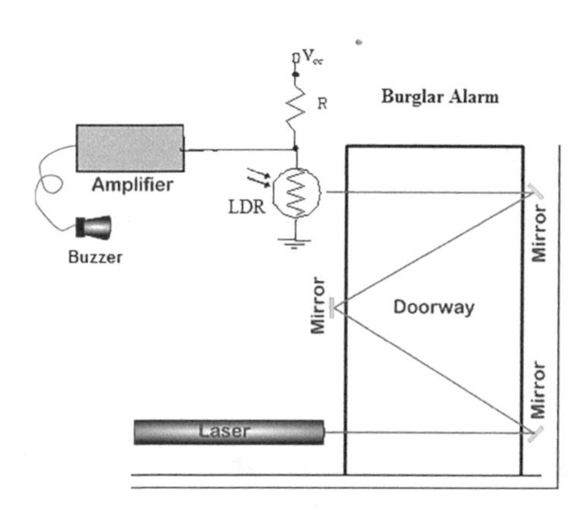

FIGURE 17.10 Laser burglar alarm.

17.2 CALIBRATION AND DATA FITTING

Calibrating a sensor means to compare its output to a known input in order to develop a relationship between what the sensor is measuring and what the sensor outputs. For example, to calibrate a force sensor, its output is recorded for a known mass applied to the sensor. After plotting the data in a spreadsheet, we can fit the data. Fitting data means finding an equation that passes through the dataset. The fit to the data will generate a formula that predicts what the input to the sensor is for a given output.

Example 17.5

A resistive force sensor is placed in a voltage divider and calibrated by recording the output voltage as mass is added on top of the sensor (Figure 17.11).

The following table was generated by measuring the output voltage as mass was added on top of the sensor (Table 17.1).

This table can be now roughly used to determine the mass on the sensor. Interpolating between data points would be necessary for readings not in the table. Rather than interpolating between data points, a better method would be to fit the data in a spreadsheet program such as Excel.

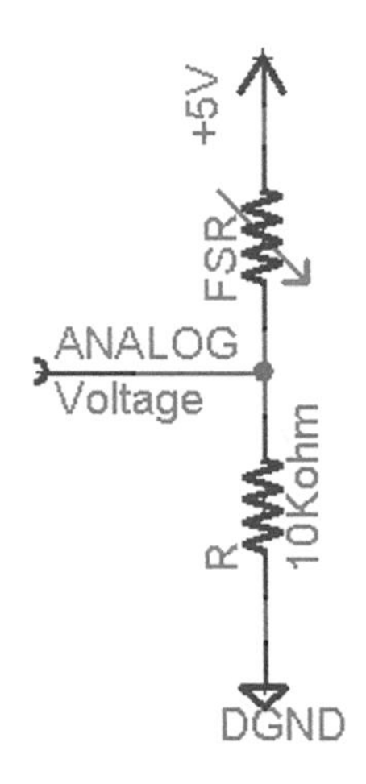

FIGURE 17.11 A force sensor resistor, FSR, placed in a voltage divider circuit, courtesy of adafruit.com

17.3 EXCEL

A spreadsheet, such as Excel, can be used to record and fit the data. Excel has a fitting function called **Trendline**, which finds the "best" line through the points and generates an equation that represents the curve. Trendline has a limited number of fitting functions. For more complicated functions, Excel has a Solver function that can be used. For this discussion, we will focus on the built-in functions in Trendline. Below are steps to plot and fit data from a force sensor to a polynomial function. The data in Table 17.1 will be used to illustrate fitting.

Steps in fitting data:

1. Label the columns in row one and fill in the data measured during calibration (Figure 17.12).
2. Click on the insert tab, then scatter as shown in Figure 17.12.
3. Click on +, then the arrow next to Trendline, then More Options (Figure 17.13).
4. Choose Polynomial and the order, then click on Display Equation on chart (Figure 17.14).
5. Now you can see the equation displayed on the graph (Figure 17.14).

TABLE 17.1
Calibration Data of Force Sensor

Force (*g*)	**Voltage Across *R***
None	0 V
50	1.5 V
100	2.7 V
200	3.2 V
300	3.6 V
400	3.8 V
500	4.0 V
600	4.1 V

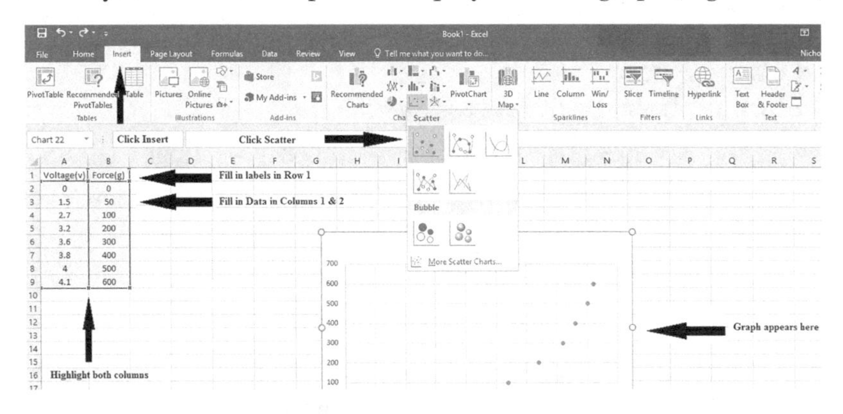

FIGURE 17.12 Enter data and click on insert, then scatter.

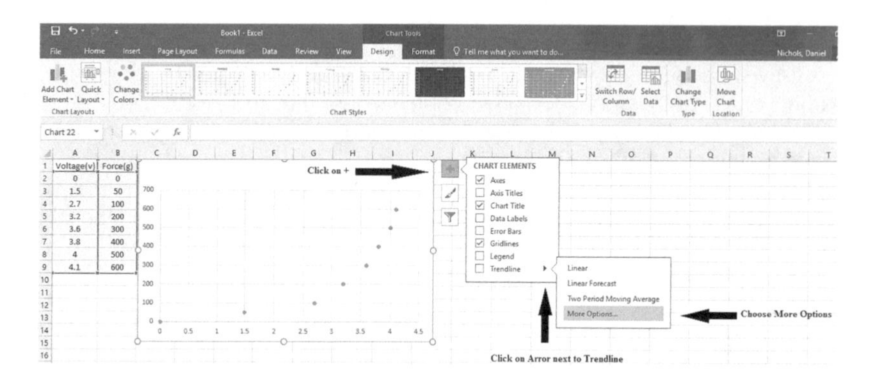

FIGURE 17.13 Click on +, then the arrow next to Trendline, then More Options to get to where to find the fitting equations.

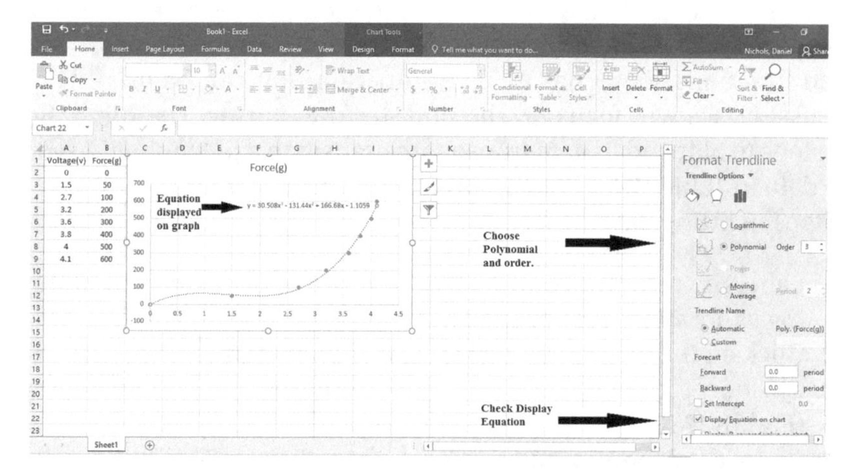

FIGURE 17.14 Choose Polynomial and the order, then click on Display Equation on chart.

The formula from the fit above (Figure 17.15) is interpreted as y is the force F and x is the voltage V.

The formula from the fit above is therefore:

$$F = 30.508 \cdot V^3 - 131.44 \cdot V^2 + 166.68 \cdot V - 1.1059$$

Checking the fitting formula from the fit at a random voltage, say $V = 3.2\,\text{V}$, gives:

$$F = 30.508 \cdot 3.2^3 - 131.44 \cdot 3.2^2 + 166.68 \cdot 3.2 - 1.1059$$

$$F = 186\,\text{g} \leftarrow \text{The actual value was } 200\,\text{g}.$$

We can see the fit is not perfect! A better fit can be found by fitting smaller sections of the data. The choice of the fitting function should be experimented with until you get a good match.

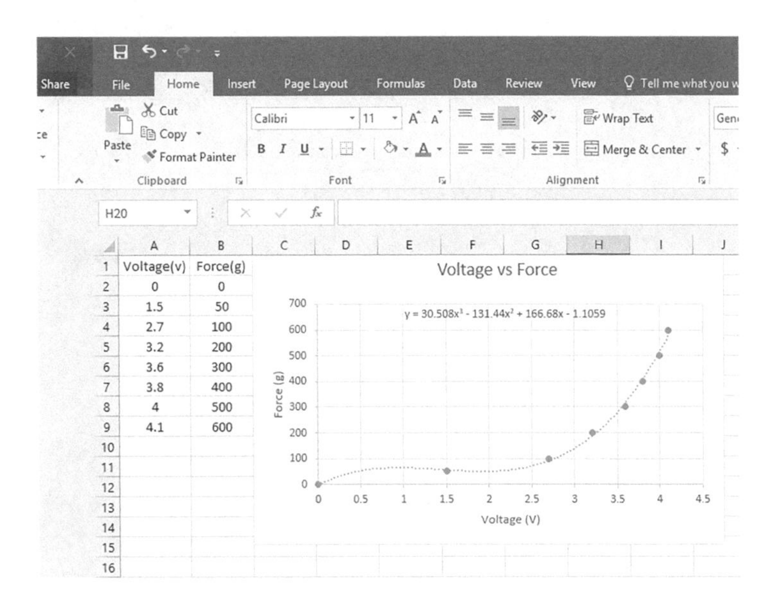

FIGURE 17.15 Graph of data and fitting equation.

17.3.1 Comparing Theory and Experiment with Data Fitting

Fitting data can be used to compare theory and experiment. For example, we found in a previous chapter that a mass dropped from rest falls with acceleration g for a distance $y - y_i$, in time t, according to:

$$y = \frac{1}{2}gt^2 + v_i t + y_i$$

By performing an experiment where we record the distance a mass falls through in time, we can find the value of g, the acceleration due to gravity.

Example 17.6

A mass is dropped, and time is recorded for how long it took to fall through a distance of 0.5, 1.0, 1.5, and 2.0 m. Find the acceleration due to gravity g, the initial velocity, and position from a data fit. Use Excel to record, plot, and fit the data to a second order polynomial (Figure 17.16).

The data fit generates the equation $y = 5.0081x^2 - 0.0073x + 0.0032$. We identify y as the distance and x as the time. Comparing the fit to theory:

We see that:

$$\frac{1}{2} \cdot g = 5.00381$$

$$v_i = -0.0073 \text{ m/s}$$

$$y_i = 0.0032 \text{ m}$$

Solving for g gives $g \approx 10\frac{\text{m}}{\text{s}^2} \leftarrow$ within 2% of the accepted value of $9.8\frac{\text{m}}{\text{s}^2}$!

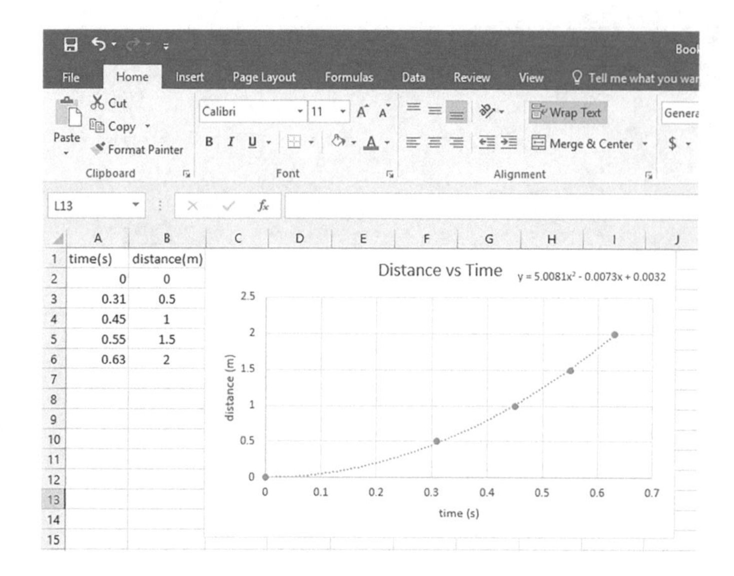

time(s)	distance(m)
0	0
0.31	0.5
0.45	1
0.55	1.5
0.63	2

FIGURE 17.16 Data fit to a mass in free fall.

Example 17.7

The temperature of a hot cup of coffee was recorded and plotted as it cooled off (Figure 17.17).

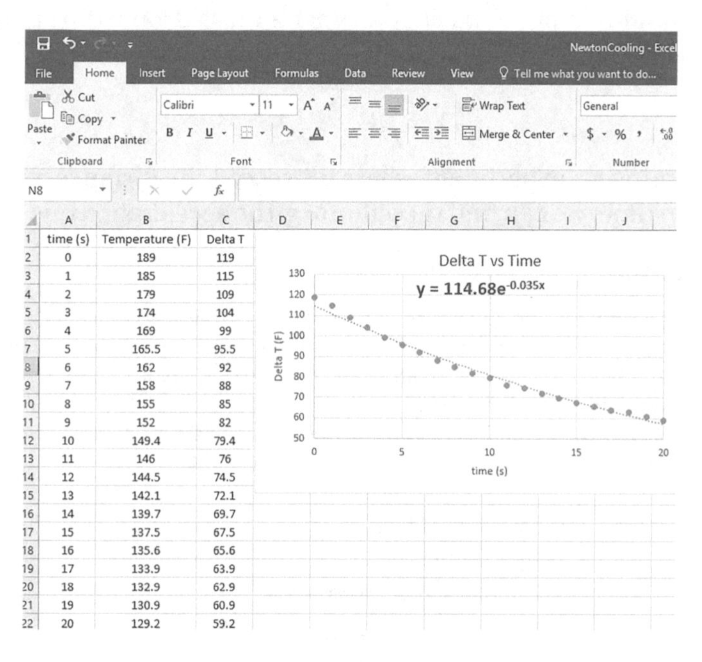

time (s)	Temperature (F)	Delta T
0	189	119
1	185	115
2	179	109
3	174	104
4	169	99
5	165.5	95.5
6	162	92
7	158	88
8	155	85
9	152	82
10	149.4	79.4
11	146	76
12	144.5	74.5
13	142.1	72.1
14	139.7	69.7
15	137.5	67.5
16	135.6	65.6
17	133.9	63.9
18	132.9	62.9
19	130.9	60.9
20	129.2	59.2

FIGURE 17.17 Plotted is a cup of hot coffee cooling down to room temperature; it has a reasonable fit to an exponential.

To measure the temperature, a calibrated RTD was used. Newton's Law of Cooling says that an object cools off at a rate proportional to the difference between the object's temperature and its surroundings, the ambient temperature.

This is described by the following equation:

$$T - T_a = (T_0 - T_a) \cdot e^{-k \cdot t}$$

where:

T = temperature
t = time
T_a = ambient temperature
T_0 = temperature at time 0
k = thermal conduction constant

This formula says that an object's temperature will decrease to room temperature at an exponential rate. Excel can fit data to an exponential function, $y = A \cdot e^{-kx}$.

Comparing this equation to a fitting equation from Excel:

$$y = A \cdot e^{-k \cdot x}$$
$$T - T_a = (T_0 - T_a) \cdot e^{-k \cdot t}$$

We identify:

$$y = T - T_a$$
$$A = (T_0 - T_a) \cdot$$
$$x = t$$

Plotted is (temperature – room temperature) vs. time along with a fit to $y = A \cdot e^{-kx}$.

From the fit we see, A = 114.68°F and k = 0.035/s.

The thermal conduction constant k is a measure of how an object is thermally connected to its environment and has a unit of 1/s.

17.3.2 Excel's Solver Function

The Trendline options in Excel are limited, for example, there is not an option to fit a sinusoidal function. Excel does have a tool called Solver that can be used to fit data to almost any function. To use Solver, the user enters a function the data is to be fit to. Solver then minimizes the difference between the declared fitting function and the data by altering the free parameters in the declared function until the difference is small.

To install Solver preform the following:

1. Click the **File tab**, click **Options**, click the **Add-ins** category.
2. In the **Manage** box, click **Excel Add-ins**, then click **Go**.
3. In the **Add-ins available** box, select **Solver Add-in** check box. If you don't see this name in the list, click the Browse button and find the folder containing Solver.xlam. Then click **OK**.
4. Now on the **Data** tab, in the **Analysis** group, you will see the **Solver** command.

Example 17.8

As an example of setting up the spreadsheet to use Solver, suppose we perform an experiment to measure the position of an oscillating object and we see if we can find an equation that describes its motion. To accomplish this, a distance sensor is placed below the oscillating mass and data is recorded (Figure 17.18). The data looks like a sinusoidal wave, so we will attempt to fit the data to the function:

$$y = A \cdot \sin(w \cdot t + \text{phi}) + \text{offset}$$

	A	B	C	D	E	F	G
1	Time(s)	Distance(mm)	D=Asin(wt+phi) + offset D= $G$3*SIN($G$4*A2+$G$5)+$G$2	chi squared (B2-C2)^2			
2	0	276	289.8289643	191.240253		offset=>	289.829
3	0.2	299	313.0634575	197.780838		A=>	26.4891
4	0.4	310	312.145474	4.60305884		w=>	5.34931
5	0.6	302	288.0292663	195.181399		phi=>	0
6	0.8	271	265.7838618	27.2080982		Sum of chi squared sum(D2:Dend) =>	1081.989
7	1	261	268.53357	56.7546764			
8	1.2	282	293.4200431	130.417385			
9	1.4	315	314.5735556	0.18185478			
10	1.6	312	310.0048301	3.98070312			
11	1.8	294	284.4631	90.9524607			
12	2	266	264.4992369	2.25228976			
13	2.2	269	270.8658663	3.48145719			
14	2.4	296	296.9448164	0.89267807			
15	2.6	312	315.6267705	13.1534642			
16	2.8	311	307.4916596	12.3084524			
17	3	286	280.9960089	25.0399271			
18	3.2	261	263.6822993	7.19472961			
19	3.4	274	273.5482967	0.20403588			
20	3.6	301	300.3382029	0.43797543			
21	3.8	319	316.2036556	7.81954221			

Fit of Distance vs Time

Distance(mm)

time(s)

Series1 Series2

FIGURE 17.18 A Scatter plot of columns 1–3 along with the fitting equation.

The following steps were performed to find the quantities, *A*, *w*, phi, and offset:

1. The collected data is entered on columns 1 and 2.
2. The fitting function is entered in column 3. The parameters to be adjusted to match the data are (offset, *A*, *w*, and phi).
3. The initial values of these parameters are entered in columns G2–G5.
4. The difference squared between the data and the fitting function is given in column D.

Now preform a Scatter Plot of columns 1–3 to see the data and the fitting equation. You can now adjust the parameters in columns G2–G5 to find good starting parameters for Solver.

17.3.3 Starting Solver

Once the data and the fitting equations and parameters are entered and plotted, open Solver (Figure 17.19).

1. Click on the Data tab.
2. Click on Solver.

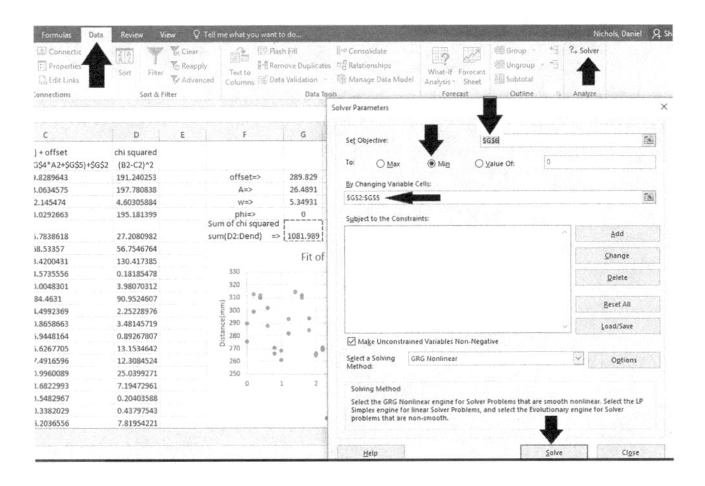

FIGURE 17.19 Opening Solver.

3. Enter the location of the Sum of chi squared in Set Objective, in this case $G$6.
4. Click on Min.
5. Enter the location of the fitting parameters, in By Changing Variable Cells, in this case $G$2:$G$5.
6. Click Solve at the bottom of the screen.
7. Look at the fit and how the values in $G$2:$G$5 have changed.
8. If the fit is not good, modify the parameters in $G$2:$G$5 again and run Solver again till the fit looks good.
9. The values at $G$2:$G$5 are now the values of (offset, A, w, and phi) in the equation,

$$y = A \cdot \sin(w \cdot t + \text{phi}) + \text{offset}$$

10. The equation that describes the position of the mass as it oscillates is therefore:

$$y = 26.49 \cdot \sin(5.35t + 0) + 289.83$$

Example 17.9

An oscillating mass on a spring is studied to determine its resonance frequency. The mass of the mass spring system is 1 kg, and the spring is stretched 33.4 cm when the mass is added to the spring. The position of the mass is measured while it is oscillating and copied into an Excel spreadsheet to be analyzed as in Figures 17.20 and 17.21.

Let us perform a theoretical calculation of the period of oscillation and compare this to the measured value. The period T of oscillation of a mass m on a spring with spring constant k is given by the formula:

$$T = 2\pi\sqrt{\frac{m}{k}}$$

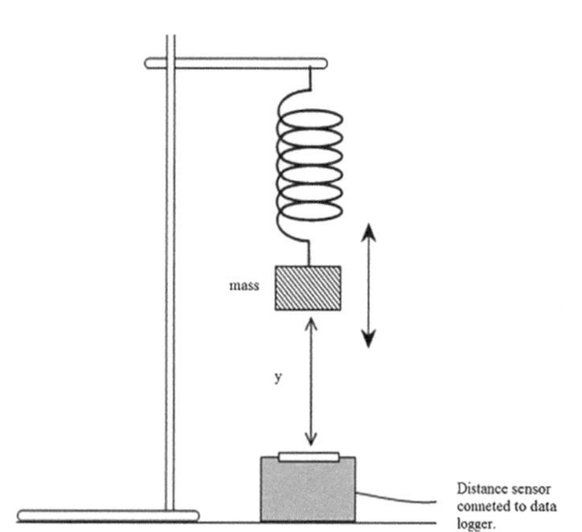

FIGURE 17.20 An oscillating mass on a spring.

	A	B	C	D
1	Time(s)	Distance(mm)	D=Asin(wt+phi) + offset D= $G$3*SIN($G$4*A2+$G$5)+$G$2	chi squared (B2-C2)^2
2	0	276	289.8289643	191.240253
3	0.2	299	313.0634575	197.780838
4	0.4	310	312.145474	4.60305884
5	0.6	302	288.0292663	195.181399
6	0.8	271	265.7838618	27.2080982
7	1	261	268.53357	56.7546764
8	1.2	282	293.4200431	130.417385
9	1.4	315	314.5735556	0.18185478
10	1.6	312	310.0048301	3.98070312
11	1.8	294	284.4631	90.9524607
12	2	266	264.4992369	2.25228976
13	2.2	269	270.8658663	3.48145719
14	2.4	296	296.9448164	0.89267807
15	2.6	312	315.6267705	13.1534642
16	2.8	311	307.4916596	12.3084524
17	3	286	280.9960089	25.0399271
18	3.2	261	263.6822993	7.19472961
19	3.4	274	273.5482967	0.20403588
20	3.6	301	300.3382029	0.43797543
21	3.8	319	316.2036556	7.81954221

F	G
offset=>	289.829
A=>	26.4891
w=>	5.34931
phi=>	0
Sum of chi squared sum(D2:Dend) =>	1081.989

Fit of Distance vs Time
Distance(mm)
time(s)
Series1 Series2

FIGURE 17.21 Data and fit of a 1 kg mass oscillating on a spring.

The spring constant k is a measure of how stiff the spring is and can be found from Hooke's law:

$$F = -k \cdot \Delta s$$

Solving for k: $k = -\dfrac{F}{\Delta s}$

We can determine k by applying a force on the spring and measuring how much it stretches.

A mass of 1 kg was added to the spring resulting in the spring stretching 33.4 cm or 0.334 m.

The spring constant k is therefore: $k = \dfrac{F}{\Delta s} = \dfrac{m \cdot g}{\Delta s}$

Plugging in for m and k, $k = \dfrac{(1\text{kg})(9.8\text{ m/s}^2)}{0.334\text{ m}}$

$$k = 29.34\ \frac{N}{m}$$

The predicted period of oscillation is therefore:

$$T = 2\pi\sqrt{\frac{m}{k}}$$

$$T = 2\pi\sqrt{\frac{1\text{kg}}{29.34\ \frac{N}{m}}}$$

$$T = 1.16\text{ s} \leftarrow \text{Predicted period}$$

Let us now calculate the measured period of oscillation.

Recall the angular velocity formula:

$$\omega = 2\pi f = \frac{2\pi}{T}$$

$$T = \frac{2\pi}{\omega}$$

The fit in Figure 17.21 gives us a value for $\omega = 5.35\ \frac{\text{rad}}{\text{s}}$ from cell G4.

$$T_{\text{measured}} = \frac{2\pi}{5.35}$$

$$T_{\text{measured}} = 1.17\ \text{s}$$

Let us now do a percent difference calculation to compare the theoretical and measured values of T.

$$\%\ \text{difference} = \frac{1.17\ \text{s} - 1.16\ \text{s}}{1.16\ \text{s}} \times 100$$

$$\%\ \text{difference} = 0.86\% \leftarrow \text{Wow so close!}$$

17.4 SENSORS THAT OUTPUT VOLTAGE

17.4.1 LM35

The LM35 is a temperature sensor whose output voltage is linearly proportional to the Fahrenheit temperature. Its supply voltage can be anywhere from 4 V→20 V and will output 10 mV/°C. The output, 10 mV/°C means for every degree Fahrenheit it measures it will output 10 mV. Illustrated in Figure 17.22, the LM35 is supplied with 9 V.

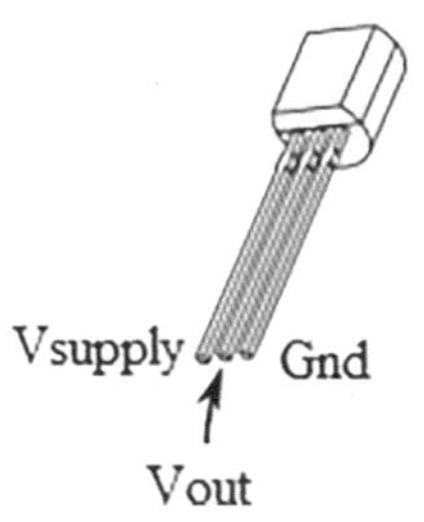

FIGURE 17.22 The LM35 temperature sensor.

Example 17.10

The LM35 temperature sensor outputs 10 mV/°C. Its output voltage reads 230 mV in Figure 17.23. What temperature is the sensor reading?

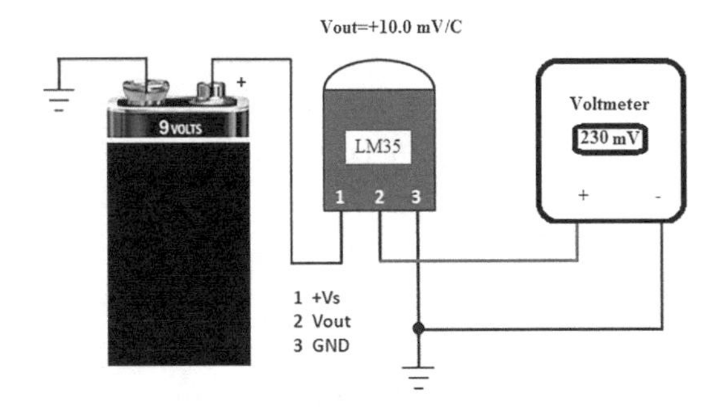

FIGURE 17.23 The LM35 wiring diagram.

$$T = \frac{1°C}{10\ mV} \cdot V_{out}$$

$$T = \frac{1°C}{10\ \cancel{mV}} \cdot 230\ \cancel{mV}$$

$$T = 23°C$$

Example 17.11

An experiment is performed on a cup of tea. The tea is made by heating 100 g of water with a 10 Ω, 5 W resistor immersed in the water and applying 5 V to it. The temperature was monitored with a LM35.

a. What is the expected temperature rise after 15 min?
b. The results of the experiment after heating the water for 15 min.
c. A comparison of the experimental results and theoretical calculation.

a) What is the expected temperature rise after 15 min?

Recall that the temperature rise of a material depends on its mass *m*, specific heat capacity *c*, and the heat *Q* added to it.

$$\Delta T = \frac{Q}{c \cdot m}$$

$$\text{Temperature change}(°C) = \frac{\text{heat}(J)}{\text{specific heat capacity}\left(\frac{J}{kg°C}\right) \cdot \text{mass}(kg)}$$

The heat *Q* is related to power *P* and *t* time by:

$$Q = P \cdot t$$

The power dissipated by a resistor is related to the voltage *V* and the resistance *R* by:

$$P = \frac{V^2}{R}$$

Plugging this power formula into the heat formula gives:

$$Q = \frac{V^2}{R} \cdot t$$

Plugging this into the heat capacity formula gives:

$$\Delta T = \frac{\frac{V^2}{R} \cdot t}{c \cdot m}$$

This equation tells us we can expect the temperature to change linearly with time; rewriting it:

$$\Delta T = \frac{\frac{V^2}{R}}{c \cdot m} \cdot t$$

Plugging in and using $c_{water} = 4{,}186 \frac{J}{kg°C}$ and 15 minutes equals 900 s, 100 g = 0.1 kg.

$$\Delta T = \frac{\frac{(5V)^2}{10\Omega}}{4{,}186 \frac{J}{kg°C} \cdot 0.1 kg} \cdot 900\ s$$

$$\Delta T = 5.37°C$$

b) The results of the experiment

The water was heated in a paper cup with a lid to minimize the heat lost to the room. Measuring the temperature with the LM35, care was taken while wrapping the sensor in a plastic wrap not to get the leads wet.

The initial temperature of the water before heating was taken with the LM35. The measured voltage out of the LM35 was 220 mV. Converting this to temperature:

$$T = \frac{1°C}{10\ mV} \cdot V_{out}$$

$$T = \frac{1°C}{10\ \cancel{mV}} \cdot 220\ \cancel{mV}$$

$$T = 22°C$$

After 15 min of heating, the water was stirred and the measured voltage out of the LM35 was 270 mV. Converting this to temperature:

$$T = \frac{1°C}{10\ mV} \cdot V_{out}$$

$$T = \frac{1°C}{10\ \cancel{mV}} \cdot 270\ \cancel{mV}$$

$$T = 27°C$$

Calculating ΔT gives us:

$$\Delta T = 5.0°C$$

c) A comparison of the experimental results and theoretical calculation.

The theoretical calculation gave: $\Delta T = 5.37°C$

The experimental result gave: $\Delta T = 5.0°C$

The measured change in temperature is within 10% of the theoretical result. The difference could be attributed to some heat loss to the room.

17.4.2 Thermocouple

A thermocouple consists of two dissimilar pieces of wires connected to each other, producing a voltage proportional to the temperature. A voltage is developed across the junction and is proportional to the temperature of the junction. A reference junction is used to obtain measurements with respect to a known reference temperature (Figure 17.24).

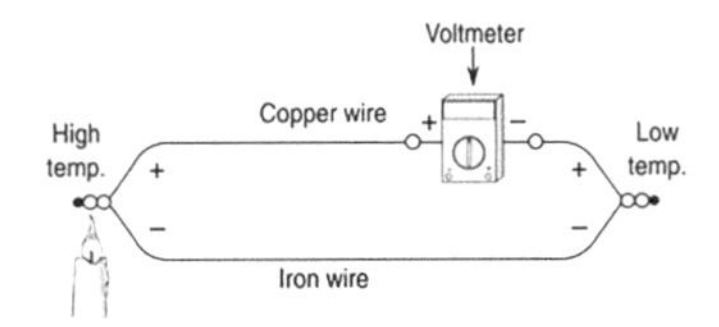

FIGURE 17.24 Thermocouple with reference junction.

Thermocouples are labeled with letters to indicate their composition. The metals that make up the thermocouple determines their sensitivity and range (Figure 17.25).

Some multimeters have thermocouples' inputs to measure temperature. The low-temperature reference junction is internal to the meter. There is an internal thermometer that measures the temperature of the isothermal junction and the meter compensates for this added voltage to display the correct temperature of the probe. Type K thermocouple probes are typically used because of their linear response (Figure 17.26).

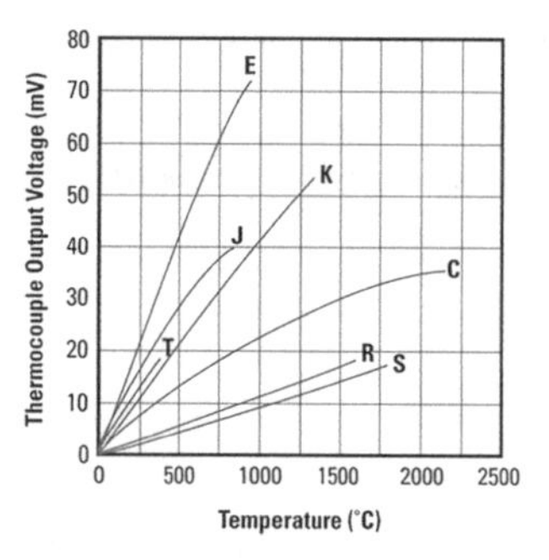

FIGURE 17.25 Thermocouples output voltage vs. temperature.

Leads at the same temperature
Volt Meter
Thermocouple Junction
Metal 1
Metal 2
Copper

FIGURE 17.26 Thermocouple with isothermal junction.

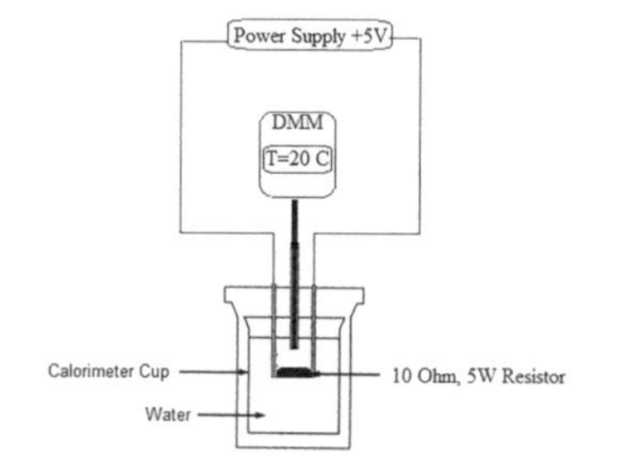

FIGURE 17.27 The calorimeter set up.

Example 17.12

Measure the specific heat capacity c of water using a thermocouple. We will measure c by heating 50 g of water in a calorimeter cup using a 10 Ω, 5 W resistor immersed in the water, and applying 5 V to it (Figure 17.27).

We found in the last example that the temperature change in a material depends on its specific heat capacity c.

$$\Delta T = \frac{\frac{V^2}{R}}{c \cdot m} \cdot t$$

Solving for c gives

$$c = \frac{\frac{V^2}{R}}{m} \cdot \frac{t}{\Delta T}$$

The time t in this equation is considered as the interval of time needed to raise the temperature by ΔT. We can replace t by Δt and obtain:

$$c = \frac{\frac{V^2}{R}}{m} \cdot \frac{\Delta t}{\Delta T}$$

When we plot the data, temperature vs. time, we can fit the data, and from the measured slope $\frac{\Delta T}{\Delta t}$ we find c. The results of the experiment are shown in (Figure 17.28).

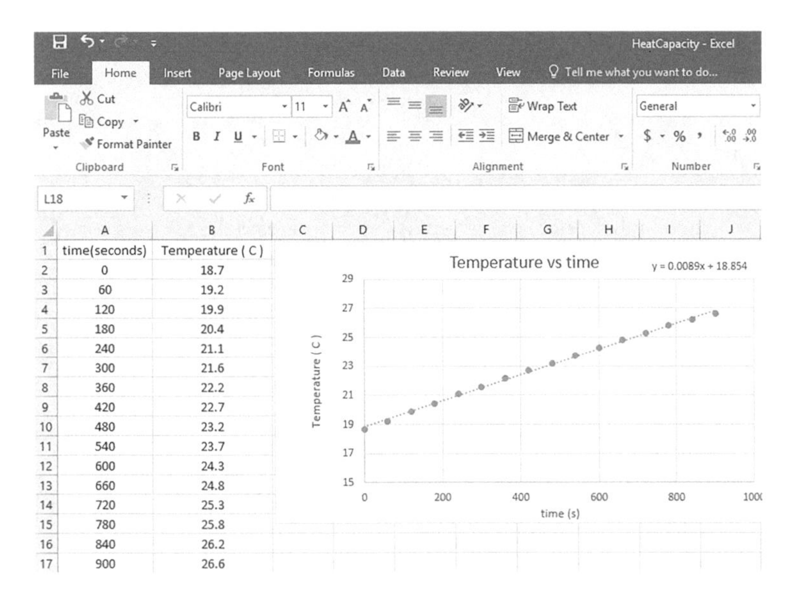

	A	B
1	time(seconds)	Temperature (C)
2	0	18.7
3	60	19.2
4	120	19.9
5	180	20.4
6	240	21.1
7	300	21.6
8	360	22.2
9	420	22.7
10	480	23.2
11	540	23.7
12	600	24.3
13	660	24.8
14	720	25.3
15	780	25.8
16	840	26.2
17	900	26.6

FIGURE 17.28 Temperature vs. time of water heated.

From the linear fit, the slope $= \frac{\Delta T}{\Delta t} = 0.0089\frac{°\text{C}}{\text{s}}$. Plugging this into the heat capacity formula:

$$c = \frac{\frac{V^2}{R}}{m} \cdot \frac{\Delta t}{\Delta T}$$

$$c = \frac{\frac{V^2}{R}}{m} \cdot \frac{1}{\frac{\Delta T}{\Delta t}}$$

Plugging in:

$$c = \frac{\frac{(5\text{V})^2}{10\text{ Ohms}}}{0.050\text{ kg}} \cdot \frac{1}{0.0089\frac{°\text{C}}{\text{s}}}$$

$$c = 5{,}618\ \frac{\text{J}}{\text{kg}°\text{C}}$$

Comparing this to the accepted value of $c_{\text{water}} = 4{,}186\ \frac{\text{J}}{\text{kg}°\text{C}}$ and calculating a % difference:

$$\%\text{ difference} = \frac{4{,}186\ \frac{\text{J}}{\text{kg}°\text{C}} - 5{,}618\ \frac{\text{J}}{\text{kg}°\text{C}}}{4{,}186\ \frac{\text{J}}{\text{kg}°\text{C}}} \times 100$$

$$\%\ \text{difference} = 34\%$$

This difference can be attributed to heat loss to the room, and the heat absorbed by the heater, and the container itself.

FIGURE 17.29 A magnifying glass concentrating the light from the sun.

Example 17.13

A magnifying glass of radius 5 cm is arranged to concentrate the sun's rays onto a spot the size of a thermocouple of radius 1 mm and mass 0.5 g. Using a multimeter with a *K*-type thermocouple what temperature can we expect if heated for 10 s? (Figure 17.29)

The intensity of sunlight is about 1 kW/m^2 at the surface of the earth. The intensity or the amount of power delivered by light per unit area is given by:

$$I = \frac{P}{A}$$

$$\text{Intensity}\left(\text{W/m}^2\right) = \frac{\text{Power}\left(\text{W}\right)}{\text{Area}\left(\text{m}^2\right)}$$

The magnifying glass will collect an amount of power given by:

$$P_{\text{collected}} = I_{\text{sun}} \cdot A_{\text{lens}}$$

$$P_{\text{collected}} = 1000\ \frac{\text{W}}{\text{m}^2} \cdot \pi \cdot \left(0.05\ \text{m}\right)^2$$

$$P_{\text{collected}} = 7.85\ \text{W}$$

This power will now be focused on a spot of smaller area to obtain a new intensity:

$$I_{\text{spot}} = \frac{P_{\text{collected}}}{A_{\text{spot}}}$$

Plugging in the numbers:

$$I_{\text{spot}} = \frac{7.85\ \text{W}}{\pi \cdot \left(0.001\ \text{m}\right)^2}$$

$$I_{\text{spot}} = 2.5 \times 10^6\ \frac{\text{W}}{\text{m}^2} \leftarrow \text{Wow!}$$

The thermocouple sensor of area A_{sensor} will receive an amount of power:

$$P_{\text{sensor}} = I_{\text{spot}} \cdot A_{\text{sensor}}$$

Plugging in the numbers:

$$P_{\text{sensor}} = 2.5 \times 10^6\ \frac{\text{W}}{\text{m}^2} \cdot \pi \cdot \left(0.001\ \text{m}\right)^2$$

$$P_{\text{sensor}} = 7.85\ \text{W}$$

$$\text{We see that } P_{\text{sensor}} = P_{\text{collected}} \text{ because } A_{\text{sensor}} = A_{\text{spot}}$$

Now let's calculate the temperature change in the sensor. A Google search shows that the specific heat capacity of a *K*-type thermocouple is $c = 0.523$ J/(g°C).

Using the heat capacity equation:

$$\Delta T = \frac{Q}{c \cdot m}$$

$$\Delta T = \frac{P \cdot t}{c \cdot m}$$

$$\Delta T = \frac{7.85\ \text{W} \cdot 10\ \text{s}}{0.523 \frac{\text{J}}{\text{g} \cdot {}^\circ\text{C}} \cdot 0.5\ \text{g}}$$

$$\Delta T = 300.2^\circ\text{C}$$

No wonder you can burn things with a magnifying glass!

17.5 MEASURING MAGNETIC FIELDS

17.5.1 Hall Probe

A Hall probe is the most common method of measuring magnetic fields. A Hall probe consists of a semiconductor material with a current running through it. If a magnetic field B penetrates the Hall probe, the current flowing inside the probe will be pushed to one side, as shown (Figure 17.30). This will produce a voltage across the probe because the charge will be concentrated on one side. This voltage is proportional to the magnetic field strength.

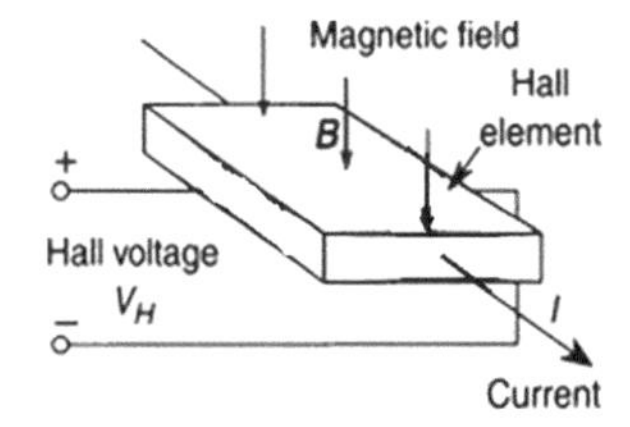

FIGURE 17.30 A Hall probe responds to a magnetic field by generating a voltage proportional to the field strength.

Example 17.14

The A1301 Hall-effect sensor can measure a magnetic field, B, with a sensitivity of 2.5 mV/gauss, when supplied with 5V. The sensor puts out half the supply voltage even when there is no magnetic field present. In the presence of a magnetic field the sensor outputs an additional voltage of 2.5 mV/gauss. For example, if the supply voltage is 5 V and the sensor sits in a magnetic field of 2 gauss (Figure 17.31), then the output voltage is:

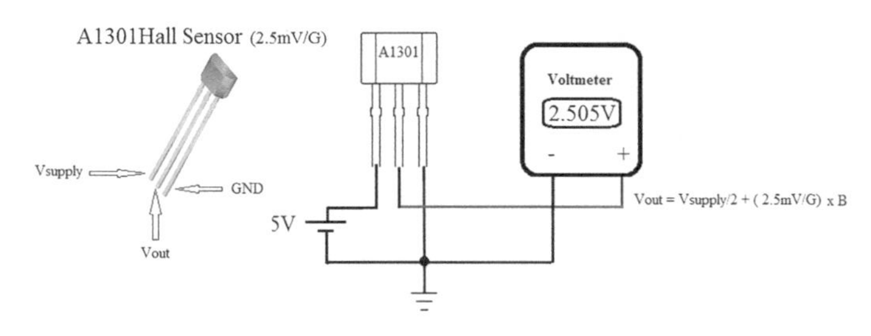

FIGURE 17.31 Wiring diagram of the A1301 Hall sensor.

$$V_{out} = \frac{V_{supply}}{2} + \left(2.5\,\frac{\text{mV}}{\text{G}}\right)\cdot B$$

$$V_{out} = \frac{5\,\text{V}}{2} + \left(2.5\times 10^{-3}\,\frac{\text{V}}{\cancel{\text{G}}}\right)\cdot 2\cancel{\text{G}}$$

$$V_{out} = 2.505\,\text{V}$$

This 5 mV is a very small change in the constant offset of $V_{supply}/2$, and therefore should be accompanied by some circuitry to eliminate the offset and amplify the remainder.

17.6 MICROCONTROLLERS AND DATA ACQUISITION

Data acquisition equipment for physics can be quite expensive. As an alternative, data can be acquired using inexpensive analog and digital sensors and interfaced to a low-cost microcontroller. A microcontroller is a small computer on a single integrated circuit chip with programmable inputs and outputs.

Analog sensors are sensors with outputs that change continuously with changing inputs. A digital sensor is an electronic sensor where data conversion and data transmission are done digitally and sent to a computer or microcontroller to be recorded, plotted, and analyzed. Examples of acquiring data from sensors, plotting and analyzing the data will be discussed in this section. This discussion is not meant as a complete guide but as a starting point with links to references (Figure 17.32).

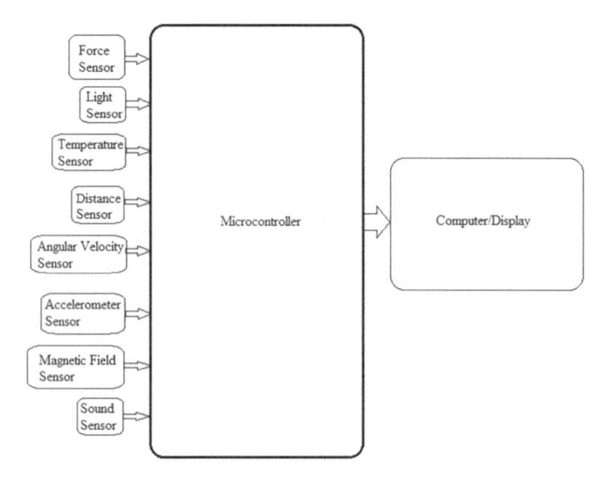

FIGURE 17.32 Data acquisition using a microcontroller.

17.6.1 The Arduino Uno Microcontroller

The Arduino Uno is a microcontroller with many digital inputs/outputs and six analog inputs with 10 bit resolution. It can be purchased for as little as $6, can be interfaced to a computer through the universal serial bus (USB). The most popular way to program it is in a language called C. There are also graphical ways of programming it similar to Scratch for the programming adverse user. The software is free and available from www.arduino.cc/en/Main/Software (Figures 17.33 and 17.34).

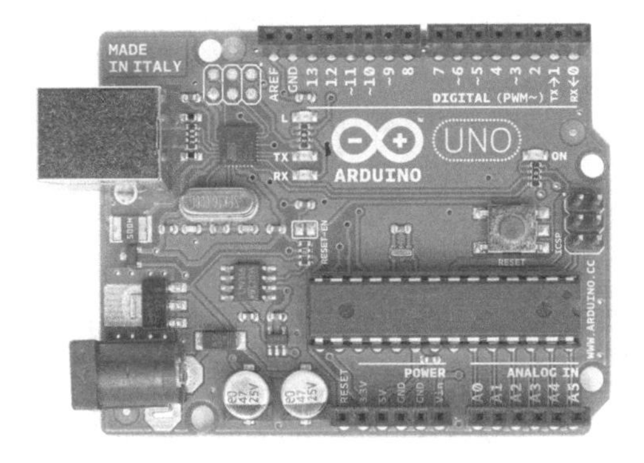

FIGURE 17.33 The Uno microcontroller.

FIGURE 17.34 Interfacing the Uno to a laptop for programming and data display.

The Integrated Development Environment, IDE, is available free from www.arduino.cc/en/Main/Software.

Contained in the IDE are many examples of code for almost any sensor you can purchase. If your device is not listed, a library for that device may be added to the IDE. The IDE can display or plot the data you collect with boards connected to it. See the figures below on how to open an example, upload it to an UNO board, and display the data on a computer connected to the UNO (Figures 17.35 and 17.36).

To view the data or the plot click on Tools > Serial Monitor or Serial Plotter (Figure 17.37). The data can be copied from the serial monitor (Figure 17.38) and then pasted into a spreadsheet for analysis.

17.6.2 Analog Sensors and Analog to Digital Conversion

The Uno has six analog input pins. These inputs are connected to an **ADC**, analog to digital converter. The ADC takes a continuous analog signal, and divides it up into finite number of levels, otherwise the memory would be overloaded. In case of Uno, input voltages from 0 to 5 V are converted to

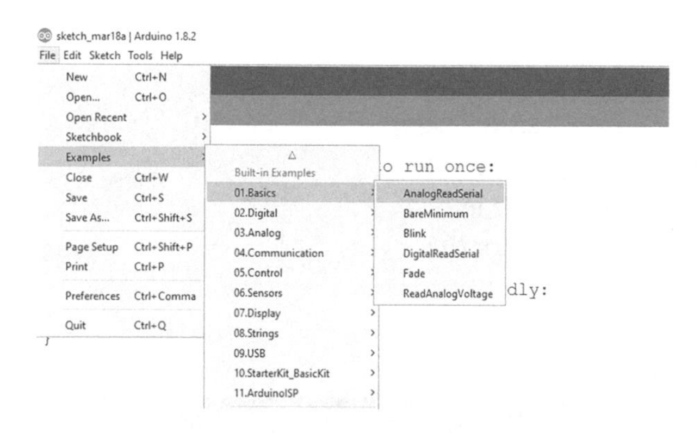

FIGURE 17.35 To find examples of code, click on File>Examples, then the category of interest.

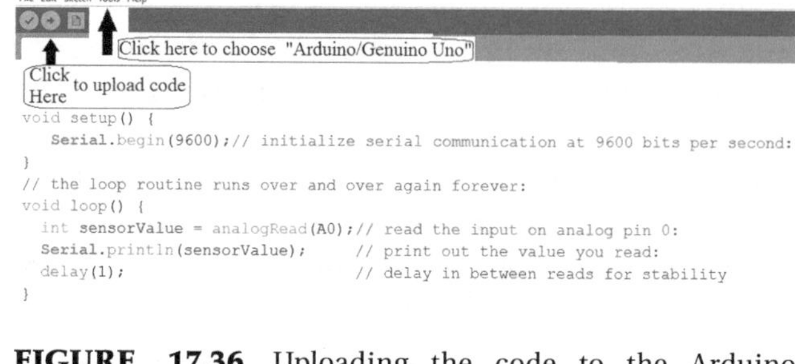

FIGURE 17.36 Uploading the code to the Arduino microcontroller.

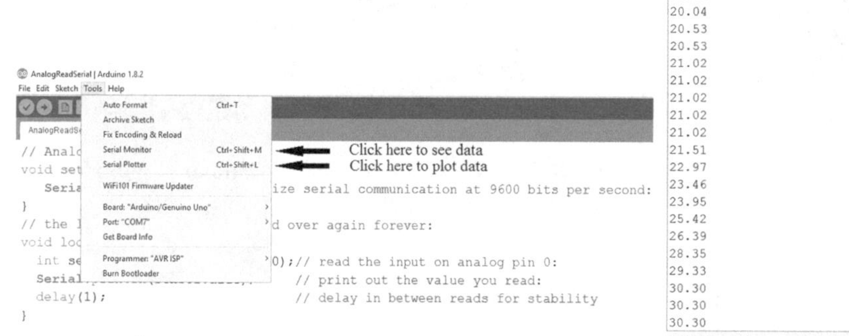

FIGURE 17.37 Serial Monitor and Plotter.

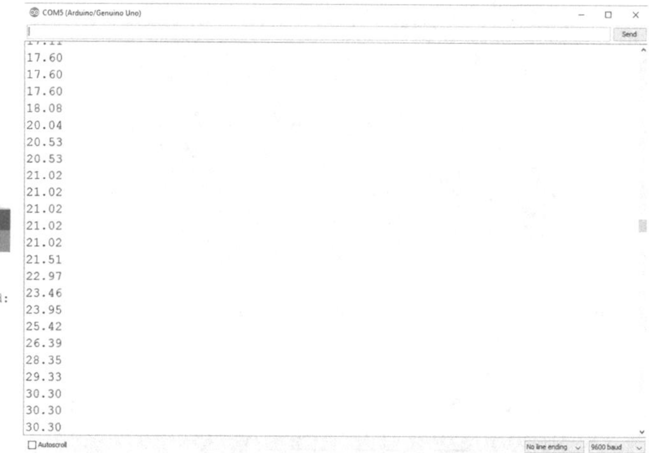

FIGURE 17.38 Data from Serial monitor.

numbers between 0 and 1,023, 0 V = 0 and 5 V = 1,023. **Resolution** is the smallest voltage change that can be measured. This Uno can measure a voltage change at input as small as 5 V/1,023 or ~4.9 mV (Figure 17.39).

For input voltages smaller than 1.1 V, there is a way to increase the resolution to 1.1 V/1,023 or ~1 mV, using a command in the code called "analogReference (INTERNAL)".

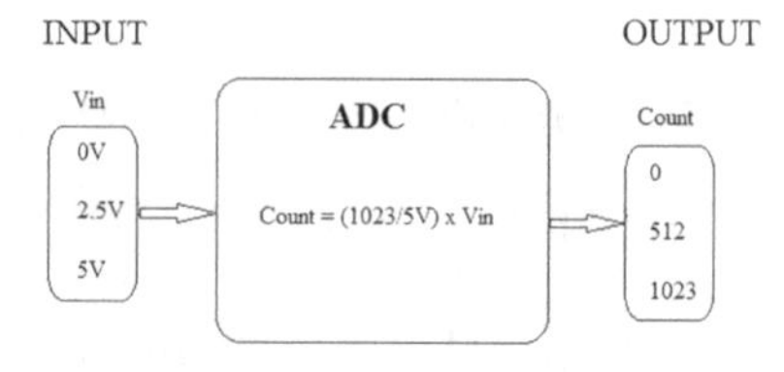

FIGURE 17.39 A 10 bit ADC divides a signal up in 2^{10} or 1,024 ways.

Example 17.15

Temperature can be measured with a LM35 temperature sensor and an Arduino Uno.

LM35 temperature sensor, which outputs 10 mV/°C, is monitored with an Arduino Uno.

The LM35 temperature sensor outputs 10 mV/°C. Assuming a voltage resolution of 4.9 mV, the temperature resolution will be:

$$\text{Temperature resolution} = \frac{4.9\text{mV}}{10\frac{\text{mV}}{°\text{C}}} = 0.49°\text{C}$$

Let us now find out how the temperature of the sensor measured will be converted to counts:

The sensitivity of the LM35 is 0.01 V/°C.

The resolution of the ADC is 1,023 counts/5 V.

Combining these two conversions: $\left(0.01\frac{V}{°\text{C}}\right)\cdot\left(\frac{1{,}023\text{ counts}}{5\,V}\right) = 2.046\frac{\text{counts}}{°\text{C}}$

This conversion says that for every degree the sensor measures, the Uno will record 2.046 counts.

To read an analog voltage the command analogRead(Input Pin Number) is used. analogRead will convert the input voltage to a count between 0->1,023. The temperature can now be calculated using:

$$T = \frac{\text{analog Read(Input Pin Number)}}{2.046\frac{\text{counts}}{°\text{C}}}$$

To measure the LM35, connect up the V_s to 5 V, GND to GND, and V_{out} to A_0 (Figure 17.40).

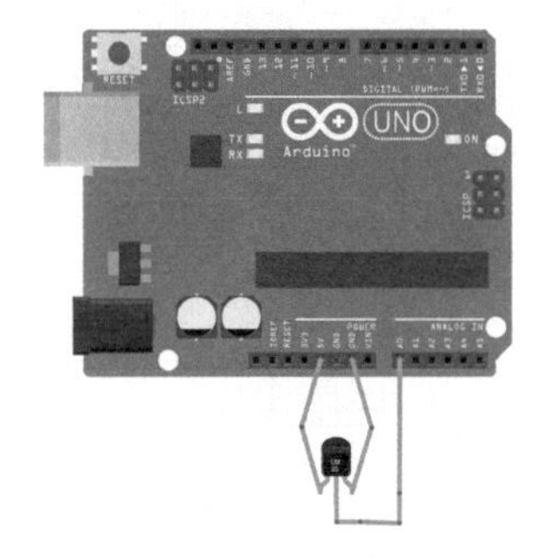

FIGURE 17.40 Connecting the sensor to the Arduino. This image was created with Fritzing.

Modify the code in an example to read temperature by following the steps:

1. Open AnalogReadSerial by clicking on File > Examples > Basics > AnalogReadSerial.
2. Change: Serial.println(sensorValue) to Serial.println(sensorValue/2.046) (Figure 17.41).

```
sketch_mar19c | Arduino 1.8.2
File Edit Sketch Tools Help
sketch_mar19c §

// the setup routine runs once when you press reset:
void setup() {
  // initialize serial communication at 9600 bits per second:
  Serial.begin(9600);
}

// the loop routine runs over and over again forever:
void loop() {
  // read the input on analog pin 0:
  int sensorValue = analogRead(A0);
  // print out the value you read:
  Serial.println(sensorValue/2.046);   <== Lines Altered!
  delay(1000);                          <==
}
```

FIGURE 17.41 Code to read LM35 every 1 s and display the data.

3. The delay(1) line sets the rate to take data every 1 mS. Change to delay(1,000) to take data every 1 s (Figure 17.41).
4. Now upload the code to the Uno and click on: Tools > Serial Monitor to see the data or Tools > Serial Plotter to plot the data (Figure 17.37).
5. Highlight and copy the data in Serial Monitor and paste it into a spreadsheet to view and analyze (Figure 17.38).

17.6.3 Digital Sensors

A digital sensor is an electronic sensor where data conversion and data transmission are done digitally and sent to a computer or microcontroller to be recorded, plotted, and analyzed. These sensors avoid many of the problems that analog sensors have in terms of noise pickup from neighboring inputs and any nonlinearities of the sensor's response. Digital sensors communicate to the microcontroller through one or two wires depending on the communication protocol they follow (Figures 17.42–17.45).

Four examples of digital sensors wired to an Uno are:

1. DHT11-Temperature/Humidity sensor. Range 0°C–50°C, 20%–80% Humidity.
2. VL53LOX- Time of flight distance sensor. Range 30–1,000 mm.
3. HC-SR04- Ultrasonic distance sensor. Range 2–400 cm.
4. ADXL345 - Accelerometer. Range adjustable from 2 to 16 g.

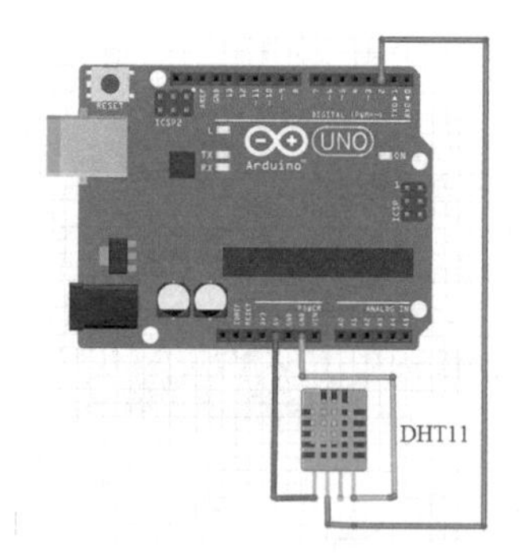

FIGURE 17.42 The DHT11 temperature/humidity sensor. This image was created with Fritzing.

A tutorial for these sensors can be found online that explains how to connect up the sensor to a microcontroller along with a library that must be installed with example code. For example, to use the VL53LOX distance sensor the library must be downloaded from adafruit.com. Then the library must be installed into the Arduino IDE. Once installed, an example of the code can be found.

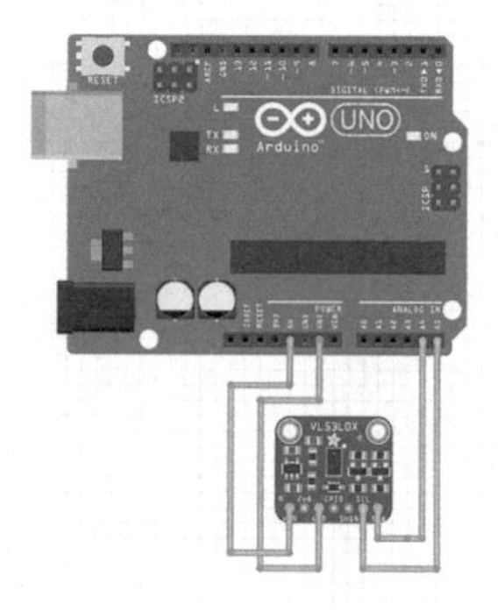

FIGURE 17.43 The VL53LOX distance sensor. This image was created with Fritzing.

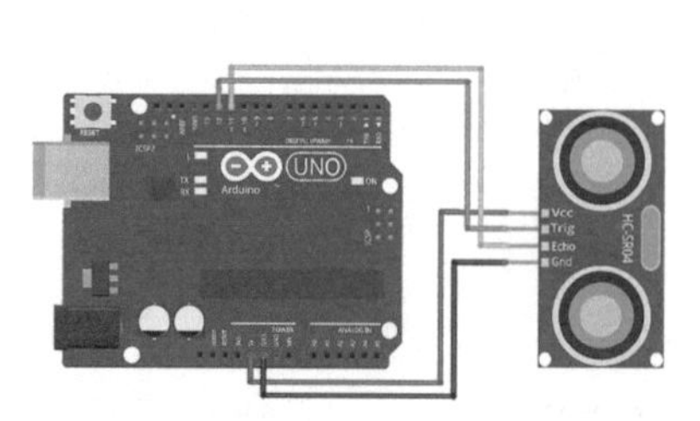

FIGURE 17.44 The HC-SR04 distance sensor. This image was created with Fritzing.

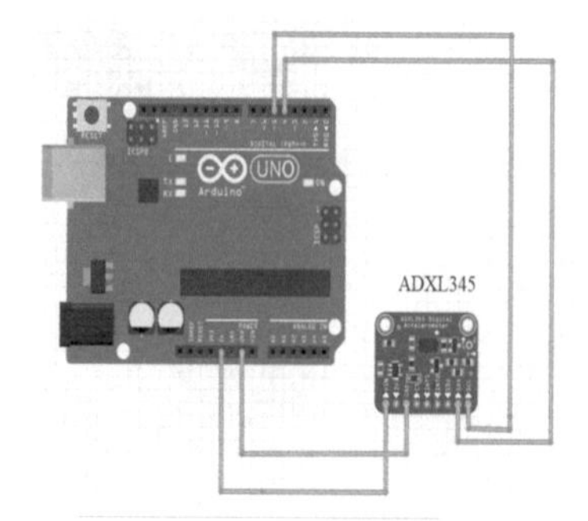

FIGURE 17.45 The ADXL345 accelerometer. This image was created with Fritzing.

Example 17.16

Monitor the motion of a mass oscillating up and down on a spring, using a VL53LOX time of flight distance sensor (Figure 17.46).

1. First download Adafruit_VL53LOX zip file from adafruit.com.
2. Click Sketch > Include Library > Add Zip library, then find library and click open (Figure 17.47).
3. Click Sketch > Include Library > Manage Library (Figure 17.48).
   Opening manage libraries: Click Sketch > Include Library > Manage Library.
4. Type in VL53LOX into the search window and click install (Figure 17.49).
5. Now open up File-> Examples-> Adafruit_VL53LOX-> vl53l0x and upload to your board (Figure 17.50).
6. Upload the code and open up the serial monitor window at 115,200 speed to see the data (Figure 17.51).

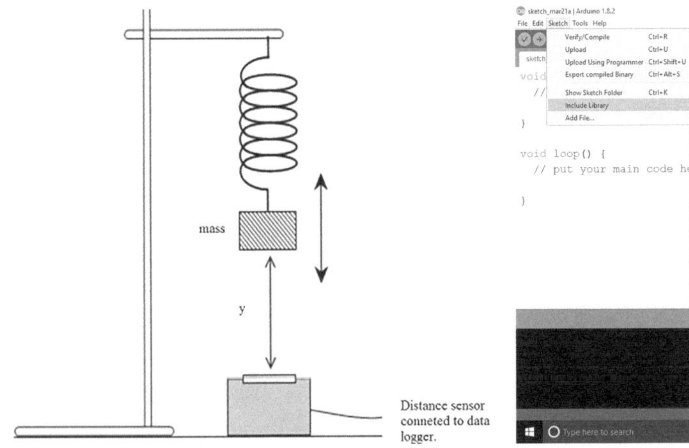

FIGURE 17.46 An oscillating mass on a spring.

FIGURE 17.47 Click Sketch > Include Library > Add Zip library, then find library and click open.

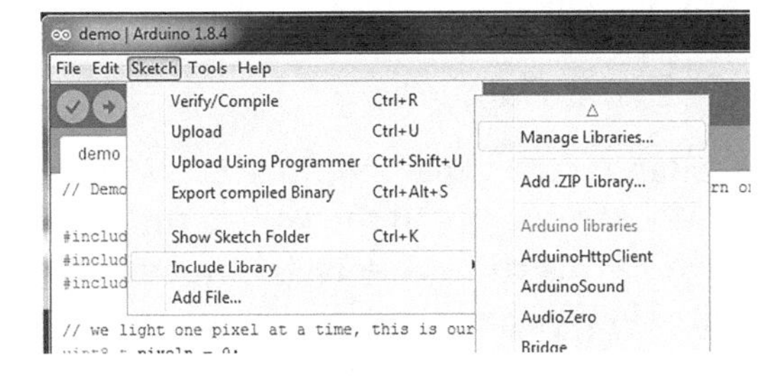

FIGURE 17.48 Click Sketch > Include library > manage library.

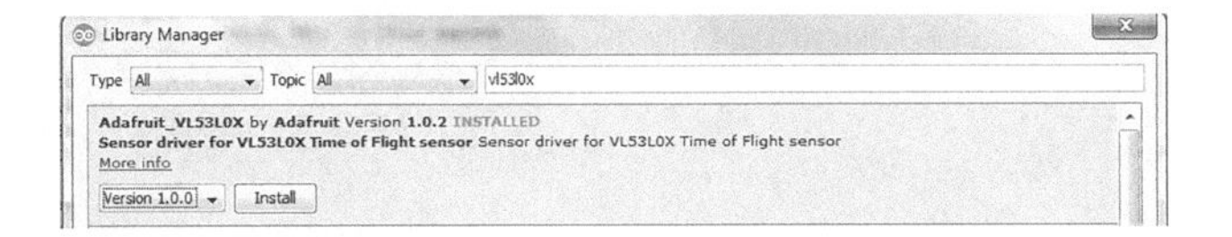

FIGURE 17.49 Type in vl53lox and click install.

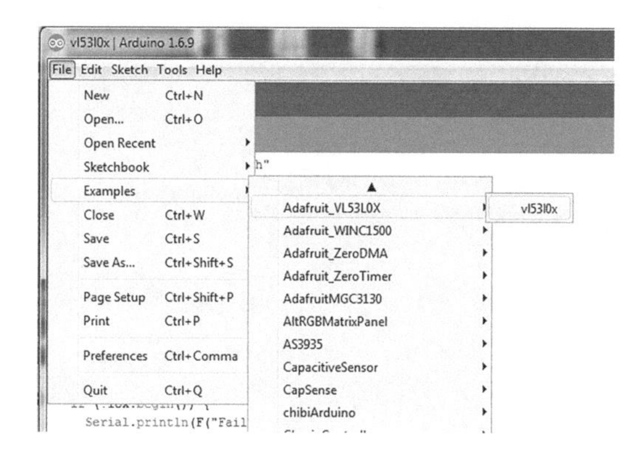

FIGURE 17.50 Click on: File > Examples > Adafruit_ VL53LOX > vl53l0x and upload to your board.

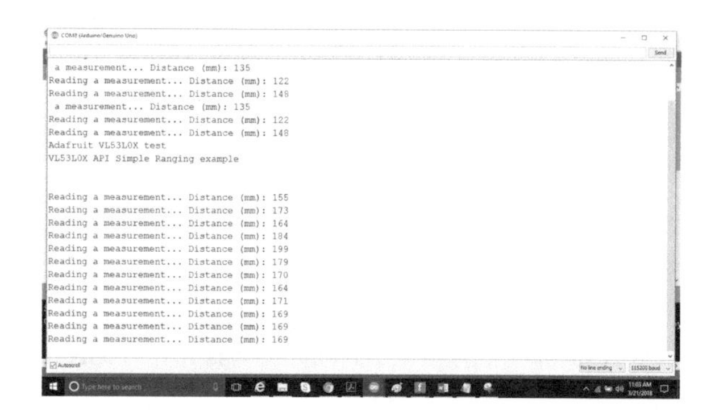

FIGURE 17.51 Serial monitor displaying data.

7. To see data plotted, open up the serial plotter, Tools > Serial Plotter, window at 115,200 speed (Figure 17.52).

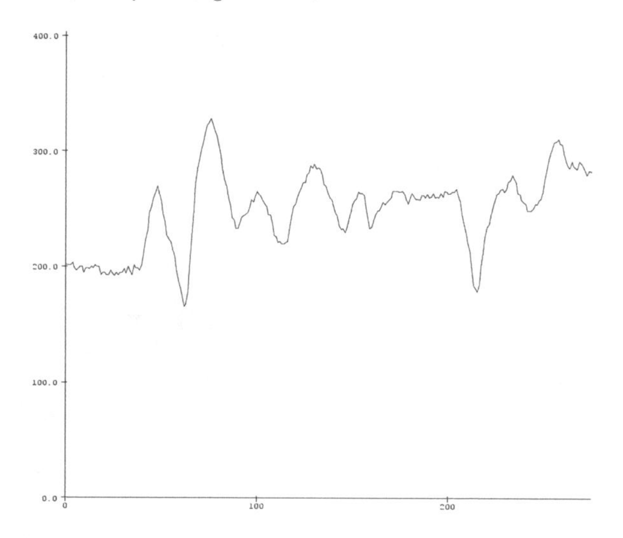

FIGURE 17.52 Serial plotter plot displaying data.

17.6.4 Sampling Data at a Fixed Rate

Measuring something that changes in time requires knowing when the measurement was taken. We can specify in the code a time interval between measurements to give us this time. A delay(x) function can be used for measurements that are slowly changing.

Example 17.17

Measuring every one second a slowly changing temperature.

```
void loop() {
  T= readTemperature();
  delay(1000); //<-- wait 1000mS
  }
```

Quantities that need to be measured more often, for example, a mass oscillating on the end of a spring, require a fast sample rate at precise intervals of measurement. We can use a built-in function called micros() that when called gives the current value of a clock in microseconds that is continuously running. By specifying that a measurement can only take place after a specified time interval has elapsed, we can record both the measurement and when it took place.

Example 17.18

Measuring something every 0.1 s (Figure 17.53).

The line "if (timeNow > lastSampleTime + samplePeriod)" means that if the current time on the clock is greater than the last time on the clock plus a sampling interval of time, then execute a measurement.

Example 17.19

Determine the acceleration due to gravity using a pendulum (Figure 17.54).

The period T of a pendulum with arm length l is given by the formula:

$$T = 2\pi\sqrt{\frac{l}{g}}$$

This formula can be rewritten as $\sqrt{\frac{g}{l}} = \frac{2\pi}{T} = \omega.$

Solving for g gives: $g = l \cdot \omega^2$

To find g, a pendulum's motion is monitored with a VL53LOX distance sensor from adafruit.com and an Arduino Uno (Figure 17.55).

From the data sheet, the Adafruit VL53LOX Time of flight Micro-LIDAR Distance Sensor Breakout sensor requires 33 ms to acquire a measurement. We cannot therefore sample faster than it. Sample times were experimented with and 0.2 s gave good results. The code from Example 17.16 was modified to take data every 200,000 μs or every 0.2 s as shown below:

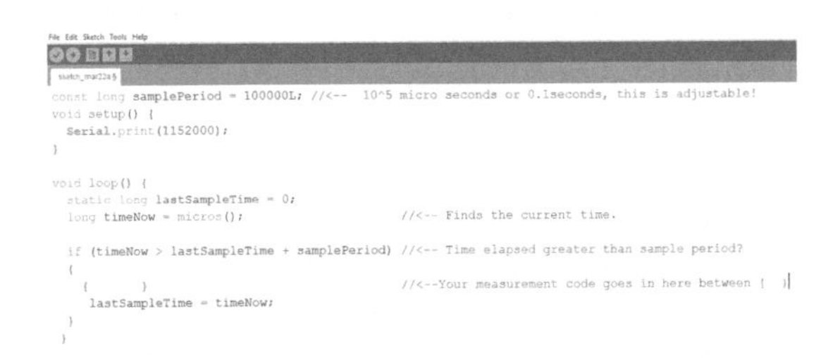

FIGURE 17.53 Arduino code to read the VL53LOX.

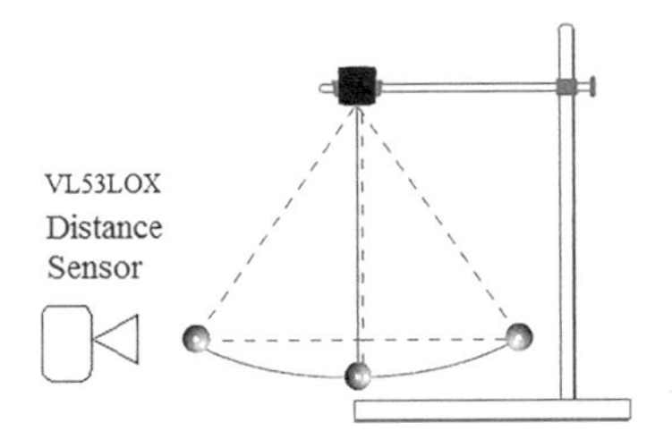

FIGURE 17.54 The VL53LOX sensor monitoring a pendulum.

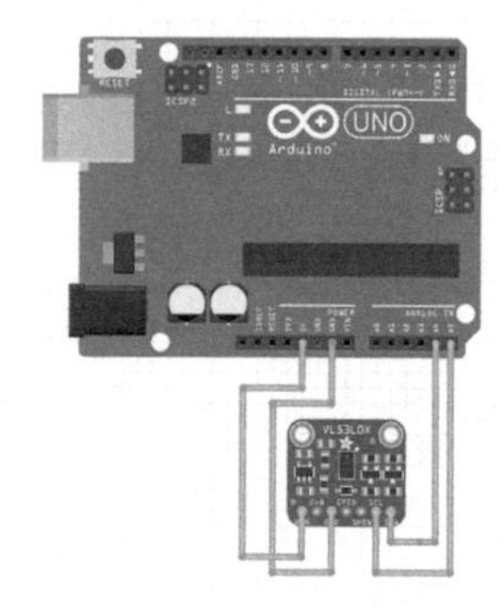

FIGURE 17.55 Wiring diagram of the sensor connected to the Arduino. This image was created with Fritzing.

```
#include "Adafruit_VL53L0X.h"
const long samplePeriod = 200000L; // micro seconds
Adafruit_VL53LOX lox = Adafruit_VL53LOX();
void setup() {

  Serial.begin(115200);
    // wait until serial port opens for native USB devices
  while (! Serial) {
      delay(1);
  }
     Serial.println("Adafruit VL53LOX test");
  if (!lox.begin()) {
    Serial.println(F("Failed to boot VL53LOX"));
      while(1);
            }
  Serial.println(F("VL53LOX API Simple Ranging
example\n\n"));
}

void loop() {
  VL53LOX_RangingMeasurementData_t measure;
     lox.rangingTest(&measure, false); // pass in 'true'
to get debug data printout!

      static long lastSampleTime = 0;
      long timeNow = micros();
      if (timeNow > lastSampleTime + samplePeriod)
      {
       if (measure.RangeStatus!= 4) { // phase failures
have incorrect data
            Serial.println(measure.RangeMilliMeter);
  } else {
      Serial.println(" out of range ");
    }
     lastSampleTime = timeNow;
  }
}
```

Data was copied from Serial Monitor and pasted into Excel for analysis (Figure 17.56).

The data is displayed in Figure 17.56, looks like a sinusoidal wave as in Example 17.8, so we will attempt to fit the data to the function:

$$y = A \cdot \sin(w \cdot t + \text{phi}) + \text{offset}$$

Again, as in Example 17.8 to find w:

1. The data is entered on columns 1 and 2 in Excel.
2. The fitting function is entered in column 3. The parameters to be adjusted to match the data are (offset, *A*, *w*, and phi).
3. The initial values of these parameters are entered in column G2–G5.
4. The difference squared between the data and the fitting function is given in column D.

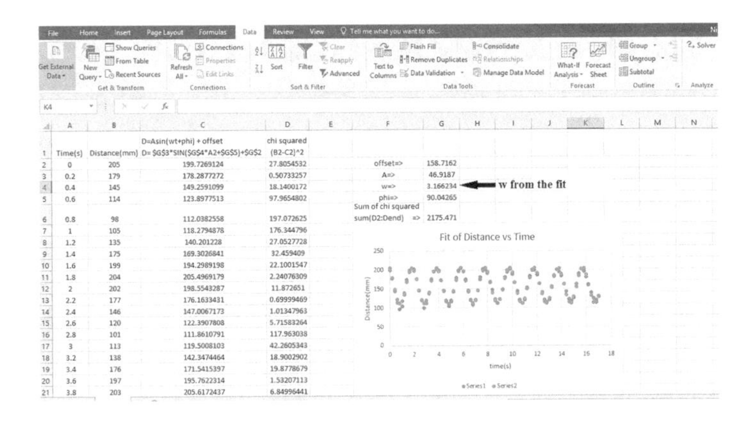

FIGURE 17.56 A plot of the data from the VL553LOX

Now preform a Scatter Plot of columns 1–3 to see the data and the fitting equation. You can now adjust the parameters in column G2–G5 to find good starting parameters for Solver.

Once the data and the fitting equations and parameters are entered and plotted, open Solver.

5. Click on the Data tab.
6. Click on Solver.
7. Enter the location of the Sum of chi squared in Set Objective, in this case $G$6.
8. Click on Min.
9. Enter the location of the fitting parameters, in By Changing Variable Cells, in this case $G$2:$G$5.
10. Click Solve at the bottom of the screen.
11. Look at the fit and how the values in $G$2:$G$5 have changed.
12. If the fit is not good, modify the parameters in $G$2:$G$5 again and run Solver again till the fit looks good.
13. The values at $G$2:$G$5 are now the values (offset, *A*, *w*, and phi) in the equation,

$$y = A \cdot \sin(w \cdot t + \text{phi}) + \text{offset}$$

From the fit, we find $\boldsymbol{w} = \mathbf{3.166}$, this is the angular velocity of the swinging pendulum, see Figure 17.56.

Let's now use this to find the acceleration due to gravity.

$$g = l \cdot \omega^2$$

$$g = (1\,\text{m}) \cdot \left(3.166 \ \text{rad/s}^2\right)^2$$

$$g = 10.02 \, \frac{\text{m}}{\text{s}^2}$$

Let's compare this to the actual value = 9.8 m/s^2, with a percent difference calculation:

$$\% \ \text{difference} = \frac{10.02 \, \frac{\text{m}}{\text{s}^2} - 9.8 \, \frac{\text{m}}{\text{s}^2}}{9.8 \, \frac{\text{m}}{\text{s}^2}} \times 100$$

$$\% \ \text{difference} = 2.2\% \leftarrow \text{Not too bad!}$$

17.7 THE NodeMcu MICROCONTROLLER

The NodeMcu is a microcontroller with an onboard WiFi chip. It can be programmed and monitored with the Arduino software just like an Arduino board, but it also has the capability of connecting to the internet using its WiFi chip, the ESP8266. With the NodeMcu you can participate with the Internet of Things (IOT), that is, sensors and devices can be monitored and controlled remotely from anywhere in the world with a web browser. There are many online tutorials on how to connect up a NodeMcu and take and post data to a website or phone app (Figure 17.57).

Example 17.20

Measure the motion of a cart moving down an inclined plane using an ultrasonic distance sensor, the HC-SR04 and the NodeMcu (Figure 17.58).

The cart has two forces acting upon it in the direction of travel, that is, gravity and friction in the opposite direction. Assume the wheels of the cart have a negligible moment of inertia. The force on the cart in the direction of travel is:

$$F_{\text{Total}} = F_{\text{gravity}} - F_{\text{friction}}$$

$$F_{\text{Total}} = m \cdot g \cdot \sin(\theta) - F_{\text{friction}}$$

$$a = \frac{F_{\text{Total}}}{m}$$

$$a = \frac{m \cdot g \cdot \sin(\theta) - F_{\text{friction}}}{m}$$

$$a = g \cdot \sin(\theta) - \frac{F_{\text{friction}}}{m}$$

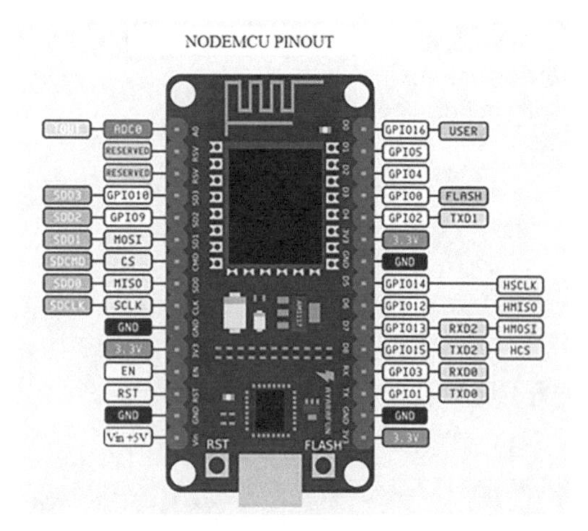

FIGURE 17.57 The NodeMcu is a microcontroller.

HC-SR04 UltraSonic Distance Sensor

F_{friction}

Cart

mgsinθ

h

mg

d

θ

FIGURE 17.58 Diagram of cart rolling down an inclined plane.

The acceleration is therefore a constant and recall that for a constant acceleration the position of an object obeys the following:

$$y = v_i t + \frac{1}{2} a t^2 + y_i$$

To measure the distance, the HC-SR04 distance sensor was wired to the NodeMcu as shown in Figure 17.59.

To use the NodeMcu with the Arduino IDE, the NodeMcu library has to be installed, it can be found online. The board has to be declared under Tools along with the port the NodeMcu is connected to. The following code was entered in Figure 17.60. The code was written to take data every 0.2 s.

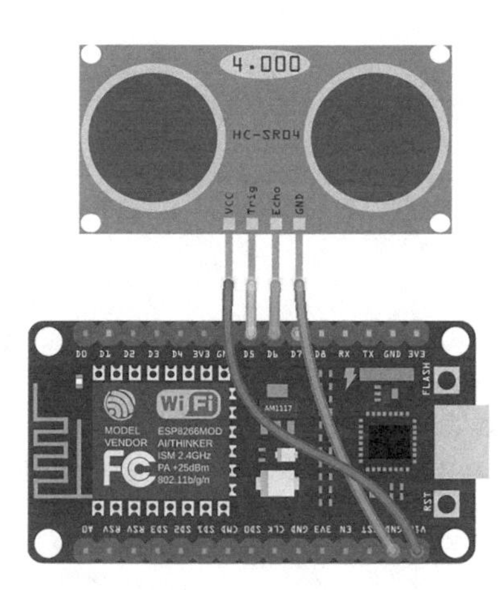

FIGURE 17.59 Wiring diagram of the NodeMcu connected to the HC-Sr04 ultrasonic range finder. This image was created with Fritzing.

```
File Edit Sketch Tools Help
NODEmcuHCSR04Ultrasonic §
//NODEMCU connected to HC-SR04 ULTRASONIC  HC-SR04 requires Vcc to be 5V
#define TRIGGERPIN D5
#define ECHOPIN    D6
const long samplePeriod = 200000L; // micro seconds or 0.2 seconds
void setup(){
Serial.begin(115200);
pinMode(TRIGGERPIN, OUTPUT);
pinMode(ECHOPIN, INPUT);
}
void loop()
{
  long duration, distance;
  static long lastSampleTime = 0;
  long timeNow = micros();
  if (timeNow > lastSampleTime + samplePeriod){
  digitalWrite(TRIGGERPIN, LOW);
  delayMicroseconds(3);
  digitalWrite(TRIGGERPIN, HIGH);
  delayMicroseconds(12);
  digitalWrite(TRIGGERPIN, LOW);
  duration = pulseIn(ECHOPIN, HIGH);
  distance = (duration/2) / 29.1;
  Serial.println(distance);
  lastSampleTime = timeNow;
  }
}
```

FIGURE 17.60 Code use to monitor the HC-Sr04 ultrasonic range finder.

The cart was released from rest from an initial distance from the sensor of 4 cm. The data along with a fit is displayed in Figure 17.61.

We can see the data fits a second order polynomial well, with the equation converted to meters instead of centimeters:

$$y = 0.15532t^2 + 0.25638t + 0.027706$$

Comparing this fit to our theoretical model:

$$y = at^2 + v_i t + y_i$$

We can identify the measured acceleration, velocity, and initial position as:

$$a = 0.15532 \text{ m/s}^2$$

$$v_i = 0.25638 \text{ m/s}$$

$$y_i = 0.027706 \text{ m}$$

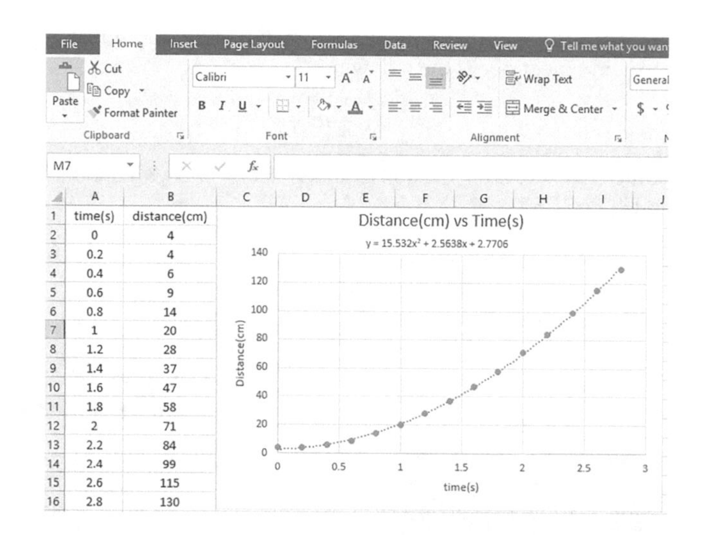

time(s)	distance(cm)
0	4
0.2	4
0.4	6
0.6	9
0.8	14
1	20
1.2	28
1.4	37
1.6	47
1.8	58
2	71
2.2	84
2.4	99
2.6	115
2.8	130

FIGURE 17.61 Plot of the data from the distance sensor module.

17.8 SUMMARY

Microcontrollers offer an inexpensive way to acquire data from analog and digital sensors. There are many online tutorials on how to get started using microcontrollers and interfacing them to sensors. Once the data from an experiment is acquired, analysis can be done with spreadsheets such as Excel.

17.8.1 Lab Ideas

Following the examples in this chapter, construct a data acquisition system to:

1. Measure the motion of a cart moving down an inclined plane at various inclines.
2. Measure the period of a pendulum of different lengths.
3. Measure the period of an oscillating mass on a spring using different masses.
4. Measure the heat capacities of different metals.
5. Measure the temperature and light level outside over a 24 h period.
6. Measure the acceleration of and the force on a mass and verify Newton's second law.
7. Measure the acceleration due to gravity of a falling object.
8. Measure the intensity of a light source as a function of distance.
9. Measure the magnetic field of a refrigerator magnet.
10. Measure the temperature of focused sunlight with a magnifying glass.

18

Smartphones and Physics

Smartphones contain a suite of sensors that allow one to measure quantities in physics without the use of any other lab equipment. Typical sensors contained onboard a smartphone are a accelerometer, gyroscope, magnetometer, GPS, barometer, microphone, and a light sensor. There are many free apps that can access these sensors, record data, and create downloadable spreadsheets. This chapter will focus on an app called **phyphox** that's available for free for both iOS and Android phones. Along with phyphox we will look at a free video analysis software package called **Tracker**, which can analyze the motion of an object in a video taken by a phone.

FIGURE 18.1 The phone app phyphox. Courtesy of phyphox.org.

18.1 phyphox

phyphox is written for physics education and will allow you to record data on the phone, which can be later downloaded, or can remotely send the data to your laptop and display in a web browser. This app is very well written, intuitive, comes with video tutorials, and will lend itself to learning other similar apps (Figure 18.1).

18.2 USING phyphox

- Download phyphox from the App Store (iOS users) or from Google Play (Android users).
- Open the phyphox app. Displayed will be the various experiments that can be performed (Figure 18.2).
- Each experiment displays the data collected along with a graph (Figure 18.3).
- Information, control, and data exporting can be controlled. Each experiment has info and videos and can be found by clicking on the menu as shown in (Figure 18.4).

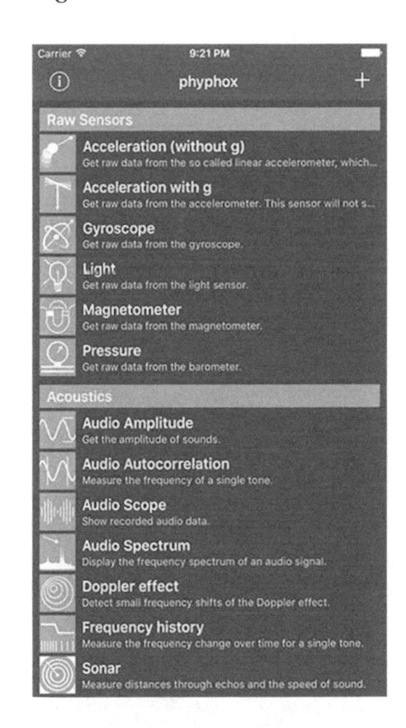

FIGURE 18.2 Some of the experiments contained in phyphox.

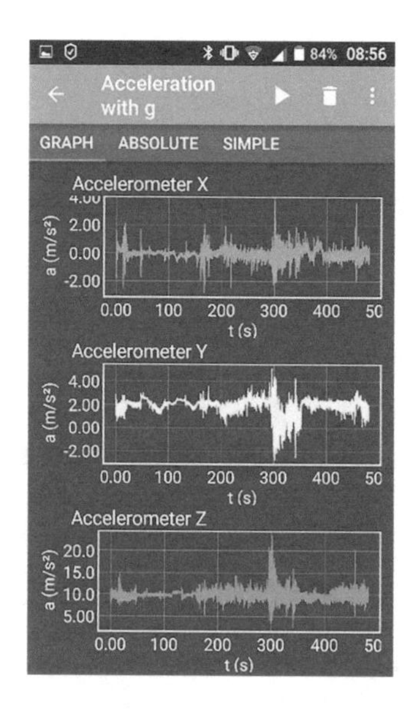

FIGURE 18.3 Each experiment displays the data collected along with a graph.

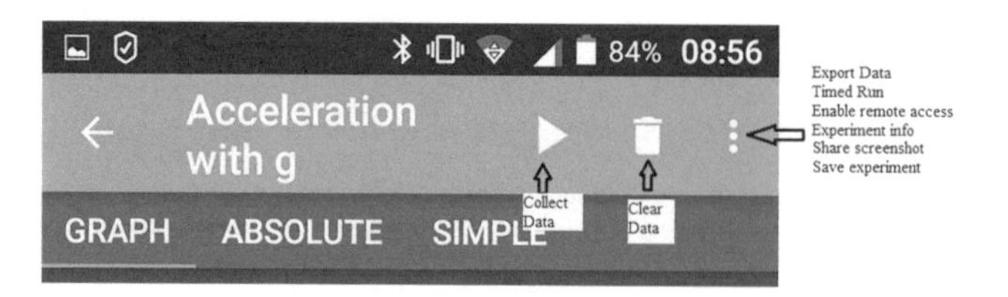

FIGURE 18.4 Each experiment has info and videos, and can be found by clicking on the menu.

- Data can be exported to a spreadsheet and downloaded to a computer via email, Google Drive, etc. Currently, phyphox supports the export formats shown in (Figure 18.5).
- Experiments can be controlled wirelessly through a web browser on your laptop and data can be exported to a spreadsheet (Figure 18.6).

18.3 MEASURING ACCELERATION AND ANGULAR VELOCITY

The accelerometer and gyroscope onboard a phone use a **MEMS** (Micro-Electro-Mechanical Systems) **Accelerometer.** A *MEMS accelerometer* is micro machined silicon wafer mass suspended by polysilicon springs. The wafer has a plate attached to it that acts as a plate of a capacitor, while the other plate is fixed. During an acceleration, the wafer is displaced and the capacitance between the plates changes. This change in capacitance produces a change in voltage that is proportional to the acceleration.

The phone's accelerometer and gyroscope are oriented as shown in Figure 18.7.

FIGURE 18.5 File formats for exporting.

FIGURE 18.6 Experiments can be controlled wirelessly through a web browser on your laptop and data can be exported to a spreadsheet.

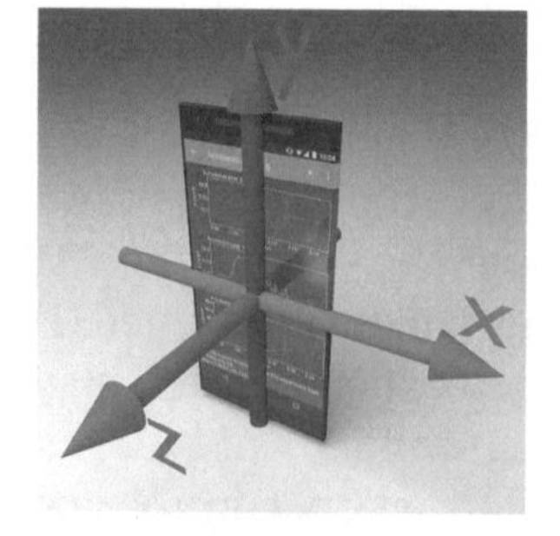

FIGURE 18.7 The phone's accelerometer and gyroscope are oriented as shown.

18.4 LAB 1: MEASURING *g*

The goal of this lab is to measure the acceleration due to gravity. There are a number of experiments in phyphox that measure acceleration. Let's measure the acceleration due to gravity using "Acceleration with *g*" and "Acceleration (without *g*)". The difference between these is that "Acceleration (without *g*)" subtracts off the constant earth's gravitational acceleration (Figure 18.8). Click on the experiment info tab or the online video for a quick overview.

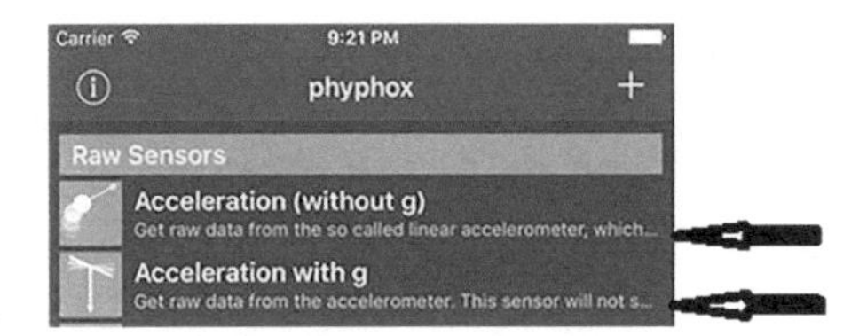

FIGURE 18.8 "Acceleration (without *g*)" experiment and "Acceleration (with *g*)" experiment in phyphox.

18.4.1 Theory

The gravitational force of attraction between two masses is:

$$F = \frac{G \cdot m_1 \cdot m_2}{r^2}$$

The gravitational force of attraction between a mass and the mass of the earth given by:

$$F = m \cdot \frac{G \cdot m_2}{r^2}$$

The force of attraction between a mass m, and the entire mass of the earth, on the surface of the earth can be calculated by entering the mass of the earth, 5.98×10^{24} kg, and the radius of the earth, 6.37×10^{6} m. This gives us a force of attraction of:

$$F = m \cdot \frac{\left(6.67 \times 10^{-11} \frac{\text{Nm}^2}{\text{kg}^2}\right) \cdot \left(5.98 \times 10^{24}\,\text{kg}\right)}{\left(6.37 \times 10^{6}\,\text{m}\right)^2}$$

$$F = m \cdot 9.8\ \frac{\text{m}}{\text{s}^2}$$

The force of gravity on a mass on the surface of the earth is given by:

$$F = m \cdot g$$

$$\text{with} \Rightarrow g = 9.8 \text{ m/s}^2$$

18.4.2 Procedure

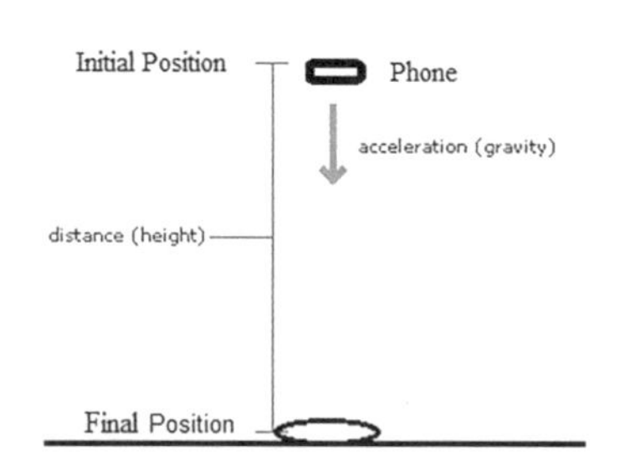

FIGURE 18.9 Position the phone over the pillow with the z axis facing upward.

1. Click on "Acceleration with g", see Figure 18.8 above.
2. Position the phone over the pillow with the z axis facing upward (Figure 18.9).
3. Press Collect Data and drop the phone onto the pillow. After the drop, press the pause button. Now click on the menu to find "Export Data" to a spread sheet (Figure 18.10).
4. Plot the acceleration in the z direction vs. time. Look for the region where the acceleration is near zero. This is where the phone is in free fall, effectively canceling out the force on the sensor (Figure 18.11).
5. Repeat the experiment using the "Acceleration (without g)" tab.
6. What is the value of g measured for "Acceleration (without g)" and "Acceleration (with g)"?

   How does it compare with the accepted value of g by performing a difference calculation for your measured g and the accepted value?

$$\%\text{ difference} = \frac{g_{\text{measured}} - g_{\text{accepted}}}{g_{\text{accepted}}} \times 100$$

$$\%\text{ difference} = \_\_\_\_\_\_\_\_\_\_$$

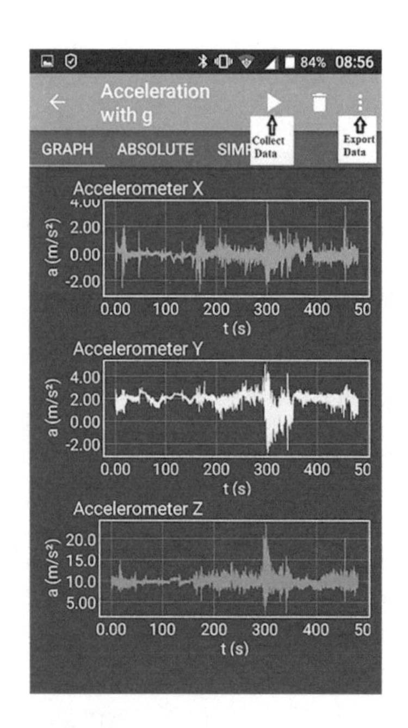

FIGURE 18.10 The menu contains "Export Data" tab to export a spread sheet.

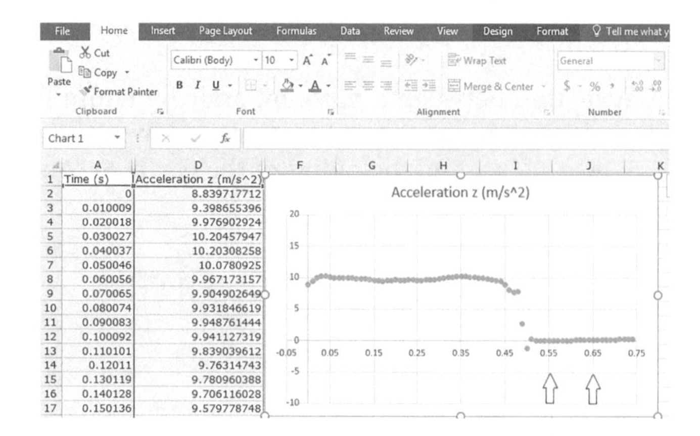

	A	D
1	Time (s)	Acceleration z (m/s^2)
2	0	8.839717712
3	0.010009	9.398655396
4	0.020018	9.976902924
5	0.030027	10.20457947
6	0.040037	10.20308258
7	0.050046	10.0780925
8	0.060056	9.967173157
9	0.070065	9.904902649
10	0.080074	9.931846619
11	0.090083	9.948761444
12	0.100092	9.941127319
13	0.110101	9.839039612
14	0.12011	9.76314743
15	0.130119	9.780960388
16	0.140128	9.706116028
17	0.150136	9.579778748

FIGURE 18.11 Plot of a_z vs time. Note where the arrows point to where the phone was in free fall.

18.5 LAB 2: MEASURING VELOCITY

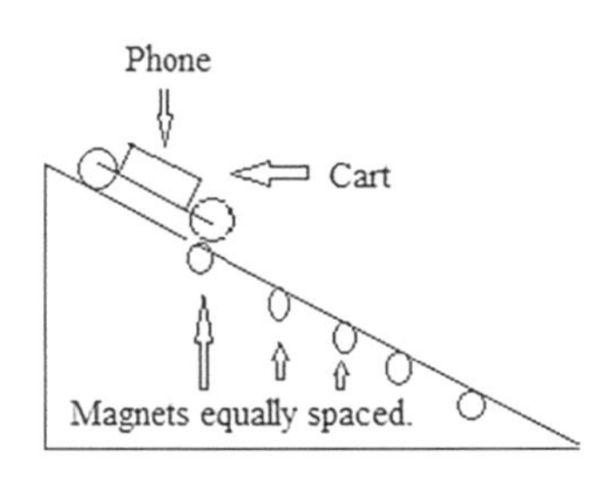

FIGURE 18.12 Magnets are equally spaced along an inclined plane.

The goal of this lab is to measure the velocity of a cart rolling down an inclined plane using the "magnetic ruler" experiment in phyphox. Click on the experiment info tab or the online video for a quick overview of the "magnetic ruler" (Figure 18.12).

18.5.1 Theory

The magnetic ruler experiment in phyphox uses the magnetic field sensor and a timer in the phone to record the time between measurements of a magnetic field. Magnets are equally spaced for a known distance *d* along an inclined plane. As the cart moves down an inclined plane a timer is running. Every time the cart passes a magnet the time is recorded. The average velocity of the cart can be determined as the cart moves down the plane by:

$$\bar{v} = \frac{d}{\Delta t} \leftarrow \frac{\text{distance between magnets}}{\text{time interval between magnets}}$$

The average acceleration can be found by plotting velocity vs. time and fitting the data with a linear fit. Recall:

$$\bar{a} = \frac{\Delta v}{\Delta t} \leftarrow \frac{\text{change in velocity}}{\text{time interval between measurements}}$$

18.5.2 Procedure

1. Place magnets at equal intervals along an inclined plane. Be sure the polarities of the magnets are all facing the same way, i.e., all the magnets have their north poles facing upward. See Figure 18.12 above.
2. Open the experiment in phyphox "magnetic ruler" and enter the distance between magnets.
3. Place the phone on a cart and the cart onto the inclined plane.
4. Press play and release the phone from rest.
5. After the phone travels down the plane press pause and export the data to a spreadsheet.
6. Plot the velocity vs. time data and perform a linear fit to find the average acceleration.

As an example, an inclined plane was set up with magnets placed 17 cm apart, at an inclined angle = 5°. The data taken and the fit are shown in the following figures (Figures 18.13 and 18.14). From the fit we see the acceleration of the cart was 0.327 m/s².

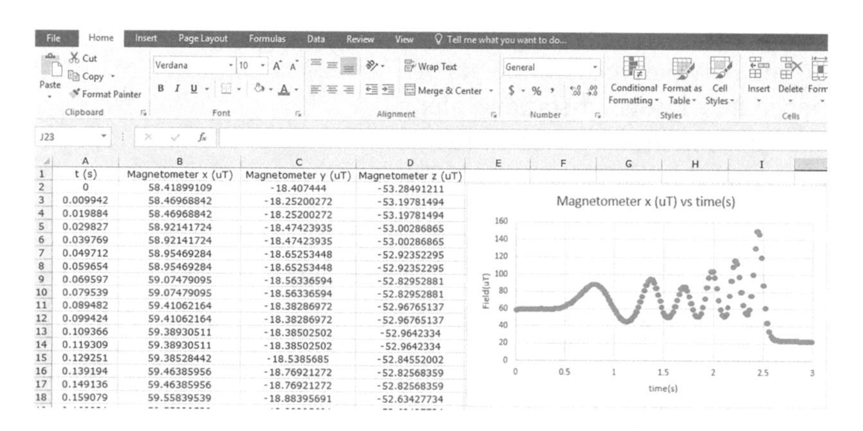

FIGURE 18.13 Sample data from the experiment.

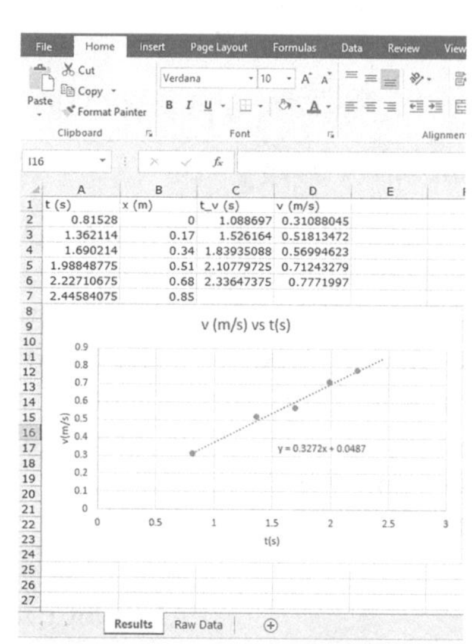

FIGURE 18.14 A fit to the data to find the acceleration.

18.6 LAB 3: MEASURING *g* USING A PENDULUM AND A SMARTPHONE'S GYROSCOPE

The goal of this lab is to measure the acceleration due to gravity using a pendulum and phyphox pendulum experiment (Figure 18.15).

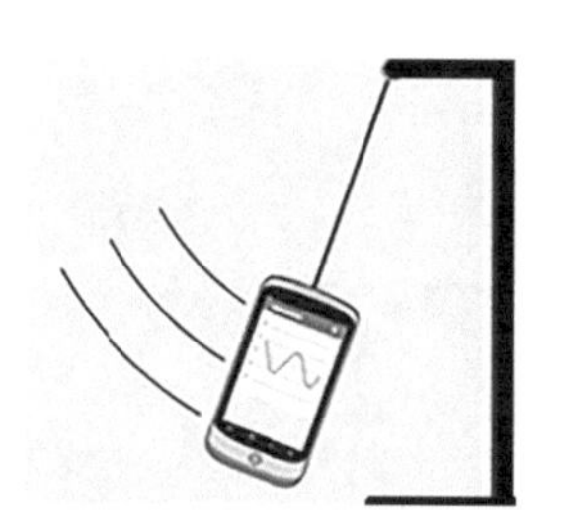

FIGURE 18.15 A pendulum made from a phone and a string.

18.6.1 Theory

The period T of a pendulum with arm length l is given by the formula:

$$T = 2 \cdot \pi \cdot \sqrt{\frac{l}{g}}$$

To measure the oscillations of the pendulum, the gyroscope in the phone will be accessed. A gyroscope measures angular velocity. The gyroscope onboard a phone uses a MEMS sensor similar to the accelerometer on the phone. The change in voltage from the sensor is proportional to the angular velocity.

18.6.2 Procedure

1. Attach a phone to a string of length *l*.
2. Scroll down in phyphox into the mechanics section and click on pendulum.
3. Click on *g* and enter the length of the pendulum (Figure 18.16).
4. Set the delay and run time by clicking on the menu and entering the times (Figure 18.17).

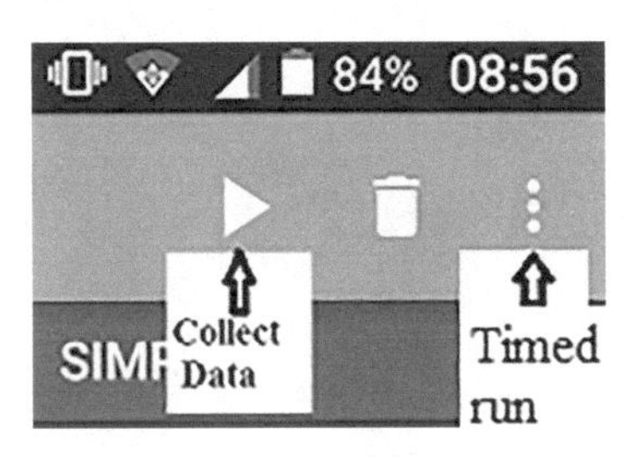

FIGURE 18.16 Set the delay and run time by clicking on the menu and entering the times.

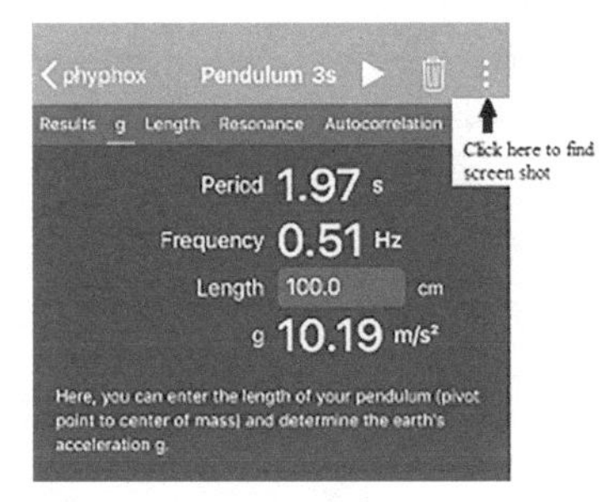

FIGURE 18.17 Click on *g* and enter the length of the pendulum.

5. Click run, set the pendulum swinging, and wait till the run time has elapsed.
6. What was your measured value for *g* and *T*. Download a screenshot from your phone as shown in Figure 18.17.

$$T_{\text{measured}} = \_\_\_\_\_$$

$$G_{\text{measured}} = \_\_\_\_\_$$

Perform a difference calculation for your measured *g* and the accepted value of 9.8 m/s^2?

$$\% \text{ difference} = \frac{g_{\text{measured}} - g_{\text{accepted}}}{g_{\text{accepted}}} \times 100$$

$$\% \text{ difference} = \_\_\_\_\_\_\_\_\_\_$$

7. What is your calculated *T* for your measured *g* and *l* and theoretical values for *g* and *T*?

18.7 LAB 4: MEASURE THE VELOCITY OF AN ELEVATOR AND HEIGHT BETWEEN FLOORS

The goal of this lab is to measure the velocity of an elevator and the height between floors using the "elevator" experiment in the phyphox app. The

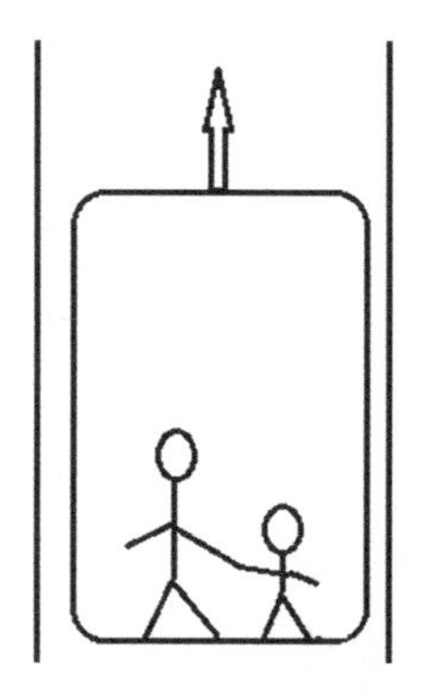

FIGURE 18.18 The elevator experiment measures the displacement, velocity, and acceleration of an elevator.

elevator experiment measures the displacement, velocity, and acceleration of an elevator (Figure 18.18). Click on the experiment info tab or the online video for a quick overview.

18.7.1 Theory

The pressure sensor onboard a phone uses a MEMS sensor similar to the accelerometer and gyroscope on the phone. Recall the formula for pressure:

$$P = \frac{F}{A}$$

$$\text{Pressure}\,(\text{Pa}) = \frac{\text{Force}\,(\text{N})}{\text{Area}\,(\text{m}^2)}$$

The unit Pa is pronounced as pascal. 1 Pa = 1 N/m^2.

Meteorologists generally measure pressures in the hectopascal (hPa) unit. The smartphone displays pressure in hPa. Some other commonly used units are:

$$1\,\text{atm} = 101.3\,\text{kPa}$$

$$1\,\text{atm} = 1013.25\,\text{hPa}$$

$$1\,\text{atm} = 1013.25\,\text{mbar}$$

$$1\,\text{hPa} = 100\,\text{Pa}$$

The atmospheric pressure decreases with an increasing height above sea level (Figure 18.19). Your height above sea level can be simply determined by measuring the barometric pressure and using the graph to determine your altitude. This pressure height dependence can be used to determine your altitude as demonstrated in this experiment.

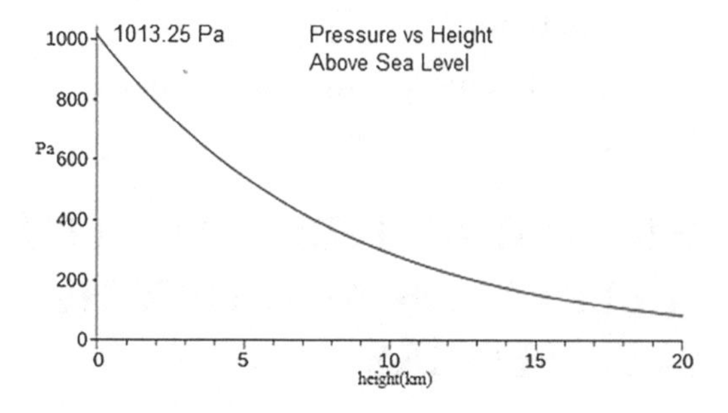

FIGURE 18.19 The atmosphere pressure decreases with an increasing height above sea level.

18.7.2 Procedure

1. Open the "elevator" experiment in phyphox.
2. Place the phone flat on the floor of the elevator and press run on the phone. As the elevator ascends or descends the pressure will change as a result of the altitude difference. This pressure change is then converted into a height change. The velocity of the elevator is determined by dividing the height change by the time between measurements. The z component of acceleration is also measured by the accelerometer and plotted.
3. Export the data and determine the average velocity of the elevator from the data by using:

$$\bar{v} = \frac{\Delta h}{t}$$

$$\text{Average velocity} = \frac{\text{distance}}{\text{time}}$$

   How does the calculated velocity compare to the velocity plotted in phyhox? Perform a difference calculation.
4. Repeat the experiment with the elevator moving downward instead.
   How is the acceleration different?
   Is this what you would expect? Why?

18.8 LAB 5: MEASURE THE PERIOD OF OSCILLATION OF A SPRING OR RUBBER BAND

The goal of this lab is to measure the period T of oscillation of a mass m of a spring using the "spring" experiment in phyphox. Click on the experiment info tab or the online video for a quick overview (Figure 18.20).

The spring experiment in phyphox uses the onboard accelerometer to measure the movement of the phone as it is oscillating. Click on the experiment info tab or the online video for a quick overview.

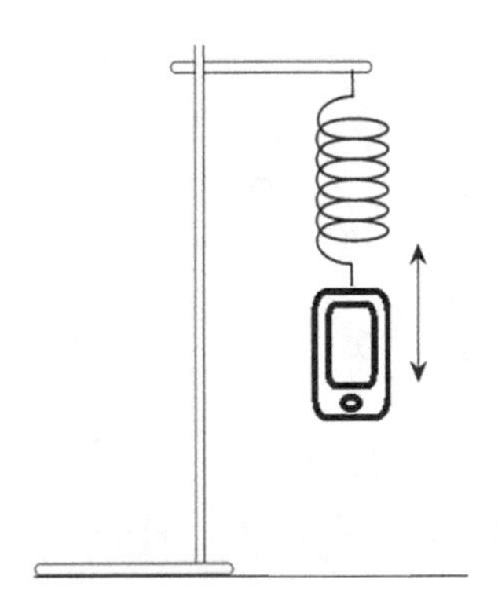

FIGURE 18.20 Smartphone oscillating up and down.

18.8.1 Theory

The period T of oscillation of a mass m on a spring with spring constant k is given by the formula:

$$T = 2 \cdot \pi \cdot \sqrt{\frac{m}{k}}$$

The spring constant k is a measure of how stiff the spring is and can be found from Hooke's law:

$$F = -k \cdot \Delta s$$

Solving for k:

$$k = -\frac{F}{\Delta s}$$

We can determine k by applying a force to the spring and measuring how much it stretches.

The spring constant k is therefore:

$$k = \frac{F}{\Delta s} = \frac{m \cdot g}{\Delta s}$$

Plug in for m and Δs to find k:

$$k = \_\_\_\_\_\_\_\_\_\_\_\_$$

The predicted period of oscillation is therefore:

$$T = 2 \cdot \pi \cdot \sqrt{\frac{m}{k}}$$

Plug in for m and k to find T:

$$T = \_\_\_\_\_\_ \leftarrow \text{Predicted period}$$

18.8.2 Procedure

Let us now measure period of oscillation with phyphox.

1. Open the spring experiment in phyphox.
2. Place the phone on the end of the spring, set the delay and end times of the experiment so that the phone is freely oscillating while the measurement is being taken.
3. Press play and record the measured period by phyphox.

$$T_{\text{measured}} = \_\_\_\_\_\_$$

4. Provide a screenshot of the experiment, see Figure 18.21 for an example.
5. Do a percent difference calculation to compare theoretical and measured values of T.

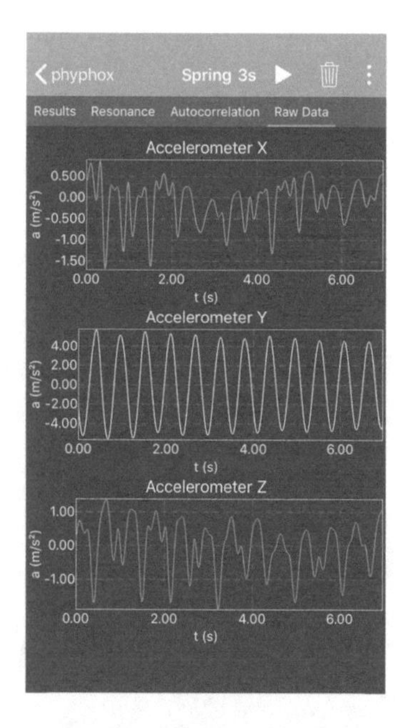

FIGURE 18.21 Example data from the experiment.

$$\% \text{ difference} = \frac{T_{\text{measured}} - T_{\text{predicted}}}{T_{\text{measured}}} \times 100$$

$$\% \text{ difference} = \_\_\_\_\_\_\_\_\_\_\_\_$$

18.9 LAB 6: MEASURE CENTRIPETAL ACCELERATION OF A ROTATING PHONE

The goal of this lab is to measure the centripetal acceleration of a phone placed on a rotating platform, such as a lazy susan, salad spinner, rotating chair, etc. using the phyphox experiment "centripetal acceleration". Determine how far the center of the phone is from the center of rotation. Click on the experiment info tab or the online video for a quick overview (Figure 18.22).

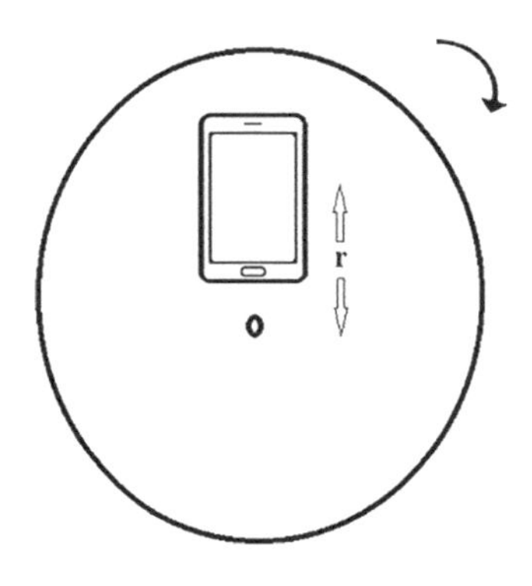

FIGURE 18.22 A smartphone rotating on a platform.

18.9.1 Theory

An object moving in a circle has a velocity vector that is constantly changing, because its direction is constantly changing. Recall, a change is velocity is an acceleration. This change in velocity is directed toward the center and is called centripetal acceleration. Centripetal acceleration is given by:

$$a_c = r \cdot \omega^2$$

$$\text{Acceleration}_{\text{centripetal}}\left(\frac{\text{m}}{\text{s}^2}\right) = \text{radius}(\text{m}) \times \text{angular velocity}\left(\frac{\text{rad}}{\text{s}}\right)^2$$

In this lab we will measure the values of a_c for various ω's and generate a plot of a_c vs. ω^2. From the plot we will calculate the slope of a fit to this data. The slope will be a measure of the radius of rotation r, since $r = a_c/\omega^2$.

18.9.2 Procedure

The centripetal acceleration experiment in phyphox uses the onboard gyroscope to measure the angular velocity and the onboard accelerometer to measure acceleration of the phone as it is rotating.

1. Open the centripetal acceleration experiment in phyphox. Place the phone on a rotating platform. Enable remote access in the experiment so that it can be controlled from a laptop. If you encounter trouble connecting, check out the phyphox website for help or just control the phone directly.
2. Click on run and begin spinning the platform at various speeds for several seconds, then pause the experiment.

3. Export the data to an Excel spreadsheet. Insert a column between the angular velocity and the acceleration columns. Into this column insert the square of the angular velocity column B.
4. Perform a scatter plot of columns C vs. D.
5. Click on the graph, then the +. Click on Trendline>Linear>More Options>Display equation on chart, see Figure 18.23 for an example. In the example data in Figure 18.23, the fit equation is: $y = 0.0792x - 0.0347$. The slope from this fit is 0.0792. We can conclude from this example that the distance the phone is from the center of rotation is: $r = ac/\omega^2 = 0.0792\,\text{m}$.

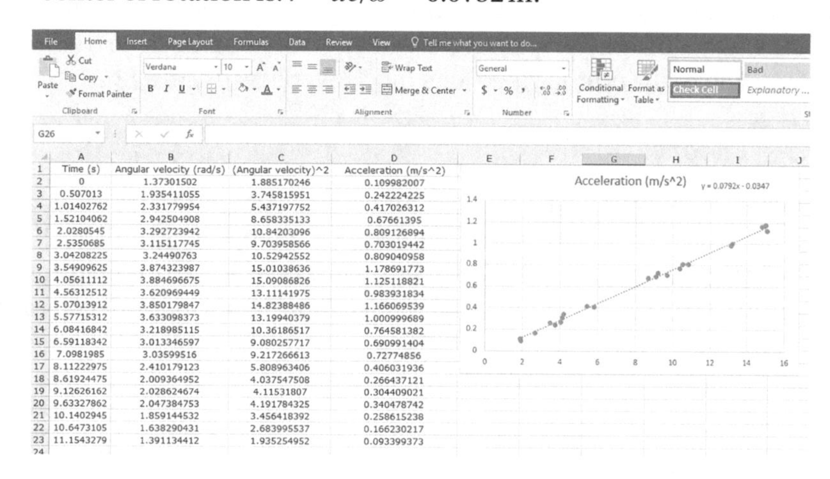

	A	B	C	D
1	Time (s)	Angular velocity (rad/s)	(Angular velocity)^2	Acceleration (m/s^2)
2	0	1.37301502	1.885170246	0.109982007
3	0.507013	1.935411055	3.745815951	0.242224225
4	1.01402762	2.331779954	5.437197752	0.417026312
5	1.52104062	2.942504908	8.658335133	0.67661395
6	2.0280545	3.292723942	10.84203096	0.809126894
7	2.5350685	3.115117745	9.703958566	0.703019442
8	3.04208225	3.24490763	10.52942552	0.809040958
9	3.54909625	3.874323987	15.01038636	1.178691773
10	4.05611112	3.884696675	15.09086826	1.125118821
11	4.56312512	3.620969449	13.11141975	0.983931834
12	5.07013912	3.850179847	14.82388486	1.166069539
13	5.57715312	3.633098373	13.19940379	1.000999689
14	6.08416842	3.218985115	10.36186517	0.764581382
15	6.59118342	3.013346597	9.080257717	0.690991404
16	7.0981985	3.03599516	9.217266613	0.72774856
17	8.11222975	2.410179123	5.808963406	0.406031936
18	8.61924475	2.009364952	4.037547508	0.266437121
19	9.12626162	2.028624674	4.11531807	0.304409021
20	9.63327862	2.047384753	4.191784325	0.340478742
21	10.1402945	1.859144532	3.456418392	0.258615238
22	10.6473105	1.638290431	2.683995537	0.166230217
23	11.1543279	1.391134412	1.935254952	0.093399373

FIGURE 18.23 Example data from the experiment along with a fit to the data.

6. Provide a screenshot from your phone of the experiment.
7. What was the fit equation and the distance the phone was from the center of rotation according to the fit?
8. What was the distance the phone was from the center of rotation according to a measurement with a ruler?
9. Perform a percent difference calculation to compare the ruler measurement and the experimental values of r.

$$\%\ \text{difference} = \frac{r_{\text{ruler}} - r_{\text{experiment}}}{r_{\text{ruler}}} \times 100$$

$$\%\ \text{difference} = \_\_\_\_\_\_\_\_\_\_\_\_$$

18.10 LAB 7: ROLLING VELOCITY

The goal of this lab is to measure the rolling velocity of a can down an inclined plane. Measure the velocity of a smartphone placed in a soda bottle rolling along an inclined plane using the "roll" experiment in phyphox. Click on the experiment info tab or the online video for a quick overview (Figure 18.24).

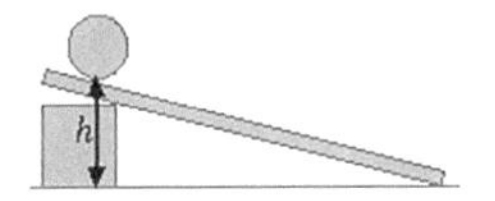

FIGURE 18.24 A smartphone placed in a soda bottle rolling along an inclined plane.

18.10.1 Theory

The velocity of a cylinder of radius r rolling along a surface is given by:

$$v = r \cdot \omega$$

$$\text{Velocity}(\text{m/s}) = \text{radius}(\text{m}) \times \text{angular velocity}(\text{rad/s})$$

where

$$\omega = \frac{\Delta\theta}{\Delta t}$$

18.10.2 Procedure

1. Open the "roll" experiment in phyphox. Enable remote access in the experiment so that it can be controlled from a laptop. If you encounter trouble connecting, check out the phyphox website for help or just control the phone directly.
2. Place the phone in a soda bottle, cut open to accept a phone, and place the bottle on an inclined plane.
3. Enter the radius of the bottle into phyphox.
4. Click on run and let the bottle roll down the plane.
5. Take a screenshot of your phone and paste it in your report as shown in Figure 18.25.
6. Can you explain the shape of your graph?

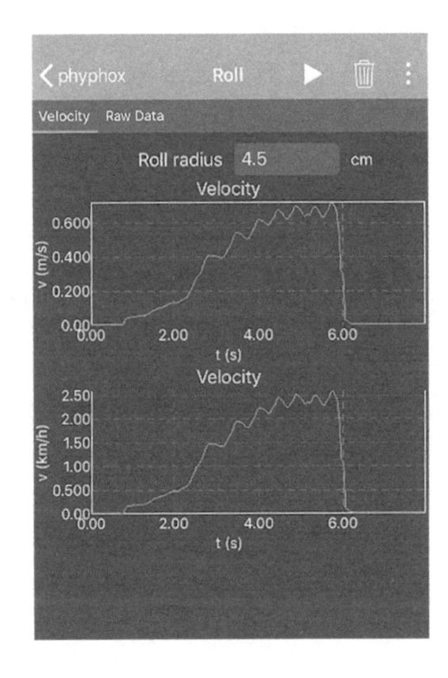

FIGURE 18.25 Example data from the experiment.

18.11 LAB 8: INELASTIC COLLISIONS

The goal of this lab is to measure the height a ball is dropped from by recording the time between bounces. We will measure the change in height and energy of a bouncing ball using the "(In)elastic collision" experiment in phyphox (Figure 18.26). Click on the experiment info tab or the online video for a quick overview.

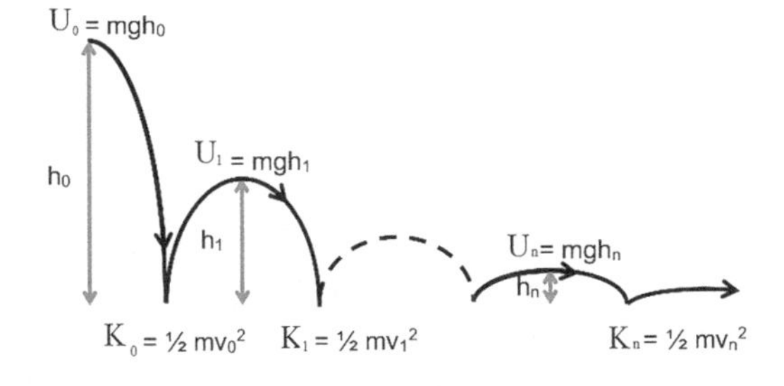

FIGURE 18.26 A ball is dropped from by recording the time between bounces.

18.11.1 Theory

Recall that a mass dropped from rest will fall a height h in a time t and is given by:

$$h = \frac{1}{2} g \cdot t^2$$

Suppose the time T between bounces is recorded. This time is the sum of time the ball moves upward plus the time the ball falls downward. The fall time is therefore $T/2$ and h becomes:

$$h = \frac{1}{2} g \cdot \left(T/2 \right)^2 \leftarrow \text{Height of each bounce, where } T \text{ is the time between bounces.}$$

Now consider the energy in the ball. The energy of the ball when it reaches its maximum bounce height is:

$$E = m \cdot g \cdot h$$

Referring to Figure 18.26, the ratio of the energies between consecutive bounces is:

$$\frac{E_1}{E_0} = \frac{h_1}{h_0} \quad \text{and} \quad \frac{E_2}{E_1} = \frac{h_2}{h_1}$$

If we assume the energy loss from the first bounce is approximately equal to the energy loss from the second bounce,

$$\frac{E_1}{E_0} \approx \frac{E_2}{E_1}$$

$$\text{or} \Rightarrow \frac{h_1}{h_0} = \frac{h_2}{h_1}$$

$$\text{Solving for } h_0 \Rightarrow h_0 = \frac{(h_1)^2}{h_2}$$

We have shown that by only measuring the time between bounces the relative energy and bounce height can be determined!

18.11.2 Procedure

1. Open the "(In)elastic collision" experiment in phyphox.
2. Place the phone on a hard floor, press play, and drop a ball from a measured height.
3. Take a screenshot of your phone and paste it in your report such as shown in Figure 18.27.

FIGURE 18.27 Example data from the experiment.

4. Weigh the ball and determine the initial energy of the ball and the energy in each successive bounce using:

$$E = m \cdot g \cdot h$$

$E_0 =$ _____

$E_1 =$ _____

$E_2 =$ _____

$E_3 =$ _____

$E_4 =$ _____

$E_5 =$ _____

18.12 MOTION ANALYSIS USING TRACKER

Tracker is a free video analysis software written for physics education and will allow you to take a video recorded on the phone and analyze the motion of an object in the video. Tracker is very well written, intuitive, and has many video tutorials. Tracker can be downloaded for free from http://physlets.org/tracker/.

To illustrate the use of Tracker let's follow step-by-step the quick start guide on the Tracker website. Record a video of an object in motion that you would like to study. Include a meter stick in the video to set a scale.

1. Download Tracker from http://physlets.org/tracker/ and open the software.
2. Click on File>Open File>choose the video to be analyzed (Figure 18.28).

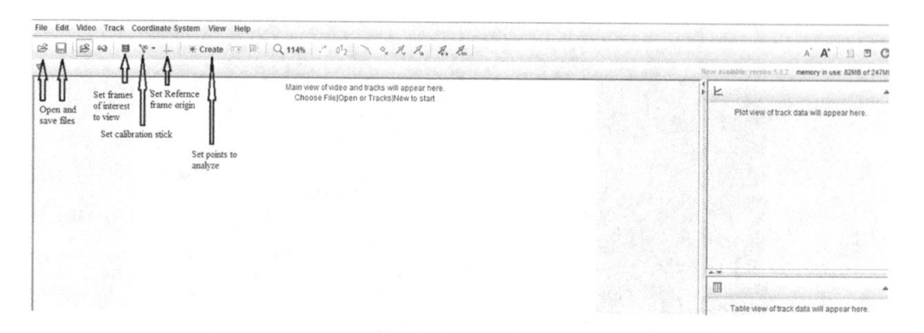

FIGURE 18.28 Various tabs in Tracker.

3. Play the video and right click on the cursor to set the start and end times of video (Figure 18.29).
4. Click on calibration stick, set length to 1, and set the stick's end points to the top and bottom of the meter stick (Figure 18.30).
5. Click on set reference frame origin, and center cursor over the projectile at the beginning of where the motion is to be analyzed (Figure 18.30).
6. Click on *Create>select point mass>press shift and click on the projectile. The cursor will advance to the next frame, then press shift click on the projectile again, and repeat this till the end is reached (Figure 18.31).

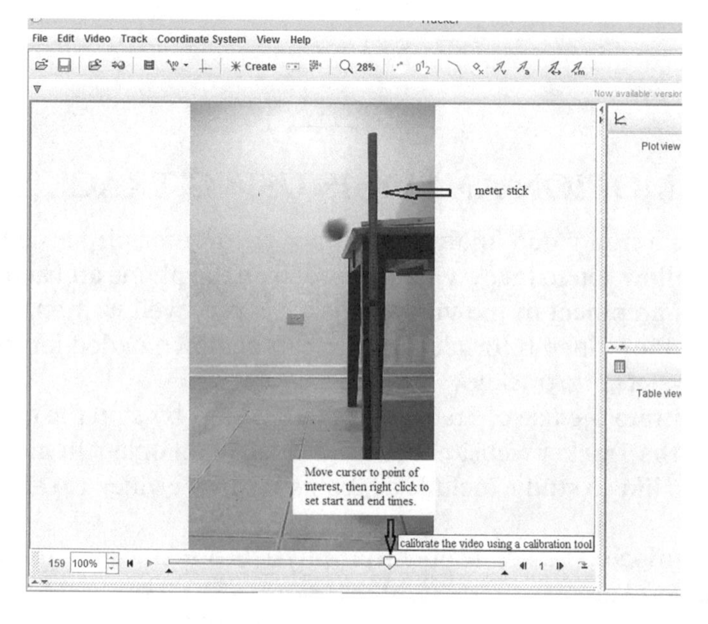

FIGURE 18.29 Play the video and right click on the cursor to set the start and end times of video.

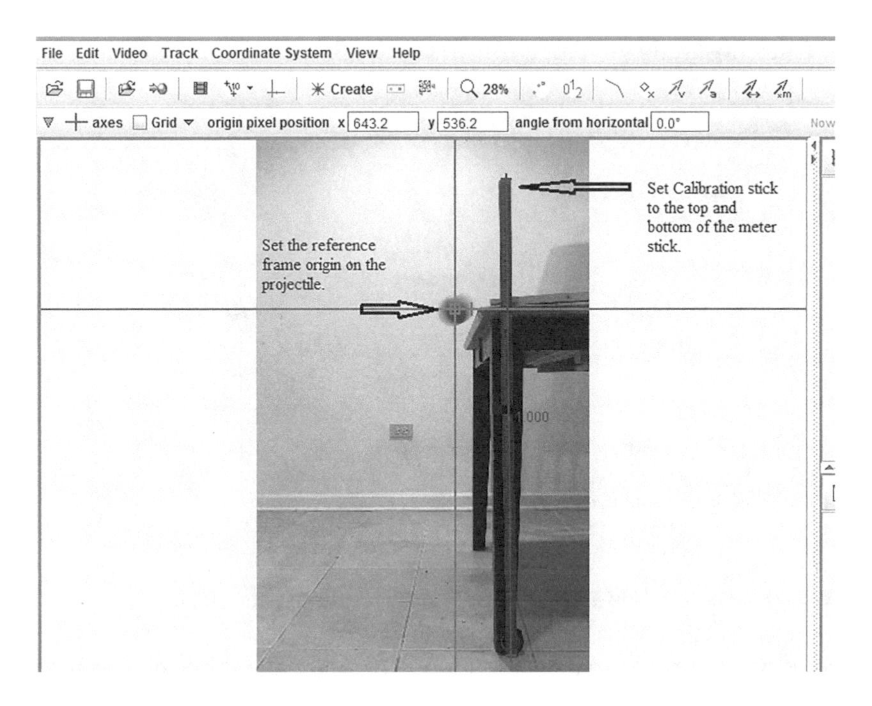

FIGURE 18.30 Click on set reference frame origin, and center cursor over the projectile at the beginning of where the motion is to be analyzed.

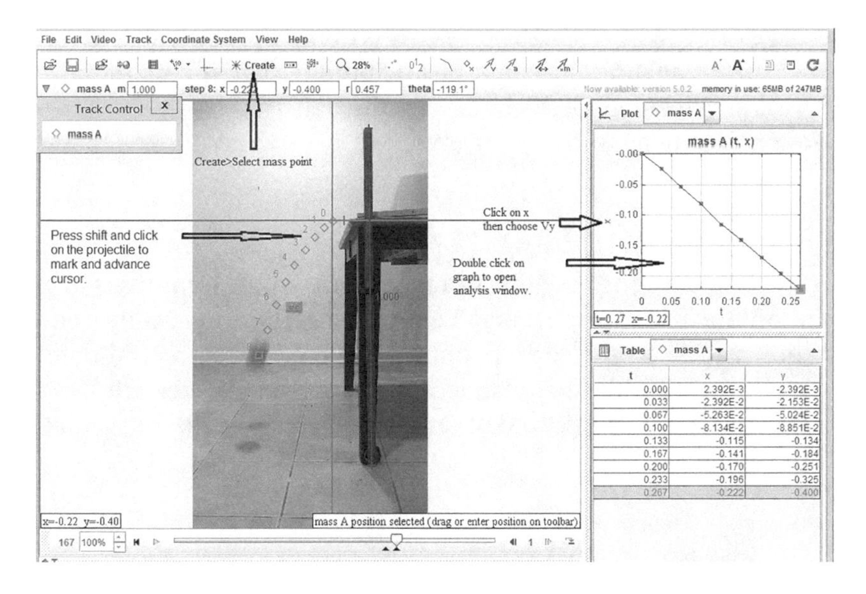

FIGURE 18.31 Click on *Create>select point mass>press shift and click on the projectile. The cursor will advance to the next frame, then press shift click on the projectile again, and repeat this until the end is reached.

7. Click on Analyze > Curve fits to see the fit to the data (Figure 18.32).

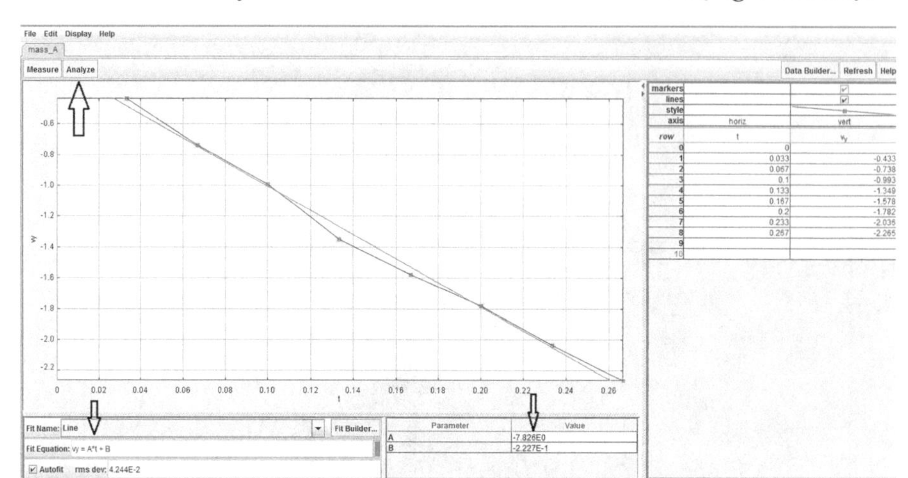

FIGURE 18.32 Click on Analyze > Curve fits to see the fit to the data.

Looking at the example data in Figure 18.32 the fit to the velocity v_y vs t is:

$$v_y = -7.826 \cdot t - 2.227 \times 10^{-1}$$

Comparing this to the free fall equation: $v_y = g \cdot t + v_i$ gives us a measured value for g of $g_{\text{measured}} = -7.826\,\text{m/s}^2$. This is about 20% off the actual value of $g = 9.8\,\text{m/s}^2$. The difference may be attributed to a non-zero initial value for v_y, or the placement of the cursor on the ball.

Better measure again!

18.13 SUMMARY

Smartphones have the capability of measuring many quantities in physics without the use of any other lab equipment. There are many free apps that allow access to the sensors onboard the phone to record data and create a downloadable spreadsheet. phyphox and Tracker are two such apps for both iOS and Android phones that measure and record data for the study of physics.

Appendix A

Trigonometry Review

Trigonometry shows us the relationship between the sides of a triangle and its angles. It is very useful in solving vector problems in two or three dimensions.

A.1 PYTHAGOREAN THEOREM AND RIGHT TRIANGLES

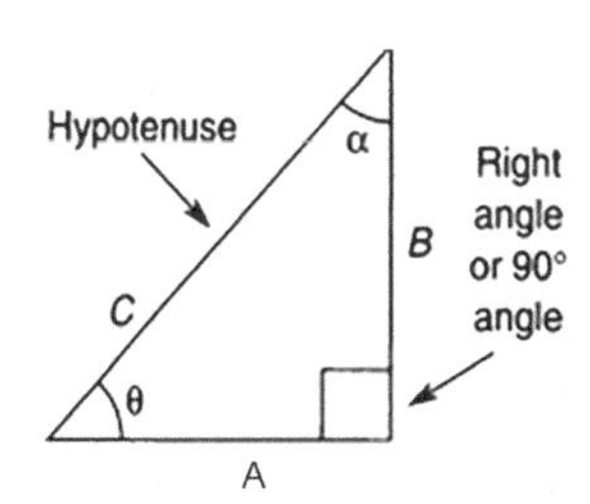

FIGURE A.1 A right triangle has one 90 degree angle.

A right triangle is a triangle with a 90° angle.

Key points about a right triangle:

- The *hypotenuse* long side of the triangle.
- $\alpha + \theta = 90°$
- **Pythagorean theorem:** The hypotenuse equals the square root of the sum of the squares of the sides.

$$C = \sqrt{A^2 + B^2}$$

$$A = \sqrt{C^2 - B^2}$$

$$B = \sqrt{C^2 - A^2}$$

Example A.1

Find the length of the sides of the following right triangles: (a) $A = 3$, $B = 4$, $C = ?$; (b) $A = 7$, $C = 9$, $B = ?$; (c) $B = 13$, $C = 18$, $A = ?$

(a) $A = 3.0$, $B = 4.0$, $C = ?$

$$C = \sqrt{A^2 + B^2}$$

$$C = \sqrt{3.0^2 + 4.0^2}$$

$$C = 5.0$$

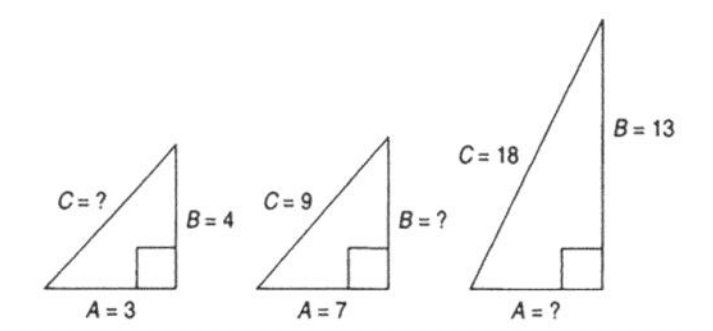

FIGURE A.2 Find the lengths of the sides of these triangles.

(b) $A = 7.0$, $C = 9.0$, $B = ?$

$$B = \sqrt{C^2 - A^2}$$
$$B = \sqrt{9.0^2 - 7.0^2}$$
$$B = 5.7$$

(c) $B = 13.0$, $C = 18.0$, $A = ?$

$$A = \sqrt{C^2 - B^2}$$
$$A = \sqrt{18.0^2 - 13.0^2}$$
$$A = 12.4$$

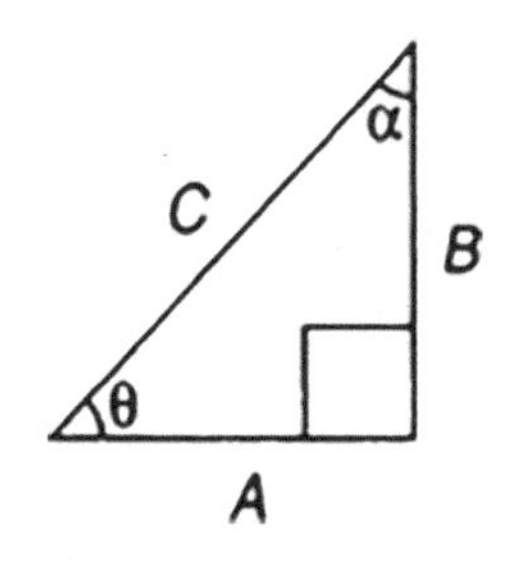

FIGURE A.3 A right triangle.

A.2 DETERMINING SIDES AND ANGLES

The hardest part of trigonometry is determining which side is opposite or adjacent to an angle.

- The *opposite side* of an angle can be found by drawing a line through the angle. This line will strike the opposite side.
- The *hypotenuse* is the long side of the triangle. It is the side opposite the 90° angle.
- The *adjacent side* is the side next to an angle. The hypotenuse may be next to an angle, but it is never called the adjacent side. It is always the hypotenuse.

***C* is the hypotenuse.**

B is the side opposite angle θ.
A is the side adjacent to angle θ.
A is the side opposite angle α.
B is the side adjacent to angle α.

A.3 Sin(θ), Cos(θ), Tan(θ)

The trigonometric functions $\sin(\theta)$, $\cos(\theta)$, $\tan(\theta)$ and $\sin(\alpha)$, $\cos(\alpha)$, $\tan(\alpha)$ are defined to be:

$$\sin(\theta) = \frac{\text{opposite side}}{\text{hypotenuse}} = \frac{B}{C}$$

$$\cos(\theta) = \frac{\text{adjacent side}}{\text{hypotenuse}} = \frac{A}{C}$$

$$\tan(\theta) = \frac{\text{opposite side}}{\text{adjacent}} = \frac{B}{A}$$

$$\sin(\alpha) = \frac{\text{opposite side}}{\text{hypotenuse}} = \frac{A}{C}$$

$$\cos(\alpha) = \frac{\text{adjacent side}}{\text{hypotenuse}} = \frac{B}{C}$$

$$\tan(\alpha) = \frac{\text{opposite side}}{\text{adjacent}} = \frac{A}{B}$$

An easy way to remember these is with the mnemonic:

SOH-CAH-TOA
Where SOH stands for $\sin(\theta) = \text{op/hy}$
Where CAH stands for $\cos(\theta) = \text{adj/hy}$
Where TOA stands for $\tan(\theta) = \text{op/adj}$

A triangle's angles can be determined using the inverse trigonometric functions:

$$\theta = \sin^{-1}\left(\frac{\text{opposite side}}{\text{hypotenuse}}\right) = \sin^{-1}(B/C)$$

$$\theta = \cos^{-1}\left(\frac{\text{adjacent side}}{\text{hypotenuse}}\right) = \cos^{-1}(A/C)$$

$$\theta = \tan^{-1}\left(\frac{\text{opposite side}}{\text{adjacent side}}\right) = \tan^{-1}(B/A)$$

$$\alpha = \sin^{-1}\left(\frac{\text{opposite side}}{\text{hypotenuse}}\right) = (A/C)$$

$$\alpha = \cos^{-1}\left(\frac{\text{adjacent side}}{\text{hypotenuse}}\right) = (B/C)$$

$$\alpha = \tan^{-1}\left(\frac{\text{opposite side}}{\text{adjacent side}}\right) = \tan^{-1}(A/B)$$

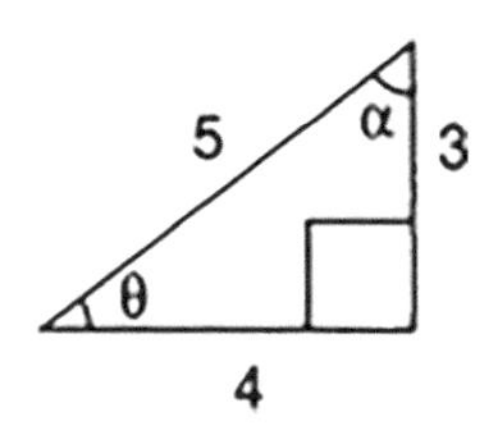

FIGURE A.4 Find the angles of this triangle.

Example A.2

Determine the sin(θ), cos(θ), tan(θ); sin(α), cos(α), tan(α); and the angles of the following triangle.

$$\sin(\theta) = \frac{3}{5}$$

$$\cos(\theta) = \frac{4}{5}$$

$$\tan(\theta) = \frac{3}{4}$$

$$\sin(\alpha) = \frac{4}{5}$$

$$\cos(\alpha) = \frac{3}{5}$$

$$\tan(\alpha) = \frac{4}{3}$$

$$\theta = \sin^{-1}\left(\frac{3}{5}\right) = 36.87°$$

$$\theta = \cos^{-1}\left(\frac{4}{5}\right) = 36.87°$$

$$\theta = \tan^{-1}\left(\frac{3}{4}\right) = 36.87°$$

$$\alpha = \sin^{-1}\left(\frac{4}{5}\right) = 53.13°$$

$$\alpha = \cos^{-1}\left(\frac{3}{5}\right) = 53.13°$$

$$\theta = \tan^{-1}\left(\frac{4}{3}\right) = 53.13°$$

Example A.3

Given $A = 12.0\,\text{cm}$ and $B = 14.0\,\text{cm}$, what are θ and β?

$A = 12.0\,\text{cm}$
$B = 14.0\,\text{cm}$

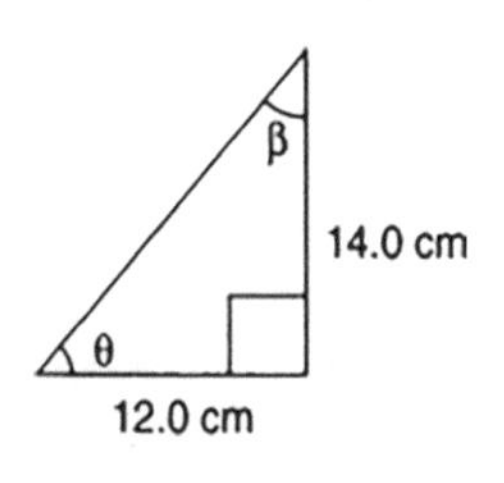

FIGURE A.5 Find the angles and the side of this triangle.

$$\theta = \tan^{-1}\left(\frac{\text{opposite side}}{\text{adjacent side}}\right)$$

$$\theta = \tan^{-1}\left(\frac{14.0\,\text{cm}}{12.0\,\text{cm}}\right)$$

$$\theta = 49°$$

$$\beta = 90° - 49° = 41°$$

Given one side and one angle of a right triangle, you can determine all other sides and angles of the triangle.

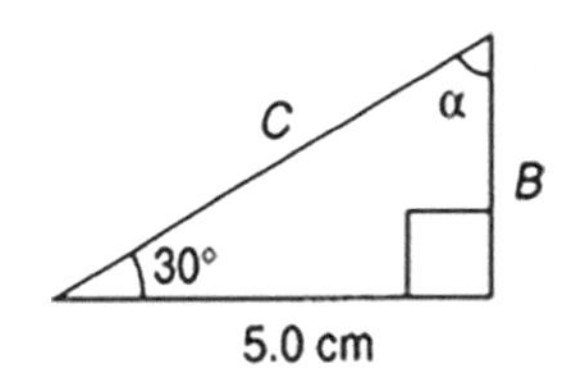

FIGURE A.6 Find the angle and the sides of this triangle.

Example A.4

Given $A = 5.0\,\text{cm}$ and $\theta = 30°$, what are (a) $\alpha = ?$; (b) $B = ?$; (c) $C = ?$

(a) $\alpha = ?$

$$\alpha + \theta = 90°$$

$$\alpha = 90° - 30°$$

$$\alpha = 60°$$

(b) $B = ?$

$$\tan(30) = B/5.0$$

$$B = 5.0 \cdot \tan(30)$$

$$B = 2.88$$

(c) $C = ?$

$$\cos(30) = \frac{5.0}{C}$$

$$C = \frac{5.0}{\cos(30)}$$

$$C = 5.77$$

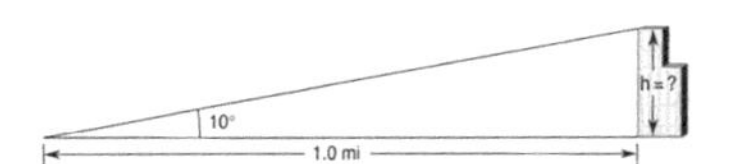

FIGURE A.7 Find the height of a building.

Example A.5

Suppose you determine the height of a building, using trigonometry. What is the height of a building when you are 1.0 mile away and measure an angle of 10.0° from the sidewalk to the top of the building?

$$\tan(10) = \frac{h}{1.0\text{mile}}$$

$$h = 1.0\text{mile} \cdot \tan(10)$$

$$h = 0.18\text{miles}$$

PROBLEM SOLVING TIPS

- Make sure your calculator is in the degree mode and not radians or gradians before calculating.
- If you have trouble determining whether to use sin, cos, or tan, write all three down. Having it on paper is easier than solving it in your head.

PROBLEMS

1. Determine the length of the hypotenuse of a right triangle with sides of 2.3 cm and 4.6 cm.
2. What are the angles that make up the triangle in problem 1?
3. Determine the length of the side of a right triangle with a side of 14.7 in. and a hypotenuse of 23.2 in.
4. What are the angles that make up the triangle in problem 3?
5. What is the height of a tree when you are 500.0 ft away and measure an angle of 25.0° from the ground to the top of the tree?
6. How long is the diagonal of a square with sides that are 6.0 cm long?
7. If an electronic signal is described by the formula $V = 2.0\text{V} \sin(2\pi ft)$ and $f = 1{,}254$ Hz, what is the value of V at $t = 12$ ms?
8. If a force of 10.0 N acts downward at a 45° angle, what component of force acts downward and horizontally?
9. At what angle are the sides of a r ight triangle equal?
10. At what angle is one side of a right triangle twice as big as the other?

Appendix B

Supplementary Lab Experiments

Contained in this appendix are ten laboratory experiments.

B.1 MEASURING LENGTH

Objective: To build and calibrate a device to measure length using a wire's resistance dependence on length.

Theory: A wire's resistance changes with its length. By measuring the resistance of a wire of a known length, we can determine the wire's resistivity ρ (the number of ohms per meter of wire). The resistance is given by:

$$R = \rho \times L$$

$$\text{Resistance}(\Omega) = \text{resistivity}(\Omega/\text{m}) \times \text{length}(\text{m})$$

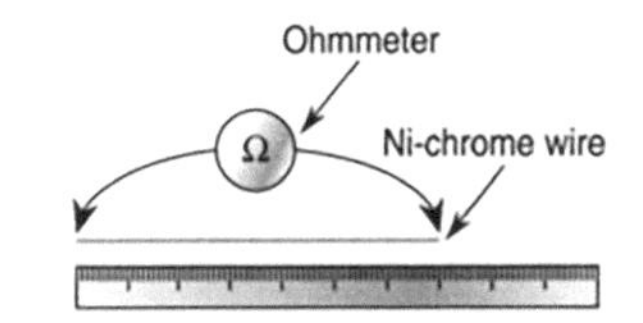

FIGURE B.1 A device to measure length using a wire's resistance dependence on length.

Once we know the wire's resistivity, we can use the wire to measure the length of an object. By measuring the resistance of the wire with the same length as the object, we can determine its length from the equation:

$$L = \frac{R}{\rho}$$

B.1.1 Procedure

1. Determine the wire's resistivity by measuring the wire's resistance for known lengths. (For best results, use a fine gauge wire made of nichrome.)

L (m)	*R* (Ω)	$\rho\left(\frac{\Omega}{\text{m}}\right) = \frac{\text{R}}{\text{L}}$
1		
0.9		
0.8		
0.7		

From these four measurements calculate the average value of ρ.

$\rho_{\text{average}} =$ ______

2. Check the accuracy of your wire length measurement device by measuring the length of some objects using the wire and compare your results with their real lengths using a meter stick.

Object #	R (Ω)	$L = \frac{R}{\rho_{average}}$	Length Measured with a Meter Stick	% Difference = $\left\lvert \frac{L_{meterstick} - L_{resistance}}{L_{meterstick}} \right\rvert \times 100$
1				
2				
3				
4				

B.2 MEASURING TIME

Objective: To make a clock using a pendulum and determine which factors influence its period of oscillation.

Theory: A pendulum's period of oscillation depends only on its length and gravity and can be approximated for small oscillations by:

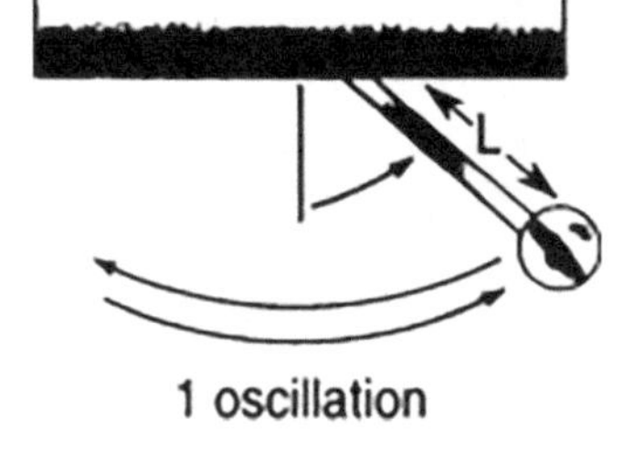

FIGURE B.2 A clock made from a pendulum.

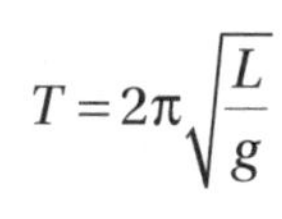

$$T = 2\pi\sqrt{\frac{L}{g}}$$

$$\text{Period of one oscillation} = 2\pi\sqrt{\frac{\text{length}}{\text{gravity}}}$$

An *oscillation* is one complete back-and-forth motion.

B.2.1 Procedure

1. Show that the period does not depend on the mass. Explain how you did this.
2. If $L = 1$ m, what is T?
3. Show by experiment that the answer you calculated in step 2 is correct. Explain what you did.
4. Use your pendulum as a clock to determine your heart rate in beats per minute. Explain how you did this. What is your rate?

B.3 MEASURING g

Objective: To measure free-fall acceleration.

Theory: Everything falls at the same rate (9.8 m/s²) independent of its mass (neglecting air friction).

The equation of motion:

$$d = v_i \cdot t + \frac{1}{2} g \cdot t^2$$

can be used to determine free-fall acceleration (g). If $v = 0$, then:

$$g = \frac{2d}{t^2}$$

B.3.1 Procedure

1. Verify that different masses fall at the same rate. Explain how you did this.
2. Time how long it takes for a mass to fall from rest from a measured distance. Do this for several trials and calculate g and the percent discrepancy from the measured and the actual $g = 9.8\,\text{m/s}^2$.

d(m)	t(s)	$g = 2d/t^2$	% Difference $= \left\lvert \frac{g_{measured} - 9.8}{9.8} \right\rvert \times 100$

3. How could you measure g more accurately?
4. Are there any real applications for measuring variations in g over the surface of the earth?

B.4 MEASURING FORCE

Objective: To build and calibrate a force sensor using conductive foam.

Theory: A simple force sensor can be constructed out of conductive foam (integrated circuit chips are stored on this foam to avoid damage done by static charge). As the foam is compressed, its resistance decreases. Using this property, we can calibrate this force sensor, developing a relationship between the force on the sensor and its resistance.

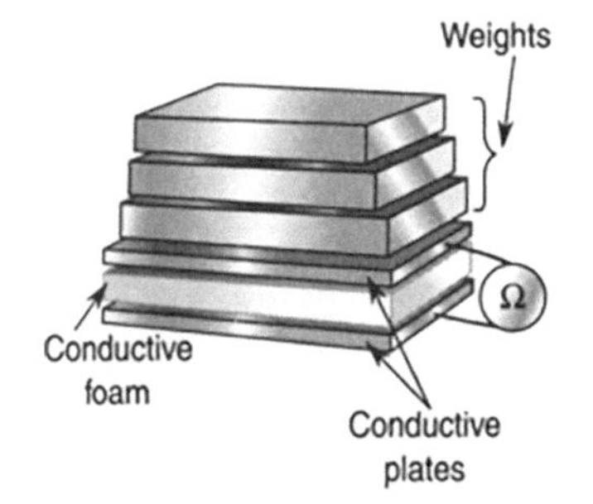

FIGURE B.3 A force sensor made from conductive foam.

B.4.1 Procedure

1. Record the resistance of the foam as weights are placed on top of the foam.

Weight	Resistance

2. Plot your data, resistance versus weight.
3. Place some unknown weights on your sensor. From your graph, determine the value of these weights and compare this value with the real value.

Weight #	Resistance	Weight Determined from Graph	Real Value of Weight	% Difference $= \left\| \frac{\text{weight}_{\text{graph}} - \text{weighty}_{\text{actual}}}{\text{weight}_{\text{actual}}} \right\| \times 100$
1				
2				
3				
4				

4. What are the problems with this simple sensor, and how could you alleviate them?

B.5 MEASURING POWER

Objective: To measure the horsepower of a small motor using a homemade dynamometer.

Theory: By applying a force to the spinning shaft of a motor and measuring its angular speed, we can determine its horsepower.

Power is given by the formula:

$$P = F \cdot v$$

$$\text{Power}(\text{ft lb/s}) = \text{force}(\text{lb}) \times \text{velocity}(\text{ft/s})$$

The tangential velocity of a spinning shaft is given by:

$$v = 2\pi \text{r} \cdot f$$

$$\text{Velocity} = 2\pi \times \text{radius}(\text{ft}) \times \text{frequency}(\text{rev/s})$$

Therefore:

$$P = F \cdot 2\pi r \cdot f$$

To convert power in ft lb/s into horsepower use:

$$\text{Hp} = P(\text{ft lb/s}) \times \left(\frac{1\,\text{hp}}{550\,\text{ft lb/s}} \right)$$

B.5.1 Procedure

Using a motor with a built-in tachometer or a means to measure angular speed, see Figure B.4, set up the following:

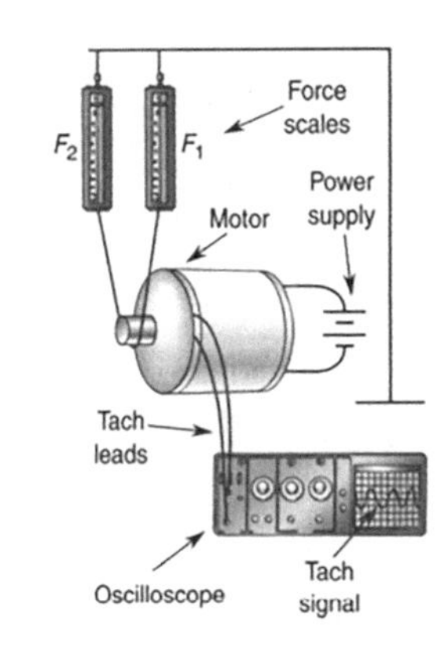

FIGURE B.4 A homemade dynamometer.

1. Arrange the motor so that it sits on the tabletop with the string looped under the motor's shaft with tension in the spring scales.
2. Apply a voltage to the motor and measure F_1, F_2, and the motor's revolutions per second.
3. Record these measurements for different applied voltages. Generate the following table.

V (I)	*I* (A)	F_1 (lb)	F_2 (lb)	*f* (rev/s)	P(ftlb/s) $= \lvert F_2 - F_1 \rvert \; 2\pi r f$	*P* (hp)

4. Electrical power in watts is given by:

$$P = I \cdot V$$
$$\text{Power(watts)} = \text{current} \times \text{voltage}$$

Calculate the electrical power consumed by the motor by measuring the current for an applied voltage. Using the conversion 1 hp = 746 W, determine the horsepower.

$$\text{Hp} = I \times V \times \left(\frac{1\text{ hp}}{746\text{ W}} \right) = \_\_\_$$

5. Determine the efficiency of the motor.

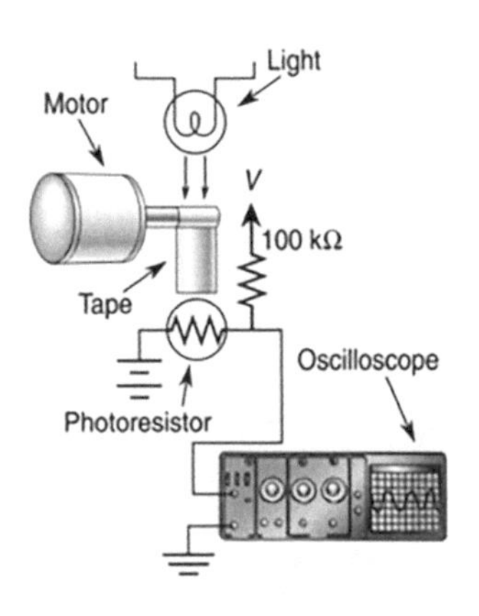

FIGURE B.5 A tachometer.

$$\text{Efficiency} = \frac{\text{power}_{\text{out}}}{\text{power}_{\text{in}}} \times 100\%$$

$$\text{Efficiency} = \frac{\text{power measured by the dynamometer}}{\text{power electrical}} \times 100\%$$

6. Is your motor very efficient? Why or why not?

B.6 MEASURING ANGULAR VELOCITY

Objective: To measure the angular velocity of a rotating shaft.

Theory: A tachometer is a device that measures angular velocity (rpm). A simple tachometer can be made from a photoresistor, a light source, and monitored with an oscilloscope. Figure B.5 shows a rotating shaft with tape attached to it.

Once every revolution, the light source is blocked from the photoresistor by the tape. The photoresistor is used in a voltage divider circuit and monitored with a scope. The period T of the waveform on the scope can be used to determine the angular velocity of the shaft.

$$\omega = \frac{1 \text{ rev}}{T}$$

$$\text{Angular velocity (rev/s)} = \frac{1 \text{ revolution}}{\text{period of one revolution (s)}}$$

To determine rpm use:

$$\text{RPM} = \frac{1 \text{ rev}}{T}\left(\frac{60 \text{ s}}{1 \text{ min}}\right)$$

B.6.1 Procedure

1. Build the setup pictured above and measure the angular velocity of the motor.
2. Determine how the motor's rpm changes with applied voltage to the motor.

Voltage (V)	T (s)	$\text{RPM} = \frac{I}{T} \times 60$

3. Plot rpm versus applied voltage.
4. How else could you measure the rpm of this motor?

B.7 MACHINES

Objective: To demonstrate several simple machines and measure their mechanical advantage.

Theory: The mechanical advantage of a machine is a measure of how much force it can generate compared with the input force into the machine. It is given by:

$$\mathrm{MA} = \frac{F_{\text{out}}}{F_{\text{in}}}$$

The mechanical advantage can also be expressed by:

$$\mathrm{MA} = \text{efficiency} \times \frac{d_{\text{in}}}{d_{\text{out}}}$$

where d_{in} and d_{out} are the input and output distances of the machine, respectively, and the efficiency is related to any friction involved in operating the machine.

We can put these two equations together to calculate the efficiency.

$$\text{Efficiency} = \left(\frac{F_{\text{out}}/F_{\text{in}}}{d_{\text{in}}/d_{\text{out}}} \right) \times 100$$

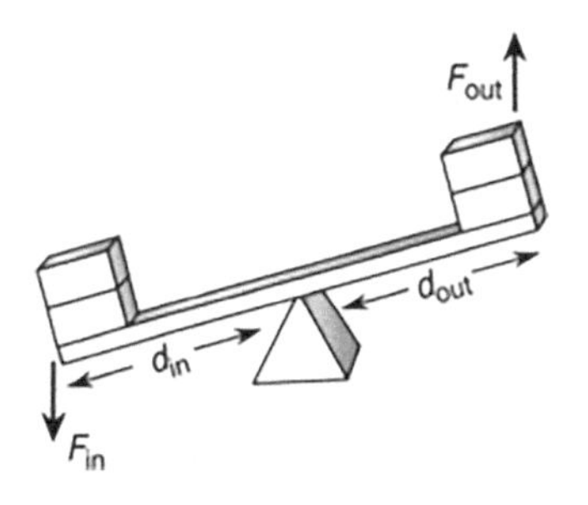

FIGURE B.6 A lever.

B.7.1 Procedure

Determine the mechanical advantage and the efficiency of the machines below.

1. **The lever.** Experiment with the lever for various lengths of d_{in} and d_{out}. See Figure B.6.

d_{in}	d_{out}	F_{in}	F_{out}	$MA = F_{out}/F_{in}$	$\text{Efficiency} = \left(\frac{F_{out}/F_{in}}{d_{in}/d_{out}} \right) \times 100\%$

2. **The pulley.** Put together the following configurations Figure B.7 and fill in the table below.

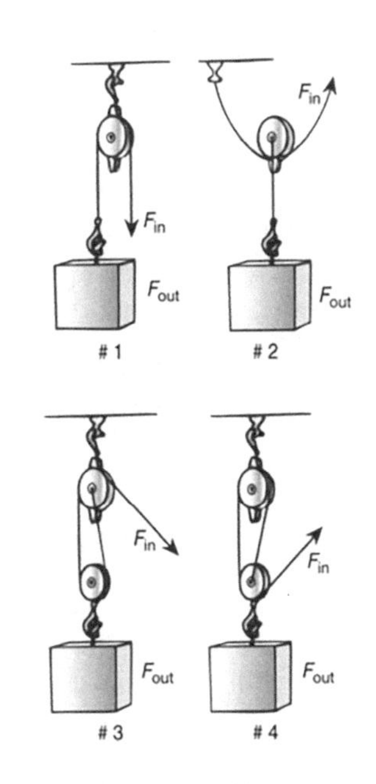

FIGURE B.7 Pulley systems.

Configuration	d_{in}	d_{out}	F_{in}	F_{out}	$MA = F_{out}/F_{in}$	$\text{Efficiency} = \left(\frac{F_{out}/F_{in}}{d_{in}/d_{out}}\right) \times 100\%$
#1						
#2						
#3						
#4						

3. **The wheel and axle.** Experiment for various radii of r_{in} and r_{out}. See Figure B.8.

r_{in}	r_{out}	F_{in}	F_{out}	$MA = F_{out}/F_{in}$	$\text{Efficiency} = \left(\frac{F_{out}/F_{in}}{r_{in}/r_{out}}\right) \times 100\%$

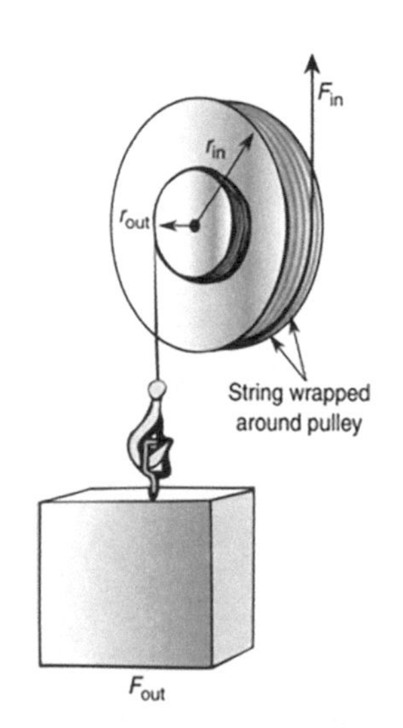

FIGURE B.8 A wheel and axle.

B.8 MEASURING TEMPERATURE

Objective: To measure and calibrate temperature sensors using a thermistor and a thermocouple.

Theory: Temperature changes the electrical properties of materials. A thermistor is a semiconductor that changes its resistance with temperature. A thermocouple consists of two dissimilar pieces of wire twisted together at one end, which when heated produces a voltage across the wires.

B.8.1 Procedure

Set up the following:

Calibrate the thermistor and thermocouple by placing them on a soldering iron along with a thermometer, see Figure B.9. Plug in the iron and while it is heating up, record the temperature of the thermometer, voltage of the thermocouple, and resistance of the thermistor. Fill in the table below.

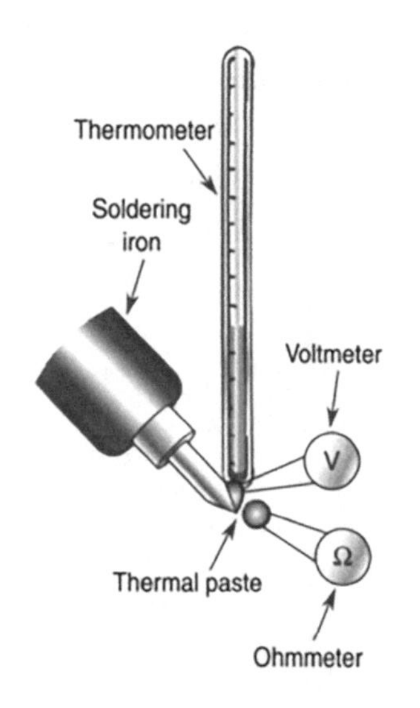

FIGURE B.9 Calibrating a thermistor and a thermocouple.

Temperature	Resistance	Voltage

- Plot resistance versus temperature.
- Plot voltage versus temperature.

Use your sensors to measure the room's temperature and your temperature. Check with your thermometer that these are correct.

B.9 MEASURING LIGHT

Objective: To measure light intensity using a photoresistor, photodiode, and solar cell, see Figure B.9.

Theory: Some materials change their electrical properties with light. A photoresistor lowers its resistance with an increasing light level. Solar cells and LEDs generate a voltage with light.

B.9.1 Procedure

Set up the following circuits using a LED of a known light output (usually this is listed as so many millicandelas, mcd, for a given applied voltage) as a source of light and measure how the sensors respond to this light, see Figure B.10.

Record how the sensors respond to various intensities of light.

Number of Light Sources	Intensity	Resistance of Photoresistor	Voltage, LED	Voltage, Solar Cell
1				
2				
3				
4				

1. Plot resistance versus intensity.
2. Plot voltage of the photodiode versus intensity.
3. Plot voltage of the solar cell versus intensity.
4. From your calibrated sensors, can you determine the intensity of the light in the room? What is this value?

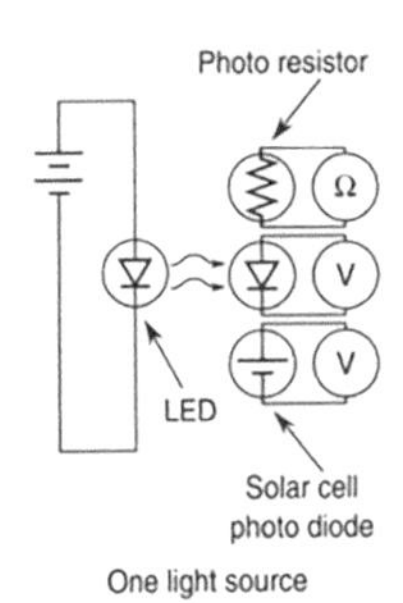

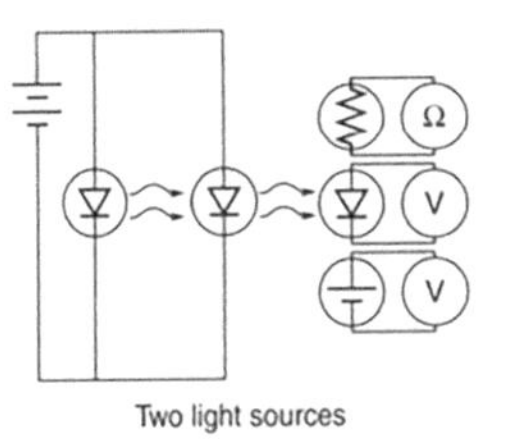

FIGURE B.10 Calibrating a light sensor.

B.10 DIRECT CURRENT MOTORS

Objective: To build a simple direct current (DC) motor, see Figure B.11.

Theory: Magnetic like poles repel, and unlike poles attract. An electric motor uses this principle to alternately repel and attract the magnetic field generated in a current-carrying coil to a stationary permanent magnet in order to produce a rotation.

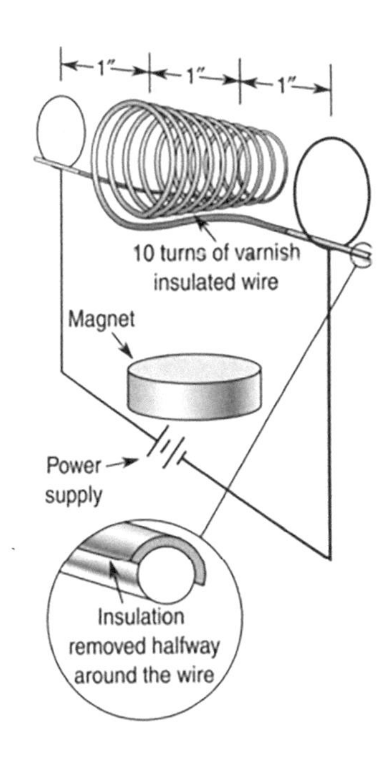

FIGURE B.11 A simple DC motor.

B.10.1 Procedure

1. Using a small-gauge varnished magnet wire such as 22-gauge or smaller, wind a small coil of approximately 10 turns and 1 in. diameter. Taking the ends of the coil, wrap them around the coil to hold the turns together. With 1 in. of wire protruding from opposite ends of the coil, sand off the varnish from each end but only halfway around the wire.
2. Using two other pieces of wire, remove the varnish insulation and create a small loop on each end. These pieces will support the coil and deliver current to it. Attach or connect these two wires to the ends of a D-cell or power supply.
3. Place the coil through the support wires and place a magnet under the coil. Spin the coil with your finger. If made correctly, the coil should continue to spin on its own.

B.10.2 Questions

1. Explain why the coil spins on its own.
2. How could you make this motor spin faster and smoother?

Appendix C

Unit Conversions

TABLE C.1
English Weights and Measures

Units of length	**Units of volume**
Standard unit—inch (in. or ″)	*Liquid*
12 inches = 1 foot (ft or ′)	16 ounces (fl oz) = 1 pint (pt)
3 feet = 1 yard (yd)	2 pints = 1 quart (qt)
$5\frac{1}{2}$ yards or 165 feet = 1 rod (rd)	4 quarts = 1 gallon (gal)
5,280 feet = 1 mile (mi)	*Dry*
Units of weight	2 pints = 1 quart
Standard unit—pound (lb)	8 quarts = 1 peck (pk)
16 ounces (oz) = 1 pound	4 pecks = 1 bushel (bu)
2,000 pounds = 1 ton (T)	

TABLE C.2
Conversion Table for Length

	cm	**m**	**km**	**in.**	**ft**	**mile**
1 cm =	1	10^{-2}	10^{-5}	0.394	3.28×10^{-2}	6.21×10^{-6}
1 m =	100	1	10^{-3}	39.4	3.28	6.21×10^{-4}
1 km =	10^5	1,000	1	3.94×10^4	3,280	0.621
1 in. =	2.54	2.54×10^{-2}	2.54×10^{-5}	1	8.33×10^{-2}	1.58×10^{-5}
1 ft =	30.5	0.305	3.05×10^{-4}	12	1	1.89×10^{-4}
1 mile =	1.61×10^5	1,610	1.61	6.34×10^4	5,280	1

TABLE C.3
Conversion Table for Area

Metric	English
1 m^2 = 10,000 cm^2	1 ft^2 = 144 in.^{2}
1 m^2 = 1,000,000 mm^2	1 yd^2 = 9 ft^2
1 cm^2 = 100 mm^2	1 rd^2 = 30.25 yd^2
1 cm^2 = 0.0001 m^2	1 acre = 160 rd^2
1 km^2 = 1,000,000 m^2	1 acre = 4,840 yd^2
	1 acre = 43,560 ft^2
	1 mile^2 = 640 acres

	m^2	cm^2	ft^2	in.^{2}
1 m^2 =	1	10^4	10.8	1,550
1 cm^2 =	10^{-4}	1	1.08×10^{-3}	0.155
1 ft^2 =	9.29×10^{-2}	929	1	144
1 in.^{2} =	6.45×10^4	6.45	6.94×10^{-3}	1

1 circular mile = 5.07×10^{-6} cm^2 = 7.85×10^{-7} in.^{2}
1 ha = 10,000 m^2 = 2.47 acres

TABLE C.4
Conversion Table for Volume

Metric	English
1 m^3 = 10^6 cm^3	1 ft^3 = 1,728 in.^{3}
1 cm^3 = 10^{-6} m^3	1 yd^3 = 27 ft^3
1 cm^3 = 10^3 mm^3	

	m^3	cm^3	L	ft^3	in.^{3}
1 m^3 =	1	10^6	1,000	35.3	6.10×10^4
1 cm^3 =	10^{-6}	1	1.00×10^{-3}	3.53×10^{-5}	6.10×10^{-2}
1 L =	1.00×10^{-3}	1,000	1	3.53×10^{-2}	61.0
1 ft^3 =	2.83×10^{-2}	2.83×10^4	28.3	1	1,728
1 in.^{3} =	1.64×10^{-5}	16.4	1.64×10^{-2}	5.79×10^{-4}	1

1 U.S. fluid gallon = 4 U.S. fluid quarts = 8 U.S. pints = 128 U.S. fluid ounces = 231 in.^{3} = 0.134 ft^3
1 L— 1,000 cm^3 = 1.06 qt = 1 fl oz = 29.5 cm^3
1 ft^3 = 7.47 gal = 28.3 L

TABLE C.5
Conversion Table for Mass

	g	kg	slug	oz	lb	T
1 g =	1	0.001	6.85×10^{-5}	3.53×10^{-2}	2.21×10^{-3}	1.10×10^{-6}
1 kg =	1,000	1	6.85×10^{-2}	35.3	2.21	1.10×10^{-3}
1 slug =	1.46×10^{4}	14.6	1	515	32.2	1.61×10^{-2}
1 oz	28.4	2.84×10^{-2}	1.94×10^{-3}	1	6.25×10^{-2}	3.13×10^{-5}
1 lb =	454	0.454	3.11×10^{-2}	16	1	5.00×10^{-4}
1 T =	9.07×10^{5}	907	62.2	3.2×10^{4}	2,000	1

1 MT = 1,000 kg = 2,205 lb

Quantities in the shaded areas are not mass units. When we write, for example, 1 kg = 2.21 lb, this means that a kilogram is a mass that weighs 2.21 lb under standard conditions of gravity ($g = 9.80\,\text{m/s}^2 = 32.2\,\text{ft/s}^2$).

TABLE C.6
Conversion Table for Density

	slug/ft³	kg/m³	g/cm³	lb/ft³	lb/in.³
1 slug/ft³ =	1	515	0.515	32.2	1.86×10^{-2}
1 kg/m³ =	1.94×10^{-3}	1	0.001	6.24×10^{-2}	3.61×10^{-5}
1 g/cm³ =	1.94	1,000	1	62.4	3.61×10^{-2}
1 lb/ft³ =	3.11×10^{-2}	16.0	1.60×10^{-2}	1	5.79×10^{-4}
1 lb/in.³ =	53.7	2.77×10^{4}	27.7	1,728	1

Quantities in the shaded areas are weight densities and, as such, are dimensionally different from mass densities.

Note that $\rho_w = \rho_m g$.

where

ρ_w = weight density

ρ_m = mass density

$g = 9.80\,\text{m/s}^2 = 32.2\,\text{ft/s}^2$

TABLE C.7
Conversion Table for Time

	year	day	h	min	s
1 year =	1	365	8.77×10^{3}	5.26×10^{5}	3.16×10^{7}
1 day =	2.74×10^{-3}	1	24	1,440	8.64×10^{4}
1 h =	1.14×10^{-4}	4.17×10^{-2}	1	60	3,600
1 min =	1.90×10^{-6}	6.94×10^{-4}	1.67×10^{-2}	1	60
1 s =	3.17×10^{-8}	1.16×10^{-5}	2.78×10^{-4}	1.67×10^{-2}	1

TABLE C.8
Conversion Table for Speed

	ft/s	km/h	m/s	mile/h	cm/s
1 ft/s =	1	1.10	0.305	0.682	30.5
1 km/h =	0.911	1	0.278	0.621	27.8
1 m/s =	3.28	3.60	1	2.24	100
1 mile/h =	1.47	1.61	0.447	1	44.7
1 cm/s =	3.28×10^{-2}	3.60×10^{-2}	0.01	2.24×10^{-2}	1

1 mile/min = 88.0 ft/s = 60.0 mile/h = 26.8 m/s

TABLE C.9
Conversion Table for Force

	N	lb	oz
1 N =	1	0.225	3.60
1 lb =	4.45	1	16
1 oz =	0.278	0.0625	1

TABLE C.10
Conversion Table for Power

	Btu/h	ft lb/s	hp	kW	W
1 Btu/h =	1	0.216	3.93×10^{-4}	2.93×10^{-4}	0.293
1 ft lb/s =	4.63	1	1.82×10^{-3}	1.36×10^{-3}	1.36
1 hp =	2,550	550	1	0.746	746
1 kW =	3,410	738	1.34	1	1,000
1 W =	3.41	0.738	1.34×10^{-3}	0.001	1

TABLE C.11
Conversion Table for Pressure

	atm	in. of H_2O	mm Hg	N/m^2 (Pa)	lb/in.^{2}	lb/ft^2
1 atm =	1	407	760	1.01×10^5	14.7	2,120
1 in. of H_2O =	2.46×10^{-3}	1	1.87	249	3.61×10^{-2}	5.20
1 mm of Hg =	1.32×10^{-3}	0.535	1	133	1.93×10^{-2}	2.79
1 N/m^2 =	9.87×10^{-6}	4.02×10^{-3}	7.50×10^{-3}	1	1.45×10^{-4}	0.021
1 lb/in.^{2} =	6.81×10^{-2}	27.7	51.7	6.90×10^3	1	144
1 lb/ft^2 =	4.73×10^{-4}	0.192	0.359	47.9	6.94×10^{-3}	1

TABLE C.12
Mass and Weight Density[a]

Substance	Mass Density (kg/m³)	Weight Density (lb/ft³)
Solids		
Copper	8,890	555
Iron	7,800	490
Lead	11,300	708
Aluminum	2,700	169
Ice	917	57
Wood, white pine	420	26
Concrete	2,300	145
Cork	240	15
Liquids		
Water	1,000[b]	62.4
Seawater	1,025	64.0
Oil	870	54.2
Mercury	13,600	846
Alcohol	790	49.4
Gasoline	680	42.0
	At 0°C and 1 atm pressure	At 32°F and 1 atm pressure
Gases[a]		
Air	1.29	0.081
Carbon dioxide	1.96	0.123
Carbon monoxide	1.25	0.078
Helium	0.178	0.011
Hydrogen	0.0899	0.0056
Oxygen	1.43	0.089
Nitrogen	1.25	0.078
Ammonia	0.760	0.047
Propane	2.02	0.126

[a] The density of a gas is found by pumping the gas into a container, by measuring its volume and mass or weight, and then by using the appropriate density formula.
[b] Metric weight density of water = 9,800 N/m³.

TABLE C.13
Specific Gravity of Certain Liquids at Room Temperature (20°C or 68°F)

Liquid	Specific Gravity
Benzene	0.90
Ethyl alcohol	0.79
Gasoline	0.68

(*Continued*)

TABLE C.13 (*Continued*)
Specific Gravity of Certain Liquids at Room Temperature (20°C or 68°F)

Liquid	Specific Gravity
Kerosene	0.82
Mercury	13.6
Seawater	1.025
Sulfuric acid	1.84
Turpentine	0.87
Water	1.000

TABLE C.14
Conversion Table for Energy, Work, and Heat

	Btu	ft lb	J	cal	kWh
1 Btu =	1	778	1,060	252	2.93×10^{-4}
1 ft lb =	1.29×10^{-3}	1	1.36	0.324	3.77×10^{-7}
1 J =	9.48×10^{-4}	0.738	1	0.239	2.78×10^{-7}
1 cal =	3.97×10^{-3}	3.09	4.19	1	1.16×10^{-6}
1 kWh =	3,410	2.66×10^{6}	3.60×10^{6}	8.60×10^{5}	1

TABLE C.15
Heat Constants

					Heat of Fusion		Heat of Vaporization	
	Melting Point (°C)	Boiling Point (°C)	Specific Heat (Btu/lb °F)	J/kg (°C)	kcal/ kg	J/kg	kcal/ kg	J/kg
Alcohol, ethyl	−117	78.5	0.58	2,400	24.9	1.04×10^{5}	204	8.54×10^{5}
Aluminum	660	2,057	0.22	920	76.8	3.21×10^{5}		
Brass	840		0.092	390				
Copper	1,083	2,330	0.092	390	49.0	2.05×10^{5}		
Glass			0.21	880				
Ice	0		0.51	2,100	80	3.35×10^{5}		
Iron (steel)	1,540	3,000	0.115	481	7.89	3.30×10^{4}		
Lead	327	1,620	0.031	130	5.86	2.45×10^{4}		
Mercury	−38.9	357	0.033	140	2.82	1.18×10^{4}	65.0	2.72×10^{5}
Silver	961	1,950	0.056	230	26.0	1.09×10^{5}		
Steam			0.48	2,000				
Water (liquid)	0	100	1.00	4,190	80	3.35×10^{5}	540	2.26×10^{6}
Zinc	419	907	0.092	390	23.0	9.63×10^{4}		

TABLE C.16
Coefficient of Linear Expansion

Material	α (metric)	α (English)
Aluminum	$2.3 \times 10^{-5}/°C$	$1.3 \times 10^{-5}/°F$
Brass	$1.9 \times 10^{-5}/°C$	$1.0 \times 10^{-5}/°F$
Concrete	$1.1 \times 10^{-5}/°C$	$6.0 \times 10^{-6}/°F$
Copper	$1.7 \times 10^{-5}/°C$	$9.5 \times 10^{-6}/°F$
Glass	$9.0 \times 10^{-6}/°C$	$5.1 \times 10^{-6}/°F$
Pyrex	$3.0 \times 10^{-6}/°C$	$1.7 \times 10^{-6}/°F$
Steel	$1.3 \times 10^{-5}/°C$	$6.5 \times 10^{-6}/°F$
Zinc	$2.6 \times 10^{-5}/°C$	$1.5 \times 10^{-5}/°F$

TABLE C.17
Coefficient of Volume Expansion

Liquid	β (metric)	β (English)
Acetone	$1.49 \times 10^{-3}/°C$	$8.28 \times 10^{-4}/°F$
Alcohol, ethyl	$1.12 \times 10^{-3}/°C$	$6.62 \times 10^{-4}/°F$
Carbon tetrachloride	$1.24 \times 10^{-3}/°C$	$6.89 \times 10^{-4}/°F$
Mercury	$1.8 \times 10^{-4}/°C$	$1.0 \times 10^{-4}/°F$
Petroleum	$9.6 \times 10^{-4}/°C$	$5.33 \times 10^{-4}/°F$
Turpentine	$9.7 \times 10^{-4}/°C$	$5.39 \times 10^{-4}/°F$
Water	$2.1 \times 10^{-4}/°C$	$1.17 \times 10^{-4}/°F$

TABLE C.18
Conversion Table for Charge

Charge on one electron = 1.60×10^{-19}C
1 C = 6.25×10^{18} electrons of charge
1 Ah = 3,600 C

TABLE C.19
Copper Wire Table

Gauge No.	Diameter (miles)	Diameter (mm)	Cross Section: Circular miles	Cross Section: in.^{2}	Ohms/1,000 ft: 25°C (77°F)	Ohms/1,000 ft: 65°C (149°F)	Weight/1,000 ft (lb)
0000	460.0		212,000	0.166	0.0500	0.0577	641.0
000	410.0		168,000	0.132	0.0630	0.0727	508.0

(Continued)

TABLE C.19 (*Continued*)
Copper Wire Table

Gauge No.	Diameter (miles)	Diameter (mm)	Cross Section: Circular miles	Cross Section: in.^{2}	Ohms/1,000 ft: 25°C (77°F)	Ohms/1,000 ft: 65°C (149°F)	Weight/1,000 ft (lb)
00	365.0		133,000	0.105	0.0795	0.0917	403.0
0	325.0		106,000	0.0829	0.100	0.116	319.0
1	289.0	7.35	83,700	0.0657	0.126	0.146	253.0
2	258.0	6.54	66,400	0.0521	0.159	0.184	201.0
3	229.0	5.83	52,600	0.0413	0.201	0.232	159.0
4	204.0	5.19	41,700	0.0328	0.253	0.292	126.0
5	182.0	4.62	33,100	0.0260	0.319	0.369	100.0
6	162.0	4.12	26,300	0.0206	0.403	0.465	79.5
7	144.0	3.67	20,800	0.0164	0.508	0.586	63.0
8	128.0	3.26	16,500	0.0130	0.641	0.739	50.0
9	114.0	2.91	13,100	0.0103	0.808	0.932	39.6
10	102.0	2.59	10,400	0.00815	1.02	1.18	31.4
11	91.0	2.31	8,230	0.00647	1.28	1.48	24.9
12	81.0	2.05	6,530	0.00513	1.62	1.87	19.8
13	72.0	1.83	5,180	0.00407	2.04	2.36	15.7
14	64.0	1.63	4,110	0.00323	2.58	2.97	12.4
15	57.0	1.45	3,260	0.00256	3.25	3.75	9.86
16	51.0	1.29	2,580	0.00203	4.09	4.73	7.82
17	45.0	1.15	2,050	0.00161	5.16	5.96	6.20
18	40.0	1.02	1,620	0.00128	6.51	7.51	4.92
19	36.0	0.91	1,290	0.00101	8.21	9.48	3.90
20	32.0	0.81	1,020	0.000802	10.4	11.9	3.09
21	28.5	0.72	810	0.000636	13.1	15.1	2.45
22	25.3	0.64	642	0.000505	16.5	19.0	1.94
23	22.6	0.57	509	0.000400	20.8	24.0	1.54
24	20.1	0.51	404	0.000317	26.2	30.2	1.22
25	17.9	0.46	320	0.000252	33.0	38.1	0.970
26	15.9	0.41	254	0.000200	41.6	48.0	0.769
27	14.2	0.36	202	0.000158	52.5	60.6	0.610
28	12.6	0.32	160	0.000126	66.2	76.4	0.484
29	11.3	0.29	127	0.0000995	83.4	96.3	0.384
30	10.0	0.26	101	0.0000789	105	121	0.304
31	8.9	0.23	79.7	0.0000626	133	153	0.241
32	8.0	0.20	63.2	0.0000496	167	193	0.191
33	7.1	0.18	50.1	0.0000394	211	243	0.152
34	6.3	0.16	39.8	0.0000312	266	307	0.120
35	5.6	0.14	31.5	0.0000248	335	387	0.0954
36	5.0	0.13	25.0	0.0000196	423	488	0.0757
37	4.5	0.11	19.8	0.0000156	533	616	0.0600

(*Continued*)

TABLE C.19 (*Continued*)
Copper Wire Table

Gauge No.	Diameter (miles)	Diameter (mm)	Cross Section Circular miles	in.^{2}	Ohms/1,000 ft 25°C (77°F)	65°C (149°F)	Weight/1,000 ft (lb)
38	4.0	0.10	15.7	0.0000123	673	776	0.0476
39	3.5	0.09	12.5	0.0000098	848	979	0.0377
40	3.1	0.08	9.9	0.0000078	1070	1230	0.0200

TABLE C.20
Conversion Table for Plane Angles

	°	′	″	rad	rev
1° =	1	60	3,600	1.75×10^{-2}	2.78×10^{-3}
1′ =	1.67×10^{-2}	1	60	2.91×10^{-4}	4.63×10^{-5}
1″ =	2.78×10^{-4}	1.67×10^{-2}	1	4.85×10^{-6}	7.72×10^{-7}
1 rad =	57.336	3,440	2.06×10^{s}	1	0.159
1 rev = 360°					

TABLE C.21
The Greek Alphabet

Capital	Lowercase	Name
Α	α	Alpha
Β	β	Beta
Γ	γ	Gamma
Δ	δ	Delta
Ε	ε	Epsilon
Ζ	ζ	Zeta
Η	η	Eta
Θ	θ	Theta
Ι	ι	Iota
Κ	κ	Kappa
Λ	λ	Lambda
Μ	μ	Mu
Ν	ν	Nu
Ξ	ξ	Xi
Ο	o	Omicron
Π	π	Pi

(*Continued*)

TABLE C.21 (*Continued*)
The Greek Alphabet

Capital	Lowercase	Name
P	ρ	Rho
Σ	σ	Sigma
T	τ	Tau
Y	υ	Upsilon
Φ	ϕ	Phi
X	χ	Chi
Ψ	Ψ	Psi
Ω	ω	Omega

TABLE C.22
Characteristics of Some Common Batteries

Battery	Voltage (V)	Ampere-Hour Rating	Mass (g)
Primary			
"pen-light"	1.5	0.58	20
D-cell	1.5	3.0	90
#6 dry cell	1.5	30	860
#2U6 battery	9.0	0.325	30
#V-60 battery	90	0.47	580
#1 mercury cell	1.4	1.0	12
#42 mercury cell	1.4	14	166
Secondary (Rechargeable)			
#CD-21 nickel-cadmium	6.0	0.15	52
#CD-29 nickel-cadmium	12.0	0.45	340
lead-acid car battery	6.0	84	13,000
lead-acid car battery	12.0	96	25,000

TABLE C.23
Coefficients of Static Friction and Sliding Friction

	Coefficient of Static Friction	
Surfaces in Contact	Dry	Well Lubricated
Hard steel and babbitt	0.42	0.17
Hard steel and hard steel	0.78	0.11
Soft steel and soft steel	0.74	0.11
Soft steel and lead	0.95	0.5
Plastic and steel	0.5	0.1
Rubber and concrete	1.5	1.0 (wet)

(*Continued*)

TABLE C.23 (*Continued*)
Coefficients of Static Friction and Sliding Friction

	Coefficient of Sliding Friction		
Surfaces in Contact	**Dry**	**Lightly Lubricated**	**Well Lubricated**
Bronze and cast iron	0.21	0.16	0.077
Cast iron and cast iron	0.4	0.15	0.064
Cast iron and hardwood	0.49	0.19	0.075
Hardwood and hardwood	0.48	0.16	0.067
Hard steel and babbitt	0.35	0.16	0.06
Hard steel and hard steel	0.42	0.12	0.03
Plastic and steel	0.35	—	0.05
Rubber and concrete	1.02	0.9 (wet)	—
Soft steel and soft steel	0.57	0.19	0.09

Values listed are for ordinary pressures and temperatures, but may still vary somewhat from sample to sample.

TABLE C.24
Elastic Limit and Ultimate Strength of Some Common Solids

	Elastic Limit		Ultimate Strength	
Solid	**Tension**	**Compression**	**Tension**	**Compression**
Aluminum	840	840	1,800	∞
Brass	630	630	2,000	∞
Brick, best hard	30	840	30	840
Brick, common	4	70	4	70
Bronze	2,800	2,800	5,300	∞
Cement, Portland				
1 month old	28	140	28	140
1 year old	35	210	35	210
Concrete, Portland				
1 month old	14	70	14	70
1 year old	28	140	28	140
Copper	700	700	2,500	∞
Douglas fir	330	330	500	430
Granite	49	1,300	49	1,300
Iron, cast	420	1,800	1,400	5,600
Lead	10	10	200	∞
Limestone and sandstone	21	630	21	630

(*Continued*)

TABLE C.24 (*Continued*)

Elastic Limit and Ultimate Strength of Some Common Solids

	Elastic Limit		Ultimate Strength	
Solid	**Tension**	**Compression**	**Tension**	**Compression**
Monel metal	6,300	6,300	7,000	00
Oak, white	310	310	600	520
Pine, white, eastern	270	270	400	345
Slate	35	980	35	980
Steel, bridge cable	6,700	6,700	15,000	∞
Steel, 1% C, tempered	5,000	5,000	8,400	8,400
Steel, chrome, tempered	9,100	9,100	11,000	11,000
Steel, stainless	2,100	2,100	5,300	5,300
Steel, structural	2,500	2,500	4,600	4,600

Units are kilogram-force per square centimeter (kgf/cm^2).
Note: $1\ kgf/cm^2 = 98.1\ kPa = 14.22\ lb/in.^2$

TABLE C.25

Hardness of Materials

Aluminum	2–2.9
Brass	3–4
Copper	2.5–3
Diamond	10
Gold	2.5–3
Iron	4–5
Lead	1.5
Magnesium	2.0
Marble	3–4
Mica	2.8
Phosphorbronze	4
Platinum	4.3
Quartz	7
Silver	2.5–4
Steel	5–8.5
Tin	1.5–1.8
Zinc	2.5

TABLE C.26
Elastic Constants for Various Materials in SI Units and USCS Units

Material	Young's Modulus (Y, MPa[a])	Shear Modulus (S, MPa)	Elastic Limit (MPa)	Ultimate Strength (MPa)
Aluminum	68,900	23,700	131	145
Brass	89,600	35,300	379	455
Copper	117,000	42,300	159	338
Iron	89,600	68,900	165	324
Steel	207,000	82,700	248	489
Material	**Young's Modulus (lb/in.^{2})**	**Shear Modulus (S, lb/in.^{2})**	**Elastic Limit (lb/in.^{2})**	**Ultimate Strength (lb/in.^{2})**
Aluminum	10×10^6	3.44×10^6	19,000	21,000
Brass	13×10^6	5.12×10^6	55,000	66,000
Copper	17×10^6	6.14×10^6	23,000	49,000
Iron	13×10^6	10×10^6	24,000	47,000
Steel	30×10^6	12×10^6	36,000	71,000

[a] 1 MPa = 10^6 Pa.

TABLE C.27
***K*-Values for Some Common Building Materials**

Material	R-value (ft^2 °F h/Btu)
Hardwood siding (1.0 in. thick)	0.91
Wood shingles (lapped)	0.87
Brick (4.0 in. thick)	4.00
Concrete block (filled cores)	1.93
Styrofoam (1.0 in. thick)	5.0
Fiberglass batting (3.5 in. thick)	10.90
Fiberglass batting (6.0 in thick)	18.80
Fiberglass board (1.0 in. thick)	4.35
Cellulose fiber (1.0 in. thick)	3.70
Flat glass (0.125 in. thick)	0.89
Insulating glass (0.25 in. space)	1.54
Vertical air space (3.5 in. thick)	1.01
Air film	0.17
Dry wall (0.50 in. thick)	0.45
Sheathing (0.50 in. thick)	1.32

TABLE C.28
Typical Heats of Combustion of Some Gaseous Fuels[a]

	Heat of Combustion				
	On Mass Basis		On Volume Basis		
Fuel	Btu/lb_m	MJ/kg	Btu/ft^3	kJ/L	Air-Fuel Ratio for Complete Combustion (ft^3/ft^3 or L/L)
Hydrogen	61,400	143.0	275	10.3	2.4
Methane	23,800	55.2	900	33	9.6
Propane	22,200	51.5	2,400	88	13.7
Natural gas	23,600	54.8	1,000	37	11.9
Coke gas	23,000	53.6	1,300	50	14

Note: 1 MJ/kg = 1,000 U/kg = 1,000 J/g.
1 U/L = 1 J/cm^3 = 1 MJ/m^3.

[a] Data applies to gases at 1.0 atm pressure and 15.6°C (60°F). Actual values vary slightly depending on geographical origin. Values do not include latent heat in any water vapor formed as a product of combustion.

TABLE C.29
Typical Heats of Combustion of Some Solid and Liquid Fuels[a]

	Heat of Combustion		
Fuel	Btu/lb_m	MJ/kg	Air-Fuel Ratio for Complete Combustion (lb_m/lb_m or g/g)
Butane	20,000	46	15.4
Gasoline	19,000	44	14.8
Crude oil	18,000	42	14.2
Fuel oil	17,500	41	13.8
Coke	14,500	34	11.3
Coal, bituminous	13,500	31	10.4

Note: 1 MJ/kg = 1,000 kJ/kg = 1,000 J/g.

[a] Actual values will vary slightly depending on geographical origin. Values do not include latent heat in any water vapor formed as a product of combustion.

TABLE C.30
Speed of Sound in Various Media

	Speed	
Medium	**m/s**	**ft/s**
Aluminum	6,420	21,100
Brass	4,700	15,400
Steel	5,960	19,500
Granite	6,000	19,700
Alcohol	1,210	3,970
Water (25°C)	1,500	4,920
Air, dry (0°C)	331	1,090
Vacuum	0	0

TABLE C.31
Indices of Refraction

Material	$n = \frac{c}{v}$
Vacuum	1.0000
Air (at STP)	1.0003
Water	1.33
Ethyl alcohol	1.36
Glass	
Fused quartz	1.46
Crown glass	1.52
Light flint	1.58
Lucite or Plexiglas	1.51
Sodium chloride	1.53
Diamond	2.42

$\lambda = 589\,\text{nm}$.

Appendix D

Answers to Odd Numbered Back of the Chapter Problems

Chapter 1: Units and Measurements

1	6,336 ft
3	1.3×10^3 ft
5	120 in.
7	560 cm
9	0.35 mm
11	64 cm
13	9.7×10^{-6} m^3
15	15 m^2
17	5.0 in.^{2}
19	740 mm^3
21	3.1×10^5 in.^{3}
23	3.78×10^7
25	38.4 s
27	0.156 cm^3
29	11.0 in.
31	47 in.^{3}
33	0.05 lb/in.^{3}
35	16 in.^{2}
37	600
39	6.3×10^7 s
41	15 lb
43	38 in.^{3}
45	13 W
47	1.2×10^2
49	9.5×10^7 s
51	3.1×10^2 lb
53	14.4 in.^{3}

Chapter 2: Linear Motion

1	1.5 m/s
3	6 s
5	10 ft/s^2
7	300 m
9	9.6 ft
11	65 mile/h
13	Acceleration is the change in velocity with time.
15	Hang the gum at the end of a hanging piece of string. If the gum moves there is an acceleration.
17	0.1 s
19	14 ft/s
21	18 s
23	4.6 g
25	2.8 s
27	0.1 s
29	8 ft/s
31	1 in./s
33	4 ft
35	5.8×10^{-3} s

Chapter 3: Force and Momentum

1	1,000 mV
3	12 N
5	38.6 ft/s^2
7	50 N
9	They are equal but opposite.
11	It states that for every force there is an equal but opposite force.
13	Use a force sensor.
15	7.5 N
17	2.0×10^3 lb
19	1.5
21	50 N
23	2.5×10^2 mV
25	31 ft/s^2
27	6.2×10^{-2} slugs
29	7.5 lb
31	1.6 ft/s^2
33	3 N
35	17.7 ft/s^2
37	2×10^{20} N

Chapter 4: Energy, Work, and Power

1	0.2 ft lb
3	17.30%
5	1.4 h
7	1,044 W
9	46.4 m/s
11	106 ft lb
13	450 ft lb
15	66.3 m/s
17	75%
19	2.0×10^2 m/s
21	2.4 kWh
23	1.4×10^2 J
25	16 miles/gal
27	10 ft lb/s
29	−100 J
31	98 J
33	If the total energy of the universe equaled 0 at one point in time, then it equaled 0 during the big bang!
35	2.2×10^3 W

Chapter 5: Rotational Motion

1	14.1 in.
3	0.76 in.
5	200 teeth
7	12.5 hp
9	6 ft lb
11	150 rev
13	2.42 m/s^2
15	66×10^3 in./min
17	12 in./s
19	1.2 ft
21	2
23	9.0×10^2 N
25	11 ft lb
27	6.9×10^{-4} rpm
29	1.5×10^2 rpm
31	0.63 W
33	0.89 Nm
35	400 rpm
37	720 N

Chapter 6: Machines

1	MA = 100
3	MA = 5.9, F = 5.9 lb
5	0.17, 330 lb
7	1.6
9	844 lb
11	MA = 2
13	Hint: Look at Figure 6.3.
15	Connect the muscle further from the joint.
17	1.2×10^2
19	1.3×10^{-2} lb
21	6.3×10^{-2} lb
23	2.4×10^2 lb
25	72%
27	80%
29	3.0×10^2
31	60 lb
33	6.0×10^2 N
35	8.4×10^3

Chapter 7: Strength of Materials

1	413 lb/in.
3	243 lb/in.^{2}, 531 ft
5	The maximum stretching or compressing without becoming permanently deformed.
7	0.01 in.
9	Aluminum
11	14.7 lb/(body area)
13	0.0002 in.
15	0.02%
17	700 lb/in.^{2}
19	12.5 lb
21	1,500 lb
23	$1.2 \times 10^4\ \Omega$
25	72 lb
27	Yes
29	Steel
31	False
33	Carbon nanotubes
35	It has no unit.

Chapter 8: Fluids

1	Cohesion is the attractive force between like molecules. Adhesion is the attractive force between unlike molecules. Glue and tape are considered adhesives.
3	Oil with a SAE rating of 5 flows more easily than an SAE rating of 30. In cold weather, oils with low SAE numbers are used, otherwise the engine would not turn over. In warm weather, high viscosity oils are used to prevent the oil from becoming too thin and leaking at the seals in the engine.
5	psia = psig + 14.7 $lb/in.^2$
7	1.6×10^6 lb
9	30:1
11	By Archimedes' principle the upward force on a body is equal to the weight of the volume of fluid displaced by the object. Air is considered as a fluid and is being displaced by the balloon, therefore there will be an upward force on the balloon.
13	–22.3°C
15	Detergent reduces the adhesive force between dirt and clothing. It can therefore more easily be washed away by the water.
17	The flow rate will increase.
19	1.9×10^2 lb
21	16.4 psi
23	10 ft
25	$2 N/m^2$
27	Three times
29	4.6
31	5 Pa
33	The cohesion of the molecules.
35	psig
37	The water is drawn up because of the adhesion between the water and the inside of the plant stem.

Chapter 9: Fluid Flow

1	$4.7 in.^3/s$
3	876 W
5	16.3 m/s
7	Four times
9	8.2×10^5 W
11	A teardrop has the lowest drag of any shape for speeds below the speed of sound.
13	5 ft/min
15	The wing is shaped so that the air moving over the top of the wing has to move faster to keep up with the air moving below the wing. As a result, the pressure is higher below the wing than above it. The difference in air pressure pushes the plane upward. When the wing is horizontal, air moving over the top of the wing has to move faster to keep up with the air moving below the wing, resulting in a pressure difference. The plane also receives a lift when the wing is tilted slightly upward. In this case, air is deflected off the bottom of the wing, pushing it upward.

(Continued)

17	Two times
19	45 cm^3/s
21	Four times
23	85.7 gpm
25	30.5 W
27	Teardrop
29	A thermal flow sensor contains a heater and an electronic thermometer placed next to one another. As the fluid flows over the heater, it cools. The greater the flow, the cooler the heater will be. The temperature is therefore a measure of the flow rate.
31	6.2×10^5 cm^3/min
33	Because the moving water creates a low-pressure system in the shower, the higher pressure outside the shower pushes the curtain inward.
35	Liters per flush
37	19.6 kPa

Chapter 10: Temperature and Heat

1	22°C, –201 K
3	68 Btu
5	1.8×10^4 Btu
7	7.8×10^3 Btu
9	A hot object gives off heat faster.
11	3.9 h
13	18°C
15	A small and large wattage soldering iron can have the same temperature, but the large wattage iron can deliver more heat because it is being delivered at a faster rate (power = energy/time).
17	Water boils at a lower temperature on a mountaintop because at lower pressures water's boiling point is lowered.
19	1.2×10^3 °F
21	3.2×10^6 J
23	2.1×10^2 J
25	3.3×10^5 J
27	8.6×10^{-2} °C
29	1.3 W
31	50 $in.^2$
33	1.5 times
35	1.6×10^2 s
37	1.1×10^5 J
39	10,884 K

Chapter 11: Thermodynamics: Heat Engines, Heat Pumps, and Thermal Expansion

1	Energy wasted = 4.95×10^5 J, 20% efficient.
3	The 1,100°C engine
5	0.03 in.
7	The expansion joint allows the bridge to thermally expand through the joint, thereby keeping the bridge from buckling.
9	A thermal switch consists of two dissimilar pieces of metal that have different coefficients of thermal expansion. When heated, the result is a bending of the strip. Heating is due to the current running through the breaker. The breakers are calibrated to open at specified current levels.
11	My bedroom becomes a mess in only one week if I do not exert any energy to clean it.
13	1,100 J, 38%
15	Energy is conserved as it moves through a thermodynamic system.
17	3.0×10^4 Btu
19	80%
21	135 Cal
23	4.6×10^{-3} in.
25	8.1 cm^2
27	Butane
29	1.9×10^3 Btu
31	The engine would melt without it.
33	It would not help.
35	A 100 ft strip would change length by 1.1 in. Not to worry, however, because the strips are not 100 ft long.

Chapter 12: Electric Force

1	2.3×10^{-8} N
3	2.8×10^{-7} C
5	1.44×10^{-5} N/C
7	+4 N in the $+x$ direction
9	10^5 V/m
11	1 m/s
13	−0.2 J
15	5,400 J
17	50×10^{-6} C
19	The capacitance increases.
21	2×10^{-11} F
23	54.45 J
25	40,000 batteries
27	It would discharge through the dielectric.
29	100 V/m

Chapter 13: Electricity

1	A voltage difference
3	0.3 W
5	Energy consumed: 4–100 W lights on for 2 h, computer in low energy mode 60 W for 2 h, refrigerator average power used 300 W. Total kWh = 1.5 kWh.
7	Measure the current and the voltage of the motor at the same time. The product will give you the apparent power.
9	66.7 kΩ
11	$304
13	2 A
15	918 miles
17	14 W
19	5.5×10^2 W
21	2.2×10^4 cents
23	1.5 h
25	48 W
27	1.6×10^3 cents
29	20 h
31	The total resistance is ½ the resistance of the individual resistors.
33	1.2×10^2 W
35	24 Ω
37	$I_{1k\Omega} = 12\,mA$, $I_{2k\Omega} = 6\,mA$, $I_{source} = 18\,mA$
39	Closed

Chapter 14: Magnetism

1	Earth's magnetic field is due to electric currents in the molten iron flow near its core.
3	A magnetic bearing can be made using electromagnets placed around a shaft designed to rotate. The magnets attract the metal shaft from all directions keeping it centered during rotation. A control system is used to adjust the current to the electromagnets, ensuring the shaft stays centered.
5	2.0×10^{-6} T
7	0.02 T
9	168 A
11	The electric meter on your house is an energy meter measuring the number of kilowatt-hours used. It works on the same principle as an alternating current (AC) induction motor. There are two electromagnets. One is connected from the source voltage to the house, and the other is connected in series with source to the house. Between the two electromagnets is an aluminum disk that is free to rotate. Because of the way the two electromagnets are connected, they are out of phase with one another by 90° and produce a rotating magnetic field. These electromagnets induce currents into the aluminum disk, which produces a magnetic field. This magnetic field interacts with the electromagnet's field, causing the disk to rotate. The number of rotations the motor makes is proportional to the current, voltage, and the time for which the motor is running. Recall $E = I \times V \times t$. A counting mechanism records this number in terms of kilowatt-hours or kWh.

(Continued)

13	See Figure 14.22
15	0.2 T
17	20 mV
19	3.1 T
21	1.0×10^{-5} T
23	1 T
25	0.52 Gauss
27	Decreases two times.
29	4.4×10^{4} J
31	1.2×10^{-5} m
33	5.6×10^{2} times
35	0 Ω
37	318.3 turns
39	$N_s/N_p = 2$

Chapter 15: Waves

1	0.2 Hz
3	3×10^{6} m
5	A speaker creates sound by moving its cone back and forth. As the speaker cone moves forward, it compresses the air; moving backward it creates a slight vacuum. The speaker is alternately compressing and decompressing the air.
7	In the old days they were modulated differently. Today, they use time division multiplexing.
9	Change in frequency = 1.9×10^{2} Hz
11	347.8 m/s
13	Warm air
15	8.8×10^{-5} W/m^2
17	3 dB
19	3.2×10^{14} Hz
21	1.7×10^{-2} s
23	343 m/s. This train must be moving at the speed of sound!
25	8×10^{-6} W
27	40 dB
29	Because there is no medium to disturb that would lead to a wave.
31	The sound waves push on a diaphragm connected to a coil that moves back and forth through a fixed magnet, which induces a current in the coil. The sound is effectively turned into a current.
33	Two times
35	High frequency
37	At resonance a system will produce a maximum response.

Chapter 16: Light

1	White light is light composed of many different colors of light. The color can be separated out using a prism. This is the part of the spectrum that we can see. At wavelengths slightly shorter than visible light, lies the ultraviolet part of the spectrum. Ultraviolet rays or UV rays cause sunburn. Welders wear goggles to protect their eyes from UV rays. At wavelengths slightly longer than visible light, lies the infrared part of the spectrum. Infrared radiation is otherwise known as thermal radiation or radiant heat.
3	545 m
5	Step 1: Electrons in their orbits around their atoms are excited to higher energy orbits. Excitation can be done in various ways using light, electrical discharge, chemical reactions etc. Step 2: Excited electrons begin to jump down to lower energy orbits. The electrons spend different amounts of time in the various orbits before jumping to the next lower orbit. If the electrons spend a lot of time in one particular orbit, soon there will be a buildup of electrons there. This is called population inversion, because a large number of the electrons of that atom now reside at that level.
7	24.9 lm/m^2
9	3,000 W, 600 W
11	Ultraviolet light has a shorter wavelength than infrared light.
13	Because the light does not get absorbed by the glass.
15	300
17	7.5×10^3 m
19	9.8×10^8 m
21	300
23	1.5 Lux
25	18 W
27	9.8×10^8 ft/s
29	1.2×10^8 m/s
31	39.8 cd
33	7.5 m
35	6 in.
37	60 cm
39	$m = -2.4$, $h_i = 2.8$ cm
41	$R = 60$ cm

Index

D

E

I

J

K

L

M

N

O